High Performance Computing in Science and Engineering '09

Wolfgang E. Nagel • Dietmar B. Kröner •
Michael M. Resch
Editors

High Performance Computing in Science and Engineering '09

Transactions of the High Performance Computing Center
Stuttgart (HLRS) 2009

Editors

Wolfgang E. Nagel
Zentrum für Informationsdienste
und Hochleistungsrechnen (ZIH)
Technische Universität Dresden
Helmholtzstr. 10
01069 Dresden
Germany
wolfgang.nagel@tu-dresden.de

Dietmar B. Kröner
Abteilung für Angewandte Mathematik
Universität Freiburg
Hermann-Herder-Str. 10
79104 Freiburg
Germany
dietmar@mathematik.uni-freiburg.de

Michael M. Resch
Höchstleistungsrechenzentrum
Stuttgart (HLRS)
Universität Stuttgart
Nobelstraße 19
70569 Stuttgart
Germany
resch@hlrs.de

Front cover figure: Interaction of dissipation elements with vortex tubes in turbulence, Institut für Technische Verbrennung, RWTH Aachen

---

ISBN 978-3-642-04664-3 e-ISBN 978-3-642-04665-0
DOI 10.1007/978-3-642-04665-0
Springer Heidelberg Dordrecht London New York

Library of Congress Control Number: 2009941206

Mathematics Subject Classification (2000): 65Cxx, 65C99, 68U20

Cover design: WMXDesign GmbH, Heidelberg

Printed on acid-free paper

Springer is part of Springer Science+Business Media (www.springer.com)

Preface

At the end of the year 2008, we have seen a strategic step towards a functioning HPC infrastructure on Tier-0 level in Germany. Based on an agreement („Verwaltungsabkommen") between the Federal Ministry of Education and Research (BMBF) and the state ministries for research of Baden-Württemberg, Bayern, and Nordrhein-Westfalen, a budget of overall 400 Million Euro had been allocated – equally shared between federal and state authorities in a five year time frame – to establish the next generation of HPC systems at the Gauss Centre for Supercomputing (GCS) – consisting of the three national supercomputing centres HLRS (Stuttgart), NIC/JSC (Jülich), and LRZ (Munich). As part of that strategic initiative, in May 2009 already NIC/JSC has installed the first phase of the GCS HPC Tier-0 resources, an IBM Blue Gene/P with roughly 300.000 Cores, this time in Jülich, With that, the GCS provides the most powerful high-performance computing infrastructure in Europe already today.

HLRS and its partners in the GCS have agreed on a common strategy for the installation of the next generation of leading edge HPC systems. Over the next few years, HLRS and LRZ as the other two GCS centers will upgrade their systems accordingly. The plan is to have a Tier-0 HPC system within GCS operating at any time in this five year period.

As an intermediate step, HLRS has replaced most of their NEC SX-8 nodes by the NEC SX-9/12M192, a system with roughly 20 TFLOPs peak. Additionally, a remarkably large Intel Nehalem cluster has been installed, bringing another 100+ TFLOPs peak to the users of HLRS. Which overall means: due to the support of the state ministry for research of Baden-Württemberg, the available cpu resources have been increased by a factor of more than ten just in this intermediate step. Further resources will be coming soon.

As part of GCS, HLRS participates in the European project PRACE (Partnership for Advances Computing in Europe) and extends its reach to all European member countries. These activities aligns well with the activities of HLRS in the European HPC infrastructure project DEISA (Distributed European Infrastructure for Supercomputing Applications) and in the European HPC support project HPC-Europa2.

Two years after the GCS has been created to coordinate the European supercomputing activities, Germany took the next step to shape the future of high performance computing (HPC) on a national level. On December 3rd 2008 the German leading supercomputing centers met in Bonn to establish the Gauß Allianz. While GCS was successfully addressing the needs on the top end, it was clear from the beginning that an additional layer of support was required to maintain the longevity with a network of competence centers across Germany. This gap is now addressed by the Gauß Allianz, where regional and topical centers team up to create the necessary infrastructure.

The 15 founding members are Gauss Centre for Supercomputing (GCS), Center for Computing and Communication of RWTH Aachen University, Norddeutscher Verbund für Hoch- und Höchstleistungsrechnen (HLRN) consisting of Konrad Zuse Center for Information Technology Berlin (ZIB) und Regionales Rechenzentrum für Niedersachsen (RRZN), Center for Information Services and High Performance Computing (ZIH) at TU Dresden, Rechenzentrum Garching (RZG) of the Max-Planck Society, Deutscher Wetterdienst (DWD), Deutsches Klimarechenzentrum (DKRZ), and Steinbuch Center for Computing (SCC) as full members and Goethe Center for Scientific Computing (G-CSC) at Frankfurt University, Forschungszentrum Computational Engineering, Kompetenzgruppe Wissenschaftliches Hochleistungsrechnen der Technischen Universität Darmstadt, Paderborn Center for Parallel Computing (PC2), Regionales Rechenzentrum Erlangen (RRZE), and the German Research and Education Network (DFN) as associated members.

The mission is to coordinate the HPC related activities of the members. With the provision of versatile computing architectures and by combining the expertise of the participating centers, this creates the ecosystem necessary for computational science. Strengthening the research and increasing the visibility to compete on an international level are further goals of the Gauß Allianz.

So far, the whole HPC community is still waiting for the second HPC call of the BMBF to enable and support Petascale applications on more than 100.000 processors. While the projects of the first funding round have started in early 2009, the follow up call got delayed by roughly one year. Nevertheless, all experts and administration authorities still see the strong need for such a program, and it is announced that the next call will follow – soon, within weeks. The strategic plan is that within the next four years another 20 Million Euro will be spend – on a yearly basis – for projects to develop scalable algorithms, methods, and tools to support massively parallel systems. As we all know, we do not only need competitive hardware but also excellent software and methods to approach – and solve – the most demanding problems in science and engineering. The success of this approach is of utmost importance for our community and also will strongly influence the development of new technologies and industrial products; beyond that, this will finally determine if Germany will be an accepted partner among the leading technology and research nations.

Since 1996, HLRS is supporting the scientific community as part of its official mission. Like in the years before, the major results of the last 12 months were reported at the Twelfth Results and Review Workshop on High Performance Computing in Science and Engineering, which has been held October 8-9, 2008 at Stuttgart University. This volume contains the written versions of the research work presented. The papers have been selected from all projects running at HLRS and at SSC Karlsruhe during the one year period beginning October 2007. Overall, more than 35 papers have been chosen from Physics, Solid State Physics, Reactive Flow, Computational Fluid Dynamics, Chemistry, and other topics. The largest number of contributions, as in many other years, came from CFD with 15 papers. Although such a small collection cannot represent a large area in total, the selected papers demonstrate the state of the art in high performance computing in Germany. The authors were encouraged to emphasize computational techniques used in solving the problems examined. This often forgotten aspect was the major focus of these proceedings, nevertheless this should not disregard the importance of the newly computed scientific results for the specific disciplines.

We gratefully acknowledge the continued support of the Land Baden-Württemberg in promoting and supporting high performance computing. Grateful acknowledgement is also due to the Deutsche Forschungsgemeinschaft (DFG): many projects processed on the machines of HLRS and SSC could not have been carried out without the support of the DFG. Also, we thank the Springer Verlag for publishing this volume and, thus, helping to position the local activities into an international frame. We hope that this series of publications is contributing to the global promotion of high performance scientific computing.

Stuttgart, October 2009

Wolfgang E. Nagel
Dietmar Kröner
Michael Resch

Contents

Physics

Prof. Dr.-Ing. Michael Resch

Höchstleistungsrechenzentrum Stuttgart (HLRS), Universität Stuttgart,
Nobelstraße 19, 70569 Stuttgart, Germany

Three selected contributions reflect the enormous progress made in supercomputer simulation techniques in the field of physics. Two of them focus on astrophysics, the third on soft matter physics.

Line-emitting gas nebulae around powerful radio galaxies in the early universe are simulated with different models by Volker Gaibler and Max Camenzind from the Landessternwarte at Heidelberg in their contribution "Numerical Models for Emission Line Nebulae in High Redshift Radio Galaxies". The formed jet cocoon is simulated in two and three dimensions with timescales of several ten million years. The simulation code runs very efficiently on the NEC SX-8 and SX-9 supercomputers and is computationally very demanding.

The paper of B. Müller, A. Marek and H.-Th. Janka from the Max-Planck-Institut für Astrophysik in Garching gives a detailed update on the scientific progress made in studying core collapse supernovae. Here, vector based supercomputers like the NEC SX machines demonstrate their extreme usefulness for supernova simulations.

Martin Hecht and Jens Harting from the Institut für Computerphysik at the University of Stuttgart simulate a suspension of a model of colloidal particles in their contribution titled "Using Computational Steering to Explore the Parameter Space of Stability in a Suspension". They use a coupled algorithm consisting of a Molecular Dynamics (MD) and a Stochastic Rotation Dynamics (SRD) part to explore the huge parameter space by a big number of simulation runs to cover the influence of physical properties and conditions. They also describe a method of steering the simulations for example to explore the regions of certain states in order to study cluster formation of particles.

Numerical Models for Emission Line Nebulae in High Redshift Radio Galaxies

Volker Gaibler[1,2] and Max Camenzind[1]

[1] Landessternwarte, Zentrum für Astronomie der Universität Heidelberg, 69117 Heidelberg, Germany
[2] Max-Planck-Institut für extraterrestrische Physik, 85748 Garching, Germany
vgaibler@mpe.mpg.de

Summary. We examine models for line-emitting gas nebulae around powerful radio galaxies in the early universe. The models assume that either the emitting gas clouds are embedded in the shocked ambient gas and driven outwards with it or that they are created and sustained by multi-phase turbulence in the jet cocoon. For this, we perform jet simulations with realistic density contrasts on large scales and on typical activity time scales of several ten million years. The employed magnetohydrodynamics code NIRVANA has been optimized for the NEC SX machines in the previous years and now runs very efficiently on the SX-6 and SX-8, allowing us simulations both in axisymmetry and full three dimensions. Future simulations on the SX-9 will benefit significantly from the increased number of shared memory processors per node for the axisymmetric runs, where an MPI-parallelization is generally inefficient for our setup due to communication overhead.

1 Introduction

Extragalactic jets are amongst the most spectacular phenomena in astrophysics: These collimated beams of dilute but highly energetic plasma are formed in the environs of active black holes, on scales of milliparsecs[3] with speeds very close to the speed of light, and propagate far outwards, eventually leaving the galaxy and pushing through the intergalactic gas at hundreds or even thousands of kiloparsecs. Over these more than seven orders of magnitude in length, they remain collimated and deposit a large fraction of their huge kinetic energies of $\sim 10^{39}$ watts on the ambient gas. Yet, due to their low densities relative to the gas in galaxy cluster, they expand only slowly and reach their largest sizes only after several tens of millions of years. While these objects have been discovered already more than 50 years ago as double radio sources, only improving observational techniques made it possible in the

[3] 1 pc = 3.26 light years = 3×10^{16} m

W.E. Nagel et al. (eds.), *High Performance Computing in Science and Engineering '09*, DOI 10.1007/978-3-642-04665-0_1,

1970s and 1980s to realize the underlying physical mechanism of supersonic beams carrying large amounts of kinetic energy.

Jets do not only occur in the local universe, but also at high redshifts [13], corresponding to a time only some billion years after the big bang, when galaxies were still actively forming. At redshifts $z > 0.5$, radio galaxies often show luminous emission regions extending far beyond the galaxy itself, which are aligned with the radio source ("alignment effect"). These gas nebulae consist of an outer quiescent component, usually outside the radio structures, and an inner component with strongly distorted kinematics and clumpy and irregular structure [15]. The alignment with the radio source as well as the large velocities involved (often exceeding 1000 km s^{-1} in line widths) indicates vigorous interaction with the jet. In the nearby universe, similar interaction between the jet and the ambient gas is observed. Morganti et al. [11] and Morganti [10] report on large outflows of neutral hydrogen gas with velocities of $\sim$ 1000 km s^{-1} in local radio galaxies seen by blue-shifted HI absorption lines. Whether these velocities are really due to jet interaction and how this interaction happens in detail is yet unclear. In our project we applied our large-scale and high-resolution magnetohydrodynamical simulations of jets to this question. Only recently, new observational studies [12] by means of integral field spectroscopy became available which now allow comparing theoretical models with observed data.

Increasing computational resources made it possible to conduct numerical simulations and to perform qualitative and quantitative tests based on hydrodynamics and magnetohydrodynamics. Due to the slow propagation of the jet in combination with its large extents, numerical studies are computationally very demanding and hence heavily rely on supercomputing facilities as the NEC SX machines at the HLRS.

Although there are many detailed observations of extragalactic jets, even the most basic parameters of these jets, as density, Mach number and the composition of the plasma, are very hard to constrain. This makes simulations difficult and enforces setups with several different parameters. Most simulations in the literature assumed less extreme density contrasts to reach reasonable jet sizes. We try to use more realistic density contrasts despite the huge computational demands because previous work in this project [4, 5] showed that these very light jets show a different behaviour than their heavier counterparts. Also, current observations suggest magnetic fields of considerable strength in jet cocoons, which makes a magnetohydrodynamical treatment of these objects necessary, even when one assumes somewhat weaker magnetic fields. As reported last year, we are now able to perform simulations of jets with realistically strong density contrasts and jet sizes. While we previously concentrated on the effects of magnetic fields on the dynamics of jets, we now apply a series of simulations to high-redshift radio galaxies [1].

2 Numerical Method

We employ the magnetohydrodynamical code NIRVANA [16], which solves the magnetohydrodynamic equations in either cartesian, cylindrical or spherical coordinates and three dimensions. It is based on a finite-differences discretization in an explicit formulation using operator splitting and uses van Leer's interpolation scheme, which is second order accurate. The advection part is solved in a conservative form and the magnetic fields are evolved using the constrained transport method, which conserves $\nabla \cdot \mathbf{B}$ up to machine round-off errors. The code is vectorized and shared-memory parallelized [2, 8] and runs very efficiently on the NEC SX machines. We also developed an MPI-parallelized version [2] which is used for fully three-dimensional simulations.

3 Computational and Scientific Results

3.1 Performance

For the simulations, we used both the SX-6 and the SX-8. For single-node axisymmetric simulations, we achieved 19.2 Gflops on the SX-6 with our optimized code version [2, 8] (8 CPUs, vector operation ratio 99.5%). A single simulation run takes $\sim$ 3 000 CPU hours for a very light jet with density contrast 10^{-4}. Multi-node runs are used for the 3D simulations on the SX-8. On 32 nodes, we reached 550–600 Gflops so far (256 CPUs, vector operation ratio 99.0%), still working on improvement of the MPI communication to increase this further before starting very long runs (we estimate that a factor of nearly 2 might be possible). However, we generally do not use this MPI-parallelized version for axisymmetric simulations, since the problem size per node becomes too small. Then vectorization becomes inefficient and the communication overhead increases. For these setups, shared-memory parallelization is more efficient. Since millions of time steps are necessary to reach realistic jet sizes, these simulations cannot be parallelized due to causality, corresponding wallclock runtimes are large and simulations have to be restarted many times due to wallclock time restrictions in the batch system. Hence we are looking forward to computations on the newly available SX-9 with its larger number of shared-memory CPUs and increased speed. Tests of the code on this machine are already underway.

3.2 High Redshift Radio Galaxies

Models

To explain the morphology and kinematics of the emission-line gas, we considered two models. In the first model, hereafter referred to as "SAG model", the line emission originates from dense clouds embedded into the shocked ambient gas. This model is similar to the model proposed by Meisenheimer & Hippelein [9], although those authors assumed that the ambient gas clouds are compressed and heated by the bow shock passing over them and only radiate with some lag (corresponding to the cooling time scale). We assume a simpler model, where emitting clouds of indefinite origin (but possibly excited by the bow shock) are pushed outwards by drag forces from the outward-moving shocked ambient gas. Clearly the cloud velocities will depend on the efficiency of momentum transfer and cloud properties as mass and size, since they are embedded in the ambient gas with a filling factor significantly less than unity. For perfect coupling, the emitting clouds will be dragged along with the shocked ambient gas with identical velocity and hence can be determined directly from the shocked ambient gas kinematics. For weaker coupling the speeds completely depend on the cloud properties and the interaction time scale. Hence, for most kinds of clouds, those will either be dragged along with the shocked ambient gas, or remain approximately at their initial position with only small changes in velocity.

In the second model, the emitting clouds are located in the jet cocoon (hereafter referred to as "COC model"). This was recently proposed by Krause & Alexander [7] based on 2D hydrodynamical simulations of multi-phase turbulence in the jet cocoon with optically thin cooling. The simulations started with a Kelvin-Helmholtz instability setup at the cocoon–ambient interface with a density contrast of 10^4 between the cocoon plasma and the ambient medium as well as dense clouds embedded in the ambient medium. The dense clouds were disrupted but the fragments remained cold and spread throughout the domain by the developing turbulence. This generated a multi-phase medium with a peak in the temperature distribution at $\sim 14\,000$ K (similar to what is derived from observations) where shock heating and increased cooling counteract and produce the line-emitting phase responsible for the observed emission-line nebulae. Krause [6] extended this to three dimensions and found a correlation between the emission luminosity and the kinetic energy of the cocoon plasma, with 10^{12} erg of kinetic energy in the cocoon corresponding to ~ 1 erg s^{-1} in line emission.

Results for the Shocked Ambient Gas Model

We use our jet simulations at a time $t = 15$ Myr to derive, which kinematic properties are expected for emission-line clouds located in the shocked ambient gas. Simulation run M3 is used here as reference for the figures. van Ojik et al. [14] find a typical cloud mass of 10^{-4} solar masses and cloud radius of 0.03 pc to be a reasonable description. Clouds with these properties are expected to be dragged along with the shocked ambient gas within a time span corresponding to typical source ages. For clouds with significantly longer acceleration time scales, the velocities derived in the following would correspond to upper limits.

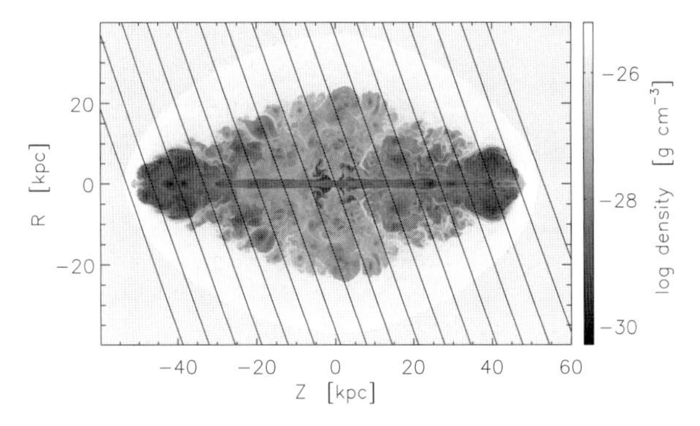

Fig. 1. Slices used for the spatial resolution of the jet system. The projection uses an inclination of $\theta = 70°$ and 16 slices in the plane through the jet beam, viewed from an observer located in the lower half of the plot. The background image is a logarithmic density map of the exemplary simulation (M3, $t = 15$ Myr)

These considerations in mind, we analyzed the projected radial velocities of the shocked ambient gas in the plane through the jet beam and the observer. Fig. 1 shows a density map of the data frame together with the borders of the 16 equidistant radial slices for an observer with inclination angle $\theta = 70°$. The slices are 6.25 kpc wide, roughly corresponding to the resolution of the observations. The cells within these slices are used to create histograms of the radial velocity (along the observer's line of sight). These histograms, in arbitrary units, are shown for some slices in Fig. 2. It shows an approaching and receding component for each slice, corresponding to the near and far side of the shocked ambient gas parts. These features typically have widths (FWHM) of 100 to 200 km s^{-1} and originate from varying velocities in the shocked ambient gas associated with the pressure waves there (for comparison: the sound speed in the shocked ambient gas is $\sim$ 1400 km s^{-1}). Since the shapes of the features depend on the pressure waves in the shocked ambient gas, it is clear that they vary significantly. However, their widths are small with respect to their offsets. If the line-emitting clouds were distributed uniformly within the shocked ambient medium, the histograms would correspond to the line profiles caused by Doppler shifting.

The position–velocity diagram in Fig. 3 shows the histograms for all slices with the counts corresponding to greyscale values in arbitrary (resolution-dependent) units. It shows two inclined and slightly curved bands with a systematic trend towards smaller (more negative) values on the side of the approaching jet, which is simply a projection effect as only the radial component of the gas motion is seen: the position of these "maximum velocity locations" is shifted towards the approaching (receding) jet head for the negative (positive) radial velocities, dependent on the inclination angle θ.

The maximum value of the radial velocity corresponds to the (unprojected) outward velocity of the shocked ambient gas, which is related with the bow

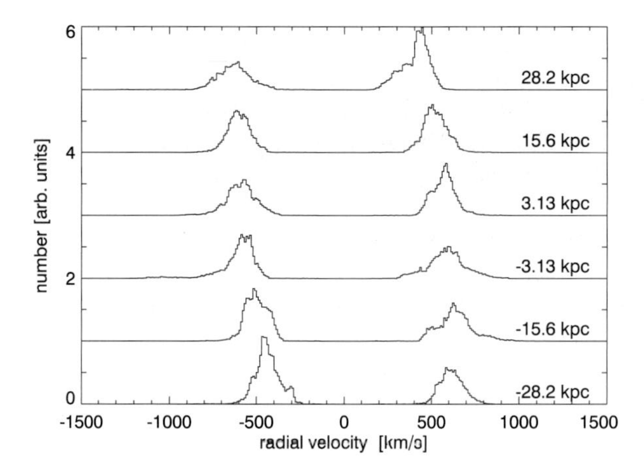

Fig. 2. SAG model: Velocity distributions for the shocked ambient gas in some slices of M3 at 15 Myr. For clarity, they are shifted by 1, respectively. Inclination angle $\theta = 70^\circ$, bin contributions ("counts") are in arbitrary units, 10 km s^{-1} bin size. The slices are labelled by their spatial offsets from core, increasing in the direction of the Z-axis. Due to projection of the shocked ambient gas velocity, the peaks shift with position

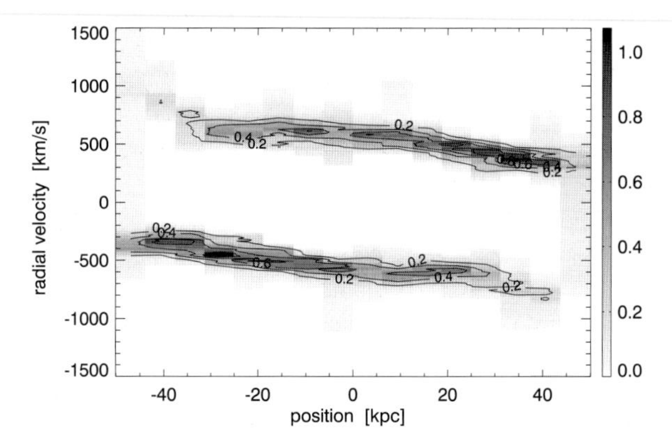

Fig. 3. SAG model: Position–velocity diagram for M3 at 15 Myr. The greyscale corresponds to the histogram counts (in arbitrary units). For clarity, contours are overlaid. The inclination angle is $\theta = 70^\circ$, velocity bin size 10 km s^{-1}. The position is measured as offset from core in plane of the sky

shock expansion speed by the Rankine–Hugoniot shock jump conditions. For the case of Fig. 3, where the bow shock is about Mach 1.5, the ambient gas outward velocity is ~ 0.6 times the ambient sound speed, corresponding to ~ 700 km s^{-1}, which is in good agreement with the maximum absolute value of the radial velocity.

At different evolutionary stages of the source, this radial velocity clearly will be different. We find that at early times, a very broad component starts at high velocities, caused by a rapidly propagating bow shock. Later, the feature becomes narrower and moves towards lower velocities (it also becomes stronger since the cell numbers along the line of sight increase). This is a consequence of a decelerating bow shock and slower shocked ambient gas motion. For heavier jets with high bow shock Mach numbers, the radial velocities generally are higher, while they are lower for lighter jets.

We conclude that the observed distribution of velocities can be understood by considering the contributions of the source evolution (decelerating bow shock), the perturbations within the shocked gas (responsible for the feature width) and projection effects.

Results for Cocoon Model

While the kinematic properties of the emission-line gas could be directly derived from the simulation data for the SAG model by assuming coupled velocities, this differs for the COC model: the cooler ($\sim 10^4$ K) phase is only modelled by the multi-phase turbulence simulation, which has a much finer spatial grid. It is not possible to include this in the large-scale simulation, since the achievable resolutions are too coarse and time steps too large. Thus our approach is to combine both simulations for the analysis. The systematic velocities and available kinetic energy in the cocoon are derived from the large-scale jet simulations ("macro physics"); the emission power, cool gas masses and turbulent velocities, in contrast, originate from the turbulence simulations ("micro physics"), with possibly additional contributions to the turbulent velocities by the large-scale models.

We will here focus on the results from the large-scale simulations, again using simulation run M3 at $t = 15$ Myr as reference. The turbulent velocities are only based on the large-scale turbulence, which cannot be easily translated into observable turbulent velocities. In Fig. 4, the cocoon motion is analyzed in the cylindrical coordinate system of the simulation (axial Z and radial R) for both contributions. This is done in 16 slices perpendicular to the full jet length. Near the jet heads ($Z \approx \pm 50$ kpc), very high velocities are found due to the backflow. More towards the midplane, the turbulent velocities level off to values of $\approx 5\,000$ km s^{-1} and the systematic velocities are around $1\,000$ km s^{-1}, with the axial component changing sign in the midplane. The systematic values are considerably distorted by large vortices (sizes of order of the slice width).

For comparison with observations, the radial velocities towards the observer are computed for 16 slices along the beam in the observer–jet beam plane for an inclination of $\theta = 70°$ (see Fig. 1). The slices are again 6.25 kpc wide, velocities are binned in 500 km s^{-1} bins and are positive for receding matter. The resulting histograms for six selected slices are shown in detail in Fig. 5, while Fig. 6 gives an overview by the position–velocity diagram. The large turbulent motion is responsible for the wide distributions, while systematic motion shifts them. There is a slight trend for positive (receding)

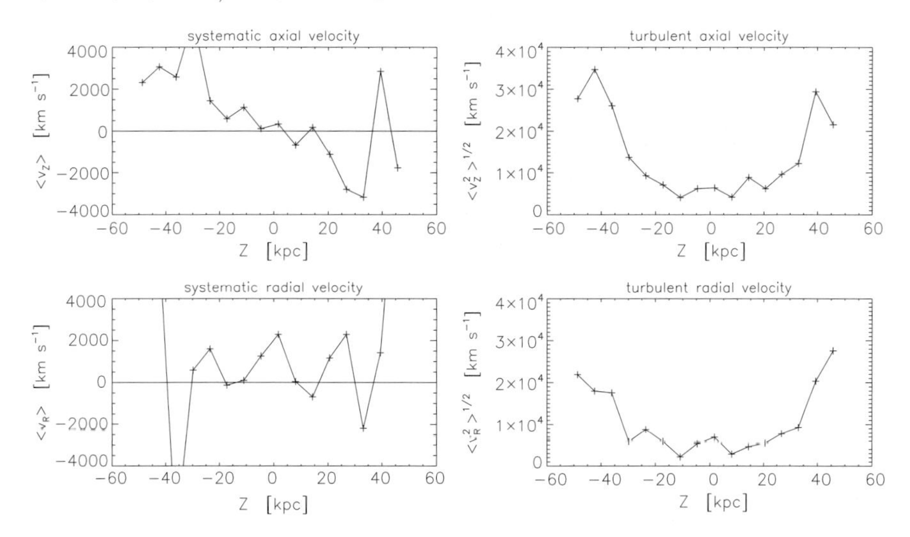

Fig. 4. COC model: Systematic and turbulent velocities in the cocoon in the jet coordinate system, spatially resolved in axial direction. Only one half of the plane is considered here (upper half of Fig. 1), no projection

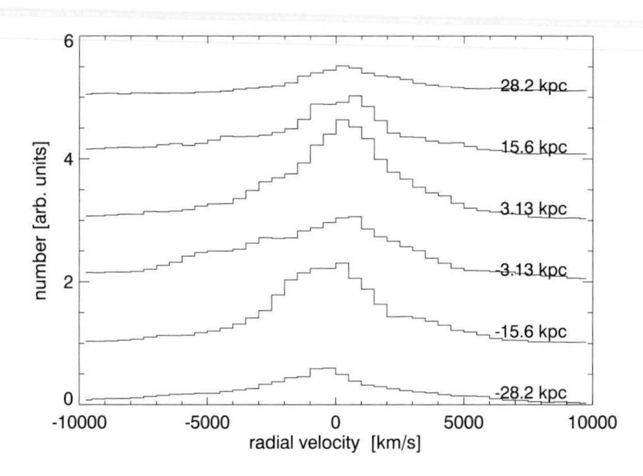

Fig. 5. COC: Velocity distributions for some slices of M3 at 15 Myr. For clarity, they are shifted by 1, respectively. Inclination angle $\theta = 70°$, counts are in arbitrary units, 500 km s^{-1} bin size. The slices are labelled by their spatial offsets from the core in the plane of the sky, increasing in direction of the Z axis

velocities on the side where the jet approaches, and negative (approaching) velocities on the receding jet side. This can also be seen in Fig. 7, where the slice averages and standard deviations are shown. This effect comes from the orientation and is somewhat stronger for smaller inclination angles, but still, the systematic shift is smaller than the turbulent width. For an inclination of $\theta = 90°$, these systematic velocities are not present, as due to the axisymmetry

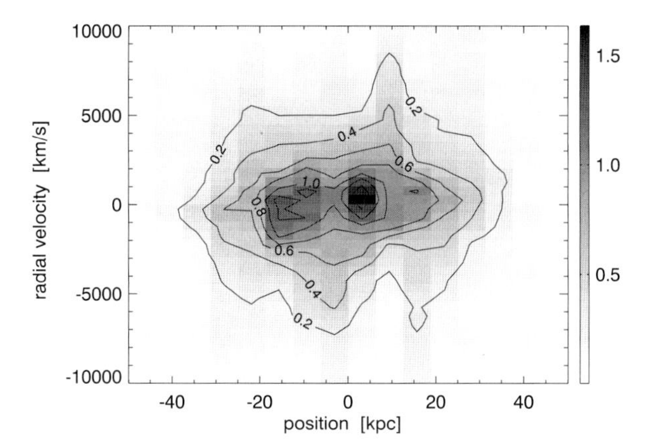

Fig. 6. COC model: Position–velocity diagram of M3 at 15 Myr. Inclination $\theta = 70°$, bin contribution are in arbitrary units, 500 km s^{-1} bin size. Positions are offsets from the core in the plane of the sky

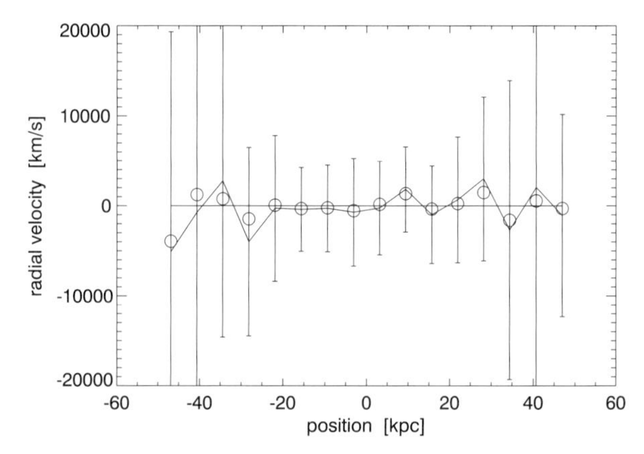

Fig. 7. COC model: Velocity averages (solid line), median (circles) and standard deviation (vertical bars) for the slices through M3 at 15 Myr. Inclination angle $\theta = 70°$, positions are offsets from the core in the plane of the sky

the systematic R-velocity is exactly balanced by the mirrored contribution (on the other side of the axis) and only broadens the distribution insignificantly.

The kinetic jet power is the major energy input into the whole system. For the COC model, we measure the fraction of the input energy that is converted to kinetic energy of the cocoon gas. This can then be further related to kinetic energy in the cool emission line gas by the microscopic simulations. Fig. 8 shows the fraction of the total jet power that is found in the cocoon kinetic energy. As underdense jets convert most of their kinetic power into thermal energy, with lighter jets showing higher thermalization, only a small percentage is found in the cocoon kinetic energy. For jets of different density

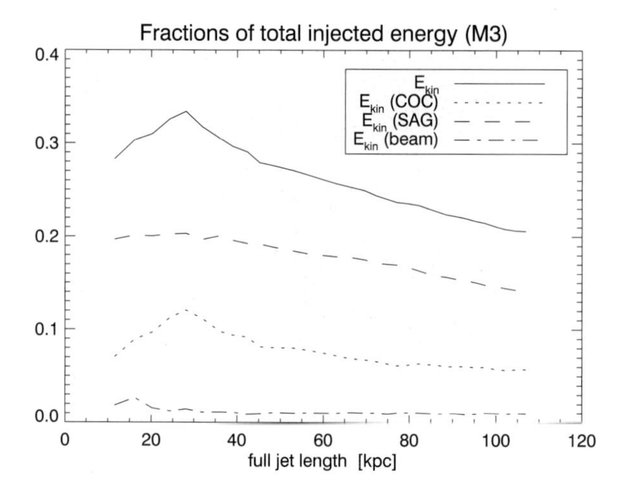

Fig. 8. Fraction of the total jet power going into kinetic energy of the cocoon and the shocked ambient gas for the jet simulation M3. The beam contribution is *included* in cocoon, but it is also shown separately

contrast but with same lengths, we find values of 21 % for $\eta = 10^{-1}$, 11 % for $\eta = 10^{-2}$, 6 % for $\eta = 10^{-3}$ and 3 % for $\eta = 10^{-4}$ (M4L run). We note that a resolution study indicated that the cocoon kinetic energy fraction somewhat decreases for higher resolution. The kinetic energy in the shocked ambient gas is similar to or up to a factor of 2 higher than the kinetic energy in the cocoon, then generally levelling off with time towards the cocoon values.

Discussion

We have analyzed our simulation data with respect to two models for the location of the emission-line gas found in high-redshift radio galaxies, which have been proposed in the literature and compared them to new integral-field spectroscopy data. In the SAG model, the emitting clouds are located in and dragged along with the shocked ambient gas, possibly emitting because they have been heated by the bow shock. The velocities of the shocked ambient gas indeed have values in the range of the observed ones for the approaching component, in particular if one considers the wider range due to the evolution of the source. Also, the gas on the approaching side shows higher approaching speeds than on the receding jet side, consistent with the observational data if only relative motion is considered. The contributions from the far side of the shocked ambient gas may be suppressed due to extinction. However, there is no sudden jump found in the systematic radial velocities near the core but only a gradual change between the two sides. Furthermore, the turbulent velocities are considerably smaller than the outward-directed systematic velocity. If the masses and densities of the emitting clouds have a wide range of masses and densities, resulting in different acceleration efficiencies, the width

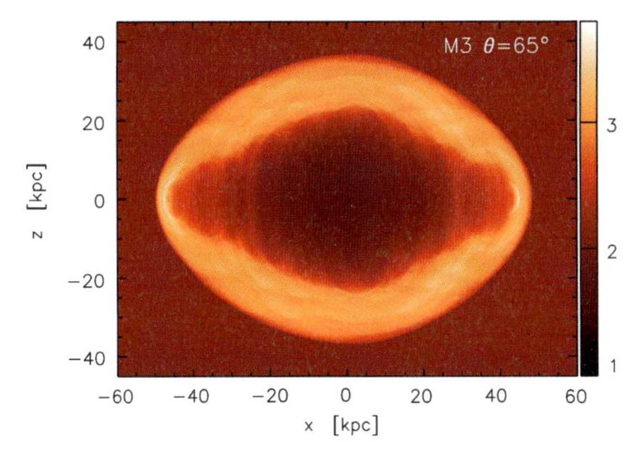

Fig. 9. Simulated X-ray bremsstrahlung emission for reference simulation M3 at $t = 15$ Myr for the energy band 1–7 keV. Assumed inclination angle $\theta = 65°$, intensity is shown in linear scaling in arbitrary units

of the velocity distributions is expected to be larger since some clouds are dragged along with the ambient gas while others are only slightly accelerated. Yet, there is still no jump in the systematic velocity expected. One might also consider the morphology of the emission-line gas as a test. If the clouds are distributed all over the shocked ambient gas, the observed spatial distribution would be similar to the regions of enhanced emission in the bremsstrahlung emission maps (Fig. 9). Shells of line emission, however, are at odds with the observations [12], where emission is concentrated towards the center.

The COC model assumes that the line-emitting gas clouds are generated in a turbulent multi-phase cocoon medium, stimulated by shearing and entrainment of ambient gas at the contact discontinuity. The magnitudes of the systematic velocities are of order 1000 km s^{-1}, however with notable variations due to large-scale vortices. The velocity distribution widths for the line-emitting gas have to be derived from the turbulence simulations of Krause & Alexander [7], since the turbulent velocities seen in the large-scale simulations are only valid for the hot cocoon phase. The cooler and denser line-emitting gas shows lower turbulent velocities, which are in the range of few hundreds to 1000 km s^{-1}, depending on the density and temperature of the ambient gas [7]. These values are in reasonable agreement with the observed values. The systematic motion is found to be receding on the side of the approaching jet and approaching on the side of the receding jet side due to the large-scale backflow towards the midplane – in contrast to the directions inferred from observations. However, extinction effects may be important and weaken the emission from the far side of the cocoon. If this is considered, the near side contributes most to the emission and the systematic radial velocity (directed away from the jet axis, Fig. 4) could dominate on a brightness-weighted distribution. The observed jump in radial velocity at the core cannot be reproduced clearly. A transition between approaching and receding motion is, however, expected for sources not lying in the plane of the sky.

Fig. 1 in our previous report [3] shows the expected morphology of the emission-line gas if it is assumed to be distributed uniformly within the cocoon. The elongation along the radio source and the distribution around the center agree with the observed morphology. The correlation between emitted power and cocoon kinetic energy described by Krause [6] allows a further check of this model. For 10^{60} erg of total injected power, 10^{58-59} erg are expected for the kinetic energy of the cocoon according to the large-scale simulations. For this energy, an emission-line luminosity of 10^{46-47} erg s^{-1} would be expected – much more than typically observed. We note, however, that this value is expected to decrease, if less ambient gas is available to the cocoon, e.g. due to suppressed entrainment.

We conclude that none of the models can explain all the kinematics and morphology of the emission-line gas at the current stage. Yet we pointed out possible explanations for the deviations. For the SAG model, the morphological differences seem to be most severe, while the COC model fails to reproduce the emission power. Both models cannot explain a distinct jump in radial velocities near the core. However, since this project is still in an early phase, significant improvement can be expected and hence, further modelling is desirable to resolve the remaining problems.

4 Summary

We use the magnetohydrodynamics code NIRVANA to simulate the evolution of powerful extragalactic jets into the surrounding intra-cluster medium, using a realistic set of parameters and taking into account the effects of magnetic fields. Due to the large extent of these objects and the large number of time steps necessary to reach the typical source ages, these simulations have to be carried out on the NEC SX vector machines. The code has been optimized for this architecture previously within this project, runs very efficiently and allows running the simulations with different parameters due to the uncertainty of the latter.

The performed simulations allowed us to test two opposing models in the literature about emitting gaseous nebulae around radio galaxies at high redshift, which show strong indication of interaction with the jets, possibly even caused by the jets. While both models qualitatively and quantitatively show some agreement with recent integral-field spectroscopy observations, neither of them is without inconsistency. We conclude that the models are still to simple to adequately describe the real situation. This may make simulations necessary which explicitly include the emitting clouds, although this is yet very much at the border of presently available computing power.

Acknowledgments. This work was also supported by the Deutsche Forschungsgemeinschaft (Sonderforschungsbereich 439).

References

1. Gaibler, V. 2008, Ph.D. Thesis
2. Gaibler, V., Vigelius, M., Krause, M., and Camenzind, M. MHD Code Optimizations and Jets in Dense Gaseous Halos. In *High Performance Computing in Science and Engineering '06, eds.: Nagel, W. E., Jäger, W., and Resch, M., Springer*, 2006
3. Gaibler, V., and Camenzind, M. Magnetic Fields in Very Light Extragalactic Jets. In *High Performance Computing in Science and Engineering '08, eds.: Nagel, W. E., Kröner, D. B., and Resch, M., Springer*, 2008
4. Krause, M. 2003, A&A, 398, 113
5. Krause, M. 2005, A&A, 431, 45
6. Krause, M. G. H. 2008, Memorie della Societa Astronomica Italiana, 79, 1162
7. Krause, M., & Alexander, P. 2007, MNRAS, 376, 465
8. Krause, M., Gaibler, V., and Camenzind, M. Simulations of Astrophysical Jets in Dense Environments. In *High Performance Computing in Science and Engineering '05, eds.: Nagel, W. E., Jäger, W., and Resch, M., Springer*, 2005
9. Meisenheimer, K., & Hippelein, H. 1992, A&A, 264, 455
10. Morganti, R. 2008, in Extragalactic Jets: Theory and Observation from Radio to Gamma Ray, eds.: Rector, T. A. and De Young, D. S., Vol. 386, 210
11. Morganti, R., Tadhunter, C. N., & Oosterloo, T. A. 2005, A&A, 444
12. Nesvadba, N. P. H., Lehnert, M. D., De Breuck, C., Gilbert, A. M., & van Breugel, W. 2008, A&A, 491, 407
13. van Breugel, W., De Breuck, C., Stanford, S. A., Stern, D., Röttgering, H., & Miley, G. 1999, ApJ, 518
14. van Ojik, R., Roettgering, H. J. A., Miley, G. K., & Hunstead, R. W. 1997, A&A, 317, 358
15. Villar-Martín, M. 2007, New Astronomy Review, 51, 194
16. Ziegler, U., & Yorke, H. W. 1997, Computer Physics Communications, 101, 54

The SuperN-Project: Current Advances in Understanding Core Collapse Supernovae

B. Müller, A. Marek, and H.-Th. Janka

Max-Planck-Institut für Astrophysik, Karl-Schwarzschild-Strasse 1, Postfach 1317, D-85741 Garching bei München, Germany
bjmuellr@mpa-garching.mpg.de

Summary. We give an overview of the problems and the current status of our two-dimensional (core collapse) supernova modelling, and present the system of equations and the algorithm for its solution that are employed in our code, and report on our continuing efforts to increase the computational efficiency of our supernova code VERTEX, which now has full MPI functionality and can be run efficiently on several nodes of the NEC SX-8. We also discuss selected results that document the progress achieved by recent simulations that are performed on the NEC SX-8 at the HLRS Stuttgart. In particular, we have studied the influence of the nuclear equation of state on the neutrino and gravitational wave signals from collapsing stellar cores and on the explosion mechanism. Another focus of our work was on the observable signatures of a subclass of supernovae originating from low-mass progenitors below about ten solar masses.

1 Introduction

A star more massive than about 8 solar masses ends its live in a cataclysmic explosion, a supernova. Its quiescent evolution comes to an end, when the pressure in its inner layers is no longer able to balance the inward pull of gravity. Throughout its life, the star sustained this balance by generating energy through a sequence of nuclear fusion reactions, forming increasingly heavier elements in its core. However, when the core consists mainly of iron-group nuclei, central energy generation ceases. The fusion reactions producing iron-group nuclei relocate to the core's surface, and their "ashes" continuously increase the core's mass. Similar to a white dwarf, such a core is stabilised against gravity by the pressure of its degenerate gas of electrons. However, to remain stable, its mass must stay smaller than the Chandrasekhar limit. When the core grows larger than this limit, it collapses to a neutron star, and a huge amount ($\sim 10^{53}$ erg) of gravitational binding energy is set free. Most ($\sim 99\%$) of this energy is radiated away in neutrinos, but a small fraction is transferred to the outer stellar layers and drives the violent mass ejection which disrupts the star in a supernova.

W.E. Nagel et al. (eds.), *High Performance Computing in Science and Engineering '09*, DOI 10.1007/978-3-642-04665-0_2,

Despite 40 years of research, the details of how this energy transfer happens and how the explosion is initiated are still not well understood. Observational evidence about the physical processes deep inside the collapsing star is sparse and almost exclusively indirect. The only direct observational access is via measurements of neutrinos or gravitational waves. To obtain insight into the events in the core, one must therefore heavily rely on sophisticated numerical simulations. The enormous amount of computer power required for this purpose has led to the use of several, often questionable, approximations and numerous ambiguous results in the past. Fortunately, however, the development of numerical tools and computational resources has meanwhile advanced to a point, where it is becoming possible to perform multi-dimensional simulations with unprecedented accuracy. Therefore there is hope that the physical processes which are essential for the explosion can finally be unravelled.

An understanding of the explosion mechanism is required to answer many important questions of nuclear, gravitational, and astro-physics like the following:

- How do the explosion energy, the explosion timescale, and the mass of the compact remnant depend on the progenitor's mass? Is the explosion mechanism the same for all progenitors? For which stars are black holes left behind as compact remnants instead of neutron stars?
- What is the role of the – poorly known – equation of state (EoS) for the proto neutron star? Do softer or stiffer EoSs favour the explosion of a core collapse supernova?
- How do neutron stars receive their natal kicks? Are they accelerated by asymmetric mass ejection and/or anisotropic neutrino emission?
- What are the generic properties of the neutrino emission and of the gravitational wave signal that are produced during stellar core collapse and explosion? Up to which distances could these signals be measured with operating or planned detectors on earth and in space? And what can one learn about supernova dynamics from a future measurement of such signals in case of a Galactic supernova?
- How do supernovae contribute to the enrichment of the intergalactic medium with heavy elements? What kind of nucleosynthesis processes occur during and after the explosion? Can the elemental composition of supernova remnants be explained correctly by the numerical simulations?

2 Numerical Models

2.1 History and Constraints

According to theory, a shock wave is launched at the moment of "core bounce" when the neutron star begins to emerge from the collapsing stellar iron core. There is general agreement, supported by all "modern" numerical simulations, that this shock is unable to propagate directly into the stellar mantle and envelope, because it looses too much energy in dissociating iron into free nucleons while it moves through the outer core. The "prompt" shock ultimately stalls. Thus the currently favoured theoretical paradigm needs to exploit the fact that a huge energy reservoir is present in the form of neutrinos, which are abundantly emitted from the hot, nascent neutron star. The absorption of electron neutrinos and anti-neutrinos by free nucleons in the post shock layer is thought to reenergize the shock, and lead to the supernova explosion.

Detailed *spherically symmetric* hydrodynamic models, which recently include a very accurate treatment of the time-dependent, multi-flavour, multi-frequency neutrino transport based on a numerical solution of the Boltzmann transport equation [1, 2, 3], reveal that this "delayed, neutrino-driven mechanism" does not work as simply as originally envisioned. Although in principle able to trigger the explosion (e.g., [4, 5, 6]), neutrino energy transfer to the post-shock matter turned out to be too weak. For inverting the infall of the stellar core and initiating powerful mass ejection, an increase of the efficiency of neutrino energy deposition is needed.

A number of physical phenomena have been pointed out that can enhance neutrino energy deposition behind the stalled supernova shock. They are all linked to the fact that the real world is multi-dimensional instead of spherically symmetric (or one-dimensional; 1D) as assumed in the work cited above:

(1) Convective instabilities in the neutrino-heated layer between the neutron star and the supernova shock develop to violent convective overturn [7]. This convective overturn is helpful for the explosion, mainly because (a) neutrino-heated matter rises and increases the pressure behind the shock, thus pushing the shock further out, and (b) cool matter is able to penetrate closer to the neutron star where it can absorb neutrino energy more efficiently. Both effects allow multi-dimensional models to explode easier than spherically symmetric ones [8, 9, 10].
(2) Recent work [11, 12, 13, 14] has demonstrated that the stalled supernova shock is also subject to a second non-radial low-mode instability, called SASI, which can grow to a dipolar, global deformation of the shock [13, 15, 16].
(3) Convective energy transport inside the nascent neutron star [17, 18, 19, 20] might enhance the energy transport to the neutrinosphere and could thus boost the neutrino luminosities. This would in turn increase the neutrino-heating behind the shock.

This list of multi-dimensional phenomena awaits more detailed exploration in multi-dimensional simulations. Until recently, such simulations have been performed with only a grossly simplified treatment of the involved microphysics, in particular of the neutrino transport and neutrino-matter interactions. At best, grey (i.e., single energy) flux-limited diffusion schemes were employed. Since, however, the role of the neutrinos is crucial for the problem, and because previous experience shows that the outcome of simulations is indeed very sensitive to the employed transport approximations, studies of the explosion mechanism require the best available description of the neutrino physics. This implies that one has to solve the Boltzmann transport equation for neutrinos.

2.2 Recent Calculations and the Need for TFlop Simulations

We have recently advanced to a new level of accuracy for supernova simulations by generalising the VERTEX code, a Boltzmann solver for neutrino transport, from spherical symmetry [21] to multi-dimensional applications [22, 23]. The corresponding mathematical model, and in particular our method for tackling the integro-differential transport problem in multi-dimensions, will be summarised in Sect. 3.

Results of a set of simulations with our code in 1D and 2D for progenitor stars with different masses have recently been published by [23, 24, 25, 26], and with respect to the expected gravitational-wave signals from rotating and convective supernova cores by [27, 28]. The recent progress in supernova modelling was summarised and set in perspective in a review article by [29]. Our collection of simulations has helped us to identify a number of aspects that are of crucial importance for the success or failure of the supernova explosion:

- The details of the stellar progenitor (i.e. the mass of the iron core and its radius–density relation) have substantial influence on the supernova evolution. Especially, we found explosions of stellar models with low-mass (i.e. small) iron cores [24], and with O-Ne-Mg cores embedded in a dilute envelope [26], whereas more massive stars resist the explosion more persistently [23, 25, 28]. Thus detailed studies with different progenitor models are necessary.
- The role of stellar rotation is ambiguous: On the one hand, it may support the expansion of the stalled shock by centrifugal forces and instigate overturn motion in the neutrino-heated post-shock matter by meridional circulation flows in addition to convective instabilities. On the other hand, the centrifugal support leads to an expansion of the proto-neutron star, which in turn reduces the neutrino luminosities; thus rotation may also be harmful for the neutrino-driven mechanism.
- The nuclear equation of state of state may play an essential role for the explosion mechanism since it affects the structure of the proto-neutron star, and hence also the neutrino emission and the heating efficiency.

All these aspects are potentially important, and some (or even all of them) may represent crucial ingredients for a successful supernova simulation. So far no multi-dimensional calculations have been performed, in which two or more of these items have been investigated simultaneously in a systematic fashion. It is clear that rather extensive parameter studies using multi-dimensional simulations are required to identify the physical processes which are essential for the explosion. Since on a dedicated machine performing at a sustained speed of about 30 GFlops already a single 2D simulation has a turn-around time of more than half a year, these parameter studies are not possible without TFlop simulations.

3 The Mathematical Model

The non-linear system of partial differential equations which is solved in our code consists of the following components:

- The Euler equations of hydrodynamics, supplemented by advection equations for the electron fraction and the chemical composition of the fluid, and formulated in spherical coordinates;
- the Poisson equation for calculating the gravitational source terms which enter the Euler equations, including corrections for general relativistic effects;
- the Boltzmann transport equation which determines the (non-equilibrium) distribution function of the neutrinos;
- the emission, absorption, and scattering rates of neutrinos, which are required for the solution of the Boltzmann equation;
- the equation of state of the stellar fluid, which provides the closure relation between the variables entering the Euler equations, i.e. density, momentum, energy, electron fraction, composition, and pressure.

In what follows we will briefly summarise the neutrino transport algorithms. For a more complete description of the entire code we refer the reader to [23], and the references therein.

3.1 "Ray-by-Ray Plus" Variable Eddington Factor Solution of the Neutrino Transport Problem

The crucial quantity required to determine the source terms for the energy, momentum, and electron fraction of the fluid owing to its interaction with the neutrinos is the neutrino distribution function in phase space, $f(r, \vartheta, \phi, \epsilon, \Theta, \Phi, t)$. Equivalently, the neutrino intensity $I = c/(2\pi\hbar c)^3 \cdot \epsilon^3 f$ may be used. Both are seven-dimensional functions, as they describe, at every point in space (r, ϑ, ϕ), the distribution of neutrinos propagating with energy ϵ into the direction (Θ, Φ) at time t (Fig. 1).

The evolution of I (or f) in time is governed by the Boltzmann equation, and solving this equation is, in general, a six-dimensional problem (as time is usually not counted as a separate dimension). A solution of this equation by direct discretisation (using an S_N scheme) would require computational resources in the PetaFlop range. Although there are attempts by at least one group in the United States to follow such an approach, we feel that, with the currently available computational resources, it is mandatory to reduce the dimensionality of the problem.

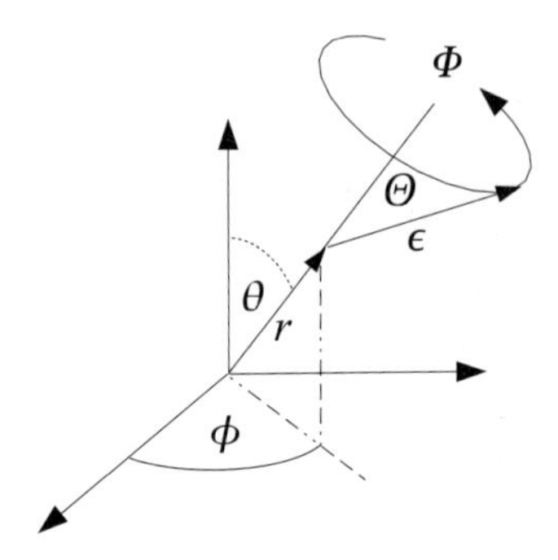

Fig. 1. Illustration of the phase space coordinates (see the main text)

Actually this should be possible, since the source terms entering the hydrodynamic equations are *integrals* of I over momentum space (i.e. over ϵ, Θ, and Φ), and thus only a fraction of the information contained in I is truly required to compute the dynamics of the flow. It makes therefore sense to consider angular moments of I, and to solve evolution equations for these moments, instead of dealing with the Boltzmann equation directly. The 0th to 3rd order moments are defined as

$$J, \boldsymbol{H}, \mathbf{K}, \mathbf{L}, \ldots (r, \vartheta, \phi, \epsilon, t) = \frac{1}{4\pi} \int I(r, \vartheta, \phi, \epsilon, \Theta, \Phi, t)\, \boldsymbol{n}^{0,1,2,3,\ldots}\, \mathrm{d}\Omega \qquad (1)$$

where $\mathrm{d}\Omega = \sin\Theta\, \mathrm{d}\Theta\, \mathrm{d}\Phi$, $\boldsymbol{n} = (\cos\Theta, \sin\Theta\cos\Phi, \sin\Theta\sin\Phi)$, and exponentiation represents repeated application of the dyadic product. Note that the moments are *tensors* of the required rank.

This leaves us with a four-dimensional problem. So far no approximations have been made. In order to reduce the size of the problem even further, one needs to resort to assumptions on its symmetry. At this point, one usually employs azimuthal symmetry for the stellar matter distribution, i.e. any dependence on the azimuth angle ϕ is ignored, which implies that the hydrodynamics of the problem can be treated in two dimensions. It also implies $I(r, \vartheta, \epsilon, \Theta, \Phi) = I(r, \vartheta, \epsilon, \Theta, -\Phi)$. If, in addition, it is assumed that I is even independent of Φ, then each of the angular moments of I becomes a *scalar*, which depends on two spatial dimensions, and one dimension in momentum space: $J, H, K, L = J, H, K, L(r, \vartheta, \epsilon, t)$. Thus we have reduced the problem to three dimensions in total.

The System of Equations

With the aforementioned assumptions it can be shown [23], that in order to compute the source terms for the energy and electron fraction of the fluid, the following two transport equations need to be solved:

$$\left(\frac{1}{c}\frac{\partial}{\partial t}+\beta_r\frac{\partial}{\partial r}+\boldsymbol{\frac{\beta_\vartheta}{r}\frac{\partial}{\partial\vartheta}}\right)J+J\left(\frac{1}{r^2}\frac{\partial(r^2\beta_r)}{\partial r}+\boldsymbol{\frac{1}{r\sin\vartheta}\frac{\partial(\sin\vartheta\beta_\vartheta)}{\partial\vartheta}}\right)$$
$$+\frac{1}{r^2}\frac{\partial(r^2H)}{\partial r}+\frac{\beta_r}{c}\frac{\partial H}{\partial t}-\frac{\partial}{\partial\epsilon}\left\{\frac{\epsilon}{c}\frac{\partial\beta_r}{\partial t}H\right\}-\frac{\partial}{\partial\epsilon}\left\{\epsilon J\left(\frac{\beta_r}{r}+\boldsymbol{\frac{1}{2r\sin\vartheta}\frac{\partial(\sin\vartheta\beta_\vartheta)}{\partial\vartheta}}\right)\right\}$$
$$-\frac{\partial}{\partial\epsilon}\left\{\epsilon K\left(\frac{\partial\beta_r}{\partial r}-\frac{\beta_r}{r}-\boldsymbol{\frac{1}{2r\sin\vartheta}\frac{\partial(\sin\vartheta\beta_\vartheta)}{\partial\vartheta}}\right)\right\}$$
$$+J\left(\frac{\beta_r}{r}+\boldsymbol{\frac{1}{2r\sin\vartheta}\frac{\partial(\sin\vartheta\beta_\vartheta)}{\partial\vartheta}}\right)+K\left(\frac{\partial\beta_r}{\partial r}-\frac{\beta_r}{r}-\boldsymbol{\frac{1}{2r\sin\vartheta}\frac{\partial(\sin\vartheta\beta_\vartheta)}{\partial\vartheta}}\right)$$
$$+\frac{2}{c}\frac{\partial\beta_r}{\partial t}H=C^{(0)},\quad(2)$$

$$\left(\frac{1}{c}\frac{\partial}{\partial t}+\beta_r\frac{\partial}{\partial r}+\boldsymbol{\frac{\beta_\vartheta}{r}\frac{\partial}{\partial\vartheta}}\right)H+H\left(\frac{1}{r^2}\frac{\partial(r^2\beta_r)}{\partial r}+\boldsymbol{\frac{1}{r\sin\vartheta}\frac{\partial(\sin\vartheta\beta_\vartheta)}{\partial\vartheta}}\right)$$
$$+\frac{\partial K}{\partial r}+\frac{3K-J}{r}+H\left(\frac{\partial\beta_r}{\partial r}\right)+\frac{\beta_r}{c}\frac{\partial K}{\partial t}-\frac{\partial}{\partial\epsilon}\left\{\frac{\epsilon}{c}\frac{\partial\beta_r}{\partial t}K\right\}$$
$$-\frac{\partial}{\partial\epsilon}\left\{\epsilon L\left(\frac{\partial\beta_r}{\partial r}-\frac{\beta_r}{r}-\boldsymbol{\frac{1}{2r\sin\vartheta}\frac{\partial(\sin\vartheta\beta_\vartheta)}{\partial\vartheta}}\right)\right\}$$
$$-\frac{\partial}{\partial\epsilon}\left\{\epsilon H\left(\frac{\beta_r}{r}+\boldsymbol{\frac{1}{2r\sin\vartheta}\frac{\partial(\sin\vartheta\beta_\vartheta)}{\partial\vartheta}}\right)\right\}+\frac{1}{c}\frac{\partial\beta_r}{\partial t}(J+K)=C^{(1)}.\quad(3)$$

These are evolution equations for the neutrino energy density, J, and the neutrino flux, H, and follow from the zeroth and first moment equations of the comoving frame (Boltzmann) transport equation in the Newtonian, $\mathcal{O}(v/c)$ approximation. The quantities $C^{(0)}$ and $C^{(1)}$ are source terms that result from the collision term of the Boltzmann equation, while $\beta_r = v_r/c$ and $\beta_\vartheta = v_\vartheta/c$, where v_r and v_ϑ are the components of the hydrodynamic velocity, and c is the speed of light. The functional dependences $\beta_r = \beta_r(r,\vartheta,t)$, $J = J(r,\vartheta,\epsilon,t)$, etc. are suppressed in the notation. This system includes four unknown moments (J, H, K, L) but only two equations, and thus needs to be supplemented by two more relations. This is done by substituting $K = f_K \cdot J$ and $L = f_L \cdot J$, where f_K and f_L are the variable Eddington factors, which for the moment may be regarded as being known, but in our case is indeed determined from a separate simplified ("model") Boltzmann equation.

The moment equations (2) and (3) are very similar to the $\mathcal{O}(v/c)$ equations in spherical symmetry which were solved in the 1D simulations of [21] (see Eqs. (7), (8), (30), and (31) of the latter work). This similarity has allowed us to reuse a good fraction of the one-dimensional version of VERTEX, for coding the multi-dimensional algorithm. The additional terms necessary for this purpose have been set in boldface above.

Finally, the changes of the energy, e, and electron fraction, Y_e, required for the hydrodynamics are given by the following two equations

$$\frac{\mathrm{d}e}{\mathrm{d}t} = -\frac{4\pi}{\rho}\int_0^\infty \mathrm{d}\epsilon \sum_{\nu\in(\nu_\mathrm{e},\bar{\nu}_\mathrm{e},\dots)} C_\nu^{(0)}(\epsilon), \tag{4}$$

$$\frac{\mathrm{d}Y_\mathrm{e}}{\mathrm{d}t} = -\frac{4\pi\, m_\mathrm{B}}{\rho}\int_0^\infty \frac{\mathrm{d}\epsilon}{\epsilon}\left(C_{\nu_\mathrm{e}}^{(0)}(\epsilon) - C_{\bar{\nu}_\mathrm{e}}^{(0)}(\epsilon)\right) \tag{5}$$

(for the momentum source terms due to neutrinos see [23]). Here m_B is the baryon mass, and the sum in Eq. (4) runs over all neutrino types. The full system consisting of Eqs. (2–5) is stiff, and thus requires an appropriate discretisation scheme for its stable solution.

Method of Solution

In order to discretise Eqs. (2–5), the spatial domain $[0, r_\mathrm{max}] \times [\vartheta_\mathrm{min}, \vartheta_\mathrm{max}]$ is covered by N_r radial, and N_ϑ angular zones, where $\vartheta_\mathrm{min} = 0$ and $\vartheta_\mathrm{max} = \pi$ correspond to the north and south poles, respectively, of the spherical grid. (In general, we allow for grids with different radial resolutions in the neutrino transport and hydrodynamic parts of the code. The number of radial zones for the hydrodynamics will be denoted by N_r^hyd.) The number of bins used in energy space is N_ϵ and the number of neutrino types taken into account is N_ν.

The equations are solved in two operator-split steps corresponding to a lateral and a radial sweep.

In the first step, we treat the boldface terms in the respectively first lines of Eqs. (2–3), which describe the lateral advection of the neutrinos with the stellar fluid, and thus couple the angular moments of the neutrino distribution of neighbouring angular zones. For this purpose we consider the equation

$$\frac{1}{c}\frac{\partial \Xi}{\partial t} + \frac{1}{r\sin\vartheta}\frac{\partial(\sin\vartheta\,\beta_\vartheta\,\Xi)}{\partial\vartheta} = 0\,, \tag{6}$$

where Ξ represents one of the moments J or H. Although it has been suppressed in the above notation, an equation of this form has to be solved for each radius, for each energy bin, and for each type of neutrino. An explicit upwind scheme is used for this purpose.

In the second step, the radial sweep is performed. Several points need to be noted here:

- terms in boldface not yet taken into account in the lateral sweep, need to be included into the discretisation scheme of the radial sweep. This can be done in a straightforward way since these remaining terms do not include derivatives of the transport variables J or H. They only depend on the hydrodynamic velocity v_ϑ, which is a *constant* scalar field for the transport problem.
- the right hand sides (source terms) of the equations and the coupling in energy space have to be accounted for. The coupling in energy is non-local,

since the source terms of Eqs. (2) and (3) stem from the Boltzmann equation, which is an integro-differential equation and couples all the energy bins
- the discretisation scheme for the radial sweep is *implicit* in time. Explicit schemes would require very small time steps to cope with the stiffness of the source terms in the optically thick regime, and the small CFL time step dictated by neutrino propagation with the speed of light in the optically thin regime. Still, even with an implicit scheme $\gtrsim 10^5$ time steps are required per simulation. This makes the calculations expensive.

Once the equations for the radial sweep have been discretized in radius and energy, the resulting solver is applied ray-by-ray for each angle ϑ and for each type of neutrino, i.e. for constant ϑ, N_ν two-dimensional problems need to be solved.

The discretisation itself is done using a second order accurate scheme with backward differencing in time according to [21]. This leads to a non-linear system of algebraic equations, which is solved by Newton-Raphson iteration with explicit construction and inversion of the corresponding Jacobian matrix with the Block-Thomas algorithm.

3.2 Parallelization

The ray-by-ray approximation readily lends itself to parallelization over the different angular zones. In order to make efficient use of the NEC SX-8 with its small shared-memory units (8 CPUs per node), distributed memory parallelism is indispensable. For this reason, a cooperation between MPA and the Teraflop Workbench at HLRS was set up in 2007 to initiate the development of an MPI version of the VERTEX code, which was completed in 2008. The development branch of VERTEX has recently been re-integrated into the production version of our code, and now provides MPI functionality not only on the NEC SX-8, but also on the IBM Power6 system at the Rechenzentrum Garching of the Max-Planck-Gesellschaft, and on the SGI Altix 4700 at LRZ.

Scaling tests demonstrate the great potential of the new code version. Fig. 2 shows the accumulated CPU time and the total wall clock time as a function of the number of processors for a test run with 256 angular rays on the NEC machines SX-8 (8 CPUs per node) and SX-9 (16 CPUs/node) at HLRS and on the IBM Power6 575 (32 CPUs/node) at the Rechenzentrum Garching (RZG) of the Max-Planck-Gesellschaft. On the SX-8, a speed-up of 7.85 could be obtained by using eight nodes instead of one, which corresponds to a parallel efficiency of more than 98%. On the IBM Power 6 575, the scaling is not quite as good; the speed-up on four nodes is 3.81 (parallel efficiency $\approx 95\%$). It should be pointed out, however, that the new MPI version is about 40% faster than the shared-memory version of the code even on a single Power 6 node, because it can better cope with the machine's ccNUMA architecture. On the new SX-9 machine at HLRS, only a few tests could be conducted so

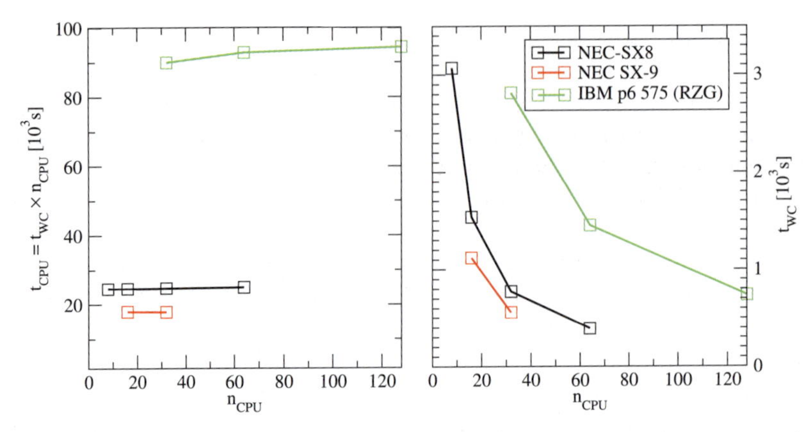

Fig. 2. Accumulated CPU time t_{CPU} (left panel) and wall clock time t_{WC} for a scaling test (20 time-steps) with 256 angular rays on different machines

far, and no statements concerning the scaling on more than two nodes can be made yet. Fig. 2 also demonstrates the extreme usefulness of the vector platforms at HLRS for supernova simulations with VERTEX. For a given number of processors, the code performance on SX-8 is almost four times as high as on the most powerful scalar platform presently available to us.

4 Recent Results and Ongoing Work

We make use of the computer resources available to us at the HLRS to address some of the important questions in SN theory (see Sect. 1) with 2D-simulations. As a full MPI version of VERTEX has become available only very recently, we typically ran our code on one node of the NEC SX-8 (8 processors, OpenMP-parallelised) with 98.3% of vector operations and up to 30000 MFLOPS per second. In the following we present some of our results from these simulations that are currently conducted at the HLRS. For the neutrino interaction rates we use the full set as described in [30], and general relativistic effects are taken into account according to [31].

4.1 Long-time Simulations of Massive Progenitor Models

During the last year, we continued our simulations of several massive progenitor models at HLRS (see [32, 33, 34] for earlier status reports) in order to study the influence of the nuclear equation of state and rotation on the explosion mechanism. As discussed in detail in Marek&Janka [25], we have obtained delayed neutrino-driven explosions with a soft nuclear equation of state of [35] for a rotating 15 $M_\odot$ progenitor. In the latter case, the explosion sets in at a rather late time , i.e. more than 500 ms after core bounce. Hydrodynamic instabilities like convection and a non-radial instability of the accretion shock front, the so-called SASI [36] were found to be of crucial importance for the explosion. Several other simulations covering the non-rotating case, and also employing the alternative stiff nuclear equation of state of [37], have been performed as well (and others are still on the machines). A comparison of the different models allows us to draw some interesting conclusions: A soft nuclear equation of state that causes a rapid contraction and a smaller radius of the forming neutron star and thus a fast release of gravitational binding energy, seems to be more favourable for an explosion. Rotation has the opposite effect because it leads to a more extended and cooler neutron star and thus lower neutrino luminosities and mean energies and overall less neutrino heating.

We have also studied the observational neutrino and gravitational wave signatures from core collapse supernovae for different nuclear equations of state [28]. Concerning the neutrino signal, we found that the absolute values of the luminosities and mean energies of the neutrinos radiated by the proto-neutron star are higher by up to 20% for a soft nuclear equation of state. Interesting equation-of-state dependent features can also be seen in the gravitational wave signal arising from non-radial mass motion and anisotropic neutrino emission (see Fig. 3 and [28]). The amplitudes of both the matter and neutrino signal differ significantly for the soft and stiff equation of state, as do the gravitational wave spectra: In the case of the soft equation of state of [35], the most prominent peak is located at significantly higher frequencies (≈ 700 Hz) than for the stiff equation of state.

4.2 Explosion Models of Oxygen-neon-magnesium Cores

In recent simulations at HLRS, we also focused on low-mass progenitors in an initial mass window between 8 and 10 solar masses. Stars in this mass range can develop cores composed of oxygen, neon, and magnesium instead of iron at the onset of core collapse, with an extremely sharp density gradient at the surface. This allows the shock front to expand continuously as it propagates into rapidly diluting infalling material, and a neutrino-powered explosion sets in within less than 100 ms after shock formation.

In [38], we provided a detailed analysis of the energetics of the explosion, the shock propagation, and the nucleosynthesis conditions in an O-Ne-Mg supernova. Our simulations allowed us to shed some light on the recent suggestion by [39], according to which the rapid neutron capture process (r-process) may operate in O-Ne-Mg supernovae, producing many of the heaviest elements in nature, such as Au and Pb. Unfortunately, the r-process scenario of [39] did not find support in our models. Instead, detailed nucleosynthesis calculations [40, 41] showed a significant overproduction of closed neutron-shell nuclei in the ($A \approx 90, N = 50$) region (like ^{90}Zr) in the early neutron-rich ejecta. The high yields of some of these nuclei may be in conflict with constraints from observed Galactic chemical abundances, but even minor variations in the models may moderate the strong overproduction (see [41]). In combination with

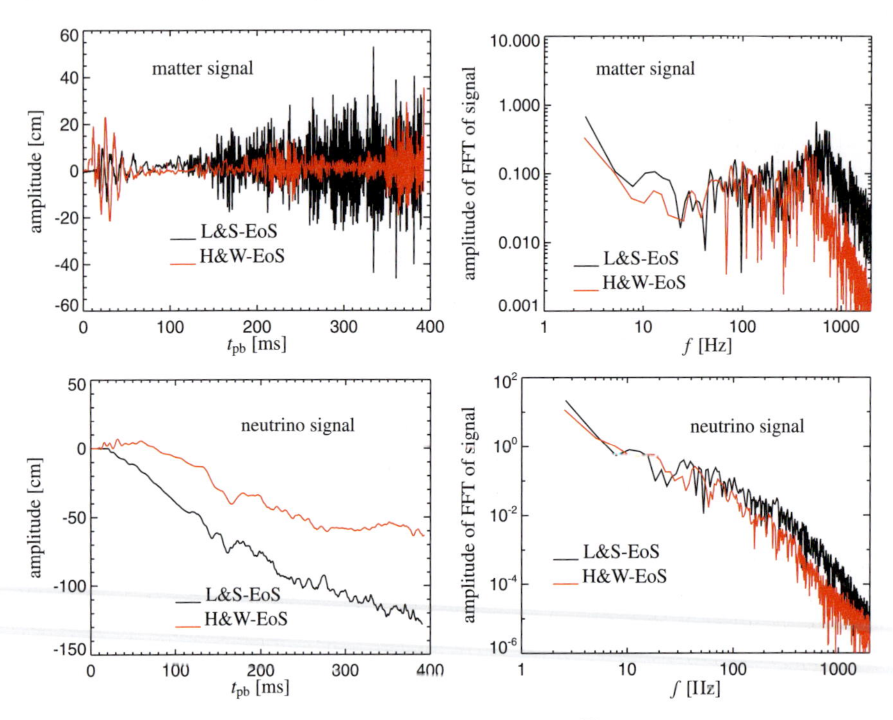

Fig. 3. Left panels: Gravitational wave amplitudes A_{20}^{E2} (see [28] for a precise definition) as a function of post-bounce time from multi-dimensional simulations of a $15\,M_\odot$ progenitor with the soft L&S EoS and the stiff H&W EoS. The signals components arising from non-radial mass motion (top) and anisotropic neutrino emission (bottom) are shown separately. Right panels: the corresponding Fourier spectra of the matter and neutrino signal

the absolute mass and relative distribution of the iron group elements, this suggests that our models are on the right track for explaining a class of subluminous and possibly underenergetic supernovae ("II-P"), and in particular the supernova of 1054, which produced the famous Crab nebula. However, the nucleosynthesis studies conducted so far [40, 41] still neglect multi-dimensional effects like convection; and we are therefore working on a detailed analysis of the nucleosynthesis conditions in a multi-dimensional simulation of an O-Ne-Mg supernova carried out at HLRS.

The neutrino signal from such successful explosion models is also of great interest, because it contains information about the structure and evolution of the newly-born proto-neutron star. In addition, a measured supernova neutrino signal may provide important clues about open questions in nuclear and particle physics, or, reversely neutrino signal features may provide information about the structure of the exploding star. Some of these related neutrino physics aspects were discussed by Lunardini et al. [42] based on data from our two-dimensional explosion model. Taking into account the effects of neutrino-

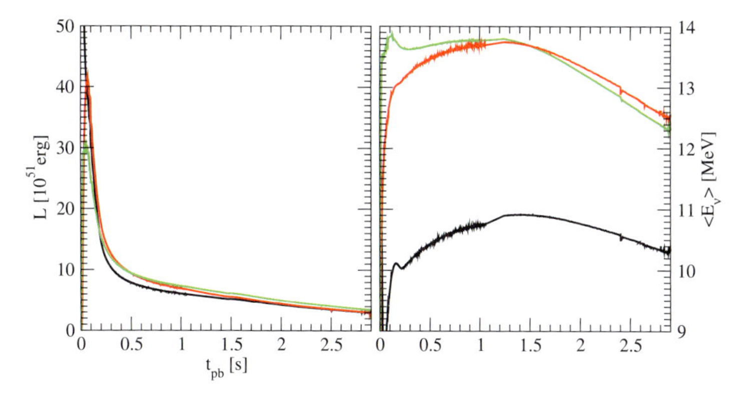

Fig. 4. Luminosities (left panel) and neutrino mean energies (right panel) from a long-time simulation of an O-Ne-Mg supernova. Electron neutrinos, electron antineutrinos and μ/τ neutrinos are shown as black, red and green lines, respectively

matter oscillations and collective $\nu\bar{\nu}$ flavour transformations in the exploding star, they found that the steep density gradient and the fast propagation of the shock lead to a detection signature clearly distinct from that of more massive stars with iron cores.

While our analysis in [42] focused on the first few hundred ms after the onset of the explosion, we have also been able to carry out long-time simulations covering the first few seconds of the Kelvin-Helmholtz cooling phase of the proto-neutron star in 1D. The neutrino signal (luminosities and mean energies) for such a model are shown in Fig. 4. It can clearly be seen that the neutrino mean energies start to drop after about 1.3 s as neutrino cooling begins to outweigh contractional heating. Such predictions for the long-time evolution of the neutrino signal are essential for determining the nucleosynthesis conditions in supernova cores and for linking the theoretical models to future neutrino observations in the event of a galactic supernova.

5 Conclusions and Outlook

We continued to simulate well-resolved 2D models of core collapse supernovae with detailed neutrino transport at the HLRS. We confirmed that non-radial hydrodynamic instabilities support the onset of supernova explosions, and for $8\,M_\odot$ – $15\,M_\odot$ progenitor models we obtained explosions powered by neutrino-heating. A comparison between different simulations of the $15\,M_\odot$ progenitor allowed us to partially assess the influence of rotation and the equation of state on the explosion mechanism, and in particular to determine the gravitational wave and neutrino signals for different nuclear equations of state. Our work on O-Ne-Mg core focused on observational signatures, namely the neutrino signal from the post-explosion phase and the nucleosynthesis yields.

However, our studies are limited to a few multi-dimensional (2D) models so far because of the prohibitively long turn-around time for supernova simulations with detailed neutrino transport. For this reason, an MPI version of the VERTEX code has been developed in a joint effort between HLRS and our group at MPA in order to exploit the parallel capabilities of the NEC SX-8 more efficiently. MPI functionality is now fully included in the production version of our code, and scaling results on the NEC SX-8 are extremely promising. With this new code version, the road towards a more thorough exploration of the parameter space (nuclear EoS, rotation, progenitor models, neutrino interactions) is now open.

Acknowledgements. Support by the Deutsche Forschungsgemeinschaft through the SFB/TR27 "Neutrinos and Beyond" and the SFB/TR7 "Gravitational Wave Astronomy", and by the Cluster of Excellence EXC 153 "Origin and Structure of the Universe" (http://www.universe-cluster.de) are acknowledged, as well computer time grants of the HLRS, NIC Jülich, and Rechenzentrum Garching are acknowledged. We thank especially K. Benkert for her extremely valuable work on the MPI version of VERTEX.

References

1. Rampp, M., Janka, H.T.: Spherically Symmetric Simulation with Boltzmann Neutrino Transport of Core Collapse and Postbounce Evolution of a 15 $M_\odot$ Star. Astrophys. J. **539** (2000) L33–L36
2. Mezzacappa, A., Liebendörfer, M., Messer, O.E., Hix, W.R., Thielemann, F., Bruenn, S.W.: Simulation of the Spherically Symmetric Stellar Core Collapse, Bounce, and Postbounce Evolution of a Star of 13 Solar Masses with Boltzmann Neutrino Transport, and Its Implications for the Supernova Mechanism. Phys. Rev. Letters **86** (2001) 1935–1938
3. Liebendörfer, M., Mezzacappa, A., Thielemann, F., Messer, O.E., Hix, W.R., Bruenn, S.W.: Probing the gravitational well: No supernova explosion in spherical symmetry with general relativistic Boltzmann neutrino transport. Phys. Rev. D **63** (2001) 103004–+
4. Bethe, H.A.: Supernova mechanisms. Reviews of Modern Physics **62** (1990) 801–866
5. Burrows, A., Goshy, J.: A Theory of Supernova Explosions. Astrophys. J. **416** (1993) L75
6. Janka, H.T.: Conditions for shock revival by neutrino heating in core-collapse supernovae. Astron. Astrophys. **368** (2001) 527–560
7. Herant, M., Benz, W., Colgate, S.: Postcollapse hydrodynamics of SN 1987A - Two-dimensional simulations of the early evolution. Astrophys. J. **395** (1992) 642–653
8. Herant, M., Benz, W., Hix, W.R., Fryer, C.L., Colgate, S.A.: Inside the supernova: A powerful convective engine. Astrophys. J. **435** (1994) 339
9. Burrows, A., Hayes, J., Fryxell, B.A.: On the nature of core-collapse supernova explosions. Astrophys. J. **450** (1995) 830

10. Janka, H.T., Müller, E.: Neutrino heating, convection, and the mechanism of Type-II supernova explosions. Astron. Astrophys. **306** (1996) 167–+
11. Thompson, C.: Accretional Heating of Asymmetric Supernova Cores. Astrophys. J. **534** (2000) 915–933
12. Blondin, J.M., Mezzacappa, A., DeMarino, C.: Stability of Standing Accretion Shocks, with an Eye toward Core-Collapse Supernovae. Astrophys. J. **584** (2003) 971–980
13. Scheck, L., Plewa, T., Janka, H.T., Kifonidis, K., Müller, E.: Pulsar Recoil by Large-Scale Anisotropies in Supernova Explosions. Phys. Rev. Letters **92** (2004) 011103–+
14. Foglizzo, T., Galletti, P., Scheck, L., Janka, H.T.: Instability of a Stalled Accretion Shock: Evidence for the Advective-Acoustic Cycle. Astrophys. J. **654** (2007) 1006–1021
15. Scheck, L., Kifonidis, K., Janka, H.T., Müller, E.: Multidimensional supernova simulations with approximative neutrino transport. I. Neutron star kicks and the anisotropy of neutrino-driven explosions in two spatial dimensions. Astron. Astrophys. **457** (2006) 963–986
16. Scheck, L., Janka, H.T., Foglizzo, T., Kifonidis, K.: Multidimensional supernova simulations with approximative neutrino transport. II. Convection and the advective-acoustic cycle in the supernova core. Astron. Astrophys. **477** (2008) 931–952
17. Keil, W., Janka, H.T., Müller, E.: Ledoux Convection in Protoneutron Stars—A Clue to Supernova Nucleosynthesis? Astrophys. J. **473** (1996) L111
18. Burrows, A., Lattimer, J.M.: The birth of neutron stars. Astrophys. J. **307** (1986) 178–196
19. Pons, J.A., Reddy, S., Prakash, M., Lattimer, J.M., Miralles, J.A.: Evolution of Proto-Neutron Stars. Astrophys. J. **513** (1999) 780–804
20. Marek, A.: Multi-dimensional simulations of core collapse supernovae with different equations of state for hot proto-neutron stars. PhD thesis, Technische Universität München (2007)
21. Rampp, M., Janka, H.T.: Radiation hydrodynamics with neutrinos. Variable Eddington factor method for core-collapse supernova simulations. Astron. Astrophys. **396** (2002) 361–392
22. Janka, H.T., Buras, R., Kifonidis, K., Marek, A., Rampp, M.: Core-Collapse Supernovae at the Threshold. In Marcaide, J.M., Weiler, K.W., eds.: Supernovae, Procs. of the IAU Coll. 192, Berlin, Springer (2004)
23. Buras, R., Rampp, M., Janka, H.T., Kifonidis, K.: Two-dimensional hydrodynamic core-collapse supernova simulations with spectral neutrino transport. I. Numerical method and results for a 15 Mȯ star. Astron. Astrophys. **447** (2006) 1049–1092
24. Buras, R., Janka, H.T., Rampp, M., Kifonidis, K.: Two–dimensional hydrodynamic core–collapse supernova simulations with spectral neutrino transport. II. Models for different progenitor stars. Astron. Astrophys. **457** (2006) 281–308
25. Marek, A., Janka, H.T.: Delayed Neutrino-Driven Supernova Explosions Aided by the Standing Accretion-Shock Instability. Astrophys. J. **694** (2009) 664–696
26. Kitaura, F.S., Janka, H.T., Hillebrandt, W.: Explosions of O–Ne–Mg cores, the Crab supernova, and subluminous type II–P supernovae. Astron. Astrophys. **450** (2006) 345–350

27. Müller, E., Rampp, M., Buras, R., Janka, H.T., Shoemaker, D.H.: Toward Gravitational Wave Signals from Realistic Core-Collapse Supernova Models. Astrophys. J. **603** (2004) 221–230
28. Marek, A., Janka, H.T., Müller, E.: Equation-of-state dependent features in shock-oscillation modulated neutrino and gravitational-wave signals from supernovae. Astron. Astrophys. **496** (2009) 475–494
29. Janka, H.T., Langanke, K., Marek, A., Martínez-Pinedo, G., Müller, B.: Theory of core-collapse supernovae. Phys. Rep. **442** (2007) 38–74
30. Marek, A., Janka, H.T., Buras, R., Liebendörfer, M., Rampp, M.: On ion-ion correlation effects during stellar core collapse. Astron. Astrophys. **443** (2005) 201–210
31. Marek, A., Dimmelmeier, H., Janka, H.T., Müller, E., Buras, R.: Exploring the relativistic regime with Newtonian hydrodynamics: an improved effective gravitational potential for supernova simulations. Astron. Astrophys. **445** (2006) 273–289
32. Marek, A., Kifonidis, K., Janka, H.T., Müller, B.: The SUPERN-project: Understanding core collapse supernovae. In Nagel, W.E., Jäger, W., Resch, M., eds.: High Performance Computing in Science and Engineering 06, Berlin, Springer (2006)
33. Marek, A., Kifonidis, K., Janka, H.T., Müller, B.: The SUPERN-project: Current progress in modelling core collapse supernovae. In Nagel, W.E., Kröner, D., Resch, M., eds.: High Performance Computing in Science and Engineering 07, Berlin, Springer (2007)
34. Müller, B., Marek, A., Janka, H.T.: The SuperN-project: Status and outlook. In Nagel, W.E., Kröner, D., Resch, M., eds.: High Performance Computing in Science and Engineering 08, Berlin, Springer (2008)
35. Lattimer, J.M., Swesty, F.D.: A generalized equation of state for hot, dense matter. Nuclear Physics A **535** (1991) 331–+
36. Foglizzo, T.: Non-radial instabilities of isothermal Bondi accretion with a shock: Vortical-acoustic cycle vs. post-shock acceleration. Astron. Astrophys. **392** (2002) 353–368
37. Hillebrandt, W., Wolff, R.G.: Models of Type II Supernova Explosions. In Arnett, W.D., Truran, J.W., eds.: Nucleosynthesis: Challenges and New Developments, Chicago, University of Chicago Press (1985) 131
38. Janka, H.T., Müller, B., Kitaura, F.S., Buras, R.: Dynamics of shock propagation and nucleosynthesis conditions in O-Ne-Mg core supernovae. Astron. Astrophys. **485** (2008) 199–208
39. Ning, H., Qian, Y.Z., Meyer, B.S.: r-Process Nucleosynthesis in Shocked Surface Layers of O-Ne-Mg Cores. Astrophys. J. **667** (2007) L159–L162
40. Hoffman, R.D., Müller, B., Janka, H.T.: Nucleosynthesis in O-Ne-Mg Supernovae. Astrophys. J. **676** (2008) L127–L130
41. Wanajo, S., Nomoto, K., Janka, H.T., Kitaura, F.S., Müller, B.: Nucleosynthesis in Electron Capture Supernovae of Asymptotic Giant Branch Stars. Astrophys. J. **695** (2009) 208–220
42. Lunardini, C., Müller, B., Janka, H.T.: Neutrino oscillation signatures of oxygen-neon-magnesium supernovae. Phys. Rev. D **78** (2008) 023016–+

Using Computational Steering to Explore the Parameter Space of Stability in a Suspension

Martin Hecht and Jens Harting

Institut für Computerphysik, Pfaffenwaldring 27, 70569 Stuttgart, Germany

Summary. We simulate a suspension of model colloidal particles interacting via DLVO (Derjaguin, Landau, Vervey, Overbeek) potentials. The interaction potentials can be related to experimental conditions, defined by the pH-value, the salt concentration and the volume fraction of solid particles suspended in the acqueous solvent. Depending on these parameters, the system shows different structural properties, including cluster formation, a glass-like repulsive structure, or a stable suspension. To explore the parameter space many simulations are required. In order to reduce the computational effort and data storage requirements, we developed a steering approach to control a running simulation and to detect interesting transitions from one region in the configuration space to another. In this article we describe the implementation of the steering in the simulation program and illustrate its applicability by several example cases.

1 Introduction

A very active field in physics is soft matter physics which has gained more and more importance during the last years. It comprises for example complex fluids, biological systems like membranes, solutions of large molecules like proteins, or suspensions of small soft or solid particles, which are commonly called "colloids". This is a very general class of materials which have in common that the particles are of a mesoscopic size, which means, they are so small that Brownian motion plays a considerable role for the behavior of the system. Soft matter has in common that the relevant energies (interaction barriers) are on the order of magnitude of the thermal energy ($k_{\mathrm{B}}T$). Since one can find examples for these materials nearly everywhere in everyday life (in medicine, food industry, paintings, glue, blood, ceramics,...), scientific research can be very relevant for practical applications. Therfore, a considerable effort has been invested to describe colloidal suspensions from a theoretical point of view and by simulations [16, 31–33, 44], as well as to understand the particle-particle interactions [1, 10, 11, 15, 59, 60] and the phase behavior [7, 8, 24, 35, 58]. Due to the small size of the particles one is usually interested in collective

W.E. Nagel et al. (eds.), *High Performance Computing in Science and Engineering '09*, DOI 10.1007/978-3-642-04665-0_3,

properties of an ensemble of many particles, e.g., the viscosity of a suspension, which is determined by the interactions among the individual particles. There are many ways to manipulate these interactions, e.g., by changing the pH-value or the ionic strength of the suspension [13, 14, 19, 20, 62], or by adding polymers [53, 54] or nanoparticles [36]. This offers the possibility to control the phase behavior of the colloidal suspension as a whole.

However, it is difficult to give quantitative predictions concerning the influence of individual changes in the details of the interactions, because in soft matter systems, typically several effects are in a subtle interplay. The large number of effects to encounter also implies that the parameter space is of high dimension and one has to perform many different simulations, each of which requiring a lot of computing resources, to gain an understanding of all inter-relationships.

Therefore, it is useful to be able to steer the simulation on-line. Steering in this context can mean changing the interaction potentials or other parameters like an externally applied shear rate. It can also mean going back in the simulation time and following a different path starting from an earlier configuration, both options induced by interaction with the user [6, 45]. The most simple case of simulation steering is just to start data acquisition or to provide output data in a running simulation as a response to a request by the user.

As an example system we have modeled Al_2O_3-particles suspended in water. The particles are represented by monodisperse spheres of 0.37μm in diameter. They interact via DLVO (Derjaguin, Landau, Vervey, Overbeek) potentials [10, 60] as well as a repulsive force ensuring excluded volume for the particles. This model system has been studied in simulations and experiments in previous works [17, 19–21, 41, 47]. In this article we discuss how simulation steering can help to explore the parameter space and which problems may occur when using steering techniques. The results have mostly been published in a proceedings paper [18], but in this article we focus more on the technical details and give a more estensive description of the simulation method. Additionally, we comment on our work in progress, which will be published elsewhere together with our collaborators [51].

In the following section we describe our simulation method. Then we describe the implementation of the steering approach. After that, we discuss some examples of simulations where steering helps to explore the parameter space, but we also highlight the limits of the steering approach in which a steered simulation might lead to different results if compared to a non-steered simulation. Finally, we give a summary of our results.

2 Simulation Method

We use a coupled algorithm consisting of a Molecular Dynamics (MD) and Stochastic Rotation Dynamics (SRD) part. The simulation method is described in detail in Refs. [17, 19, 21]. The MD algorithm solves the equations of motion for the colloidal particles, whereas the SRD algorithm is a mesoscopic simulation method to take into account the influence of the fluid solvent: it calculates the hydrodynamic interactions and intrinsically includes Brownian fluctuations.

In the MD part we include effective electrostatic interactions and van der Waals attraction, known as DLVO potentials [10, 60]. The repulsive term results from the surface charge of the suspended particles

$$V_{\mathrm{Coul}} = \pi \varepsilon_r \varepsilon_0 \left[\frac{2+\kappa d}{1+\kappa d} \cdot \frac{4k_{\mathrm{B}}T}{ze} \tanh\left(\frac{ze\zeta}{4k_{\mathrm{B}}T} \right) \right]^2 \times \frac{d^2 \mathrm{e}^{-\kappa[r-d]}}{r}, \tag{1}$$

where d denotes the particle diameter, r the distance between the particle centers, e the elementary charge, T the temperature, k_{B} the Boltzmann constant, and z is the valency of the ions of added salt. ε_0 is the permittivity of the vacuum, $\varepsilon_r = 81$ the relative dielectric constant of the solvent, κ the inverse Debye length defined by $\kappa^2 = 8\pi \ell_B I$, with I being the ionic strength and the Bjerrum length $\ell_B = 7\,\text{Å}$. The effective surface potential ζ of the particles depends on the distinct experimental conditions. Apart from the material and surface structure of the particles it is mainly determined by the pH-value and the ionic strength of the solvent. In our previous work [19] we have developed a charge regulation model to describe these dependencies based on the adsorption and desorption of the so-called charge determining ions. The equillibrium of this process depends on two influences: firstly, the electrostatic potential is determined by the surface charge density, i.e., by the amount of ions currently bound to the colloidal particle, and secondly, the electrostatic potential influences the equillibrium of the adsorption process. The ζ potential of a charged spherical colloidal particle of radius R immersed in an electrolytic environment of relative dielectric constant ε_r and ionic strength I is given within Debye-Hückel theory [9, 40] by

$$\zeta = \frac{R\sigma}{\varepsilon_0 \varepsilon_r (1+\kappa R)} . \tag{2}$$

The second part of the model which describes the equillibrium of the adsorption of protons (H^+) on surface leads to the relation [19]

$$\frac{\sigma}{eN_S} = \frac{\delta \sinh(\psi_N - \zeta e/k_{\mathrm{B}}T)}{1+\delta \cosh(\psi_N - \zeta e/k_{\mathrm{B}}T)} , \tag{3}$$

with the Nernst potential $\psi_N := \ln(10)(p\mathrm{H}_z - p\mathrm{H})$ and $p\mathrm{H}_z$ being the point of neutrally charged colloids, the so-called isoelectric point. The equations (2) and (3) can be solved numerically for ζ as a function of pH using an iteration scheme. The number density of surface sites N_S, $p\mathrm{H}_z$ and the parameter δ are adjusted by fitting the resulting dependencey $\zeta(p\mathrm{H})$ to experimental data for the specific materials [19]. This calibration procedure is required once for a known ionic strength. Afterwards the charge regulation model allows to calculate the ζ-potential, and thus the electrostatic interactions among the particles, for different experimental conditions, i.e., a different pH-value and different ionic strength.

The second term of the DLVO potentials which does not depend on the pH-value or the ionic strength is the attractive van der Waals interaction

$$V_{\mathrm{VdW}} = -\frac{A_{\mathrm{H}}}{12}\left[\frac{d^2}{r^2-d^2} + \frac{d^2}{r^2} + 2\ln\left(\frac{r^2-d^2}{r^2}\right)\right]. \tag{4}$$

$A_{\mathrm{H}} = 4.76 \cdot 10^{-20}$ J is the Hamaker constant [22]. The attractive contribution V_{VdW} competes with the repulsive term and is responsible for the cluster formation one can observe for conditions in which the attraction dominates.

Since DLVO theory is based on the assumption of large particle separations, it does not correctly reproduce the primary minimum in the potential, which should appear at particle contact. Therefore, we cut off the DLVO potentials and model the minimum by a parabola as described in Refs. [17, 21]. The only free parameter in this model is the depth of the primary minimum, which in principle can be adjusted to experimental data. However, due to surface roughness and lubrication effects, an exact measurement is very difficult. On the other hand, for the simulation it is sufficient to choose the minimum deep enough so that clusters of particles do not break up at the primary minimum.

Once the forces are calculated, the integration of the equations of motion is accomplished in the MD framework using a velocity verlet algorithm [3, 57],

$$\mathbf{x}_i(t+\delta t) = \mathbf{x}_i(t) + \delta t\, \mathbf{v}_i(t) + \delta t^2\, \frac{\mathbf{F}_i(t)}{m}, \tag{5}$$

$$\mathbf{v}_i(t+\delta t) = \mathbf{v}_i(t) + \delta t\, \frac{\mathbf{F}_i(t) + \mathbf{F}_i(t+\delta t)}{2m}. \tag{6}$$

To ensure excluded volume of the particles we use a repulsive (Hertzian) potential. Below the resolution of the SRD algorithm short range hydrodynamics is corrected by a lubrication force within the MD framework as explained in Refs. [17, 19, 21].

For the simulation of a fluid solvent, many different simulation methods have been proposed: direct Navier Stokes solvers [28, 29, 52, 61], Stokesian Dynamics (SD) [4, 5, 44], Accelerated Stokesian Dynamics (ASD) [55, 56], pair drag simulations [31], Brownian Dynamics (BD) [22, 23], Lattice Boltzmann (LB) [30, 32–34], and Stochastic Rotation Dynamics (SRD) [21, 27, 42]. These mesoscopic fluid simulation methods have in common that they impose certain approximations to reduce the computational effort. Some of them include thermal noise intrinsically, or it can be included consistently. They scale differently with the number of embedded particles, and the complexity of the algorithm differs largely. In particular, there are big differences in the concepts how to couple the suspended particles to the surrounding fluid.

We apply the Stochastic Rotation Dynamics method (SRD) introduced by Malevanets and Kapral [37, 38]. It intrinsically contains fluctuations, is easy to implement, and has been shown to be well suitable for simulations of colloidal and polymer suspensions [2, 19, 21, 27, 42, 48, 63]. The method is also known as "Real-coded Lattice Gas" [27] or as "Multi-Particle-Collision Dynamics" (MPCD) [49, 50]. It is based on so-called fluid particles with continuous positions and velocities. A streaming step and an interaction step are performed alternately. In the streaming step, each particle i is moved according to

$$\mathbf{r}_i(t+\tau) = \mathbf{r}_i(t) + \tau\, \mathbf{v}_i(t), \tag{7}$$

where $\mathbf{r}_i(t)$ denotes the position of particle i at time t, and τ is the time step. In the interaction step, the fluid particles are sorted into cubic cells of a regular lattice, and only the particles within the same cell interact with each other according to an artificial interaction. The interaction step is designed to exchange momentum among the particles, but at the same time to conserve total energy and total momentum within each cell, and to be very simple, i.e., computationally cheap. Each cell j is treated independently: first, the mean velocity $\mathbf{u}_j(t') = \frac{1}{N_j(t')}\sum_{i=1}^{N_j(t')} \mathbf{v}_i(t)$ in cell j is calculated. $N_j(t')$ is the number of fluid particles contained in cell j at time $t' = t + \tau$. Then, the velocities of each fluid particle in this cell are rotated according to

$$\mathbf{v}_i(t+\tau) = \mathbf{u}_j(t') + \boldsymbol{\Omega}_j(t') \cdot [\mathbf{v}_i(t) - \mathbf{u}_j(t')]. \tag{8}$$

$\boldsymbol{\Omega}_j(t')$ is a rotation matrix, which is independently chosen at random for each time step and each cell. We use rotations about one of the coordinate axes by an angle $\pm\alpha$, with α fixed. The coordinate axis as well as the sign of the rotation are chosen at random, resulting in 6 possible rotation matrices. However, there is a great freedom to choose the rotation matrices. Any set of rotation matrices satisfying the detailed balance for the space of velocity vectors could be used here. To remove anomalies introduced by the regular grid, one can either choose the mean free path to be sufficiently large or shift the whole grid by a random vector once per SRD time step [25, 26].

For the coupling of the SRD and the MD simulation, different methods have been introduced in the literature. Inoue *et al.* proposed a way to implement no slip boundary conditions on the particle surface [27]. To achieve full slip boundary conditions, Lennard-Jones potentials can be applied for the interaction between the fluid particles and the colloidal particles [38, 43]. A more coarse grained method was originally designed to couple the monomers of a polymer chain to the fluid [12, 39], but in our previous work [19–21] we have demonstrated that it can also be applied to colloids, as long as no detailed spatial resolution of the hydrodynamics is required. We use this coupling method in our simulations and describe it shortly in the following.

To couple the colloidal particles to the fluid, they are sorted into the SRD cells and included in the SRD interaction step. The stochastic rotation is performed in momentum space instead of the velocity space to take into account the difference of inertia between light fluid and heavy colloidal particles. We have described the simulation method in more detail in Refs. [17, 19, 21].

3 Simulation Steering

In this section we describe the technical realization of the steering interface in our simulation code. The program is an object oriented code written in C++. Each object contains virtual routines `save()` and `load()`, which write the data of the object to a buffer, or load it from there, respectively. The buffer contains a plain text description of all variables contained in the object including their values. This is similar to the C++ source code one would write to initialize an instance of that class. This plain text format is used in the simulation input file, too. The simulation program can dump its whole data structure in the same format, so that the automatically generated files can be used as checkpoints to restart the simulation from this point. The work flow of a simulation, including the actual MD loop as well as data input/output tasks is described in this manner using specialized "`workstep`-classes". Each `workstep`-class provides a specialized `work()`-routine, which performs different actions depending on the actual class type of the respective object. All specialized `workstep`-class objects are derived from the same (virtual) `workstep`-class. One of the `workstep`-classes is designed to change a specified object by using its `save()` and `load()` routine. First, the current values are stored to a temporary buffer, then one or several variables may be overwritten by new values, and finally the object to be modified is loaded again from the temporary buffer. The description of the changes may be read from standard input or from a file, similar to the simulation setup, which is read during the initialization of the simulation. `workstep`-class objects can also be included into the work flow at later times, or may even be disconnected from the usual work flow and bound to standard UNIX system signals. By default a workstep writing particle positions to standard output and a workstep to change objects getting its buffer from standard input are bound to the `SIGTTOU` ("terminal output") and `SIGTTIN` ("terminal input") signal, respectively. This allows to embed the simulation program into a framework of shell scripts which generate the appropriate input, redirect the output to a visualization tool and send the signals according to the user interaction. This can be realized in a client-server fashion, even on different hosts and platforms using appropriate scripts and TCP/IP connections. Another way to realize the steering is to drive the system along predefined trajectories in the parameter space, e.g., to continuously change the pH-value and the ionic strength. In this case a workstep which changes objects is included in the standard work flow and regularly reads its instructions from a so-called steering file: in each time step a new set of parameters for the interaction potentials is read in and when the simulation has reached the end of the predefined trajectory, a workstep triggering the end of the simulation and final output procedures is loaded from the steering file and added to the simulation protocol.

4 Results

We now apply the steering approach just presented to several example cases of computer simulations and discuss its advantages compared to traditional, non-steered simulations. We highlight some pitfalls resulting from the physical background in the context of steering. This section is subdivided into subsections each of them focusing on a particular example to illustrate the steering approach in practice.

4.1 Pressure Filtration

Filtration processes are widely used in industry to separate the solid fraction of a suspension from the solvent. In a detailed study we investigate how the interactions among the particles determine the structure of the so-called filter cake [51], i.e., the solid fraction accumulated at the filter plate. The porosity, and especially the permeability of the filter cake strongly determines the efficiency of the filtration process.

In our model the suspended particles cannot pass the filter, whereas the fluid passes through the filter without resistance in an idealized filtration process. The suspended particles agglomerate in front of the filter and form the filter cake. Since the dynamics of the particles is not only governed by the hydrodynamics of the fluid, but also by their DLVO interaction, the density and the structure of the filter cake depend on the pH-value and the ionic strength [51].

In a preliminary study [18] we check if this dependence can be seen in our simulations. We also study how an equillibration prior to the filtration process influences the structure of the filter cake [51]: if attractive forces are present in dilute suspensions, clusters form, which afterwards aggregate at the filter plane. In our simulations the filter is modeled by a closed boundary condition for the particles, which is fully permeable to the fluid. We use fully periodic boundaries for the fluid and drive the system either by applying a force to the fluid [18] or by moving the boundaries for the particles [51]. In the first case the force is applied to a limited region of the fluid so that the fluid is accelerated downwards and drags the particles with the flow towards the filter plane. In the latter case the pressure acting in the process can be controlled by monitoring the total force exerted to the particles by the boundary condition.

In Fig. 1 we plot the density profile for for several pH-values at a constant ionic strength of $I = 3\,\mathrm{mmol/l}$ and a constant total volume fraction of $\Phi = 5\%$. To gain good statistics for the structure, large simulations are needed. Together with low volume fractions and long simulation runs until the filter cake has formed the question raises how to handle the output data.

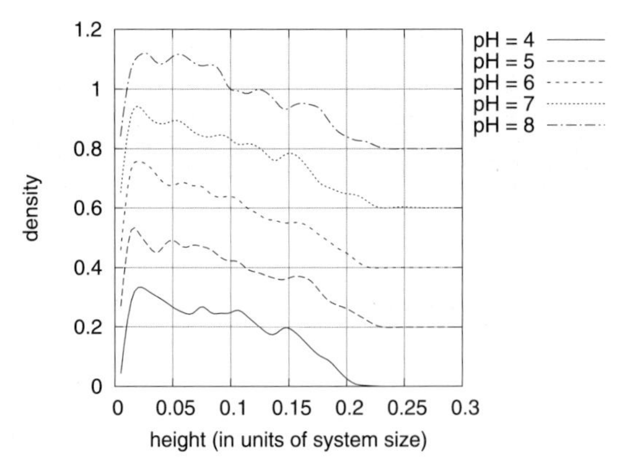

Fig. 1. Local density in terms of solid volume fraction depending on the pH-value (from Ref. [18]). The total volume fraction is kept constant at $\Phi = 5\%$ and the ionic strength at $I = 3\,\mathrm{mmol/l}$. The plots are shifted against each other vertically by 0.2 for better visibility. The shape of the profile differs due to the different interactions in each of the cases

Regularly writing the whole configuration and afterwards discarding all configurations in which the simulation has not yet settled is definetly not the optimal procedure. This is a typical problem, in which one would like to observe the running simulation occasionally to check if the filter cake is already formed. When the structure does not change significantly anymore, one would like to measure pressure profiles, local streaming velocities, the shape of the clusters, or simply the final density profile of the filter cake. In this context steering means to initiate data acquisition by user interaction. It is useful to start it manually, when the required conditions are difficult to check automatically. In the present case the density and structure of the filter cake and the time until the filter cake has formed are a priori unknown. This makes it difficult to automatically detect if the desired state has been reached.

4.2 Observation of Cluster Formation

Another application of simulation steering is to drive the system along a trajectory in parameter space to explore the regions of certain states or to check if transitions are reversible or not. Especially in soft matter systems, in which the parameter space is very large due to many competing influences, performing such a "steered voyage in parameter space" may help to gain understanding the phase behavior of the system of interest.

In Fig. 2 we show several snapshots of a simulation in which the parameters are constantly varied in this sense of a steered simulation. As indicated in Sec. 3 the trajectory in this simulation was fixed before the simulation was started, but it could as well be delivered to the program "on the fly", such that a steering tool provides the values for the interactions to the program

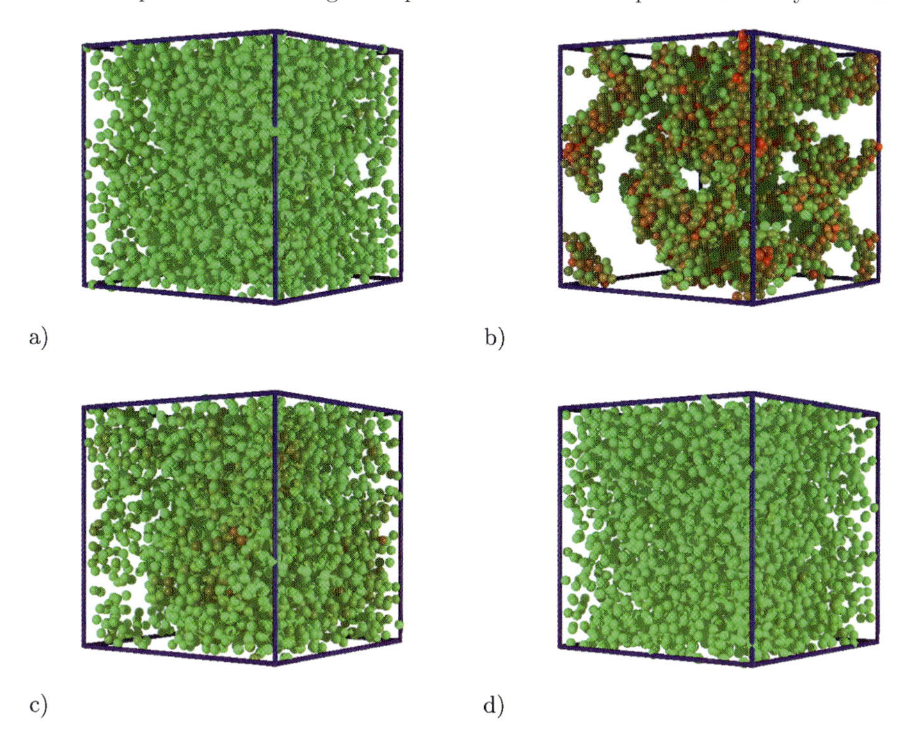

Fig. 2. Selected snapshots during a steered simulation to depict "memory" effects: the color scale denotes the coordination number (Ref. [18]). Particles closer than 2.4 radii are considered to be neighbors. The number of neighbors in this sense is depicted by red color for large numbers, or high local density, and green color for low numbers of neighbors in dilute regions. a) right after initialization, homogeneous distribution of the particles, b) cluster formation after tuning the potentials to be attractive, c) remaining inhomogeneities after steering the simulation back to the region of stabilized suspensions, d) by diffusion the system becomes homogeneous again. The steering path and the points at which the snapshots are taken are illustrated in Fig. 3

each time the user changes them. However, the changes have to be sufficiently smooth, so that it might be necessary to generate intermediate steps.

In Fig. 2 the parameter sub-space of pH-value and ionic strength is explored, whereas the volume fraction is kept constant and external driving forces are not applied. In our previous work [17, 20] we have explored the parameter space by performing numerous individual simulations. However, with the newly implemented steering approach one can detect boundaries between different regions of the stability diagram more quickly and more sensitively: When the parameters are modified such that the system crosses the border from a stable suspension (a) to the clustered regime (b), suddenly the system becomes inhomogeneous and cluster formation starts. This inhomogeneity remains (c) when the interactions become repulsive again, but on the charac-

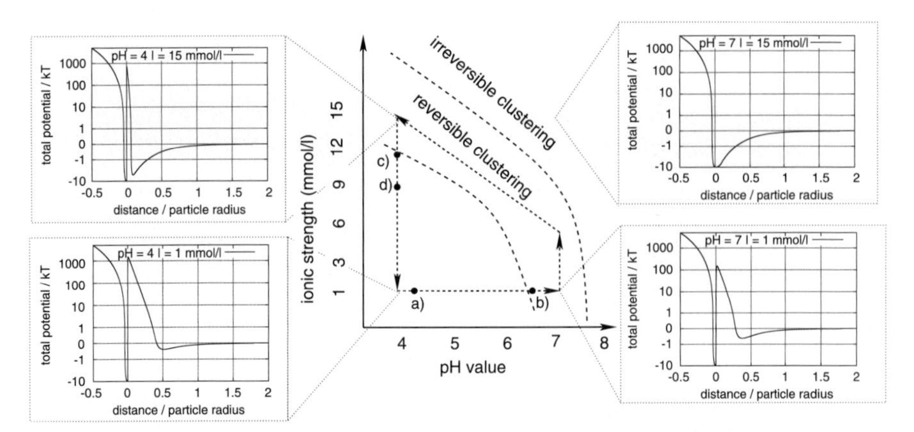

Fig. 3. Trajectory in parameter space for the steering (Ref. [18]): the simulation starts at low pH-value and low ionic strength and follows the dashed path in parameter space. After a short time of equillibration snapshot (a) of Fig. 2 is taken, snapshot (b) after crossing the boundary to the reversibly clustered region, and (c) and (d) after crossing back to the suspended region. The insets show the total potential at some selected extreme points of the parameter space. (see text)

teristic time scale of the diffusion the homogeneity of the system is restored (d).

The steering path in the configuration space is shown in Fig. 3. In the insets of Fig. 3 the effective potential is plotted for several cases. The steering path is chosen such that the cluster formation is reversible and the barrier between the primary and the secondary minimum in the DLVO potential does not go below 5 $k_{\mathrm{B}}T$. A case in which the barrier vanishes and irreversible aggregation would appear is also shown in the upper right inset of Fig. 3. The observation in our simulation confirms that the cluster formation is reversible for the chosen trajectory.

However, the simulation "remembers" the trajectory of the steering interaction. Even if the steering path is selected carefully inside the reversible range, the simulated suspension needs a characteristic time to relax after changing the interactions. The interaction can be seen as a perturbation in the physical sense and the system needs time to adopt to the new situation. Using steered simulations one can design and simulate processes in which the interactions are changed by chemical reactions that shift the pH-value. In the steered simulation the response of the simulated system to the perturbation can be observed directly. In any case, a careful interpretation of the simulation data is adviced, because the freedom to change the conditions "on the fly" is bought for the price of having to take into account the response time the system requires to react on the changes.

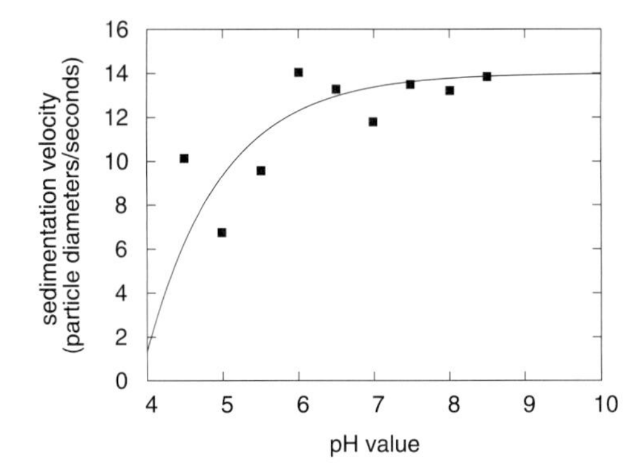

Fig. 4. Sedimentation velocity of a $\Phi = 5\%$ suspension in a closed vessel (Ref. [18]). The sedimentation velocity of the particles in the upper part of the system is averaged over several time steps. The particles at the bottom of the system are not taken into account. The velocity depends on the particle-particle interaction. Clusters settle down faster than individual particles. Symbols denote simulation results, the line is a guide to the eye

4.3 Sedimentation: Hydrodynamic Interaction

As we noticed in the previous section, the simulated system may "remember" the steering interaction. Therefore special care is advised when steering a simulation. We illustrate this by another example. Especially, when hydrodynamic interactions are important for the behavior of the system, memory effects become important and might dominate the systems behavior. Then, simulation steering cannot be applied.

Let us consider a sedimenting suspension of particles. If cluster formation is induced by modifying the electrostatic interactions among the particles, the sedimentation proceeds much faster. In Ref. [21] we have confirmed that the sedimentation velocity in our simulation depends on the volume fraction, as it is well-known from sedimentation theory [46]. But even at constant volume fraction we have observed different sedimentation velocities [18] depending on the pH-value: If the particles form clusters, they settle down and the fluid streams around the whole cluster. The resistance is much smaller compared to the case when the particles are distributed homogeneously and the fluid streams around each of them separately (compare Ref. [17] and references therein). In Fig. 4 we have plotted the sedimentation velocity evaluated in several simulations, which only differ by the potentials. The volume fraction is the same for all of them. However, as one can see in the figure, the sedimentation velocity is smaller for low pH-values. Around $p\mathrm{H} = 4$ the sedimentation velocity of an isolated particle of 0.6 diameters/s is reached [17] because the interactions among the particles are repulsive for this case and cluster formation is suppressed. This is a typical case in which one should avoid changing the interactions during a simulation, because the behavior of the whole system strongly depends on the distribution of the particles and changes in the interactions are "remembered" by the system in form of correlations due to the hydrodynamic interactions.

5 Summary

We have described the implementation of our steering approach in the simulation program and we have illustrated its application in two cases and given an example in which simulation steering is not advisable: if the system is sensitive to correlation effects and contains a "long time memory" which may be the case when hydrodynamic interactions are present, steering should be avoided. On the other hand it can save a large amount of computing time when exploring the parameter space and when the position of transition lines between different regions is investigated. We have shown this by the onset of cluster formation due to modifying the interaction potentials in a model suspension. Steering can be used as well to start data acquisition after a transient at the beginning of a simulation or to adjust the frequency for the data acquisition according to the current state of the simulation. This is especially interesting for large simulations with long transient times, and when a criterion to automatically start data acquisition is difficult to formulate. We have illustrated this by the simulation of filter flow.

Acknowledgments. The High Performance Computing Center Stuttgart and the Scientific Supercomputing Center Karlsruhe are highly acknowledged for providing the computing time and the technical support needed for our research. The authors thank the "Landesstiftung Baden-Württemberg" and the DFG for financial support within the the "Eliteprogram for Postdocs", priority program "nano- and microfluidics", SFB 716, and "Exzellenzakademie Materialwissenschaft und Werkstofftechnik".

References

1. S. Alexander, P. M. Chaikin, P. Grant, G. J. Morales, P. Pincus, and D. Hone. Charge renormalization, osmotic pressure, and bulk modulus of colloidal crystals: Theory. *J. Chem. Phys.*, 80:5776–5781, 1984.
2. I. Ali, D. Marenduzzo, and J. M. Yeomans. Dynamics of polymer packaging. *J. Chem. Phys.*, 121:8635–8641, 2004.
3. M. P. Allen and D. J. Tildesley. *Computer simulation of liquids.* Oxford Science Publications. Clarendon Press, Oxford, 1987.
4. J. F. Brady. The rheological behavior of concentrated colloidal suspensions. *J. Chem. Phys.*, 99:567–581, 1993.
5. J. F. Brady and G. Bossis. Stokesian dynamics. *Ann. Rev. Fluid Mech.*, 20:111–157, 1988.

6. J. Chin, J. Harting, S. Jha, P. Coveney, A. Porter, and S. Pickles. Steering in computational science: mesoscale modelling and simulation. *Journal of Contemporary Physics*, 44:417–434, 2003.
7. D. Costa, P. Hansen, and L. Harnau. Structure and equation of state of interaction site models for disc-shaped lamellar colloids. *Mol. Phys.*, 103:1917–1927, 2005.
8. A. de Candia, E. del del Gado, A. Fierro, N. Sator, and A. Coniglio. Colloidal gelation, percolation and structural arrest. *Physica A*, 358:239–248, 2005.
9. P. Debye and E. Hückel. Zur theorie der elektrolyte. I. Gefrierpunktserniedrigung. *Phys. Z.*, 24:185, 1923.
10. B. V. Derjaguin and L. D. Landau. Theory of the stability of strongly charged lyophobic sols and of the adhesion of strongly charged particles in solutions of electrolytes. *Acta Physicochimica USSR*, 14:633, 1941.
11. J. Dobnikar, Y. Chen, R. Rzehak, and H. H. von Grünberg. Many-body interactions in colloidal suspensions. *J. Phys.: Condensed Matter*, 15:S263, 2003.
12. E. Falck, J. M. Lahtinen, I. Vattulainen, and T. Ala-Nissila. Influence of hydrodynamics on many-particle diffusion in 2d colloidal suspensions. *Eur. Phys. J. E*, 13:267–275, 2004.
13. T. J. Graule, F. H. Baader, and L. J. Gauckler. Shaping of ceramic green compacts direct from suspensions by enzyme catalyzed reactions. *Ceramic Forum International*, 71:314–322, 1994.
14. T. J. Graule, F. H. Baader, and L. J. Gauckler. Casting uniform ceramics with direct coagulation. *CHEMTECH*, 25(6):31–37, 1995.
15. M. J. Grimson and M. Silbert. A self-consistent theory of the effective interactions in charge-stabilized colloidal dispersions. *Molecular Physics*, 74:397–404, 1991.
16. L. Harnau and S. Dietrich. Depletion potential in colloidal mixtures of hard spheres and paltelets. *Phys. Rev. E*, 69:051501, 2004.
17. M. Hecht. *Simulation of Peloids*. PhD thesis, Universität Stuttgart, Germany, 2007.
18. M. Hecht and J. Harting. Computational steering of cluster formation in Brownian suspensions. *Computers and Mathematics with Applications*, 58:995–1002, 2009. Proceedings of the Fourth International Conference on Mesoscopic Methods in Engineering and Science, ICMMES (Munich, Germany).
19. M. Hecht, J. Harting, M. Bier, J. Reinshagen, and H. J. Herrmann. Shear viscosity of clay-like colloids in computer simulations and experiments. *Phys. Rev. E*, 74:021403, 2006.
20. M. Hecht, J. Harting, and H. J. Herrmann. Stability diagram for dense suspensions of model colloidal Al2O3-particles in shear flow. *Phys. Rev. E*, 75:051404, 2007.
21. M. Hecht, J. Harting, T. Ihle, and H. J. Herrmann. Simulation of claylike colloids. *Phys. Rev. E*, 72:011408, 2005.
22. M. Hütter. *Brownian dynamics simulation of stable and of coagulating colloids in aqueous suspension.* PhD thesis, Swiss Federal Institute of Technology Zurich, 1999.
23. M. Hütter. Local structure evolution in particle network formation studied by Brownian dynamics simulation. *J. Colloid Interface Sci.*, 231:337–350, 2000.

24. A.-P. Hynninen, M. Dijkstra, and R. van Roij. Effect of three-body interactions on the phase behavior of charge-stabilized colloidal suspensions. *Phys. Rev. E*, 69:061407, 2004.
25. T. Ihle and D. M. Kroll. Stochastic rotation dynamics I: formalism, Galilean invariance, Green-Kubo relations. *Phys. Rev. E*, 67:066705, 2003.
26. T. Ihle and D. M. Kroll. Stochastic rotation dynamics II: transport coefficients, numerics, long time tails. *Phys. Rev. E*, 67:066706, 2003.
27. Y. Inoue, Y. Chen, and H. Ohashi. Development of a simulation model for solid objects suspended in a fluctuating fluid. *J. Stat. Phys.*, 107:85–100, 2002.
28. W. Kalthoff, S. Schwarzer, and H. J. Herrmann. Algorithm for the simulation of particle suspensions with inertia effects. *Phys. Rev. E*, 56:2234–2242, 1997.
29. W. Kalthoff, S. Schwarzer, G. Ristow, and H. J. Hermann. On the application of a novel algorithm to hydrodynamic diffusion and velocity fluctuations in sedimenting systems. *Int. J. Mod. Phys. C*, 7:543–561, 1996.
30. A. Komnik, J. Harting, and H. J. Herrmann. Transport phenomena and structuring in shear flow of suspensions near solid walls. *Journal of Statistical Mechanics: theory and experiment*, P12003, 2004.
31. L. E. Silbert, J. R. Melrose, and R. C. Ball. Colloidal microdynamics: Pair-drag simulations of model-concentrated aggregated systems. *Phys. Rev. E*, 56:7067–7077, 1997.
32. A. J. C. Ladd. Numerical simulations of particulate suspensions via a discretized Boltzmann equation. Part 1. Theoretical foundation. *J. Fluid Mech.*, 271:285–309, 1994.
33. A. J. C. Ladd. Numerical simulations of particulate suspensions via a discretized Boltzmann equation. Part 2. Numerical results. *J. Fluid Mech.*, 271:311–339, 1994.
34. A. J. C. Ladd and R. Verberg. Lattice-Boltzmann simulations of particle-fluid suspensions. *J. Stat. Phys.*, 104:1191, 2001.
35. Y. Levin, T. Trizac, and L. Bocquet. On the fluid-fluid phase separation in charged-stabilized colloidal suspensions. *J. Phys.: Condensed Matter*, 15:S3523, 2003.
36. J. Liu and E. Luijten. Stabilization of colloidal suspensions by means of highly-charged nanoparticles. *Phys. Rev. Lett.*, 93:247802, 2004.
37. A. Malevanets and R. Kapral. Mesoscopic model for solvent dynamics. *J. Chem. Phys.*, 110:8605, 1999.
38. A. Malevanets and R. Kapral. Solute molecular dynamics in a mesoscale solvent. *J. Chem. Phys.*, 112:7260, 2000.
39. A. Malevanets and J. M. Yeomans. Dynamics of short polymer chains in solution. *Europhys. Lett.*, 52:231, 2000.
40. D. A. McQuarrie. *Statistical mechanics.* Univ. Science Books, Sausalito, 2000.
41. R. Oberacker, J. Reinshagen, H. von Both, and M. J. Hoffmann. Ceramic slurries with bimodal particle size distributions: Rheology, suspension structure and behaviour during pressure filtration. In N. C. S. Hirano, G. L. Messing, editor, *Ceramic Processing Science VI*, volume 112, pages 179–184. Am. Cer. Soc., 2001. ISBN 1574981048.
42. J. T. Padding and A. A. Louis. Hydrodynamic and Brownian fluctuations in sedimenting suspensions. *Phys. Rev. Lett.*, 93:220601, 2004.

43. J. T. Padding and A. A. Louis. Hydrodynamic interactions and Brownian forces in colloidal suspensions: Coarse-graining over time and length-scales. *Phys. Rev. E*, 74:031402, 2006.
44. T. N. Phung, J. F. Brady, and G. Bossis. Stokesian dynamics simulation of Brownian suspensions. *J. Fluid Mech.*, 313:181–207, 1996.
45. J. Prins, J. Hermans, G. Mann, L. Nyland, and M. Simons. A virtual environment for steered molecular dynamics. *Future Generation Computer Systems*, 15:485–495, 1999.
46. J. F. Richardson and W. N. Zaki. Sedimentation and fluidisation: Part 1. *Trans. Instn. Chem. Engrs.*, 32:35–53, 1954.
47. S. Richter and G. Huber. Resonant column experiments with fine-grained model material - evidence of particle surface forces. *Granular Matter*, 5:121–128, 2003.
48. M. Ripoll, K. Mussawisade, R. G. Winkler, and G. Gompper. Low-Reynolds-number hydrodynamics of complex fluids by multi-particle-collision dynamics. *Europhys. Lett.*, 68:106–112, 2004.
49. M. Ripoll, K. Mussawisade, R. G. Winkler, and G. Gompper. Dynamic regimes of fluids simulated by multiparticle-collision dynamics. *Phys. Rev. E*, 72:016701, 2005.
50. M. Ripoll, R. G. Winkler, and G. Gompper. Star polymers in shear flow. *Phys. Rev. Lett.*, 96:188302, 2006.
51. B. Schäfer, J. Harting, M. Hecht, and H. Nirschl. Agglomeration and filtration of colloidal suspensions with DLVO interactions in simulation and experiment. *J. Colloid Interface Sci.*, 2009. (submitted) arxiv:0907.1551.
52. S. Schwarzer. Sedimentation and flow through porous media: Simulating dynamically coupled discrete and continuum phases. *Phys. Rev. E*, 52:6461–6475, 1995.
53. S. A. Shah, L. Chen, S. Ramakrishnan, K. S. Schweizer, and C. F. Zukoski. Microstructure of dense colloid-polymer suspensions and gels. *J. Phys.: Condensed Matter*, 15:4751–4778, 2003.
54. S. A. Shah, S. Ramakrishnan, L. Chen, K. S. Schweizer, and C. F. Zukoski. Scattering studies of the structure of colloid-polymer suspensions and gels. *Langmuir*, 19:5128–5136, 2003.
55. A. Sierou and J. F. Brady. Accelerated Stokesian dynamics simulations. *J. Fluid Mech.*, 448:115, 2001.
56. A. Sierou and J. F. Brady. Shear-induced self-diffusion in non-colloidal suspensions. *J. Fluid Mech.*, 506:285, 2004.
57. W. C. Swope, H. C. Andersen, P. H. Berens, and K. R. Wilson. A computer simulation method for the calculation of equilibrium constants for the formation of physical clusters of molecules: application to small water clusters. *J. Chem. Phys.*, 76:637–649, 1982.
58. V. Trappe, V. Prasad, L. Cipelletti, P. N. Segre, and D. A. Weitz. Jamming phase diagram for attractive particles. *Nature*, 411:772–774, 2001.
59. R. van Roij and P. Hansen. Van der Waals-like instability in suspensions of mutually repelling charged colloids. *Phys. Rev. Lett.*, 79:3082–3085, 1997.
60. E. J. W. Vervey and J. T. G. Overbeek. *Theory of the Stability of Lyophobic Colloids*. Elsevier, Amsterdam, 1948.
61. B. Wachmann, W. Kalthoff, S. Schwarzer, and H. J. Herrmann. Collective drag and sedimentation: comparison of simulation and experiment in two and three dimensions. *Granular Matter*, 1:75–82, 1998.

62. G. Wang, P. Sarkar, and P. S. Nicholson. Surface chemistry and rheology of electrostatically (ionically) stabillized alumina suspensions in polar media. *J. Am. Ceram. Soc.*, 82:849–856, 1999.
63. R. G. Winkler, K. Mussawisade, M. Ripoll, and G. Gompper. Rod-like colloids and polymers in shear flow: a multi-particle-collision dynamics study. *J. Phys.: Condensed Matter*, 16:S3941–S3954, 2004.

Solid State Physics

Prof. Dr. Werner Hanke

Institut für Theoretische Physik und Astrophysik, Universität Würzburg,
Am Hubland, 97074 Würzburg, Germany

A variety of large-scale computer projects in the area of solid-state physics center around topical applications of density functional theory.

One example is the study of the adsorption of the amino-acid cysteine on the Au(110) surface by the Jena theory group around F. Bechstedt, lead by B. Höffling. This project is aiming at a recently very much studied surface problem, namely the adsorption of organic molecules on specific crystal surfaces. This is due to the fact, that the metal/organic interfaces are of particular interest for the development of molecular electronics. Here, a central question is the specific microscopic role of the underlying substrate (metal) which gives rise to an additional material-dependent degree of freedom. This freedom may result in a material-dependent change of the reconstruction of the absorbed atoms on a metal surface. The project uses the so-called Vienna-Ab-Initio Simulation Package, within the framework of density functional theory (DFT) employing a specific approximation (LDA) for the exchange and correlation functional, as well as further exchange correlation corrections (GGA). The results include a study of stable geometries of the Au (110) surface and present conversion criteria and results, in particular for their surface energies. On the basis of these calculations most interesting is the prediction of a specific cysteine dimer formation on the gold surfaces to play an important role in the adsorbtion process.

Technically, somewhat related is a second project of the group in Jena under R. Leitsmann and F. Bechstedt, studying the characterization of colloidal semiconductor quantum dots. Here, the main point is again to make density functional theory based predictions guiding the efficiency of light sources in the mid-ifrared spectral region. This topic is crucial for many applications, e.g. molecular spectroscopy and gas-sensor systems for environmental monitoring or medical diagnostics. Both the above projects and this quantum dot project give a nice summary concerning the computational cost. A limiting factor in the latter project is the communication overhead required for the redistribution of the wave-functions between all nodes, which is necessary during the orthogonalization procedure of the DFT eigenstates. Therefore, the scalability of the code depends strongly on the size of the actual problem. This is shown nicely in the Leitsmann-Bechstedt summary. On the basis of this computational scheme a structural optimization has been carried through, which allows for discussion of the electronic and optical properties with respect to the nano-crystal size and chemical trap.

A very interesting project concerns the understanding of long-ranged indirect interactions between surfaces adsorbed molecules by W. G. Schmidt et al. from the University of Paderborn. Experiments and first principle (DFT) calculations have been very successful in rationalizing the adsorption of single- as well as directly interacting molecules on surfaces. However, in addition to direct molecular interactions, such as e.g. H-bonds, indirect substrate-mediated interactions between the adsorbates may occur. The corresponding long-ranged character prevented for a long time an accurate modeling of the respective forces. However, recent progress, especially on the large scale computational front with the availability of parallel vector systems allow now for studying the origin of these indirect interactions in detail. This is a nice demonstration of how up to date DFT calculations in combination with large scale computational efforts can explain details in the atomic and molecular adsorption on metal surfaces, in particular the long-range order of these adsorbed species.

The study of W. L. Yim and Th. Klüner from the University of Oldenburg is directed towards microscopic chemical modeling of charging and discharging processes and case studies on small carbon- and boron nitride nanotubes. This project is of relevance for improvements of the use of electro-chemical equipments as fuel-cells, in which hydrogen and oxygen would be transformed into water and electrical power. Recently, based on both experimental and theoretical progress, the fuel-cell-technology has been advanced, but there are still challenging problems to be addressed, such as the hydrogen storage, which is in the center of interest in the Yim-Klüner work. Again, this project uses the Vienna-Ab-Intio-DFT program to perform calculation within density functional theory. On top of the density functional simulations, thermodynamic models are used to describe the chemical equilibrium of the various states which emerge during the hydrogen storage process. An important insight gained by the project is, that the hydrogen carrying ability is predominantly due to physisorption rather then chemisorption.

A project by D. V. Bachurin and P. Gumbsch from the University of Karlsruhe and the Fraunhofer Institut für Werkstoffmechanik Freiburg is concerned with nanocrystalline metals, which are polycrystals with a mean grain size below 100nm. Their outstanding properties are mostly explained by the peculiarities of the microstructure of nanocrystals: the large number of strain boundaries and the presence of porosity relating to the synthesis technique. In this collaboration, a molecular dynamic study of plastic deformation of nanocrystalline palladium has been performed at room temperature and different strain grades. The main conclusion of this interesting work is that the molecular dynamic simulations show, that the deformation behavior exhibits different deformation mechanisms depending on the specific straining process at room temperature.

Last, not least, another University of Karlsruhe paper by A. Branschädel, T. Ulbricht und P. Schmitteckert is aiming at insight into the signal transport in and conductance of correlated nanostructures. This project is centered around a very topical issue in modern solid state physics, namely transport properties of strongly interacting or correlated quantum systems, which are at the same time of great practical use. The so-called density matrix renormalization group method (DMRG) is applied to study transport properties and quantum devices attached to metallic leads. Whereas a large number of results are already available, partly also with DMRG, for the calculation of equilibrium transport properties, a new aspect of the present project is the focus on the calculation of non-equilibrium transport properties. The authors studied, in particular, the competition between interference due to multiple single-particle paths and Coulomb interactions in a simple model of an Anderson-like impurity with local magnetic field and used levelsplitting coupled to ferromagnetic leads. This model, along with its potential experimental relevance in the field of spintronics, serves as a non-trivial benchmark system, where various quantum transport approaches can be tested and compared. For this relevant model, but also for others discussed in the paper, it was shown, that the DMRG method combined with large scale computing resources was able to provide the linear conductance for arbitrary strength of the on-site Coulomb interaction and arbitrary level splitting. The key result is here the strong suppression of conductance with the increase of the on-site Coulomb interaction, if the magnetic field on the dot is opposite to the lead-polarization.

Adsorption of Cysteine on the Au(110)-surface: A Density Functional Theory Study

B. Höffling, F. Ortmann, K. Hannewald, and F. Bechstedt

European Theoretical Spectroscopy Facility (ETSF) and
Institut für Festkörpertheorie und -optik
Friedrich-Schiller-Universität Jena
Max-Wien-Platz 1, 07743 Jena, Germany
benjamin.hoeffling@uni-jena.de

Summary. We present ab initio studies towards the adsorption of the amino acid cysteine on the Au(110) surface. By means of density functional theory calculations and using the repeated-slab supercell method, we investigate three main aspects relevant for the adsorption process. First, in order to estimate the slab width required for an accurate description of the gold surface, we calculate the surface energies for both the unreconstructed and the missing row (1×2) reconstructed Au(110) surface for varying slab widths. Then, we determine the formation energies for vacancies in the salient row on the missing-row reconstructed surface. This allows us to estimate the energy cost for a local lifting of the missing-row reconstruction upon molecular adsorption. Finally, we examine the formation of cysteine dimers via carboxyl-carboxyl hydrogen bonds and investigate the changes in the bond energy caused by intermolecular strain. We predict the cysteine dimer formation in Au surface vacancies to play an important role in the adsorption process.

1 Introduction

In recent years, there has been considerable interest in the adsorption of organic molecules on crystal surfaces. Special attention has been devoted to metal/organic interfaces [1, 2, 3, 4, 5, 6, 7, 8, 9], which are of particular interest for the development of molecular electronics [10, 11, 12]. The existence of different functional groups leads to numerous practical configurations for the organic functionalization of the substrates, each of which is characterized by a unique electronic signature as evident from single molecule conductance experiments [13]. However, also the underlying substrate gives rise to additional degrees of freedom, which may result in a change of the reconstruction upon adsorption [14]. For example, the complexity of the adsorption process is evident from the vivid discussion about the alkanethiol adsorption on Au(111) which is a prototypical system for a gold/organic interface [15].

W.E. Nagel et al. (eds.), *High Performance Computing in Science and Engineering '09*, DOI 10.1007/978-3-642-04665-0_4,

In order to tailor the surface functionalization, one resorts to a large pool of functional groups such as the amino group which, for instance, allows for modifications of the metal work functions. From this viewpoint, cysteine is of particular interest and its interaction with metal surfaces has been the subject of intense investigations both experimental [14, 16, 17, 18, 19] and theoretical [20, 21]. Since cysteine not only contains a -SH thiol head group, but also a NH_2 amino group, its adsorption behavior on Au surfaces is rather complex. Recent scanning tunneling microscopy findings have observed highly ordered adsorption structures such as chirally homogenous dimers [14], directional one-dimensional rows [16] and large homochiral cysteine islands growing from chiral kink sites [17] on Au (1×2) missing-row reconstructed surfaces. Adsorption-induced surface reconstruction has been proposed as the driving forces for dimer formation and one-dimensional row growth. In this picture, the cysteine dimers form in four-atom vacancies in the top-rows of the missing-row reconstructed (110)-surface. The directional cysteine rows grow along unidirectional vacancies. The basic idea for the driving force of both homochiral dimer adsorption and unidirectional row-growth proposed by Kuehnle *et al.* is the creation of vacancies on the Au (1×2) reconstructed surface.

While the bonding of cysteine on Au has been investigated in terms of bonding configuration and changes in the density of states (DOS) for adsorption on the Au(111) surface [20, 21], only rudimentary studies have been conducted for the Au(110) surface. Especially the possible contribution of a bonding via the amino group and its interplay with the thiolate bond in the formation of a more flat or perhaps a more vertical adsorption geometry needs a deeper understanding. Closely related to this question are two other aspects of the adsorption process: the cysteine-cysteine interaction and the role of vacancy creation upon adsorption-induced rearrangement of the Au surface. These two points form the focus of the present paper.

After the introduction of the methodology in Sec. 2, we present the results in Sec. 3. First, we study in Sec. 3.1 two stable geometries of the Au(110) surface and present convergence criteria and results for their surface energies. Second, in Sec. 3.2, we calculate the vacancy formation energies for the missing-row reconstructed surface. Third, in Sec. 3.3, we investigate the structural and energetical properties of cysteine dimers which may form in the vacancies created on the Au surface. Finally, in Sec. 4, we conclude by a short outlook.

2 Computational Methods

The calculations are carried out using the in the *Vienna ab initio simulation package* (VASP) [22, 23] within the framework of density functional theory (DFT) using the local density approximation (LDA) for the exchange and correlation functional as well as the generalized gradient approximation (GGA) as parametrized by Perdew and Wang [24, 25]. The GGA treatment has been shown by Maul et al. [26] to be superior to LDA for the description of amino acid bonds. The explicit inclusion of long-range correlations as in previous adsorption studies [27, 28] has not been considered because such contributions are expected to be of minor importance in the present case. The pseudopotentials and wave functions are represented within the projector-augmented wave (PAW) method [29]. The electron wave functions in the regions between the cores are expanded in a plane wave basis set up to a cutoff energy of 500 eV. This allows for an accurate treatment of the first-row elements as well as the Au 5d electrons. The Brillouin zone (BZ) integrations are represented by a sum over Monkhorst-Pack points [30]. Energy convergence with respect to these parameters is carefully checked. The supercell sizes for the various systems studied is given in the next chapter.

3 Results and Discussion

3.1 Au(110) Surface Energy

There are two different stable geometries for the Au(110) surface (see Fig. 1). In comparison to the unreconstructed surface, the missing-row reconstructed surface exhibits a stronger corrugation due to the removal of every second top row. To obtain the Au surface energy of the unreconstructed and the missing-row Au(110) surface we perform a series of ionic relaxations of gold slabs with varying thickness. The supercells are constructed as (1×2) surface cells laterally. The gold slab thickness is varied from 5 to 40 layers with a vacuum width equivalent to 10 atomic Au layers in [110] direction. The three central layers of the slab are kept fixed, all other ions are left free to move until the Hellmann-Feynman forces acting on each ion were below 0.01 eV/Å. We follow previous work on Au surface energy [31, 32] and perform the calculations using the LDA parametrization of exchange and correlation in order to make our results comparable to literature. However, the main goal of this investigation

(a) 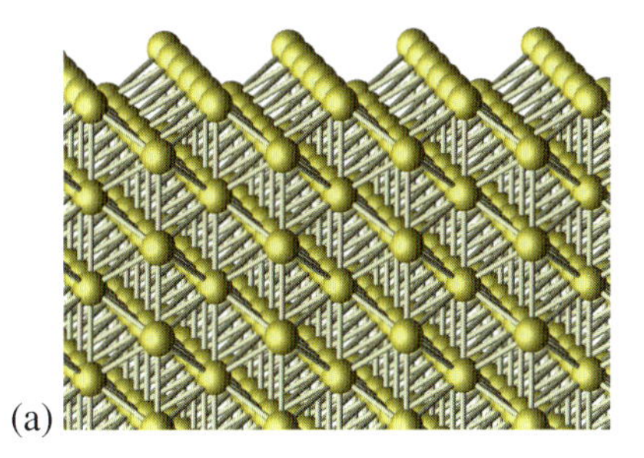

(b) 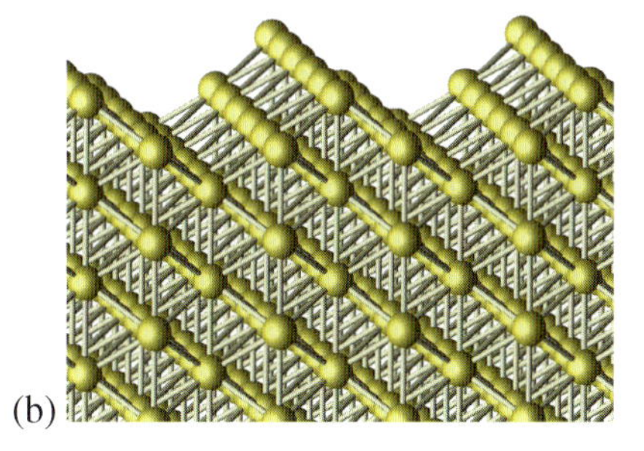

Fig. 1. Spatial view of the Au(110) surface: (a) Unreconstructed surface. (b) (1×2) missing-row reconstructed surface

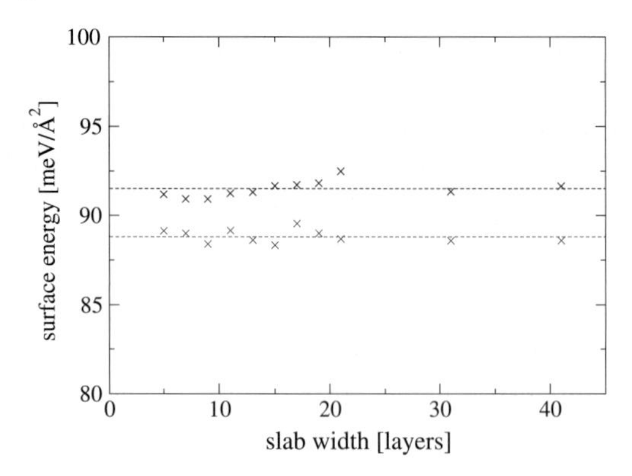

Fig. 2. Surface energy of the unreconstructed (black) and of the (1×2) missing-row reconstructed (red) Au surface

is to estimate the slab width required for an accurate representation of a real surface rather then a precise value for the surface energy. Since this should not depend on the applied XC functional, this is the most efficient way to obtain this result.

The surface energy is calculated as

$$E_{SF} = \frac{E_{SC} - NE_{bulk}}{2A}, \tag{1}$$

where E_{SC} and E_{bulk} are the total energy of the supercell and the energy of a bulk atom, respectively. n is the number of atoms in the supercell and A is the surface area of the supercell. The factor 2 reflects the two surfaces in the Au-supercell. Fig. 2 shows the surface energy plotted against the slab width. The surface energy is already converged for slab widths of just five layers, but seven (nine) layers are needed for the unreconstructed (missing-row reconstructed) slab to bring the residual forces in the three central layers below 0.01 eV/Å. Independent of the thickness, the (1×2) missing-row reconstruction is found to be energetically favoured over the unreconstructed surface. The energy gain through missing row reconstruction is 2.7 meV/Å^2. The mean energies are 91.5 meV/Å^2 and 88.8 meV/Å^2 for unreconstructed and missing-row reconstructed surface, respectively. These values are in very good agreement with recent DFT LDA results (91.5 meV/Å^2 and 88.5 meV/Å^2, respectively, of Ref. [31]) and very similar to the values obtained in Ref. [32] (90.6 meV/Å^2 and 85.9 meV/Å^2).

3.2 Au (110)-(1 × 2) Vacancy Formation Energy

Experimental observations [14, 16] indicate that cysteine does not adsorb on the strongly corrugated but most stable missing-row reconstructed Au(110) surface. The molecules rather remove the salient rows, thus lifting the (1×2)-reconstruction. The energy required for such a surface modification is directly accessible via the vacancy formation energies. Since the creation of a vacancy requires considerably more energy at an unbroken surface than at a kink site, the vacancies grow from step edges over the Au(110) terraces. In order to quantify this effect, a series of ionic relaxations is carried out, using both GGA and LDA XC-functionals. Slab width and surface size are varied in order to check for energy convergence. As in Sec. 3.1, the three central layers of the slab are kept fixed and all other atoms are free to relax. The formation energy Ω of a defect was calculated using the formula of Zhang and Northrup [33]

$$\Omega = E(N-1) + E_{bulk} - E(N), \tag{2}$$

where $E(N-1)$, E_{bulk}, and $E(N)$ are the energy of the supercell with the created vacancy, the energy of the bulk atom, and the energy of the slab without that vacancy, respectively. Since the equivalent vacancies are created symmetrically on both sides of the slab, Eq. (2) is modified to

$$\Omega = \frac{E(N-2) + 2E_{bulk} - E(N)}{2}. \tag{3}$$

Table 1. Vacancy formation energies (in eV) of the Au(110) missing-row reconstructed surface upon successive removal of atoms

XC	layers	cell surface	1st atom	2nd atom	3rd atom	4th atom
LDA	13	(1×2)	0.06			
	13	(4×2)	0.47	0.06	-0.01	-0.30
	13	(4×4)	0.46	0.04	0.02	-0.30
	7	(12×2)	0.47	0.04	0.06	0.05
GGA	13	(1×2)	0.08			
	7	(12×2)	0.36	0.06	0.07	0.06

The energy cost for the removal of the first, second, third, and fourth top row atom is determined. Table 1 shows the vacancy formation energies for various slab widths and cell surfaces. One important question, with respect to the adsorption of cysteine (and other potential adsorbates that lift the missing-row reconstruction), arises as to how fast the vacancy formation energy converges. How does the energy cost for the removal of a top-row atom change with the number of removed atoms? As can be seen, the vacancy formation energy converges already at the second vacancy. To investigate the influence of neighbouring top-rows, a series of calculations is performed using a supercell with lateral cell size (4×4), so that two salient gold rows are represented in the cell. Vacancies are created in only one of these two top-rows. The influence of the next top-rows is found to be negligibly small as can

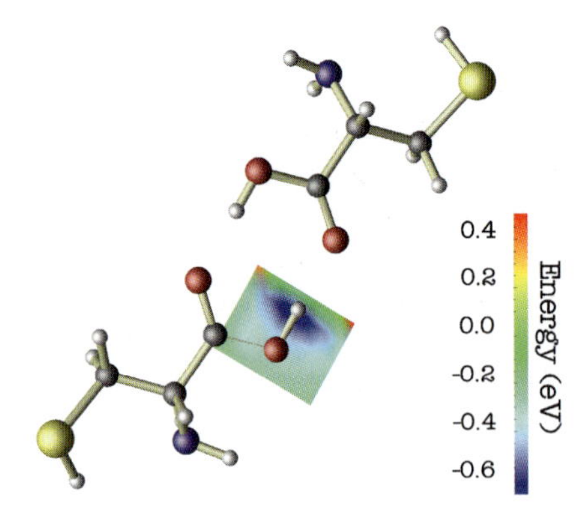

Fig. 3. Position dependent binding energy of the carboxyl-carboxyl double hydrogen bond, displayed for the position of the hydrogen atom belonging to the carboxyl group of the lower cysteine, here shown at its most favoured site

be seen by comparing the (4×2) results with the (4×4) results. The small and/or negative energy costs for the third and fourth vacancy formations for this cell sizes are due to the fact that creating these vacancies amounts to returning to the (1×1) reconstructed surface, that - though less stable then the missing-row reconstruction - is energetically favoured to an only half-filled missing-row reconstructed surface. Table 1 also shows that GGA calculations give significantly lower vacancy formation energies than LDA calculations for the removal of the first atom and slightly higher values for the following atoms.

3.3 Cysteine Dimers: The Carboxyl-Carboxyl Hydrogen Bond

As a next step we investigate gas-phase cysteine dimers, and, in particular, the molecule-molecule interaction via the carboxyl group hydrogen bonds. For the intermolecular dimer bond, such hydrogen bonds of the O-H$\cdots$O type between the cysteine carboxyl groups have been proposed [14]. To analyze the bonding properties a series of calculations are carried out using GGA XC functionals. The cell-size is chosen as $26 \times 19 \times 24$ Å^3. A cell of this size is not easily treatable with common workstation computers but requires larger clusters. Such a large cell size is necessary to obtain the same molecule-molecule distance in all three directions as in the cubic (17 Å)^3 supercell which was found suitable for the description of the single gas-phase cysteine molecule [34]. The position of one of the two molecules is varied on a two-dimensional 5×6 grid of size 2.5 Å $\times$ 3 Å in the plane described by the carboxyl groups.

Figure 3 displays the position dependent dimer bonding energy due to the carboxyl-carboxyl double hydrogen bond. The optimal binding geometry is obtained for an O-H$\cdots$O bonding angle of 176.8° and a bond length of 2.67 Å. Comparing the bonding geometry to typical literature values [35], we find very similar results. For the discussion of possible dimer adsorption geometries on the gold surface we are mainly interested in the energy cost for stretching or deforming the bond out of its optimal configuration. The energy surface shows, that while a certain stretching ($\lessapprox 0.5$ Å) perpendicular to the bonding direction can be done with little costs, further increasing of the carboxyl-carboxyl distance to more than 0.5 Å gives rise to an energy difference of the order of 0.2 eV. Displacement along the hydrogen bond axis requires even larger energies. Earlier studies have shown [26], that hydrogen bonding energies are systematically underestimated in DFT. However, the used PW91 functional performs well for the description of such a bond. The error was estimated by Maul *et al.* as approximately 0.1 eV, which is only a minor contribution to the bonding energy of the molecule. Consequently, the variation of the bonding energy due to spatial relocalization of the bonding partners should be reproduced accurately.

4 Outlook

After having studied the isolated systems of the Au surface (Sec. 3.1 and Sec. 3.2) and cysteine dimers (Sec. 3.3), we are planning to combine them to model the actual adsorption process with the amino acid placed at various locations on a regular grid above the Au(110) surface. This will allow us to gain deeper insight into the characteristics of the Au-thiolate bond and the Au-amino bond by performing a series of DFT calculations analogous to the carboxyl-carboxyl bond investigation in Sec. 3.3. The results can then be interpreted in terms of position dependent bonding energy, electron rearrangement (analogous to the treatment in Ref. [36]) and changes in the density of states. The results might be combined to predict likely adsorption geometries for cysteine on Au(110), for which ionic relaxations can then be performed. The question of a possible interplay between the bonds established by the two functional groups can be tackled by comparing the bonding characteristics of the molecule adsorbed via two bonds with those of the isolated bonds.

Acknowledgement. We gratefully acknowledge fruitful discussions with J. Furthmüller and M. Preuss. The work was financially supported by the EU NANOQUANTA Network of Excellence. Grants of computer time from the Höchstleistungsrechenzentrum Stuttgart are gratefully acknowledged.

References

1. A. Nilsson and G. M. Petterson, Surf. Sci. Rep. **55**, 49 (2004).
2. H. Ishii, K. Sugiyama, I. Eisuke, and K. Seki, Adv. Mater. **11**, 605 (1999).
3. G. Heimel, L. Romaner, J.-L. Brédas, and E. Zojer, Phys. Rev. Lett. **96**, 196806 (2006).
4. H. Vásquez, Y. J. Dappe, J. Ortega, and F. Flores, J. Chem. Phys. **126**, 144703 (2007).

5. I. G. Hill, A. Rajagopal, A. Kahn, and Y. Hu, Appl. Phys. Lett. **73**, 662 (1998).
6. W. G. Schmidt, K. Seino, M. Preuss, A. Hermann, F. Ortmann, and F. Bechstedt, Appl. Phys. A **85**, 387 (2006).
7. C. Vericat, M. E. Vela, and R. C. Salvarezza, Phys. Chem. Chem. Phys. **7**, 3258 (2005).
8. V. De Renzi, R. Rousseau, D. Marchetto, R. Biagi, S. Scandolo, and U. del Pennino, Phys. Rev. Lett. **95**, 046804 (2005).
9. E. Rauls, S. Blankenburg, and W. G. Schmidt, Surf. Sci. **602**, 2170 (2008).
10. C. Joachim, J. K. Gimzewski, and A. Aviram, Nature **408**, 541 (2000).
11. H. B. Akkerman, P. W. M. Blom, D. M. de Leeuw, and B. de Boer, Nature **441**, 69 (2006).
12. C. P. Collier, E. W. Wong, M. Belohradsky, F. M. Raymo, J. F. Stoddart, P. J. Kuekes, R. S. Williams, and J. R. Heath, Science **285**, 391 (1999).
13. S. Y. Quek, L. Venkataraman, H. J. Choi, S. G. Loule, M. S. Hybertsen, and J. B. Neaton, Nano Letters **7**, 3477 (2007).
14. A. Kühnle, T. R. Linderoth, B. Hammer, and F. Besenbacher, Nature **415**, 891 (2002).
15. M. Yu, N. Bovet, C. S. Satterley, S. Bengiō, K. R. J. Lovelock, P. K. Milligan, R. G. Jones, D. P. Woodruff, and V. Dhanak, Phys. Rev. Lett. **97**, 166102 (2006).
16. A. Kühnle, L. M. Molina, T. R. Linderoth, B. Hammer, and F. Besenbacher, Phys. Rev. Lett. **93**, 086101 (2004).
17. A. Kühnle, T. R. Linderoth, and F. Besenbacher, J. Am. Chem. Soc. **128**, 1076 (2005).
18. A. Kühnle, T. R. Linderoth, and F. Besenbacher, J. Am. Chem. Soc. **125**, 14680 (2003).
19. R. R. Nazmutdinov, J. Zhang, T. T. Zinkicheva, I. R. Manyurov, and J. Ulstrup, Langmuir **22**, 7556 (2006).
20. R. Di Felice, A. Selloni, and E. Molinari, J. Phys. Chem. B **107**, 1151 (2003).
21. R. Di Felice and A. Selloni, J. Chem. Phys. **120**, 4906 (2004).
22. G. Kresse and J. Furthmüller, Comp. Mater. Sci **6**, 15 (1996).
23. G. Kresse and J. Furthmüller, Phys. Rev. B **54**, 11169 (1996).
24. J. P. Perdew, *Electronic Structure of Solids '91*, p. 11, Akademie-Verlag, Berlin (1991).
25. J. P. Perdew and Y. Wang, Phys. Rev. B **45**, 13244 (1992).
26. R. Maul, M. Preuss, F. Ortmann, K. Hannewald, and F. Bechstedt, J. Phys. Chem. A **111**, 4370 (2007).
27. F. Ortmann, W. G. Schmidt, and F. Bechstedt, Phys. Rev. Lett. **95**, 186101 (2005).
28. F. Ortmann, W. G. Schmidt, and F. Bechstedt, Phys. Rev. B **73**, 205101 (2006).
29. G. Kresse and D. Joubert, Phys. Rev. B **59**, 1758 (1999).
30. H. J. Monkhorst and J. D. Pack, Phys. Rev. B **13**, 5188 (1976).
31. A. Nduwimana, X. G. Gong, and X. Q. Wang, Appl. Surf. Sci. **219**, 129 (2003).
32. S. Olivier, G. Tréglia, A. Saúl, and F. Williaime, Surf. Sci. **600**, 5131 (2006).
33. S. B. Zhang and J. E. Northrup, Phys. Rev. Lett. **67**, 2339 (1991).
34. R. Maul, F. Ortmann, M. Preuss, K. Hannewald, and F. Bechstedt, J. Comp. Chem. **28**, 1817 (2007).
35. D. Troitiño, L. Bailey, and F. Peral, Theochem **767**, 131 (2006).
36. M. Preuss, W. G. Schmidt, and F. Bechstedt, Phys. Rev. Lett. **94**, 236102 (2005).

Ab-initio Characterization of Colloidal IV-VI Semiconductor Quantum Dots

R. Leitsmann and F. Bechstedt

European Theoretical Spectroscopy Facility (ETSF) and
Institut für Festkörpertheorie und -optik
Friedrich-Schiller-Universität Jena
Max-Wien-Platz 1, 07743 Jena, Germany
roman@ifto.physik.uni-jena.de

Summary. We investigate structural, electronic and optical properties of colloidal IV-VI semiconductor quantum dots (QDs) using an *ab initio* pseudopotential method and a repeated super cell approximation. In particular rhombo-cubo-octahedral quantum dots (QDs) with a pseudo-hydrogen passivation shell consisting of PbSe, PbTe and SnTe are investigated for different QD sizes. The obtained dependence of the confinement energy on the QD size questions the use of 3-dimensional spherical potential well models for very small QD structures.

1 Introduction

The availability of light sources in the mid-infrared spectral region is crucial for many applications, e.g. molecular spectroscopy and gas-sensor systems for environmental monitoring or medical diagnostics. In the last decade a large progress in synthesizing colloidal IV-VI semiconductor nanorcystals (NCs) made of PbS, PbSe, PbTe has been achieved [1, 2, 3]. Typically the band gaps of these materials can be tuned between 0.5 and 1.5 eV, which covers the entire near-infrared spectral region. On the other hand, the synthesis of NCs with band gaps smaller than 0.5 eV is still a challenging task [4]. But also the theoretical description of IV-VI materials is ambitious for different reasons. There is a remarkable ionic contribution to the chemical bonding. As a consequence most of the common IV-VI semiconductors crystallize in rocksalt (*rs*) structure with the space group Fm3m (O_h). In addition, PbTe exhibits shallow Pb-derived d states [5]. Therefore, we have to treat (beside the s, p states) the outermost d states of the group IV elements as valence electrons. On the other hand for a precise description of heavy elements like Pb the inclusion of relativistic effects (e.g. spin-orbit coupling) is necessary [6]. This increases the computational cost of those calculations considerable. In addition we have to use hugh supercells to be able to compare our results to

W.E. Nagel et al. (eds.), *High Performance Computing in Science and Engineering '09*, DOI 10.1007/978-3-642-04665-0_5,

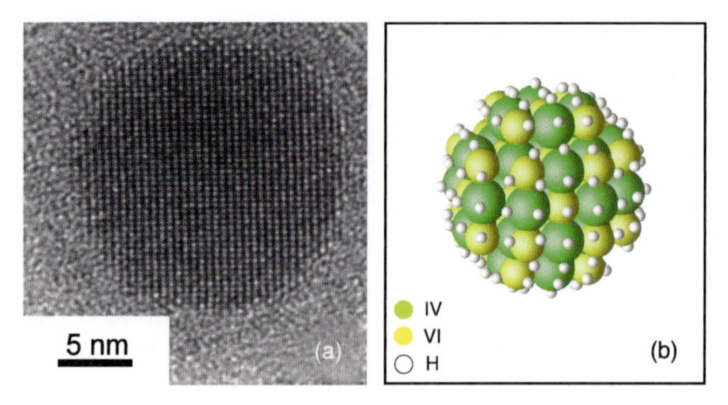

Fig. 1. (a) Experimentally observed HRXTEM image (private comunication W. Heiss, University Linz) of a small colloidal SnTe quantum dot. (b) Schematic picture of the atomic structure of a passivated IV-VI semiconductor quantum dot

experimental observations. In the present article we focus on the description of rhombo-cubo-octahedral IV-VI semiconductor NCs with a pseudo-hydrogen passivation shell. In particular PbSe, PbTe and SnTe quantum dots will be investigated. After a structural optimization we will discuss the electronic and optic properties with respect to NC size and chemical trend.

2 Computational Method

2.1 Theoretical Background

Our calculations are based on the density functional theory (DFT) [7] within the local density approximation (LDA) [8]. In this framework the ground state properties of the investigated structures are obtained by the minimization of the total energy, which is a functional of the ground-state electron density. This variational property leads to an one-particle equation, the Kohn-Sham (KS) equation, that maps the many-electron problem onto a system of non-interacting electrons $\{\Psi_i\}_i^N$ that has the same ground state density as the system of interacting electrons [9]:

$$\left\{-\frac{\hbar^2}{2m}\nabla^2 + V_{\mathbf{ion}}(\mathbf{r}) + V_{\mathbf{H}}[n](\mathbf{r}) + V_{\mathbf{XC}}[n](\mathbf{r})\right\}\Psi_i(\mathbf{r}) = \varepsilon_i\Psi_i(\mathbf{r}). \tag{1}$$

Here $V_{\mathbf{H}}$ is the classical Hartree potential, while the exchange-correlation (XC) potential is defined as the density variation of the XC energy functional $\delta E_{\mathbf{XC}}/\delta n(\mathbf{r}) = V_{\mathbf{XC}}[n](\mathbf{r})$. For the numerical solution of the Kohn-Sham equation we expand the electronic wave functions into plane waves

$$\Psi_i(\mathbf{r}) \equiv \Psi_{n\mathbf{k}}(\mathbf{r}) = \langle\mathbf{r}\,|\,n\mathbf{k}\rangle = \sum_{\mathbf{G}} c_{n\mathbf{k}}(\mathbf{G})e^{i(\mathbf{k}+\mathbf{G})\mathbf{r}} \tag{2}$$

with the band-index n and the reciprocal vector $\mathbf{k}$, which is an element of the first Brillouin zone. This leads to very efficient algorithms for periodic systems because it renders the kinetic energy operator diagonal, and symmetries in reciprocal space can be exploited. The convergence is easily controlled by adjusting the energy cutoff of the plane-wave components in the basis. The representation of the rapid oscillations of the wave functions near the nuclei demands many plane waves. However, the valence electrons determine most of the interesting physical properties, and in the interstitial region the wave functions are rather smooth and can be described by comparatively few plane waves. We therefore employ the projector-augmented wave (PAW) method [10] which establishes a one-to-one relation between the exact wave functions and a set of smooth pseudo wave functions that match the exact ones outside a certain radius around each nucleus. The application of the plane wave expansion and the PAW pseudopotential method to the Kohn-Sham equation results in a generalized eigenvalue problem that can be efficiently solved with iterative algorithms like the Residual Minimization Method with Direct Inversion in Iterative Subspace (RMM-DIIS) [11] as implemented in the Vienna Ab-initio Simulation Package (VASP) [12, 13]. Parallelization is done using the Message Passing Interface (MPI).

The electronic structure obtained within DFT-LDA is used to calculate the optical response: within independent-particle approximation the microscopic dielectric function is obtained from Bloch band eigenfunctions $|n\mathbf{k}\rangle$ using the Ehrenreich-Cohen Formula. For semiconducting systems the microscopic dielectric tensor is given by

$$\epsilon_{ij}(\omega) = \delta_{ij} + \frac{16\pi e^2\hbar^2}{\Omega} \sum_{\mathbf{k},c,v} \frac{\langle c\mathbf{k}|v_i|v\mathbf{k}\rangle\langle v\mathbf{k}|v_j|c\mathbf{k}\rangle}{[\varepsilon_c(\mathbf{k}) - \varepsilon_v(\mathbf{k})]([\varepsilon_c(\mathbf{k}) - \varepsilon_v(\mathbf{k})]^2 - \hbar^2\tilde{\omega}^2)}, \quad (3)$$

where $\tilde{\omega} = \omega + i\eta$ is the photon frequency provided with a small imaginary part η describing finite lifetime effects. The matrix elements $\langle c\mathbf{k}|v_i|v\mathbf{k}\rangle$ of the velocity operator v_i with the valence and conduction states are evaluated by calculating the corresponding Bloch integrals, which allows for an efficient parallelization. Of particular importance for the predictive power of the calculated spectra is their full numerical convergence. For the investigated nanocrystals we use up to 4000 electronic states and a Γ-point approximation.

2.2 Computational Cost

Since the Kohn-Sham matrix is diagonal in the index n of the eigenstate ("inter-band-distribution") the diagonalization can be efficiently parallelized. If there are enough nodes available, the diagonalization for the n-th state may be parallelized as well ("intra-band-distribution"). A limiting factor is, however the communication overhead required for the redistribution of the wavefunctions between all nodes, which is necessary during the orthogonalization procedure of the eigenstates. Therefore, the scalability of the VASP-code

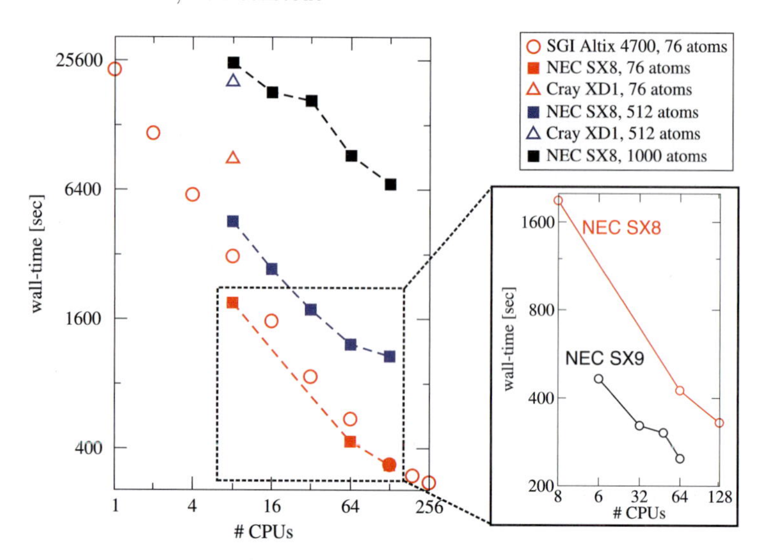

Fig. 2. Scaling behavior of differently sized jobs (red 76 atoms, blue 512 atoms, black 1000 atoms) on the NEC SX-8 (inset - SX-9), SGI Altix 4700, and a local Cray XD1 cluster

depends strongly on the size of the actual problem as can be seen in Fig. 2. In this graph we compare three systems containing 76 (red), 512 (blue), and 1000 (black) valence electrons, which correspond to Kohn-Sham matrices of the size $(168 \times 15399)^2$, $(2810 \times 108383)^2$, and $(5445 \times 211441)^2$, respectively. The calculations carried out on the NEC SX-8 and the NEC SX-9 (subset Fig. 2) system demonstrating the very good scaling behavior of the code, which is in few cases biased by different numbers of electronic iteration cycles per ionic step. In addition the performance of a singe CPU on the SX-9 system is enhanced by about 60 % compared to the SX-8 machine. Our calculations are dominated by complex matrix-matrix multiplications (CGEMM). The sustained iteration performance for both cases exceeds 1 TFLOPS already on 16 nodes NEC SX-8. The sustained efficiency is between 79 and 50 % [14]. Compared to other high-computation facilities and local machines the implementation on the NEC SX-8/9 system is the most efficient one. This fact is a consequence of the vector architecture, which can handle CGEMM operations in a very efficient way.

3 Nanocrystal Construction Using Supercells

The IV-VI semiconductor quantum dots (QDs) are modeled using the so-called supercell method, [15] i.e., periodic boundary conditions are applied. Therefore, a three-dimensional arrangement of nanocrystals embedded in a vacuum region is assumed. We use a simple-cubic arrangement of supercells.

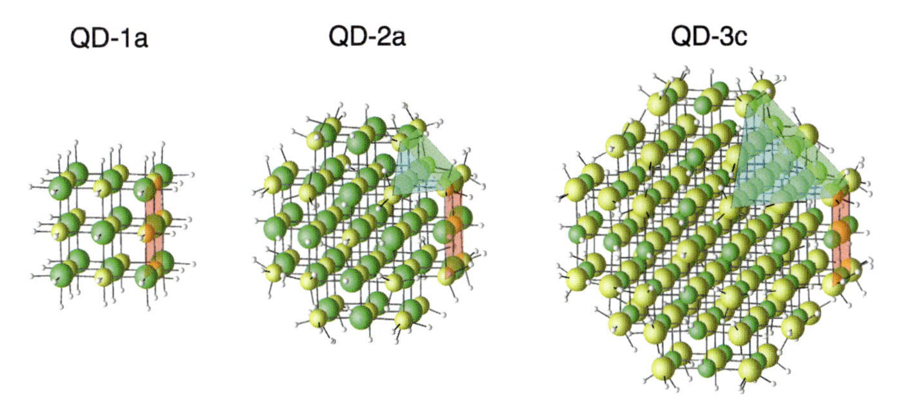

Fig. 3. Optimized structure of three differently sized PbTe quantum dots. The green, yellow, and white balls represent Pb, Te and pseudo-hydrogen atoms, respectively. The occurring surface facets are indicated by red $\{001\}$, green $\{110\}$, and blue $\{111\}$ faces

Each of them contains one single QD and a certain amount of vacuum. The size of the simple-cubic supercells vary with the diameter of the IV-VI QDs and the size of the vacuum.

The shell wise constructed nanocrystals can have either an anion or a cation in the center. This leads to different surface terminations and different QD shapes: cubes and rhombo-cubo-octahedrons. They differ with respect to the relative contributions from the $\{100\}$, $\{110\}$, and $\{111\}$ facets. In the case of the cubes only $\{100\}$ facets occur. The rhombo-cubo-octahedrons are regular octahedrons with $\{111\}$ facets truncated at each apex by $\{100\}$ planes perpendicular to the cube axis. They are distorted by additional $\{110\}$ facets between the other ones. The areas of the six $\{100\}$ facets, the twelve $\{110\}$ facets, and the eight $\{111\}$ facets occur in two different ratios, $1 : \sqrt{2} : \frac{1}{3}$ $[d = 2a_0]$ and $1 : 2\sqrt{2} : \frac{4}{3}$ $[d = 3a_0]$. The differently sized NCs are denoted as QD-da or QD-dc, where d is the diameter in units of the bulk lattice constant a_0 and a/c are describing the ionic character (anion/cation) of the central atom.

4 Results and Discussion

4.1 Structural Properties

Independent of the character of the central ion the structural optimization of the IV-VI semiconductor quantum dots yield only minimal deviations from the ideal rocksalt structure of these materials. This is illustrated in Fig. 3 where we show the optimized structure of a 0.64 nm (1a), 1.28 nm (2a), and 1.92 nm (3c) PbTe QD. Even at the surfaces the six-fold coordinated rs-structure remains stabile. Therefore, the original O_h symmetry of the bulk phase is kept. The

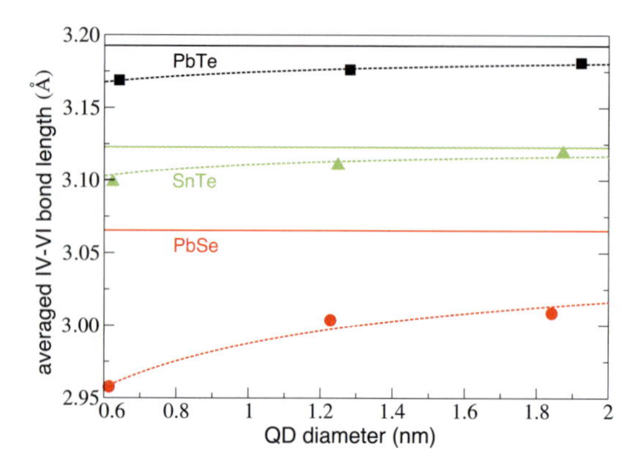

Fig. 4. Obtained IV-VI bond length for different NC diameter and materials. The filled symbols represent averaged bond length. Squares (black), triangles (green), and circles (red) represent PbTe, SnTe, and PbSe respectively. The vertical lines represent the theoretical bulk bond length. Dashed lines are fits using equation (4)

occurring surface facets are indicated in Fig. 3 by red (100), green (110) and blue (111) planes. While the (100) and (110) planes contain the same number of anions and cations, the (111) planes lead to differently anion- or cation-terminated surfaces. However, due to the O_h symmetry the appearance of surface-induced electrostatic fields is suppressed. For a quantitative analysis of the QD structure the IV-VI bond lengths are a good measure. Therefore we have plotted the averaged IV-VI bond length for different QD diameters and materials in Fig. 4. Compared to the bulk bond length (horizontal solid lines) we obtain a small reduction which increases with decreasing NC diameter D_{NC}. This behavior may be modeled via

$$d^{NC}_{IV-VI}[\text{Å}] = d^{bulk}_{IV-VI}[\text{Å}] - \frac{\alpha}{D^{\beta}_{NC}[\text{Å}]}, \tag{4}$$

where we have determined the parameter α to 0.08 (PbTe), 0.12 (SnTe), and 0.34 (PbSe) and β to 0.72 (PbTe), 0.99 (SnTe), and 0.62 (PbSe). Since α can be interpreted as a measure for the strength of the lattice deviations, the largest differences from the ideal rocksalt structure are obtained in PbSe NCs. This is a consequence of the strong ionic character of PbSe ($f_i = 0.76$ on the Phillips scale) compared to PbTe ($f_i = 0.65$) and SnTe ($f_i = 0.63$) [16]. The larger electrostatic forces between the anions and cations in PbSe lead to larger adjustments of the lattice structure at the NC surfaces and hence to a larger bond length deviation. In general the reduction of the IV-VI bond length corresponds to a compression of the material, which becomes larger with increasing influence of the surface tension.

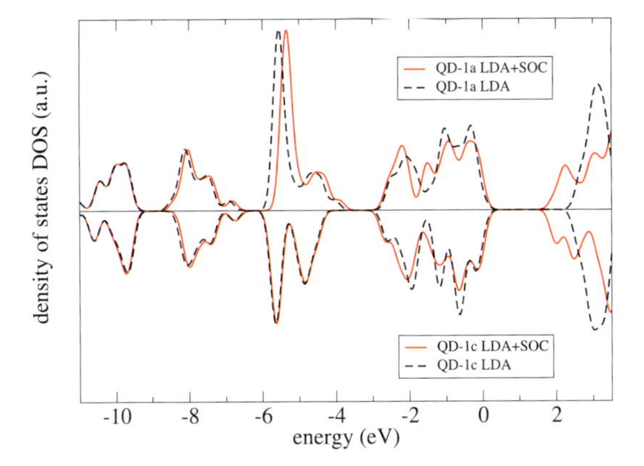

Fig. 5. Density of states of two 0.62 nm PbTe NCs with a central anion (QD-1a) or a central cation (QD-1c). The black dashed line corresponds to LDA and the red solid line to LDA+SOC. The HONO is taken as energy zero

4.2 Electronic Properties

As demonstrated in Fig. 5 at the example of a 0.62 nm PbTe NC the electronic properties are more or less independent of the ionic character of the central NC atom. This is in good agreement with the results for the NC geometries. The small remaining differences can be traced back to the different surface terminations. Therefore, in the ongoing discussion we will concentrate on NCs with a central anion.

As expected all investigated materials PbTe, PbSe, and SnTe show quite similar electronic properties. The main difference is the different size of the fundamental band gap $E^{\mathrm{KS}}_{\mathrm{gap}}(\mathrm{PbSe}) > E^{\mathrm{KS}}_{\mathrm{gap}}(\mathrm{PbTe}) > E^{\mathrm{KS}}_{\mathrm{gap}}(\mathrm{SnTe})$, see Fig. 6. This series is consistent with the expectations according to the ionicities $f_i(\mathrm{PbSe}) > f_i(\mathrm{PbTe}) > f_i(\mathrm{SnTe})$ at the Phillips scale[1]. The material with the strongest ionic character (PbSe) shows the largest band gap, because in this case the electrons are more strongly localized at the atomic cores. With increasing quantum dot size the differences, however, decrease until for $D_{\mathrm{NC}} \rightarrow \infty$ the bulk limit is reached. Taking into account the spin-orbit coupling, in the bulk limit even a change in the band gap series appears $E^{\mathrm{KS}}_{\mathrm{gap}}(\mathrm{SnTe}) > E^{\mathrm{KS}}_{\mathrm{gap}}(\mathrm{PbSe}) > E^{\mathrm{KS}}_{\mathrm{gap}}(\mathrm{PbTe})$[2]. This fact is owed by the somewhat incorrect description of the band characters close to the fundamental band gap in LSDA+SOC [17]. However, in the QD systems under investigation this problem can be neglected. Due to the confinement effect the character of the highest occupied nanocrystal orbital (HONO) and the lowest unoccupied nanocrystal orbital (LUNO) are correctly described.

[1] f_i =0.76 (PbSe), 0.65 (PbTe), and 0.64 (SnTe)

[2] The bulk band gaps in LSDA+SOC can be calculated to: 0.21 eV (PbSe), 0.13 eV (PbTe), and 0.35 eV (SnTe).

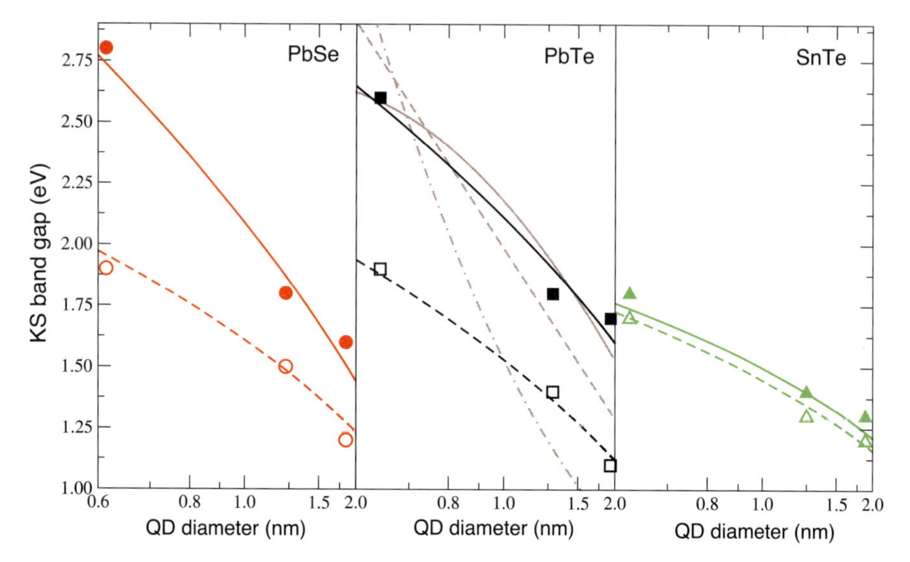

Fig. 6. KS-band gap of differently sized PbSe(red)-, PbTe(black)- and SnTe(green) QDs with central anion in LDA (filled symbols) and LSDA+SOC (empty symbols). The solid (LDA) or dashed (LSDA+SOC) lines represent the fitted size dependence according to Δ_1^{Conf}. In brown the alternative dependencies Δ_2^{Conf} with $\alpha = 1.4$ (dashed line), Δ_3^{Conf} with $\alpha = 0.92$ (dash-dotted line) and Δ_4^{Conf} with $\alpha = 2.1$ and $\beta = 0.54$ (solid line) for PbTe QDs in LDA are shown

In addition we observe a large influence of the spin-orbit interaction in PbSe and PbTe, while in SnTe QDs this effect can be neglected. In PbSe and PbTe 1a-QDs for example the inclusion of spin-orbit coupling yields decrease of the LDA band gap of 0.9 eV and 0.7 eV, respectively. The reason for this behavior is the weight of the considered cations [3]. The Sn atoms with atomic number 50 are much lighter than Pb atoms with atomic number 82, hence, the SOC in tin salts is much weaker than in lead salts.

The common confinement effect, i.e. the increase of the band gap with decreasing QD size, may be modeled using different potential well models. The size of the gap opening usually called confinement energy is the difference between the QD and the bulk band gap

$$\Delta^{\text{Conf}} = E_{\text{gap}}^{\text{KS,QD}} - E_{\text{gap}}^{\text{KS,bulk}}. \tag{5}$$

The confinement energy depends on the diameter of the nanocrystal D_{NC} and the shape of the confinement potential. In the case of a 3d spherical potential well model (Δ_3^{Conf}), e.g., a $1/D_{\text{NC}}^2$ dependence can be obtained, while 3d parabolic potentials (Δ_2^{Conf}) yield a $1/D_{\text{NC}}$ behavior. Sometimes

[3] The expectation value and the strength of the orbital induced magnetic field increases with increasing atomic number. Therefore the spin-orbit coupling is particularly large for heavy elements.

also combinations of these two models (Δ_4^{Conf}) are considered [18]. Recent *ab initio* results, however, propose a $D_{\text{NC}}^{-\beta}$ proportionality of the confinement energy (Δ_1^{Conf}) with $\beta < 1$ [19, 20, 21]. Therefore, we tried to model our results using those functional dependencies

$$\Delta_1^{\text{Conf}}[\text{eV}] = \alpha / D_{\text{NC}}^{\beta}[\text{nm}], \tag{6}$$
$$\Delta_2^{\text{Conf}}[\text{eV}] = \alpha / D_{\text{NC}}[\text{nm}], \tag{7}$$
$$\Delta_3^{\text{Conf}}[\text{eV}] = \alpha / D_{\text{NC}}^{2}[\text{nm}], \tag{8}$$
$$\Delta_4^{\text{Conf}}[\text{eV}] = \alpha / D_{\text{NC}}^{1}[\text{nm}] + \beta / D_{\text{NC}}^{2}[\text{nm}]. \tag{9}$$

As can be seen in Fig. 6 for PbTe QDs (in LDA), the best agreement is achieved for Δ_1^{Conf} with the parameters α, β in Tab. 1. This behavior suggests that theoretical descriptions of QD systems using 3d spherical or parabolic potential well models should be understood only as a fist approximation. In very small QDs with $D_{\text{NC}} \sim a_0$ the QD-vacuum interface is no longer an abrupt transition, but more likely a continuous potential curve. Therefore, in such a case the parabolic potential well model yields a better (in the sense of $\beta < 2$) description of the confinement energy as 3d spherical potentials.

Table 1. Confinement energy parameter α and β according to Δ_1^{Conf} of PbSe, PbTe and SnTe QDs

	PbSe		PbTe		SnTe	
	α	β	α	β	α	β
LDA	1.9	0.60	1.5	0.61	1.3	0.36
LSDA+SOC	1.4	0.46	1.4	0.53	1.1	0.44

Among the band gaps of PbTe QDs in Fig. 6 the band gaps of PbSe and SnTe QDs are displayed. The solid (LDA) or dashed (LSDA+SOC) lines correspond, thereby, to the band gap model with the aid of the confinement energy Δ_1^{Conf}. The respective parameters α and β are collected in Tab. 1. Taking into account SOC effects, the dependence of the band gap on the NC size and its absolute value is very similar in all three materials.

Since the KS band gap is already a good approximation for the true excitation energies in Si QD systems [22] we compare in Fig. 7 the theoretically predicted (6) and optically measured confinement energies for different QD sizes and materials. Taking into account that the parameters α and β used are determined with only three different data points, the good agreement is remarkable. In average the theoretical values overestimate the experiment by about 100 meV (dashed line). Only for very small QD sizes the experimental value exceeds the theoretical predictions for the confinement energy (3.9nm-PbSe QD). One possible reason for the small remaining discrepancies between

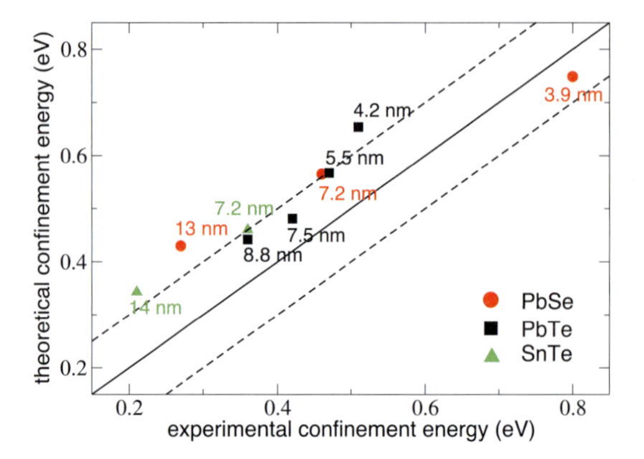

Fig. 7. Comparison of experimental and theoretical confinement energies according to equation (6). The experimental values for PbSe, PbTe and SnTe QDs are taken from the references [2], [3] and [4], respectively

experiment and theory is the influence of different surface passivations. For example the confinement energy of hydrogen or oxygen passivated Si QDs differs considerably [23, 24]. In addition, the exact measurement of the QD sizes is difficult and therefore often associated with relatively large error bars.

4.3 Optical Properties

To describe the optical properties of the considered NC systems we have calculated the dielectric tensor ϵ_{ij} according to equation (3). Since the considered QDs are central symmetric the dielectric tensor becomes isotropic $\epsilon(\omega) = \epsilon_{xx}(\omega) = \epsilon_{yy}(\omega) = \epsilon_{zz}(\omega)$. The imaginary part of the remaining dielectric function is the central quantity describing the optical absorption spectra

$$\alpha(\omega) \sim \mathrm{Im}[\epsilon(\omega)]. \tag{10}$$

In Fig. 8 we have plotted $\epsilon(\omega)$ for a PbSe, PbTe, and SnTe QD with $d = 1a_0{}^4$. As expected in all three cases the inclusion of relativistic effects (SOC) leads to a red shift of the optical spectra. In accordance with the similar electronic properties also the optical spectra of all three materials are comparable. However, in PbTe and SnTe QDs the strength of first optical transition (marked with an arrow) is enhanced including SOC effects, while in PbSe QDs this transition remains less pronounced. This is the result of the character and the spatial localization of the contributing HONO and LUNO states. In PbSe 1a-QDs the HONO state consist mainly of Se p states distributed in the whole QD and Pb s states localized at the QD center. The LUNO state, however, consist mainly of Pb p states at the QD surface. Taking into account the selection rule $\Delta l = \pm 1$, this leads to an almost forbidden HONO $\rightarrow$ LUNO

[4] a_0 - bulk lattice constant

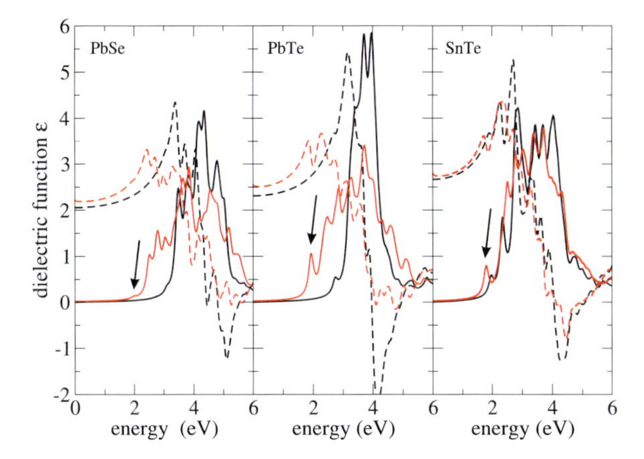

Fig. 8. Optical dielectric function (dashed line - real part; solid line - imaginary part) of a PbSe, PbTe and SnTe QD-1a. In black DFT-LDA and in red DFT-LSDA+SOC results are shown. The black arrow indicates the first optical transition

transition. On the other hand, in PbTe and SnTe 1a-QDs the selection rule for the HONO → LUNO transition is fulfilled, yielding an apparent absorption peak.

Concerning the static electronic dielectric constant $\epsilon_\infty \equiv \epsilon(0)$ we have to take into account that the calculated values are effective dielectric constants

$$\epsilon(0) = \epsilon_\infty^{\mathrm{QD}} \frac{V_{\mathrm{QD}}}{V_{\mathrm{SC}}} + \frac{V_{\mathrm{vac}}}{V_{\mathrm{SC}}} \tag{11}$$

normalized with the QD volume V_{QD}, the vacuum volume V_{vac}, and the supercell volume $V_{\mathrm{SC}} = V_{\mathrm{QD}} + V_{\mathrm{vac}}$. Therefore, we observe the following trend for the static electronic dielectric constant of small QD systems: $(\epsilon_\infty^{\mathrm{QD,PbSe}} = 33.4) < (\epsilon_\infty^{\mathrm{QD,PbTe}} = 41.5) < (\epsilon_\infty^{\mathrm{QD,SnTe}} = 48.3)$. This compares quite well with the results for the corresponding bulk materials of: $(\epsilon_\infty^{\mathrm{bulk,PbSe}} = 25.2) < (\epsilon_\infty^{\mathrm{bulk,PbTe}} = 36.9) < (\epsilon_\infty^{\mathrm{bulk,SnTe}} = 45)$ [25].

5 Summary and Outlook

We have calculated the structural, electronic, and optical properties of IV-VI semiconductor quantum dots using *ab-initio* methods. Beside the almost unperturbed rocksalt structure we found a $1/\mathrm{D}^\beta$ dependence of the confinement energy on the QD diameter with $\beta \sim 0.5$. Therefore, 3d spherical potential well models, which are commonly used in the literature, are not suitable for a precise description of the confinement energy of small QD systems. In addition we have calculated the optical absorption spectra and observed a forbidden HONO → LUNO transition in PbSe 1a-QDs. The obtained static dielectric constants are in good agreement with experimental data.

In further investigation the results concerning the optical properties have to be extended to larger QD sizes and emission spectra. For this purpose one has to calculate the influence of electronic excitations on the QD structure, i.e., the so called Stokes shift.

Acknowledgement. We acknowledge valuable discussions with colleagues of our group J. Furthmüller and F. Fuchs. In addition we thank Prof. Heiss and Prof. Schäffler from the University of Linz for providing us the experimental HRXTEM- and PL-data. The work was financially supported through the Fonds zur Förderung der Wissenschaftlichen Forschung (Austria) in the framework of SFB25, Nanostrukturen für Infrarot-Photonik (IR-ON) and the EU NANOQUANTA network of excellence (NMP4-CT-2004-500198). Grants of computer time from the Höchstleistungsrechenzentrum Stuttgart are gratefully acknowledged.

References

1. R. E. de Lamaestre, H. Bernas, D. Pacifici, G. Franzo, and F. Priolo, Appl. Phys. Lett. **88**, 181115 (2006).
2. M. V. Kovalenko, D. V. Talapin, M. A. Loi, F. Cordella, G. Hesser, M. I. Bodnarchuk, and W. Heiss, Angewandte Chemie Int. **47**, 3029 (2008).
3. J. J. Urban, D. V. Talapin, E. V. Shevchenko, C. R. Kagan, and C. B. Murray, Nature Mat. **6**, 115 (2007).
4. M. V. Kovalenko, W. Heiss, E. V. Shevchenko, J.-S. Lee, H. Schwinghammer, A. P. Alivisatos, and D. V. Talapin, J. Am. Chem. Soc. **129**, 11354 (2007).
5. R. Leitsmann and F. Bechstedt, Phys. Rev. B **76**, 125315 (2007).
6. E. A. Albanesi, C. M. I. Okoye, C. O. Rodriguez, E. L. P. y Blanca, and A. G. Petukhov Phys. Rev. B **61**, 16589 (2000).
7. P. Hohenberg and W. Kohn, Phys. Rev. **136**, B864 (1964).
8. D. M. Ceperley and B. J. Adler, Phys. Rev. Lett. **45**, 566 (1980).
9. W. Kohn and L. J. Sham, Phys. Rev. **140**, A1133 (1965).
10. P. E. Blöchl, Phys. Rev. B **50**, 17953 (1994).
11. P. Pulay, Chem. Phys. Lett. **73**, 393 (1980).
12. G. Kresse and J. Furthmüller, Comp. Mat. Sci. **6**, 15 (1996).
13. G. Kresse and J. Furthmüller, Phys. Rev. B **54**, 11169 (1996).
14. S. Haberhauer, NEC - High Performance Computing Europr GmbH SHaberhauer@hpce.nec.com (2006).
15. M. C. Payne, M. P. Teter, D. C. Allan, T. A. Arias, and J. D. Joannopoulos, Rev. Mod. Phys. **64**, 1045 (1992).
16. D. Schiferl, Phys. Rev. B **10**, 3316 (1974).
17. K. Hummer, A. Grüneis, and G. Kresse, Phys. Rev. B **75**, 195211 (2007).
18. L. Cademartiri, E. Montanari, G. Calestani, A. Migliori, A. Guagliardi, and G. A. Ozin, J. Am. Chem. Soc. **128**, 10337 (2006).
19. J. Heitmann, F. Müller, L. Yi, M. Zacharias, D. Kovalev, and F. Eichhorn, Phys. Rev. B **69**, 195309 (2004).
20. H.-C. Weissker, J. Furthmüller, and F. Bechstedt, Phys. Rev. B **65**, 155328 (2002).
21. L. E. Ramos, J. Furthmüller, and F. Bechstedt, Phys. Rev. B **70**, 033311 (2004).

22. L. E. Ramos, H.-C. Weissker, J. Furthmüller, and F. Bechstedt, Phys. Stat. Sol. (B) **15**, 3053 (2005).
23. L. E. Ramos, J. Furthmüller, and F. Bechstedt, Appl. Phys. Lett. **87**, 143113 (2005).
24. L. E. Ramos, J. Furthmüller, and F. Bechstedt, Phys. Rev. B **72**, 045351 (2005).
25. E. Burnstein, S. Perkowitz, and M. Brodsky, Journal de Physique **29**, C4 (1968).

Understanding Long-range Indirect Interactions Between Surface Adsorbed Molecules

W.G. Schmidt, S. Blankenburg, E. Rauls, S. Wippermann, U. Gerstmann, S. Sanna, C. Thierfelder, N. Koch, and M. Landmann

Lehrstuhl für Theoretische Physik, Universität Paderborn, 33095 Paderborn, Germany

Summary. Large-scale *first-principles* calculations are used to rationalize the formation of well-separated (~10 Å) molecular rows of phenylglycine upon co-adsorption of adenine and phenylglycine on Cu(110) [Chen and Richardson, Nature Materials 2, 324 (2003)]. It is found that the molecular adsorption leads to longwave oscillations of the charge density at the Cu(110) surface. The experimentally observed indirect interaction between the molecular rows is mediated by these charge fluctuations. Strain effects, in contrast, are of minor importance.

1 Introduction

The self-assembly of highly organized systems from molecular building blocks opens new avenues for realizing and exploring nanodevice concepts [1]. Typically, metal substrates serve as platforms for the engineering of such supramolecular structures [2]. For the handling of complex molecules on metals, one needs to understand their lateral interactions in dependence on the substrate and co-adsorbed species. Experiment and *first-principles* calculations have been very successful in rationalizing the adsorption of single as well as directly interacting molecules on surfaces [3, 4, 5, 6, 7, 8]. However, in addition to direct molecular interactions, such as, e.g., H bonds, indirect substrate-mediated interactions between the adsorbates may occur. Although typically smaller than 0.1 eV, they often decisively influence molecular self-assembly and may act at distances beyond 100 Å [2, 9]. This long-range character prevented for a long time an accurate modeling of the respective forces. The recent progress in computational power, in particular the availability of parallel vector systems that address both the memory and speed requirements of large-scale *first-principles* calculations allow now for studying the origin of these indirect interactions in detail [10].

Theoretically, long-range adsorbate interactions mediated by substrate electrons were predicted already decades ago [11]. Thereby the interaction

W.E. Nagel et al. (eds.), *High Performance Computing in Science and Engineering '09*, DOI 10.1007/978-3-642-04665-0_6,

energy shows damped oscillations with wavelength π/k_F, where k_F is the Fermi wave vector. This seems to be in accord with a series of observations of indirect adsorbate interactions on the (111) surfaces of Au, Ag, and Cu, e.g., Refs. [12, 1, 9]. At these surfaces partially filled, strongly surface localized electronic states with short Fermi wave vectors occur. This is not the case, however, for numerous further systems where similar interactions are observed, see, e.g., Refs. [13, 14, 15, 16]. The molecular row separation of about $\sim$10 Å on Cu(110) detected in Ref. [15], for example, cannot be explained by Cu Fermi wave vectors between about 1.36 (bulk) and 2 $Å^{-1}$ (thin films) [17]. Is the long-range interaction in such cases still due to the response of the substrate electrons?

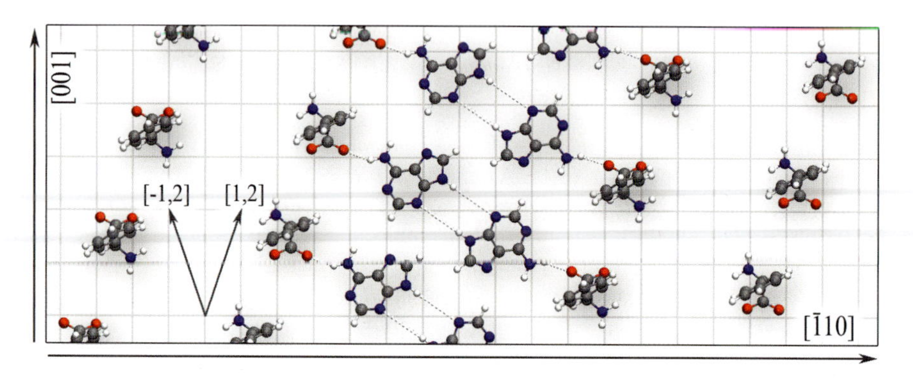

Fig. 1. Molecular model derived in Ref. [15] for phenylglycine co-adsorbed with adenine forming dimer rows along the [1,2] direction on Cu(110). Hydrogen bonds and the Cu(110) surface unit cells are indicated with dashed and gray lines, respectively

In order to answer this question, one prototypical system is modeled here on the basis of *first-principles* calculations. We employ density functional theory (DFT) for simulations of adenine and phenylglycine adsorbed on Cu(110) [15]. In this case the presence of indirect interactions between the admolecules is obvious from the adsorption configuration. The present DFT calculations yield a spatial modulation of the molecular adsorption energy that is suitable to explain the experimental findings. The modulation is indeed traced to adsorption-induced oscillations of the surface charge density that in turn lead to a local variation in the substrate-adsorbate bond strength.

Let us start by a brief description of the experimental findings [15]. Adenine deposited on Cu(110) at room temperature forms ordered one-dimensional molecular dimer chains. Co-adsorbed phenylglycine forms double rows that run parallel to the adenine dimer chains, see Fig. 1. The interaction between the adenine dimer row and the neighboring first phenylglycine row is understood in terms of hydrogen bonds, substrate locking and Coulomb repulsion [18]. Its interaction with the second row of phenylglycine molecules

– separated from the first row by ∼10 Å – is the subject of the present study. The authors of the experimental study proposed "metal-mediated, dipole-dipole interactions" to cause the molecular spacing.

2 Computational Method

Here the system is modeled using DFT within the generalized gradient approximation (GGA) [19], as implemented in the Vienna Ab Initio Simulation Package (VASP) [20]. The electron-ion interaction is described by the projector-augmented wave (PAW) method [21], which allows for an accurate treatment of the first-row elements as well as the Cu 3*d* electrons with an energy cutoff of 340 eV.

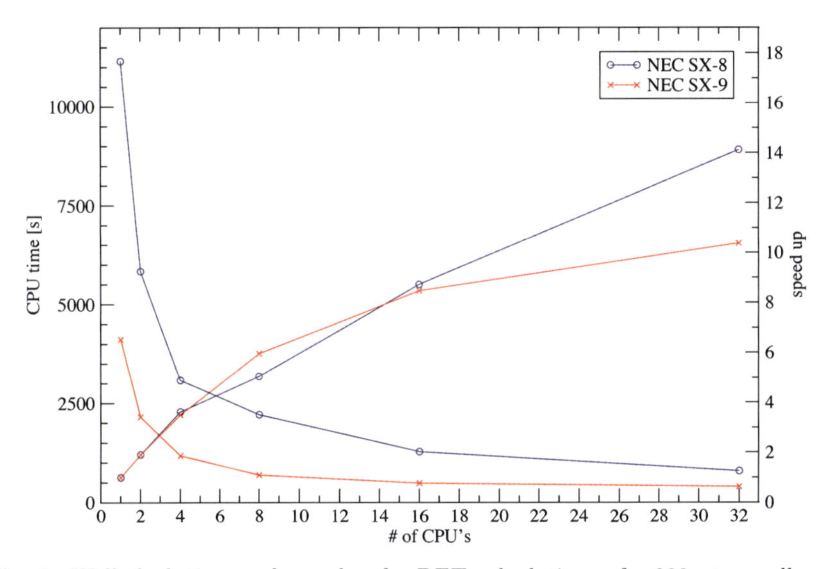

Fig. 2. Wall clock time and speedup for DFT calculations of a 200 atom cell used for surface modelling. The calculations were performed with the Stuttgart optimized VASP version on the HLRS NEC SX-8 and SX-9 machines

Within the DFT calculations, the system of Kohn-Sham equations

$$\left\{-\frac{\hbar^2}{2m}\triangle + V_{ext}(\mathbf{r}) + \int \frac{n(\mathbf{r}')}{|\mathbf{r}-\mathbf{r}'|}d\mathbf{r}' + V_{xc}(\mathbf{r})\right\}\psi_{n\mathbf{k}}(\mathbf{r}) = \varepsilon_{n\mathbf{k}}\psi_{n\mathbf{k}} \quad (1)$$

$$n(\mathbf{r}) = \sum_{n,\mathbf{k}} f_{n\mathbf{k}}|\psi_{n\mathbf{k}}|^2 \quad (2)$$

is solved iteratively until self-consistency in the total electron density $n(\mathbf{r})$ is reached. The ground-state DFT calculations were parallelised for different bands and sampling points in the Brillouin zone using the message passing interface (MPI). Fig. 2 shows benchmark calculations to determine the electronic ground state of a 200 atom cell used for surface modelling. The calculations within this project were performed on the NEC SX-8 of the Höchstleistungs-Rechenzentrum Stuttgart. As can be seen, a reasonable scaling is achieved for using up to 32 CPUs. Very recently test runs were performed on the NEC SX-9. While a substantial decrease of the CPU time (up to a factor of three) is observed, we also noted a somewhat worse scaling if more than 8 CPSs are used. We are presently investigating the origin of this problem.

Turning to the numerical parameters for the simulations, the surface Brillouin zone is sampled using a 1×2×1 mesh. Periodically repeated slabs, containing six atomic Cu layers plus the adsorbed molecules and a vacuum region equivalent in thickness to about 17 atomic Cu layers describe the adsystem. This methodology reproduces the measured geometry for phenylglycine adsorbed on Cu(110) [22].

3 Results

Basically, two mechanisms are conceivable that – possibly in combination – could be responsible for the measured spacing between the two phenylglycine rows: substrate-mediated interactions via long-range strain fields [23] or adsorption-induced Friedel oscillations of the electron density at the metal surface [24, 11, 25]. By performing calculations where the substrate atoms are either frozen at ideal bulk positions or fully relaxed according to the adsorption induced forces, one can identify the magnitude of the respective contributions to the spatial modulation of the adsorption energy.

In a first step, we neglect strain effects and vary the distance between the molecular rows without taking the structural relaxations into account. For the calculations we use a model system that includes one adenine and two phenylglycine molecules adsorbed on Cu(110) within a translational symmetry that varies from $\left(\begin{smallmatrix} 1 & 2 \\ 22 & 0 \end{smallmatrix}\right)$ to $\left(\begin{smallmatrix} 1 & 2 \\ 30 & 0 \end{smallmatrix}\right)$. The measured scanning tunneling microscopy (STM) data do not allow to conclude unambiguously on the position and orientation of the phenylglycine in the second molecular row with respect to the molecules in the first row. However, given the highly corrugated potential energy surface (PES) as well as rotational profile for phenylglycine adsorbed on the Cu(110) surface [22], only two molecular orientations are likely to be relevant. The molecules in the two phenylglycine rows may either be parallel or antiparallel, as shown in the upper and lower panel of Fig. 3, respectively.

The adsorption energy for the 2nd row phenylglycine in the presence of the primary row of phenylglycine/adenine is calculated according to

$$E_{ads} = E_{tot} - E_{adn+1pgl} - E_{pgl} + \frac{1}{2} E_{H_2}, \tag{3}$$

where E_{tot}, $E_{adn+1pgl}$, E_{pgl} and E_{H_2} refer to energies of the total system, the adsystem containing adenine and one phenylglycine, phenylglycine and hy-

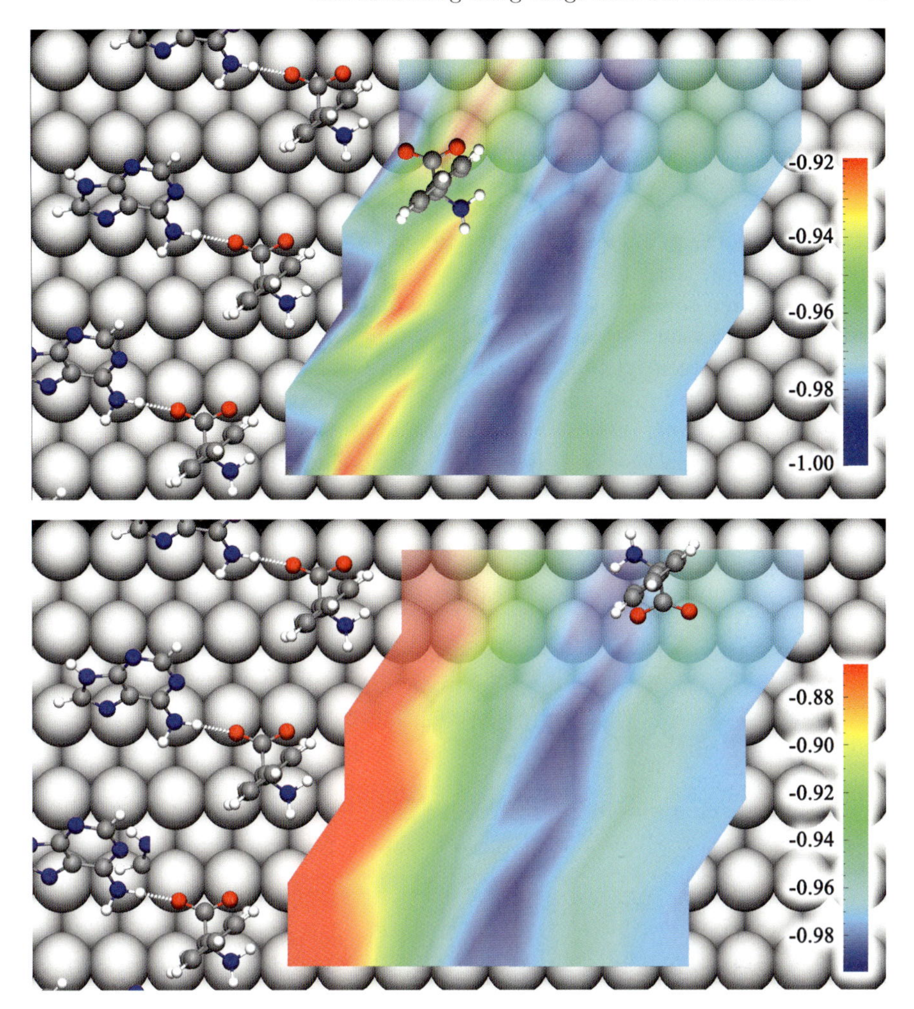

Fig. 3. Calculated adsorption energies (in eV, see text) of the 2nd row phenylglycine molecules for parallel (upper panel) and antiparallel (lower panel) orientation in the presence of pre-adsorbed adenine and phenylglycine. Thereby the leftmost atom of the 2nd row phenylglycine serves as molecular point of reference

drogen in gas phase, respectively. The corresponding energy surface is shown in Fig. 3. For clarity, the fine structure of the PES [22] is suppressed and only the minimum energy values within the respective Cu(110)(1×1) unit cells are used for the color coding. Obviously, the adsorption energies show damped oscillations with increasing molecular row distance. Depending on the relative orientation of the amino acids, local minima occur for a row distance between about 10 and 14 Å. These seem well suited to explain the experimental finding that "the molecular centres could be closer than 1.28 nm but probably not less than 800 pm." [15]. Moreover, the calculated magnitude of the adsorption energy oscillations of $\sim$ 100 meV and their wave length of $\sim$ 10 Å are compatible with many experimental examples for indirect interactions [13, 12, 14]. Still, on closer inspection, the adsorption energies calculated here do not perfectly reproduce the experimental findings [15]: The calculated global energy minimum does not correspond to separated phenylglycine rows, but to the geometry shown in the upper panel of Fig. 3. Here, the phenylglycine molecules are sufficiently close to form hydrogen bonds, similar to the case of phenylglycine monolayers adsorbed on Cu(110) [22].

There are several explanations that possibly account for the discrepancy between experiment and theory in the present case. The perhaps most likely one is related to the modeling of the H bonds. It is discussed in a number of studies, see, e.g., Refs. [26, 27], that DFT-GGA noticeably overestimates the energy of H-bonds. Based on the same methodology as used here, the cohesive energy of ice Ih was found to be overestimated by 120 meV per molecule [28]. This is of the same order of magnitude as the energy differences relevant in the present case. Further, we cannot exclude that the limited size of our supercell introduces numerical artefacts. In particular the superposition of charge density waves due to artificial image molecules will affect the accuracy of the calculations. However, the apparent discrepancy between experiment and theory may also be related to the interpretation of the STM micrographs. Additional species not resolved experimentally and not included in the simulations might alter the surface energetics. The adsorption of benzoate on Cu(110) for example, was found to be strongly influenced by naturally occurring or deliberately deposited Cu adatoms [29].

Regardless of the reasons for the incomplete reproduction of the experimental findings, the calculated adsorption energies compiled in Fig. 3 prove the presence of indirect interactions not related to strain effects. To probe their possible relation to adsorption-induced Friedel oscillations of the charge density, we calculated the difference between the self-consistently obtained charge densities of the adenine and phenylglycine (1st row) adsorbed system and the clean Cu surface. The resulting density difference is plotted in Fig. 4. If a Friedel-like behavior of the lateral density oscillations along $[\bar{1}10]$ is assumed, $\rho(d) \sim \rho_0 \cos(2k_F d + \Phi)/d^2$, a surface Fermi wave vector $k_F = 0.35$ Å^{-1} – substantially smaller than the bulk value – is obtained from the calculated data. This indicates a strong modification of the Cu surface electronic structure upon molecular adsorption.

The magnitude of the adsorption induced charge-density oscillations is rather small, of the order of $10^{-3}e$Å^{-3}. Is this sufficient to explain adsorption energy differences of $\sim$ 100 meV? To estimate the influence of the Friedel oscillations on the surface energetics, we follow Harrison [30] and perform MO-LCAO (molecular orbitals from a linear combination of atomic orbitals) calculations. The adsorption of phenylglycine on Cu(110) leads to the formation of covalent bonds between the amino-group nitrogen and the carboxyl-group oxygen with the copper substrate atoms, see Ref. [22]. We approximate

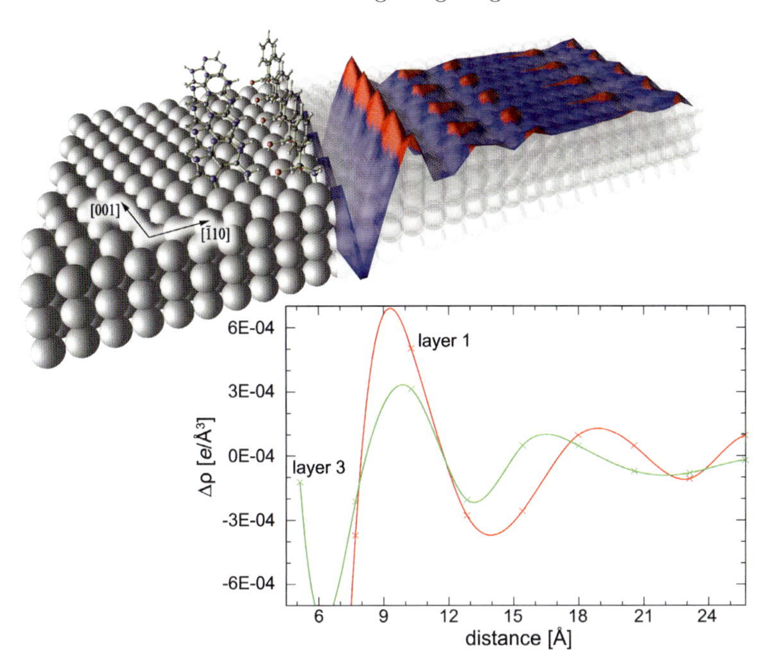

Fig. 4. Charge density difference of adenine and phenylglycine (1st row) adsorbed and clean Cu(110) surface. In the upper panel charge depletion and accumulation in the uppermost atomic layer are indicated blue and red, respectively. The lower panel shows the density differences in the first and third atomic layer vs the distance from the phenylglycine N atom along the [$\bar{1}$10] direction

the variation of the respective bond energies by

$$\Delta E = \Delta q \left(E^* - \frac{\epsilon_1 + \epsilon_2}{2} \right), \tag{4}$$

where E^* and ϵ_i are the energies of the molecular and atomic orbitals, respectively. The local variation of the charge Δq was calculated from

$$\Delta q = \frac{4}{3} \pi r_{at}^3 \Delta \rho, \tag{5}$$

where an radius of $r_{at} = 1.278$ Å has been used to approximate the size of the Cu atom. The O-Cu and N-Cu interaction parameters were calculated within the two center approximation using the semi-relativistic code described in Ref. [31]. The resulting bond energy difference vs lateral spacing is shown in Fig. 5. Obviously, the adsorption of phenylglycine on Cu(110) is not perfectly described within the simple tight-binding scheme, as seen from the phase shift between the oscillations calculated within the MO-LCAO model and DFT-GGA. However, given the simplicity of the approximation, the variation of the

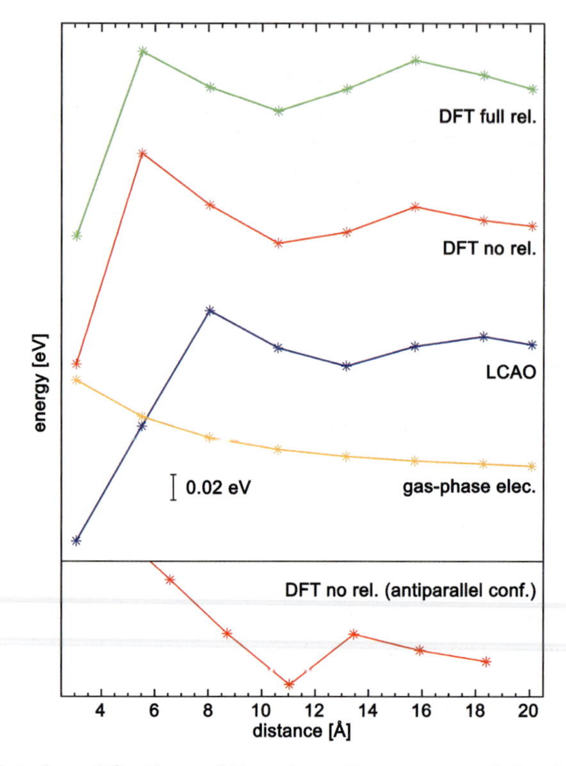

Fig. 5. Calculated modifications of the adsorption energy vs lateral spacing resulting from DFT-GGA calculations including structural relaxations, DFT-GGA calculations for frozen adsorbate and substrate, MO-LCAO calculations, and gas-phase dipole-dipole repulsion for the parallel molecular configuration are shown in the upper panel. The lower panel shows the corresponding DFT-GGA data for antiparallel molecular configuration (cf. Fig. 3)

bond energies calculated within the tight-binding approach agrees surprisingly well with the *first principles* data. For small distances, the MO-LCAO model tends to overestimate the molecular attraction. This can be explained by the neglect of the repulsive dipole-dipole interaction between the phenylglycine molecules, see lowest but one curve in Fig. 5.

Till now we have completely neglected the influence of atomic relaxations. How important are strain fields in the context of indirect interactions? The uppermost curve in Fig. 5 contains DFT-GGA results obtained upon fully relaxing both adsorbate and substrate. The differences to the adsorption energies obtained under the assumption that the Cu surface as well as the phenylglycine molecules are frozen are minor, and occur basically for molecular distances smaller than $\sim$ 10 Å. This shows that electronic effects are far more important for long-range indirect interactions than structural effects, at least in the present case. We cross-checked this finding by comparative calculations for glutamic acid adsorbed on Ag(110). Here strain effects are slightly more important: While the positions of the local energy minima are still dominated by the electron density oscillations, the inclusion of structural relaxations is required to obtain the relative depths of the minima in accordance with the measured molecular row spacing of more than 30 Å [16].

4 Summary

In summary, we used DFT calculations to rationalize the frequently observed long-range order in atomic and molecular adsorption on metal surfaces. Using the model system of adenine and phenylglycine adsorbed on Cu(110), we find local minima in the PES that are suitable to explain the molecular distances observed experimentally. The electronic origin of the indirect interaction in the case studied here is confirmed by (i) MO-LCAO model calculations for the bond strength variation due to charge fluctuations and (ii) calculations with and without structural relaxations that allow for quantifying the influence of long-range strain fields. Our results suggest that mainly surface charge density oscillations are responsible for the long-range order observed for many atomic and molecular adsorbates on metal substrates, even if no highly surface localized electronic states – as observed at noble metal (111) surfaces – exist at the clean substrate. The mainly electronic origin of the indirect interactions suggests their tuning by modifying the substrate-adsorbate charge transfer and its screening by the choice of specific adsorbate functional groups and substrates.

Acknowledgments. Generous grants of computer time from the Höchstleistungs-Rechenzentrum Stuttgart (HLRS) and the Paderborn Center for Parallel Computing (PC^2) are gratefully acknowledged. We thank the Deutsche Forschungsgemeinschaft for financial support.

References

1. Y. Pennec, W. Auwärter, A. Schiffrin, A. Weber-Bargioni, A. Riemann, and J. V. Barth, Nature Nanotechnol. **2**, 99 (2007).
2. J. V. Barth, Annu. Rev. Phys. Chem. **58**, 375 (2007).
3. A. Nilsson and L. G. M. Pettersson, Surf. Sci. Rep. **55**, 49 (2004).
4. A. Hauschild, K. Karki, B. C. C. Cowie, M. Rohlfing, F. S. Tautz, and M. Sokolowski, Phys. Rev. Lett. **94**, 036106 (2005).
5. M. Preuss, W. G. Schmidt, and F. Bechstedt, Phys. Rev. Lett. **94**, 236102 (2005).
6. W. G. Schmidt, K. Seino, M. Preuss, A. Hermann, F. Ortmann, and F. Bechstedt, Appl. Phys. A **85**, 387 (2006).
7. A. Ferretti, C. Baldacchini, A. Calzolari, R. Di Felice, A. Ruini, E. Molinari, and M. G. Betti, Phys. Rev. Lett. **99**, 046802 (2007).

8. N. Nyberg, M. Odelius, A. Nilsson, and L. G. M. Petterson, J. Chem. Phys. **119**, 12577 (2003).
9. A. Schiffrin, A. Riemann, W. Auwärter, Y. Pennec, A. Weber-Bargioni, D. Cvetko, A. Cossaro, A. Morgante, and J. V. Barth, Proc Natl Acad Sci USA **104**, 5279 (2007).
10. S. Blankenburg and W. G. Schmidt, Phys. Rev. B. **78**, 233411 (2008).
11. K. H. Lau and W. Kohn, Surf. Sci. **75**, 69 (1978).
12. J. Repp, F. Moresco, G. Meyer, K.-H. Rieder, P. Hyldgaard, and M. Persson, Phys. Rev. Lett. **85**, 2981 (2000).
13. T. T. Tsong, Phys. Rev. Lett. **31**, 1207 (1973).
14. S. Lukas, G. Witte, and C. Wöll, Phys. Rev. Lett. **88**, 028301 (2001).
15. Q. Chen and N. V. Richardson, Nature Materials **2**, 324 (2003).
16. T. Jones, C. Baddeley, A. Gerbi, L. Savio, M. Rocca, and L. Vattuone, Langmuir **21**, 9468 (2005).
17. .T Balcerzak, Materials Science-Poland **24**, 719 (2006).
18. S. Blankenburg and W. G. Schmidt, Phys. Rev. Lett. **99**, 196107 (2007).
19. J. P. Perdew, J. A. Chevary, S. H. Vosko, K. A. Jackson, M. R. Pederson, D. J. Singh, and C. Fiolhais, Phys. Rev. B **46**, 6671 (1992).
20. G. Kresse and J. Furthmüller, Comp. Mat. Sci. **6**, 15 (1996).
21. G. Kresse and D. Joubert, Phys. Rev. B **59**, 1758 (1999).
22. S. Blankenburg and W. G. Schmidt, Phys. Rev. B **74**, 155419 (2006).
23. V. Humblot, S. Haq, C. Muryn, W. A. Hofer, and R. Raval, J. Am. Chem. Soc. **124**, 503 (2002).
24. J. Friedel, Nuovo Cimento Suppl. **7**, 287 (1958).
25. P. T. Sprunger, L. Petersen, E. W. Plummer, E. Laegsgaard, and F. Besenbacher, Science **275**, 1764 (1997).
26. S. Tsuzuki and H. P. Lüthi, J. Chem. Phys. **114**, 3949 (2001).
27. G. Jones, S. J. Jenkins, and D. A. King, Surf. Sci. **600**, L224 (2006).
28. C. Thierfelder, A. Hermann, P. Schwerdtfeger, and W. G. Schmidt, Phys. Rev. B **74**, 045422 (2006).
29. C. C. Perry, S. Haq, B. G. Frederick, and N. V. Richardson, Surf. Sci. **409**, 512 (1998).
30. W. A. Harrison, *Electronic Structure and the Properties of Solids* (Freeman, San Francisco, 1980).
31. V. Heera, G. Seifert, and P. Ziesche, J. Phys. B: At. Mol. Phys. **17**, 519 (1984).

H_2 Carrying Capacity by Considering Charging and Discharging Processes – Case Studies on Small Carbon- and Boron Nitride Nanotubes

Wai-Leung Yim* and Thorsten Klüner

Institut für Reine und Angewandte Chemie, Theoretische Chemie,
Carl von Ossietzky Universität Oldenburg,
Carl-von-Ossietzky-Str. 9-11, 26129 Oldenburg, Germany

1 Introduction

A fuel-cell is an electrochemical equipment in which hydrogen and oxygen would be transformed into water and electric power. As a result, no carbon-containing gas pollutant would be emitted. In the past two decades, the fuel-cell technology has been advanced,[1, 2, 3] but still there are some challenging problems to be addressed, e.g. the durability of the electrode, oxygen conduction efficiency, and the limited hydrogen storage capacity. The later issue – hydrogen storage – is in the center of interest in this work. The targeted hydrogen-carrying capacity is 6.5 %wt by 2010, as designated by the US Department of Energy.[4]

Nowadays, several kinds of materials receive much attention, including graphitic carbon materials,[5, 6, 7] boron nitride materials,[5, 6, 7] metal-organic frameworks,[11, 12, 13, 14] and metal hydrides.[8, 9, 10] Hydrogen chemisorption on graphene-like surfaces is not likely, because of the high dissociation barrier of H_2. Therefore, it is widely accepted that hydrogen gas is stored in its physisorbed state on the surfaces.[11] Recent materials design was motivated by the fact that hydrogen might adsorb on charged species, such as charged fullerenes or C_{60}-supported alkaline metal.[12, 13] They claimed that the hydrogen storage can be as high as 8 wt% on charged fullerenes and C_{60}-supported alkaline metals. However, the operational temperature-pressure conditions are not known. So, the price/performance for these systems cannot be estimated at the moment.

In addition to the physisorbed state of H_2 on carbon-materials, many attempts have been made to investigate the hydrogen storage using their dissociative chemisorption state. On bare graphene-systems, the dissociative

* Present address: Institute of High Performance Computing, 1 Fusionopolis Way, #16-16 Connexis, Singapore 138632

W.E. Nagel et al. (eds.), *High Performance Computing in Science and Engineering '09*, DOI 10.1007/978-3-642-04665-0_7,

chemisorption of hydrogen is not likely because of its large dissociation barrier. By chemical intuition and recent detailed theoretical analyses on hydrogenation of benzene, the [2+2] cycloaddition of H_2 on graphene-like system is symmetry-forbidden and such a reaction would experience a very high barrier.[14] To avoid this, hydrogen-spillover catalysts, such as Pt-particles, have been used to generate hydrogen atoms.[15, 16, 17] On some transition metal particles, it is almost barrierless for H_2 dissociation to occur, and H_2 can then be stored by atomic migration from the catalyst to the carbon surfaces.[18, 19] However, there are two major technical problems that limit the use of such a hydrogen-spillover scheme: 1) slow hydrogen migration on graphene-like substances in the chemisorption state; 2) aggregation of transition metal catalysts, which lowers the surface area of the catalyst. As a result the weight of the support increases while the weight percentage of hydrogen storage decreases.

Boron nitride nanotube (BNNT) surfaces may serve as important alternative to store H_2.[5, 6, 7] Boron nitride has a similar molecular mass compared to carbon-based materials. The chemical bonding network, however, is very different between these two kinds of materials. The boron nitride network is linked together by polar covalent bonds, while the graphene network is formed by non-polar covalent bonds.

Concerning the H_2 dissociation mechanism, it is noteworthy that some diatomic dissociation reactions can be favored by polar interactions.[20] Recently, in another work of oxygen-reduction reaction on transition metal surfaces, we found that the dissociating O_2-species has considerable polar character. Furthermore, the transition state of O_2 can be stabilized by introducing some electropositive transition metal elements around the reaction center. Similarly, we can also analyze the bonding character for H_2 on single-walled carbon nanotubes and boron nitride nanotube surfaces.

We have investigated the complete hydrogenation cycles on (3,3)-, (4,4)- and (5,5)- BNNTs, and (3,3)-single-walled carbon nanotubes (SWCNT), respectively. The Woodward-Hoffmann (W-H) rule has been widely used to rationalize the stereochemistry of pericyclic reactions, like 1,3-dipolar interactions between O_2, NO_2 and SWCNT.[21] For H_2 cycloaddition, it involves 2 electrons from H_2, and 2 electrons from the C-C p-bond, i.e. a [2+2] cycloaddition. By the W-H rule, such an addition requires an antarafacial interaction between H_2 and C=C, which is symmetry-forbidden. On the other hand, the boron nitride network constitutes of partially covalent and partially ionic bonds, and therefore, the W-H rule may not be applicable. Note, that 1,3- and 1,4-cycloadditions are not considered in this work because they are geometrically disfavored.[14]

In this work, the only assumption is that we calculate the direct dissociative chemisorption of H_2 on SWCNT and BNNT, and no atomic hydrogen migration would be considered. We believe that this is a reasonable assumption, as previous experimental work suggested that hydrogen migration on SWCNT is limited.

2 Computational Details

Energy Profile Calculations

We used the Vienna *Ab Initio* Simulation Package (VASP) to perform calculations within density functional theory (DFT).[22, 23, 24, 25] Throughout the study, the PBE exchange-correlation functional was used.[26] Projected augmented wave (PAW) pseudopotentials were adopted for B, N and C atoms,[27] and the planewave energy cutoff and the augmentation charge cutoff were set to 400 and 645 eV, respectively. Geometry optimizations were carried out using the conjugate gradient minimization scheme in VASP. The convergence threshold for electronic structure calculations and geometry optimizations were set to $1x10_{-4}$ eV. We used a $1\times1\times6$ Monkhorst-Pack (MP) grid for k-space integration,[28] and Gaussian smearing with a smearing parameter of 0.1 eV was chosen. Systematic convergence tests showed that the computed results were sufficiently accurate. Furthermore, the Climbing Nudged Elastic Band scheme was used in order to locate the transition structures of H_2 dissociative chemisorption.[29, 30] At the transition structures, the convergence threshold of the atomic forces was set to 0.02 eV/Å.

Thermodynamic Models

We set up thermodynamic models to describe the chemical equilibrium of various states during the hydrogen storage process. In our simulations, we presume that the migration process of atomic hydrogen in the chemisorption state is slow, which was evidenced by previous studies.[18] Thus, we considered a series of elementary processes as listed below.

Physisorption process:

$$\mathrm{Tube} - 2n\mathrm{H} \ldots m\mathrm{H}_2 + \mathrm{H}_2(g) \rightarrow \mathrm{Tube} - 2n\mathrm{H} \ldots (m+1)\mathrm{H}_2 \tag{1}$$

To predict the equilibrium conditions, we make use of fundamentals of thermodynamics. During the physisorption process,

$$\Delta H = \Delta U + \Delta nRT = \Delta U - RT \tag{2}$$

Assume $S^\circ = 0$ for an immobilized phase, we have

$$\Delta G = \Delta H - T\Delta S$$

$\Delta G^\circ = \Delta U - RT - T\left(-S^\circ_{H_2}\right) = \Delta U - RT + TS^\circ_{H_2}$, $S^\circ_{H_2}$= 130.7 J mol K^{-1} (experimental value)

$$\log K^\circ = -\frac{\Delta G}{2.303RT} = \frac{-\left(\Delta U - RT + TS^\circ_{H_2}\right)}{2.303RT}$$

i.e.,

$$\log \frac{[\text{Tube} - 2n\text{H} \ldots (m+1)\,\text{H}_2]}{[\text{Tube} - 2n\text{H} \ldots m\text{H}_2]\, p_{\text{H}_2}} = -\frac{\Delta U}{2.303RT} + \frac{1}{2.303} - \frac{S^{\circ}_{H_2}}{2.303R} \tag{3}$$

Chemisorption process:

$$\begin{aligned} \text{Tube} - 2n\text{H} \ldots (m+1)\text{H}_2 &\rightarrow [\text{Tube} - 2(n+1)\text{H} \ldots m\text{H}_2]^* \\ &\rightarrow \text{Tube} - 2(n+1)\text{H} \ldots m\text{H}_2 \end{aligned} \tag{4}$$

For chemisorption process, we can split the hydrogen addition into two elementary processes. We have

$$\log \frac{[\text{Tube} - 2\,(n+1)\,\text{H} \ldots m\text{H}_2]^*}{[\text{Tube} - 2n\text{H} \ldots (m+1)\,\text{H}_2]} = -\frac{\Delta U}{2.303RT} + \frac{1}{2.303} \tag{5}$$

$$\log \frac{[\text{Tube} - 2\,(n+1)\,\text{H} \ldots m\text{H}_2]}{[\text{Tube} - 2\,(n+1)\,\text{H} \ldots m\text{H}_2]^*} = -\frac{\Delta U}{2.303RT} + \frac{1}{2.303} \tag{6}$$

In this study, we assume that hydrogen molecules can be saturated up to one monolayer coverage. Our calculations are based on two periodic unit cells. So, hydrogen physisorption and chemisorption are restricted to satisfy the following conditions:

$n + m \leq 12$, for (3,3)-SWCNT and (3,3)-BNNT

$n + m \leq 16$, for (4,4)-BNNT

$n + m \leq 20$, for (5,5)-BNNT

Let $P_{n,m}$ be the probability of the tube having n-pairs of chemisorbed H atoms and m-physisorbed H_2. We have

$$\sum_{n+m\leq 12} P_{n,m} = 1, \text{for (3,3)-tubes} \tag{7}$$

$$\sum_{n+m\leq 16} P_{n,m} = 1, \text{for (4,4)-tubes} \tag{8}$$

$$\sum_{n+m\leq 20} P_{n,m} = 1, \text{for (5,5)-tubes} \tag{9}$$

We can extract the wt% H_2 information from $P_{n,m}$.

$$wt\%\text{H}_2 = \frac{weight\,(\text{H}_2)}{weight\,(\text{tube})} = \frac{2\sum_{n+m\leq 12} P_{n,m}}{weight\,(\text{tube})}, \text{for (3,3)-tubes.} \tag{10}$$

$$wt\%\text{H}_2 = \frac{weight\,(\text{H}_2)}{weight\,(\text{tube})} = \frac{2\sum_{n+m\leq 16} P_{n,m}}{weight\,(\text{tube})}, \text{for (4,4)-tubes.} \tag{11}$$

$$wt\%\text{H}_2 = \frac{weight\,(\text{H}_2)}{weight\,(\text{tube})} = \frac{2\sum_{n+m\leq 20} P_{n,m}}{weight\,(\text{tube})}, \text{for (5,5)-tubes.} \tag{12}$$

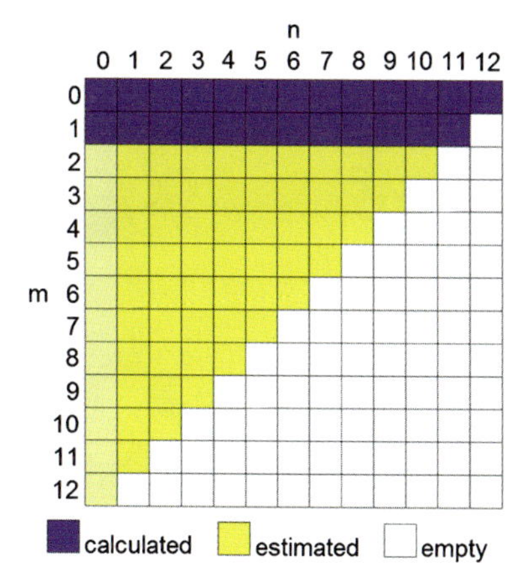

Fig. 1. Investigated configuration space for (3,3)-nanotubes. X-axis: n – number of pairs of chemisorbed H-atoms. Y-axis: m – number of physisorbed H_2. Purple squares: the structures were optimized explicitly. Yellow squares: energetics are estimated using constant physisorption energy. Empty squares: invalid states

3 Results and Discussions

We have performed DFT calculations and obtained energetic and structural information for hydrogen adsorption on armchaired (3,3)-, (4,4)- and (5,5)-BNNTs. For comparison, we also studied hydrogen adsorption on (3,3) SWCNT. In this section, we will report the hydrogen adsorption mechanisms, and estimate the hydrogen storage capacity using a simple thermodynamic model. By this method, we are able to find out the underlying factor governing the storage capacity and determine the optimal working conditions on this system.

It is worth mentioning that we have calculated a subset of the whole configuration space. Figure 1 illustrates the configuration space for H_2 adsorption on (3,3) nanotubes. We considered hydrogen addition cycles, first by hydrogen physisorption on the tube surfaces, followed by dissociative chemisorption forming two atomic hydrogen adatoms. Due to the huge number of combinations of forming physisorbed and chemisorbed hydrogen, we considered the hydrogen addition cycles following a selected pattern.

(3,3) SWCNT and (3,3) BNNT

The structures for hydrogen adsorbed on (3,3)-SWCNT and (3,3)-BNNT are depicted in Figures 2–3. Figures 4a and 4b show the potential energy surface of hydrogen physisorption and chemisorption processes on (3,3)-SWCNT and (3,3)-BNNT, respectively.

On small (3,3)-BNNT, we studied physisorption processes on 12 adsorption sites. The reaction energy of these processes range from -0.14 eV to

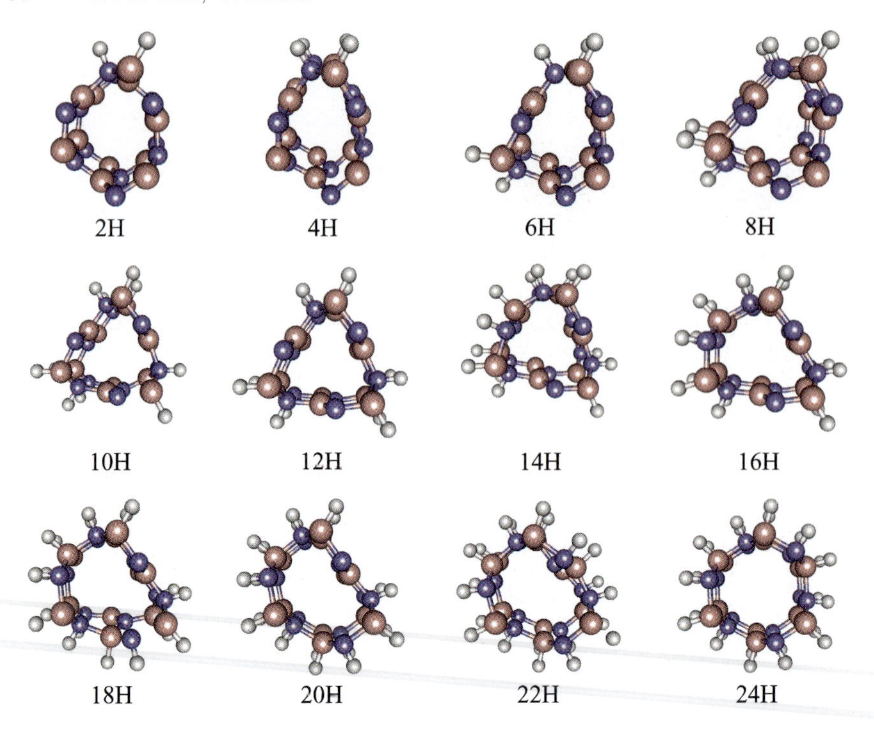

Fig. 2. Optimized geometry of hydrogenated (3,3)-BNNTs. The supercell units, each containing two primitive unit cells, are shown. Labels of the structures indicate the number of hydrogen adatoms in the supercells

-0.11 eV, therefore spanning 0.03 eV which is within the relative accuracy of current DFT, although the absolute accuracy is much worse. So, it is virtually found that the physisorption energy is not sensitive to the morphology of the substrate, and we will use the average value of the physisorption energy ($E_{physi,ave}$) to estimate the energetics of the chemical systems up to monolayer coverage of physisorbed H_2. For (3,3)-BNNT, we obtained $E_{physi,ave}$ of -0.13 eV.

We have computed the hydrogen addition cycles on small (3,3)-SWCNTs. The physisorption energy of H_2 ranges from -0.10 eV and -0.12 eV, and $E_{physi,ave}$ is -0.11 eV. So, the hydrogen physisorption strength is only slightly weaker for H_2 adsorbed on SWCNTs when compared to H_2 adsorbing on BNNTs ($E_{physi,ave}$: -0.11 eV vs -0.13 eV). Thereby, the hydrogen storage capacity due to physisorption is thought to be similar on both (3,3)-BNNTs and (3,3)-SWCNTs. Moreover, our computed results fall in a reasonable range which has been determined by previous theoretical calculations. It should be mentioned that currently available pure DFT methods underestimate the dispersion interaction which may affect the prediction of our calculations.

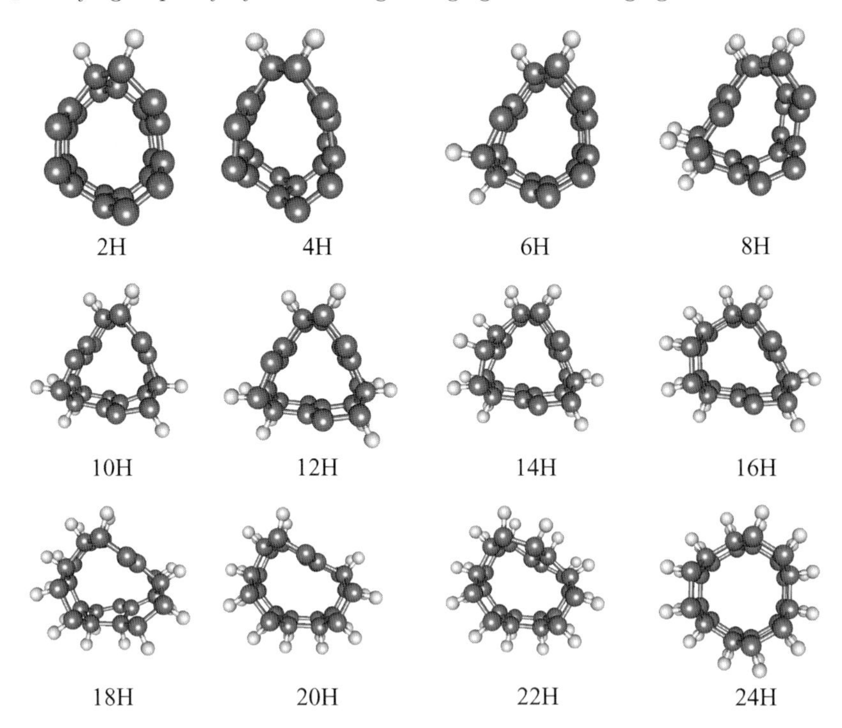

Fig. 3. Optimized geometry of hydrogenated (3,3)-SWCNTs. The supercell units, each containing two primitive unit cells, are shown. Labels of the structures indicate the number of hydrogen adatoms in the supercells

Chen *et al.* studied hydrogen adsorption on (5,5)-SWCNT, whereas only the first chemisorption cycle was taken into account.[31] In their study, hydrogen dissociative adsorption experienced a large activation barrier of around 2.4-3.1 eV, despite of the high reactivity of the highly curved carbon nanotube surfaces. The most plausible reason according to the classical Woodward-Hoffman rule is that the [2+2]-cycloaddition of H_2 on the C=C double bond is symmetry-forbidden. Recent detailed computational study also illustrated the importance of the symmetry consideration on the energetics of hydrogenation of the benzene systems.[14]

In this work, we have investigated the chemisorption of H_2 on (3,3)-BNNTs and (3,3)-SWCNTs. Firstly, we find that the chemisorption of the first H_2 experiences an activation barrier of 2.47 eV, which is in excellent agreement with previous theoretical work.[31] Note, that the accuracy of the calculated barrier is sensitive to the convergence threshold during the optimization of the transition state. Here, 0.015 eV/Å is found to be sufficiently fine to obtain conclusive results.

Except for the 4-th chemisorption of H_2, the chemisorption barriers on the (3,3)-SWCNT range from 1.74 eV and 4.3 eV, and the average value of the

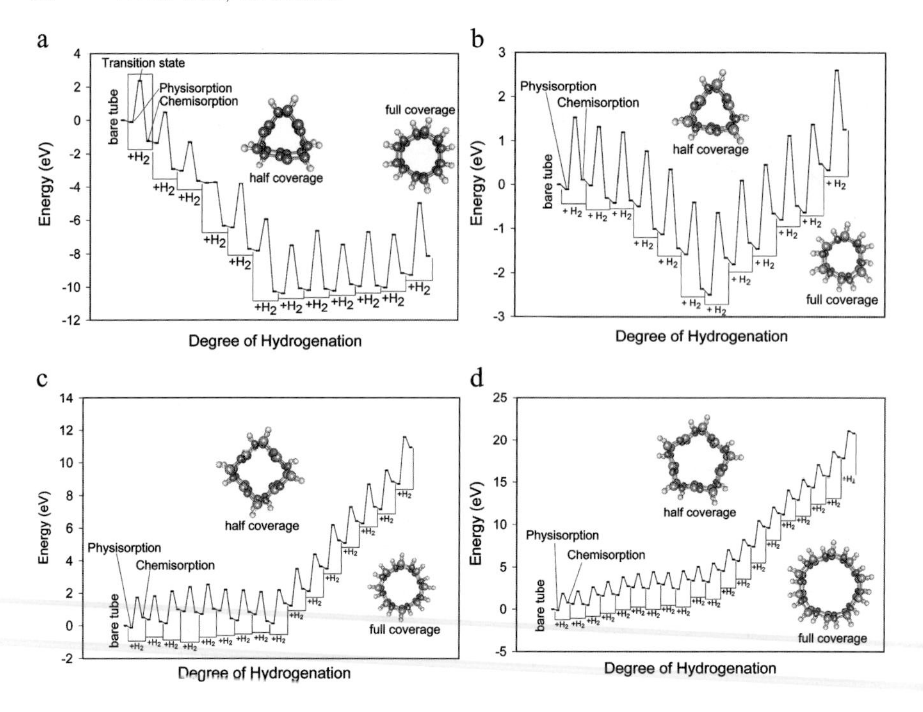

Fig. 4. Reaction energy profile for hydrogen additions on: a) (3,3)-SWCNTs; b) (3,3)-BNNTs; c) (4,4)-BNNTs; d) (5,5)-BNNTs

barriers is 2.54 eV. The first chemisorption of H_2 would activate the tube surface and the chemisorption barriers for the subsequent chemisorption events are largely reduced (1st E_a: 2.47 eV; 2nd E_a: 1.84 eV). This raises an interesting question about the feasibility of H_2 chemisorption as a storage pathway. Since the carbon nanotubes can be partially functionalized by other chemical functional groups,[32] such as fluorine, the hydrogen chemisorption may be made possible on functionalized carbon nanotubes at lower temperature (< 600 K). In general, the activation barriers are smaller for the early stage of hydrogen chemisorption.

In sharp contrast to the case of (3,3)-SWCNTs, the H_2 chemisorption barrier is significantly lower for the (3,3)-BNNTs. In the first H_2 chemisorption on (3,3)-BNNTs, the barrier is 1.63 eV (vs 2.47 eV on SWCNTs). The barriers range from 1.17 eV (the 5th addition) to 2.26 eV (the 12th addition). Similar to the SWCNT case, the activation barriers are generally smaller at the early stage of chemisorptions. The average value of the chemisorption for (3,3)-BNNTs is 1.70 eV. Therefore, our calculations unambiguously show that the (3,3)-BNNT is more vulnerable to the hydrogen chemisorption from the chemical kinetic point of view, when compared to the case of the (3,3)-SWCNT.

Figures 4a and 4b show the energy landscape of the hydrogen physisorption and chemisorption on the (3,3)-SWCNTs and (3,3)-BNNTs. Both (3,3)-SWCNT and (3,3)-BNNT exhibit an energy minimum for half-monolayer coverage of hydrogen adatoms. Moreover, it is obvious that hydrogen chemisorption on SWCNT releases a great amount of energy. At half-monolayer coverage, the system releases -10 eV by (3,3)-SWCNTs, while (3,3)-BNNT releases about -2.4 eV. The huge difference of the exothermicity can be traced back to the strengths of C-H, B-H and N-H bonds, which can also rationalize the fact that the full coverage of chemisorbed H on (3,3)-SWCNTs is exothermic while the same bonding connectivity on (3,3)-BNNTs turns out to be an endothermic reaction. We have also determined the energetics of half- and full-coverage of chemisorbed H nanotubes of different size, and confirmed that the H_2 chemisorption on SWCNTs was more exothermic as compared with the analogous case of BNNTs.

(4,4) BNNT

As mentioned above, H_2 chemisorption on both (3,3)-SWCNTs and (3,3)-BNNTs are exothermic, in particular, the reaction is kinetically more favorable on (3,3)-BNNTs. On nanotubes of larger diameter, however, H_2 chemisorption might be less probable. As generally accepted carbon nanotubes of larger diameter are less reactive towards chemical functionalization. To evaluate the hydrogen storage contributed by chemisorption, we have also calculated H_2 physisorption and chemisorption on (4,4)- and (5,5)-BNNTs until full coverage.

For the hydrogen adsorption on (4,4)-BNNTs, $E_{physi,ave}$ is found to be virtually identical to that for (3,3)-BNNTs. In detail, E_{physi} spans a range of -0.11 eV and -0.15 eV.

The energy landscape of hydrogen addition on (4,4)-BNNTs is depicted in Figure 4c. Interestingly, the influence of tube size is not pronounced when comparing the chemisorption barriers on (3,3)- and (4,4)-BNNT surfaces, respectively. Before half-monolayer coverage, the average value of the chemisorption barrier is 1.42 eV and 1.64 eV for (3,3)- and (4,4)-BNNT, respectively. However, along the hydrogen chemisorption steps, the reaction is largely endothermic, especially beyond the half-monolayer coverage. Thereby, although the chemisorption barrier is relatively small on the BNNTs, hydrogen storage due to chemisorption on (4,4)-BNNTs is not likely because large amount of energy is required to enter the chemical system in order to form chemisorbed hydrogen species.

(5,5) BNNT

For a BNNT of larger diameter, we find that the $E_{physi,ave}$ is -0.14 eV on (5,5)-BNNTs. In detail, the physisorption energies E_{physi} range from -0.11 eV to -0.16 eV. The enhancement of physisorption strength on BNNTs is not significant, but still implies a gradual increase of dispersion interaction for larger nanotubes. This is consistent with chemical intuition that dispersion forces would be strengthened with increasing molecular mass, i.e. increasing number of electrons.

The energy landscape of H_2 additions on (5,5)-BNNTs is illustrated in Figure 4d. As mentioned in the previous subsection, the influence of tube size on the chemisorption barriers is not significant. Before half-monolayer coverage, the average value of the chemisorption barriers is 1.80 eV. Moreover, as found in (4,4)-BNNTs in which H_2 chemisorption experiences relatively small activation barriers but turns out to be highly endothermic, a similar situation occurs on (5,5)-BNNTs. As shown in Figure 4d, the total energy of the system increases with the increasing degree of H_2 chemisorption. After half-monolayer coverage, the total energy increases much more rapidly with the degree of chemisorption. So, from a thermodynamical point of view, the H_2 chemisorption, especially up to saturation, is not likely because of the raise of the enthalpy and the entropy loss of the molecular hydrogen.

Energy Decomposition Analyses

To understand the influence of tube size on the activation barrier, we apply an energy decomposition analyses on (3,3)-, (4,4)- and (5,5)-BNNTs. We decompose the change of the energy attributed to several terms, namely, deformation energy of H_2 [E(dis-H_2)], deformation energy of nanotubes [E(dis-tube)], and interaction energy [E(int)]. These energy terms are related by the following formula.

$$E_a = E(\text{dis-H}_2) + E(\text{dis-tube}) + \text{E(int)} \tag{13}$$

Figures 4–6 show the results of energy decomposition analysis. It is obvious that the results for (3,3)-BNNTs are different from those for (4,4)- and (5,5)-BNNTs. As shown in Figure 5a, Ea is reduced with increasing E(dis-H_2) which indicates an increase of H-H bond distance. Thus, on (3,3)-BNNTs, a late transition state would favor the dissociative adsorption. Before half-monolayer coverage, the data points show a nice correlation for E(dis-H_2) and E(dis-tube). On (3,3)-BNNTs, the data points before half-monolayer coverage and after half-monolayer coverage correlate very well, separately. However, for (4,4)- and (5,5)-BNNTs (cf. Figures 6a & 7a), Ea is reduced with decreasing E(dis-H_2). Therefore, hydrogen chemisorption is favored by early transition states for the larger nanotubes.

We have compared the relative importance of the three energy terms, E(dis-H_2), E(dis-tube) and E(int). As shown in Figures 5d, 6d and 7d, we found that the chemisorption barrier is predominately dependent on the H_2 distortion and the tubular distortion. The interaction energy term only plays a minor role, particularly for nanotubes of larger diameter; as the tube size increases, the magnitude of the interaction energy decreases, which goes along with the fact that the chemical reactivity decreases with surfaces of smaller curvature.

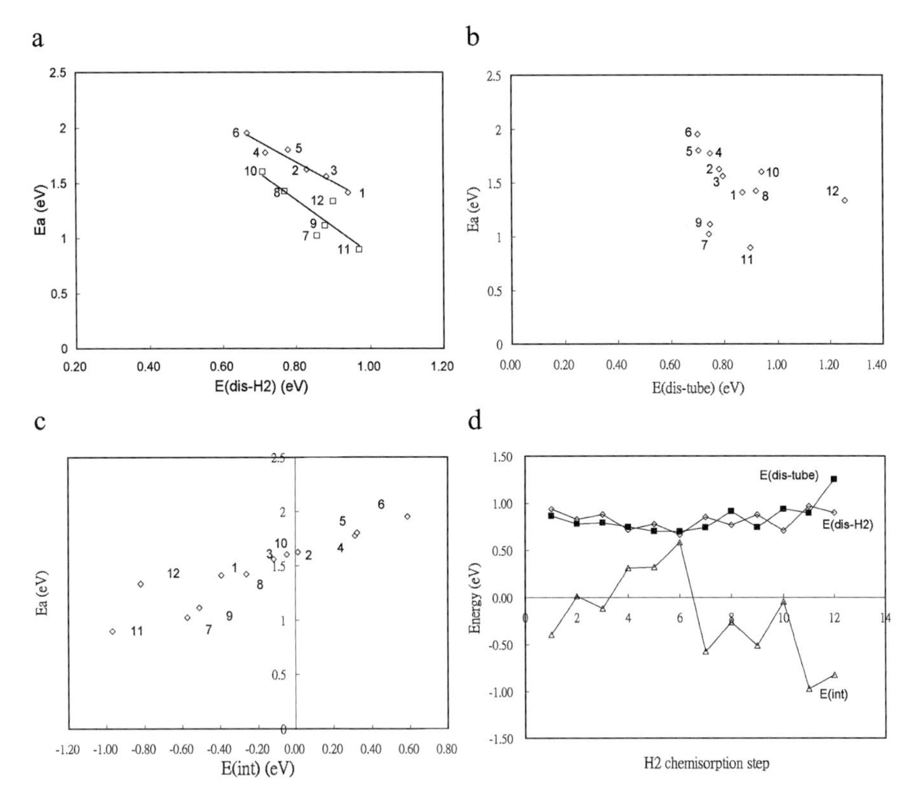

Fig. 5. Energy decomposition analysis for H_2 chemisorption barriers on (3,3)-BN-NTs. The Y-axis represents the magnitude of the activation barrier. Remarks: a) effect of H_2 deformation; b) effect of tubular deformation; c) effect of interaction energy; d) relative importance of E(dis-H_2), E(dis-tube) and E(int). Numberings in Figure a–c reveal the number of chemisorption cycles

Thermodynamics and Hydrogen Storage Capacity

We have used the calculated thermodynamic information to predict the hydrogen storage capacity based on equations 10–12. In general, we have found that the physisorption energy of hydrogen on the nanotube surfaces fall in a narrow range. Within the accuracy of DFT, we assume that each physisorption step releases a constant amount of energy, i.e. the average value of the calculated physisorption energy. Based on this assumption, we could investigate a significantly larger configuration space which is marked in yellow color in Figure 1. Note, in the physisorption calculation, H_2-chemisorbed-H interaction has already been encountered.

The equilibrium distributions of the products are calculated according to the formulae 3, 5 and 6. In this work, we have obtained physical pictures of the hydrogen storage mechanism and studied the influence of pressure, temperature and strength of H_2 physisorption on the storage capacity. The

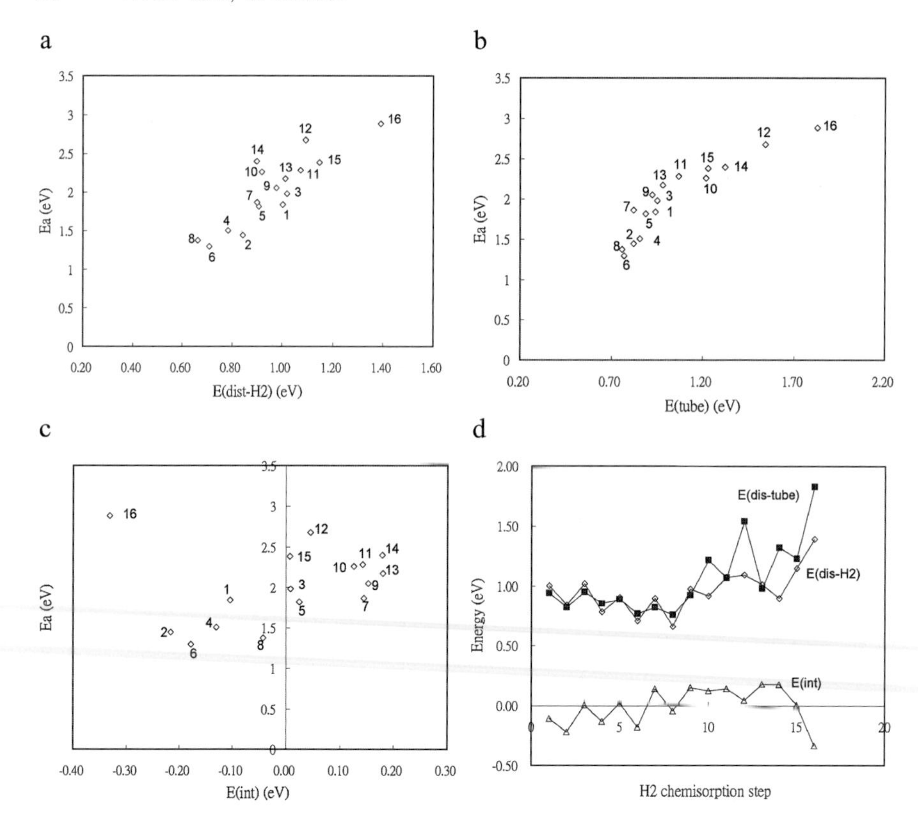

Fig. 6. Energy decomposition analysis for H_2 chemisorption barriers on (4,4)-BNNTs. The Y-axis represents the magnitude of the activation barrier. Remarks: a) effect of H_2 deformation; b) effect of tubular deformation; c) effect of interaction energy; d) relative importance of E(dis-H_2), E(dis-tube) and E(int). Numberings in Figure a–c reveal the number of chemisorption cycles

temperature range from 0 K to 1000 K and the pressure range from 0 atm to 5 atm have been investigated.

(3,3)-SWCNTs and (3,3)-BNNTs

We have plotted the 3-dimensional surfaces to illustrate %wt H_2 as a function of temperature and pressure. The results for (3,3)-SWCNTs and (3,3)-BNNTs are displayed in Figure 8.

As aforementioned, (3,3) nanotubes are subject to H_2 chemisorption. For (3,3)-SWCNTs, H_2 chemisorption is exothermic up to monolayer saturation coverage. For (3,3)-BNNTs, half-monolayer coverage is exothermic. These energetic results imply that the formation of H adatoms on these nanotube surfaces are thermodynamically more stable than forming physisorbed H_2. In consequence, these nanotube surfaces can most probably covered by a layer of

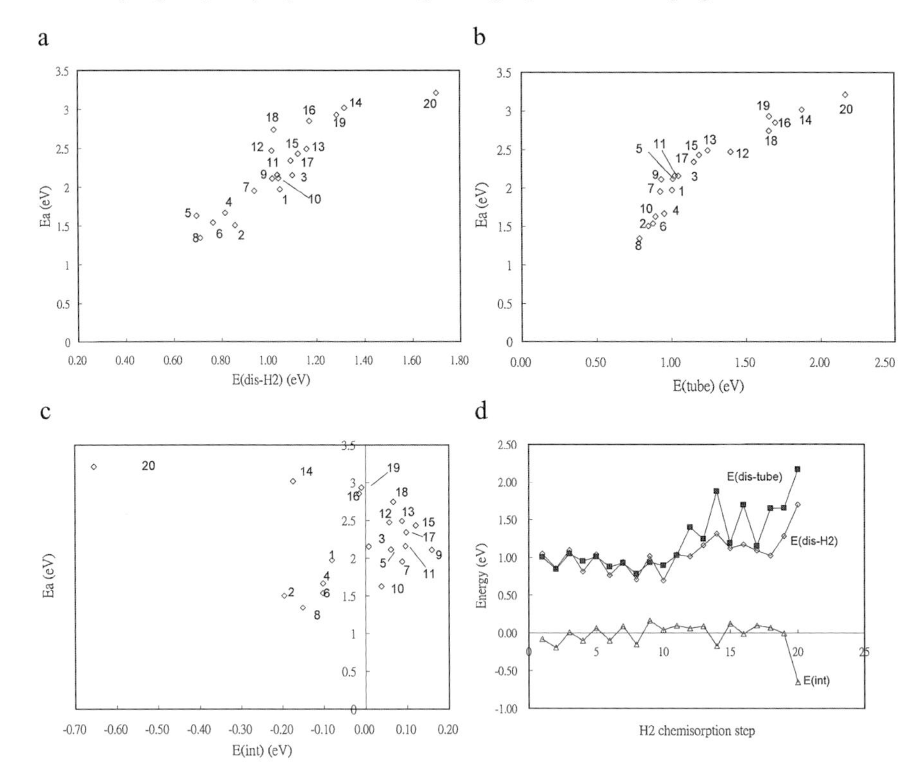

Fig. 7. Energy decomposition analysis for H_2 chemisorption barriers on (5,5)-BN-NTs. The Y-axis represents the magnitude of the activation barrier. Remarks: a) effect of H_2 deformation; b) effect of tubular deformation; c) effect of interaction energy; d) relative importance of E(dis-H_2), E(dis-tube) and E(int). Numberings in Figure a–c reveal the number of chemisorption cycles

H adatoms. Such a scenario is shown in Figure 8. On the left panel of Figure 8, we can see that about 4.15-4.5 wt% H_2 arises from the H adatoms on the (3,3)-SWCNTs under a wide range of temperature and pressure. On (3,3)-BNNTs, the nanotube is covered by about 4 %wt with H adatoms. These calculated results suggest that these nanotube surfaces are "poisoned" by chemisored H species, as they can hardly be removed by applying a change of temperature or pressure.

In the central panel of Figure 8, we can observe the hydrogen storage capacity due to physisorption, while the combined effects of chemisorption and physisorption are shown in the right panel of Figure 8. It is observed that a transition from high %wt H_2 to low %wt H_2 takes place around ambient conditions due to physisorption processes. In addition, as mentioned above, the nanotube surfaces are poisoned by chemsisorbed H, therefore, the overall H_2 carrying ability is originating from the physisorption steps.

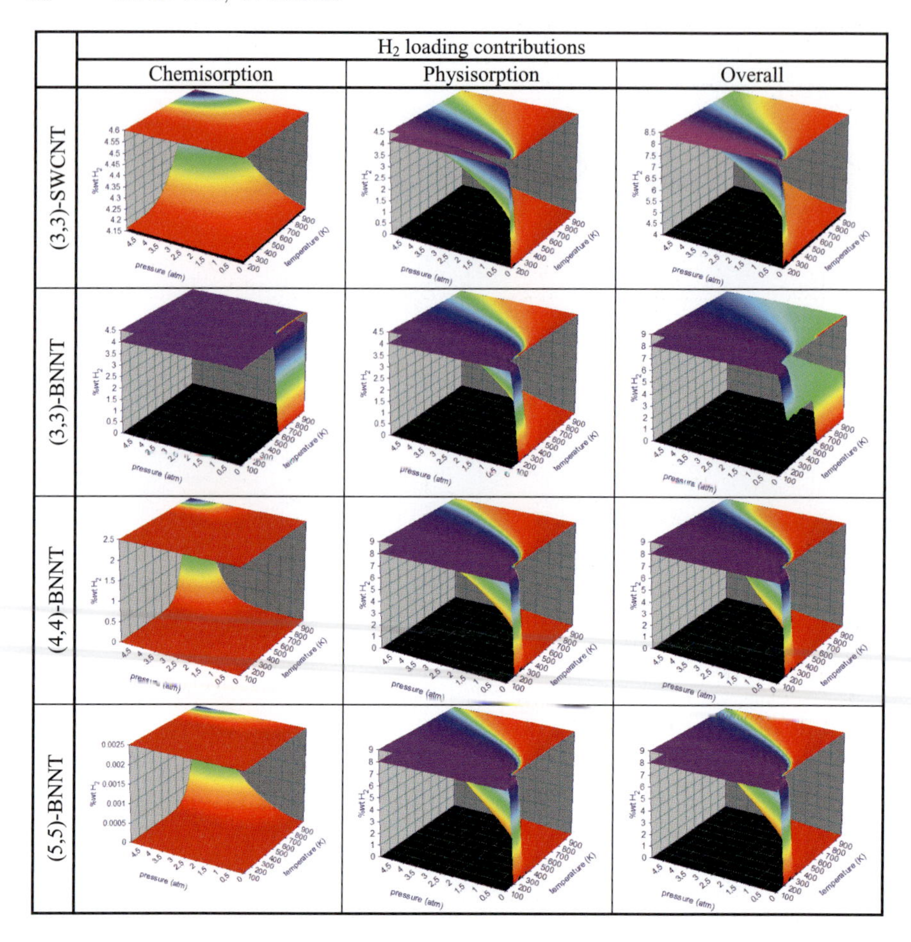

Fig. 8. Dependence of %wt-H_2 on temperature and pressure on selective nanotubes. Remarks: Left panel: contribution by chemisorption; Central panel: contribution by physisorption; Right panel: overall %wt-H_2

Our results provide important and interesting insight for designing nanotubes for H_2 storage purposes. Various approaches have been proposed and reported in the literature. One of them was to increase the hydrogen storage capacity by enhancing the H_2-surface interactions. For example, Chandrakumar *et al.* suggested a maximum loading of 9.5 % by H_2 adsorption on alkaline-metal doped fullerenes12 and Yoon *et al.* suggested an improvement of storage capacity by charged fullerenes. Another method was to use hydrogen spillover technique in which the hydrogen was stored as H adatoms on storage media, like carbon nanotubes.[15, 16, 17] However, the working conditions for such systems are still unknown.

For the first time, we predict the %wt H_2 as a function of temperature and pressure. This allows us to monitor the maximum H_2 storage capacity from

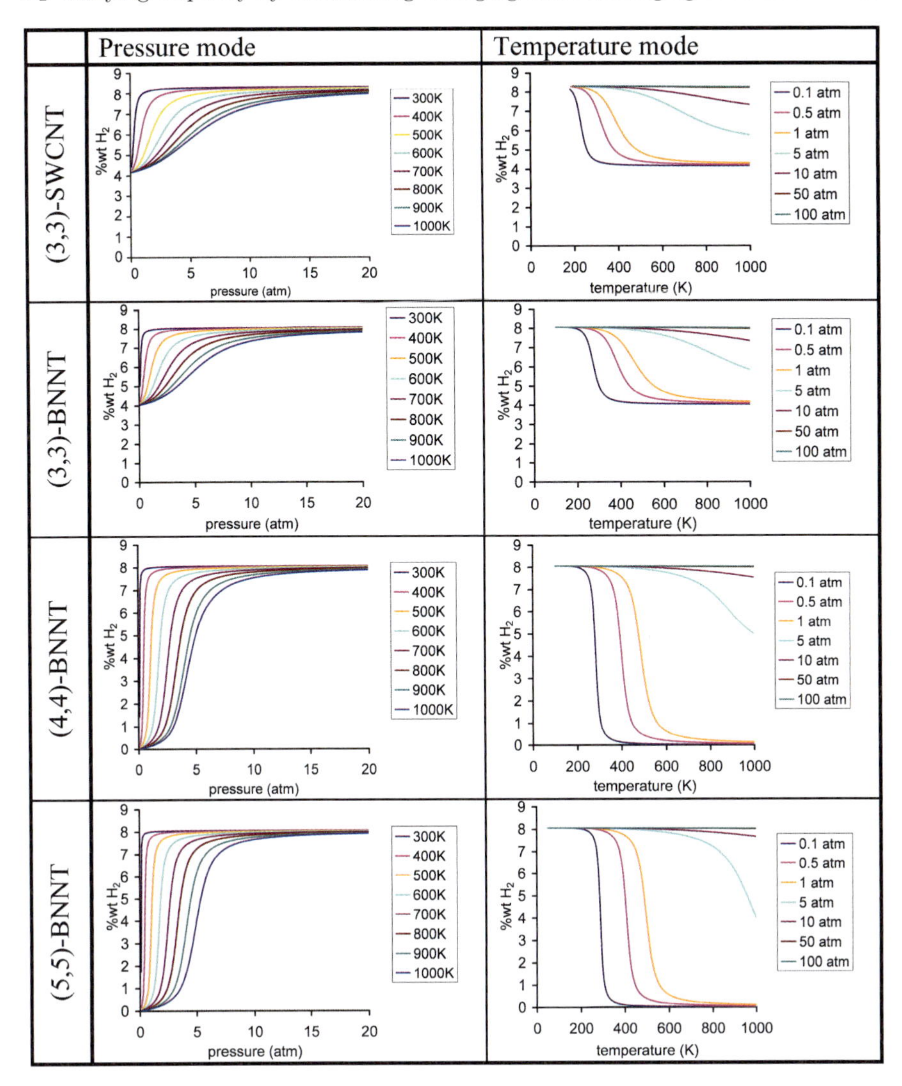

Fig. 9. Variation of %wt-H_2 due to pressure (left panel) and temperature (right panel). The colored lines refer to their selected temperature (left panel) or pressure (right panel) in the figure legends

the initial conditions (P1, T1) to the final conditions (P2, T2). The results are also projected in planes and displayed in Figure 9. It is seen that under ambient conditions (300 K and 1 atm), about 8 %wt H_2 can be achieved. However, discharge of the hydrogen loading by either reducing pressure or increasing temperature, 4 %wt H_2 is still obtained on both (3,3)-SWCNTs and (3,3)-BNNTs. As a result, the net H_2 loading is about 4 %wt H_2 only.

Again, this is due to poisoning of these small tube surfaces by chemisorbed H species.

The poisoning of the tube surfaces by chemisorbed H might be implied by previous experiments. Tang *et al.* reported a maximum H_2 loading of 4 %wt on collapsed boron nitride nanotubes.[33] On the collapsed nanotubes, buckled and kinked sites might be available because of the structural distortions. On these distorted surfaces, dissociative chemisorption of H_2 is favoured. Further computational studies are required to determine the energetics of H_2 chemisorption on these collapsed BN nanotubes.

(4,4)-BNNTs and (5,5)-BNNTs

(4,4)-BNNTs and (5,5)-BNNTs are more resistant to the hydrogen chemisorption because these reactions are highly endothermic, therefore, the chemisorption states are thermodynamically unstable. As seen in the left panel of Figure 8, the chemisorption states play a minor role in a wide range of temperature and pressure. Only at high temperature and pressure, %wt H_2 becomes significant due to H_2 chemisorption. On (5,5)-BNNTs, the %wt H_2 contribution by chemisorption is even negligible: at 1000 K and 5 atm, chemisorption only contributes 0.0025 %wt H_2-loading.

Thus, on the (4,4)-BNNTs and (5,5)-BNNTs, the hydrogen storage mechanism is predominately due to the physisorption process. At low temperature, H_2 molecules are physisorbed on the BNNTs, and a H_2 loading of 8 %wt can be obtained. These physisorbed H_2-molecules can be discharged by raising temperature. At 1 atm, H_2 loading can be changed from ˜8 %wt (at 300 K) to ˜0.5 %wt (at 600 K) on (4,4)-BNNTs. As a result, a net H_2 carrying capacity of $>$ 7 %wt can be obtained, which is almost doubled as compared to the experimental results on collapsed BNNTs.

On (5,5)-BNNTs, similar results to those on (4,4)-BNNTs are found. Under ambient conditions, H_2 chemisorption is negligible. The hydrogen loading is exclusively due to the physisorption processes. On these thicker nanotubes, the hydrogen loading is seemingly not sensitive to the tubes, but to the physisorption strength. Since the physisorption energy is similar for both cases, -0.13 eV and -0.14 eV for (3,3)-BNNTs and (4,4)-BNNTs, respectively, the hydrogen loading on these two surfaces under various conditions are very similar. So, on (5,5)-BNNTs, the net hydrogen carrying capacity can be as high as 8 %wt .

Temperature Mode and Pressure Mode

We elaborate and describe the effects of temperature and pressure in detail. From this, we can estimate the performance of BNNTs which are operated under temperature and pressure modes.

We plotted a series of curves which illustrate %wt H_2 as a function of pressure in Figure 9. The data is presented from 300 K to 1000K, incremented by 100 K. We observe that at 300 K, %wt transition takes place very rapidly at the low pressure regime. This feature would disfavor the use of BNNTs at low temperature, as it might be difficult to control the gas flow in a narrow range of pressure.

As the operational temperature increases, the %wt(P) curves become broadened. The broad spectra enable us to control the H_2 loading in a wider range of pressure. For example, at 600 K, %wt H_2 changes from 1 %wt (at 1 atm) to above 7 %wt (at 4 atm). Such a pressure change is very desirable for hydrogen carrying purpose.

In addition to the hydrogen carrying process by pressure mode, we have also projected the results in order to study the temperature-mode operation. As shown in Figure 9, we plotted the curves keeping pressure as constant. The %wt curves are represented at 0.1 atm, 0.5 atm, 1 atm, 5 atm, 10 atm, 50 atm and 100 atm. For a fixed amount of pressure, we can see that the nanotubes can carry about 8 %wt H_2 for (4,4)-BNNTs and (5,5)-BNNTs at low temperature, say 200 K. As the temperature increases, physisorbed H_2 escapes from the nanotube surfaces and %wt H_2 would approach 0 % at a characteristic temperature. Temperature and pressure exhibit opposite effects on the storage: on the one hand, higher temperature would decrease %wt; and on the other hand, higher pressure would increase %wt. From the figures, we can see that at low pressure of 0.1 atm, %wt changes sharply around 300 K. However, this operational pressure may not be desirable for porting H_2 in vechicles. At higher pressure of 1 atm, the transition temperature increases to about 500 K. So, under ambient conditions of 1 atm, %wt H_2 would change from 8 %wt (300 K) to ˜1 %wt (600 K), thus, a net change of about 7 %wt can be obtained.

Figure 9 show that at very high pressure, say 100 atm, the nanotube surfaces might be saturated with H_2. A net change of H_2 could not be observed according to our present calculations. However, Tang et al. reported that a 4 %wt loss of H_2 was found under the same pressure.[33] The underlying reason cannot be explained at this moment. But one plausible reason for the difference is that we assume the hydrogen adsorption taking place on isolated nanotubes in our calculations. But in reality, the nanotubes might form bundles and aggregate together in spaghetti form. This would decrease the outer surface area and modify the H_2 diffusion rate significantly.

Effects of H_2 Physisorption Strength

In this work, we find that the hydrogen loading process is directly related to the physisorption strength. Both our computational work and the previous experimental work by Tang et al. show that H_2 can be released around 500 K.[33] By chemical intuition one might argue that, the transition temperature could be reduced by decreasing the H_2-surface adsorption strength.

To investigate the effect of H_2 physisorption strength, we artificially adjust the average value of the physisorption energy, while the energy landscape for the transition state and product state of the H_2 chemisorption remain unchanged. In particular, for (5,5)-BNNTs, H_2 chemisorption plays a very minor role, and the hydrogen storage capacity would then be controlled by the adjusted physisorption energy. In this work, we studied a range of physisorption strength, from -0.01 eV to -0.19 eV.

The results are depicted in Figures 10–13. With very weak physisorption strength of -0.01 eV (cf. Figure 10), physisorption does not contribute to the hydrogen loading in the ranges of temperature and pressure studied, and the total H_2 loading is exclusively due to H_2 chemisorption. When H_2 physisorption strength increases, higher H_2 loading (8 %wt on (4,4)- and (5,5)-BNNTs) can be obtained. The transition temperature or the transition pressure does depend on the physisorption strength: higher physisorption strength leads to higher transition temperature. This feature is universal and applicable for all nanotubes studied in this work. In analogy to (3,3)-SWCNTs, (3,3)-BNNTs are also half-saturated by chemisorbed H (cf. Figure 11). It should be noted that both (3,3)-SWCNTs and (3,3)-BNNTs are poisoned by H adatoms and their hydrogen storage capacity is limited to about 4 %wt H_2.

Our calculated results provide important insight into hydrogen storage on nanotube systems. In our calculations, we find that the original energetic data results in a relatively high transition temperature, say 500 K, under a constant pressure of 1 atm. To lower the transition temperature, the physisorption strength is required to be weakened. However, it is out of scope of our current work to suggest a method to achieve this effect. However, one of our objectives is to point out the direction of optimizing the nanotubes in order to operate the hydrogen storage materials in an economical way.

For the BNNT system, the optimal H_2 physisorption strength is -0.07 eV. Using this physisorption strength, %wt can be changed from 8%wt (4 atm) to $\sim$ 0 %wt (1 atm), at a constant temperature of 300 K. Such ambient conditions are very suitable for high performance hydrogen storage media. It was previously suggested that higher physisorption strength, up to -0.2 eV – -0.4 eV, was very desirable for hydrogen storage. However, as shown in our work, the high physisorption strength would greatly increase the transition temperature (cf. Figures 12 & 13), which would not be an appropriate tuning because of additional consumption of energy.

4 Conclusion

In summary, we have established the theoretical basis for hydrogen storage on nanotube surfaces. We have investigated the hydrogen storage on (n,n)-BNNTs (n=3-5). Both chemisorption and physisorption were taken into account. For comparison, we also studied the (3,3)-SWCNT.

We have found that H_2 chemisorption on small (3,3)-nanotubes is highly exothermic. The nanotubes can be activated by the first chemisorption cycle. (3,3)-BNNTs are found to be more reactive towards H_2 chemisorption

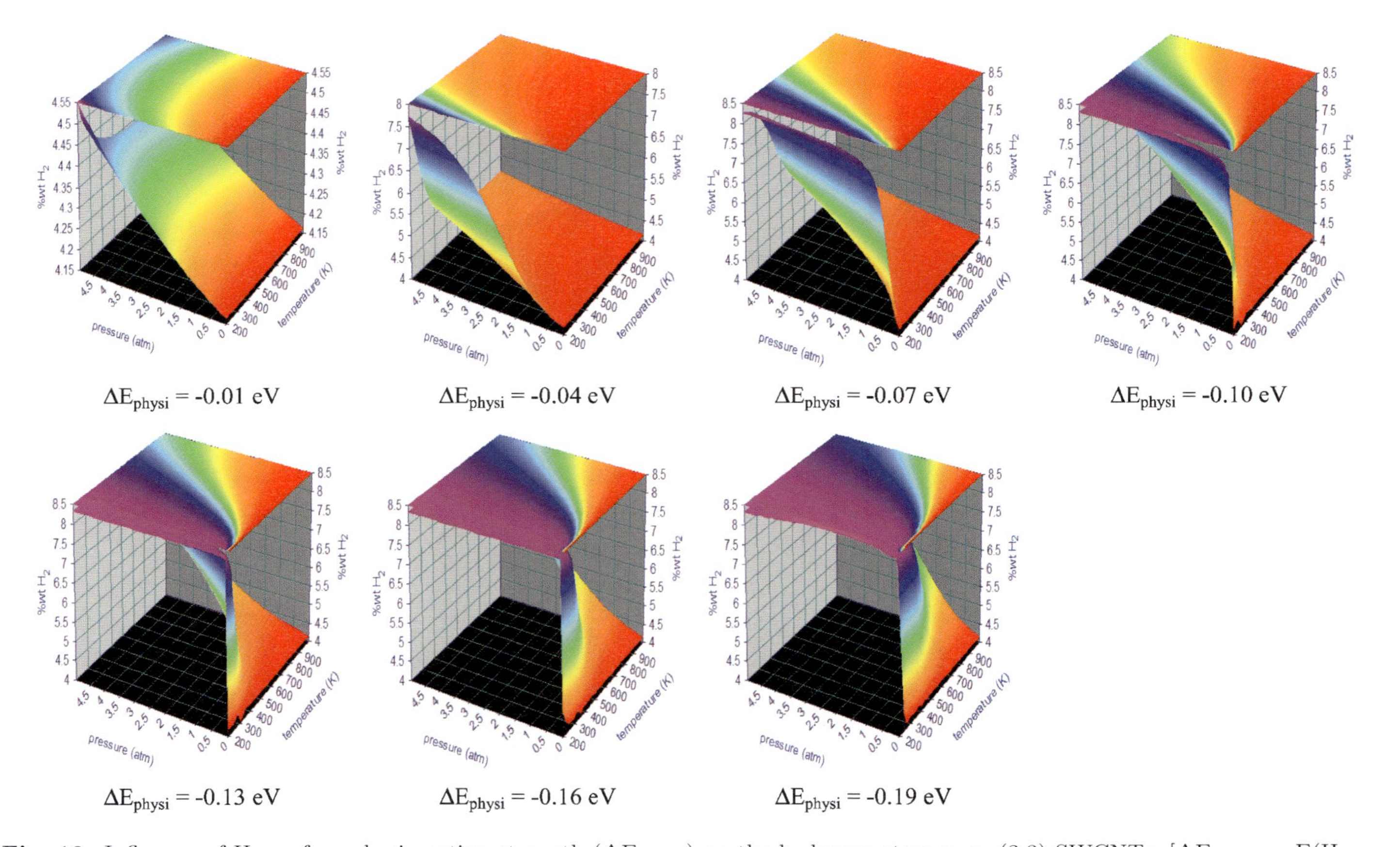

Fig. 10. Influence of H_2-surface physisorption strength (ΔE_{physi}) on the hydrogen storage on (3,3)-SWCNTs. [$\Delta E_{physi} = E(H_2 \ldots tube) - E(tube) - E(H_2)$]

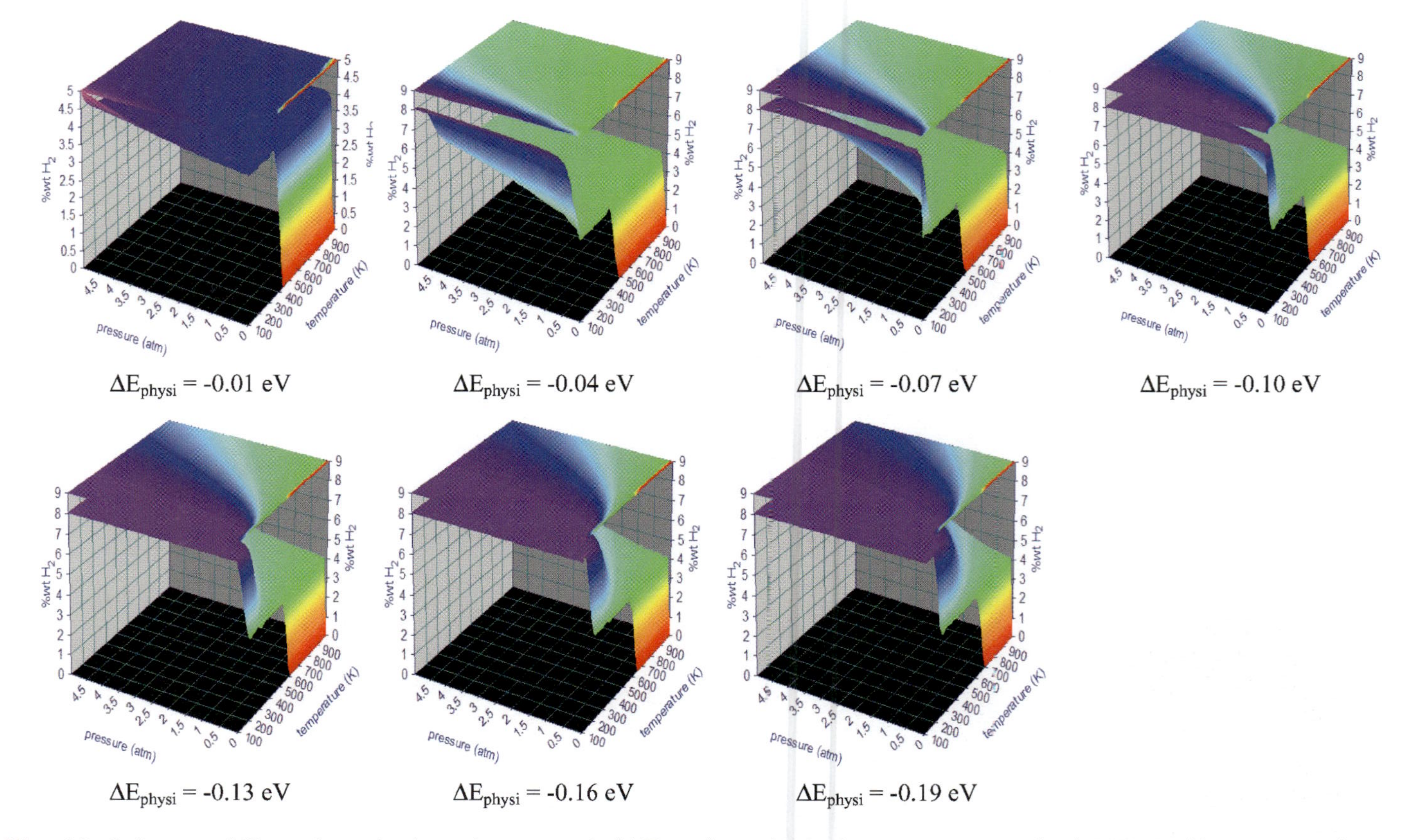

Fig. 11. Influence of H_2-surface physisorption strength (ΔE_{physi}) on the hydrogen storage on (3,3)-BNNTs. [ΔE_{physi} = E(H_2 ... tube) − E(tube) − E(H_2)]

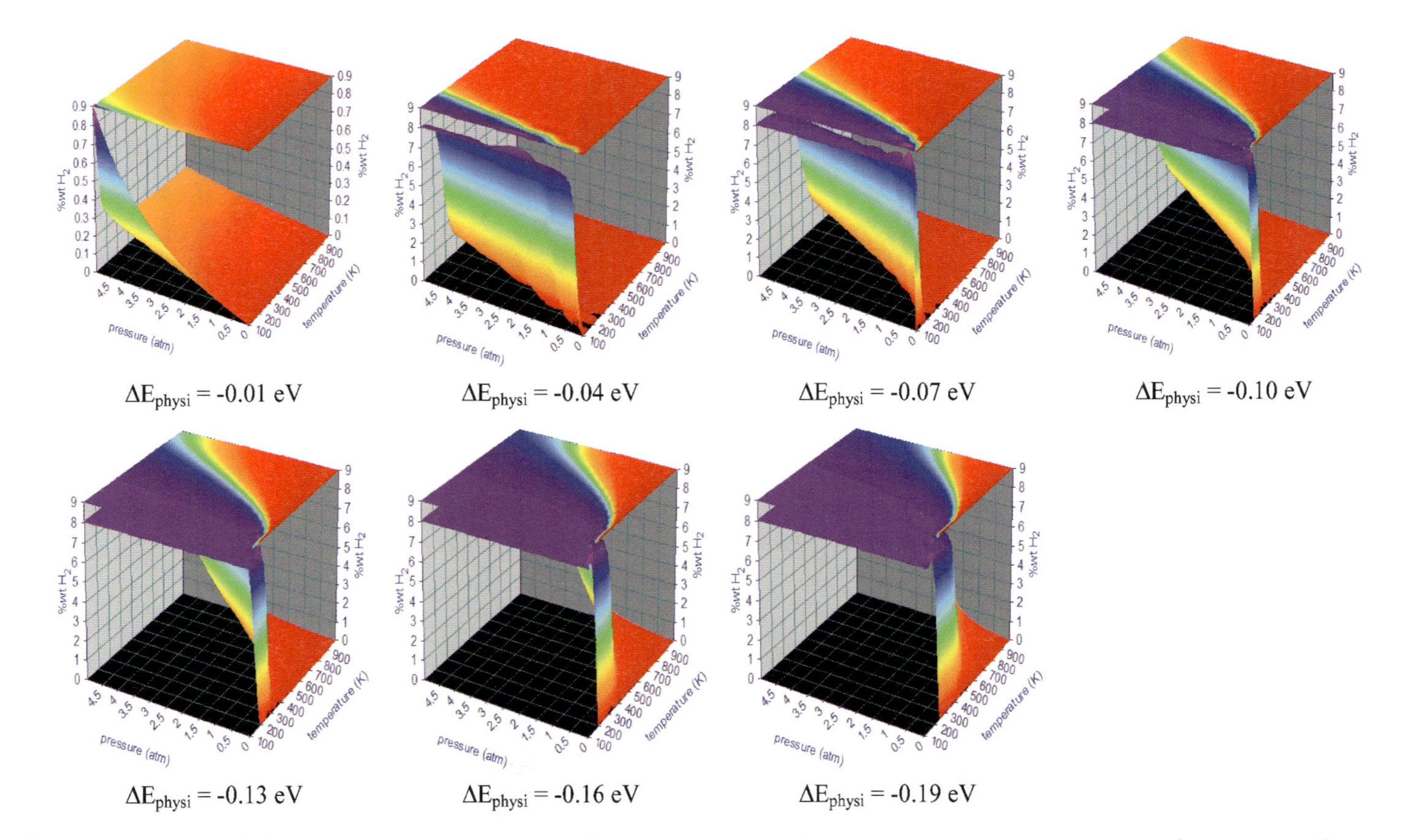

Fig. 12. Influence of H_2-surface physisorption strength (ΔE_{physi}) on the hydrogen storage on (4,4)-BNNTs. [ΔE_{physi} = E(H_2 ... tube) − E(tube) − E(H_2)]

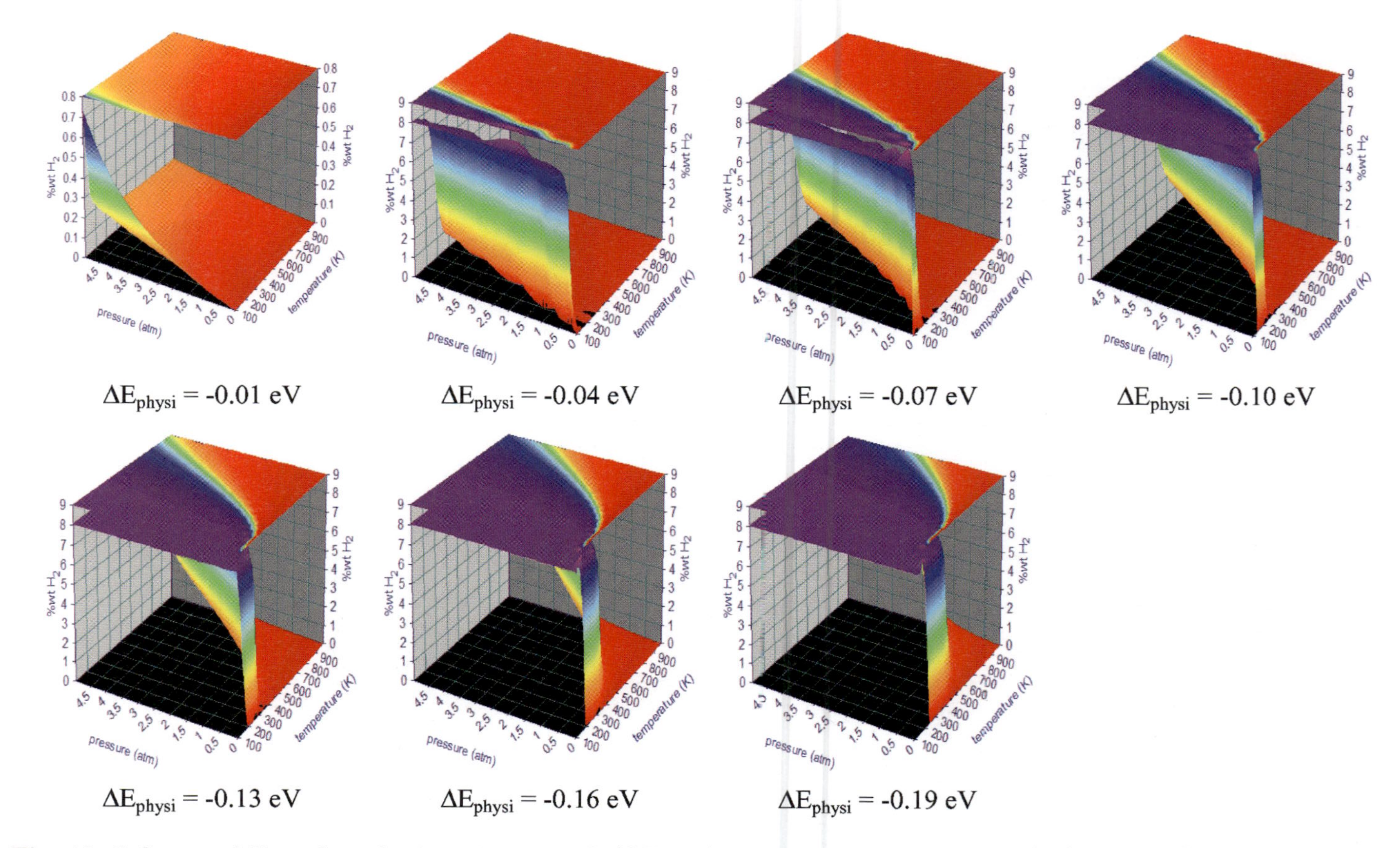

Fig. 13. Influence of H_2-surface physisorption strength (ΔE_{physi}) on the hydrogen storage on (5,5)-BNNTs. [$\Delta E_{physi} = E(H_2 \ldots tube) - E(tube) - E(H_2)$]

from chemical kinetics point of view, when compared to the (3,3)-SWCNTs counterpart. The lower chemisorption barrier is attributed to a different reaction mechanism of H_2 linking to BN surfaces, as such surfaces exhibit a partial ionic character. This may also be the reason for the insensitivity of H_2 chemisorption barriers with respect to the diameter of the nanotubes.

We have found that the hydrogen carrying ability is predominately due to physisorption rather than chemisorption. On small (3,3)-nanotubes, H_2 chemisorption would poison the tube surfaces and decrease the net hydrogen carrying capacity. Both (3,3)-SWCNTs and (3,3)-BNNTs have limited H_2 loading up to about 4 %wt H_2. On larger boron nitride nanotubes, hydrogen chemisorptions play a minor role because such reactions are highly endothermic, discouraging the formation of H adatoms. In these cases, the net storage capacity can be as large as $\sim$ 8 %wt H_2.

We have also estimated the %wt(P,T) relationship based on the fundamentals of thermodynamics. The phase diagram provides important information about the working conditions for hydrogen carrying purposes. In agreement with available experimental data, our calculations suggest that hydrogen can be displaced around the transition temperature of 500 K under atmospheric pressure. Moreover, if the hydrogen storage is underwent by pressure mode at 300 K, a sharp transition for H_2 displacement is observed at low pressure (about 0.1 atm), which is not practical in real applications.

To understand the role of the H_2 physisorption strength, we adjust the H_2 physisorption strength artificially. H_2 is found to be stored/displaced very easily in the pressure range of 1-4 atm at 300 K, when the H_2-surface interaction is about -0.07 eV. This is in contrast to the previous suggestions implying that a stronger H_2 physisorption interaction was more desirable; however, in that case, the ease of H_2 evolution has not been taken into account. Nonetheless, despite of important insight observed in this work, further experimental works will be required to consolidate our proposals.

5 Publications Associated with the NEC SX-8 Account Under Grant Number WLYIM

We have published three papers by using the titled account. Some published works have been noted in the previous progress report, and there are several manuscripts (requiring sophisticated analysis methods) pending to be submitted.

Published papers include:

1. Wai-Leung Yim* and Thorsten Klüner*. "Atoms-In-Molecules Analysis For Planewave DFT Calculations – A Numerical Approach on a Successively Interpolated Charge Density Grid." ***J. Comput. Chem.*** 29: 1306-1315, 2008.
2. Wai-Leung Yim and Thorsten Klüner*. "Promoting O_2 Activation on Noble Metal Surfaces." ***J. Catal.*** 254: 349-354, 2008.

3. Wai-Leung Yim and Thorsten Klüner*. "Role of Electrostatic Interactions on Engineering Reaction Barriers: The Case of CO Dissociation on Supported Cobalt Particles." ***J. Chem. Theory Comput.*** 4: 1709-1717, 2008.

6 Performance of VASP on Different Platforms

We mainly use VASP in our projects. The VASP performance benchmark has been performed and reported in the previous annual report in 2008.

Typical Number of Processors Used

We usually performed calculations using 8 or 16 CPUs for VASP. For transition structure calculations 16 CPUs were used.

CPU Time

In total, we received a grant of 190.000 CPU hours.

Acknowledgment. We gratefully acknowledge financial support from Fonds der Chemischen Industrie, Alexander von Humboldt Foundation, and Hanse Wissenschaftskolleg (WLY). Calculations were performed using the NEC SX-8 supercomputer at the High Performance Computing Center Stuttgart, and the Opteron cluster at the chemistry department of the University of Oldenburg.

References

1. Steele, B. C. H., Heinzel, A., Nature 2001, 414, 345.
2. Steele, B. C. H., Nature 1999, 400, 619.
3. Bashyam, R., Zelenay, P., Nature 2006, 443, 63.
4. US Department of Energy.
5. Shevlin, S. A., Guo, Z. X., Phys. Rev. B 2007, 76, 024104.
6. Wu, X. J., Yang, J. L., Hou, J. G., Zhu, Q. S., J. Chem. Phys. 2004, 121, 8481.
7. Han, S. S., Lee, S. H., Kang, J. K., Lee, H. M., Phys. Rev. B 2005, 72, 113402.
8. Luo, W. F., Gross, K. J., J. Alloys Compd. 2004, 385, 224.
9. Løvvik, O. M., Opalka, S. M., Phys. Rev. B 2005, 71, 054103.
10. Alapati, S. V., Johnson, J. K., Sholl, D. S., J. Phys. Chem. B 2006, 110, 8769.
11. Wang, Q. Y., Johnson, J. K., J. Phys. Chem. B 1999, 103, 277.
12. Chandrakumar, K. R. S., Ghosh, S. K., Nano Lett. 2008, 8, 13.
13. Yoon, M., Yang, S. Y., Wang, E., Zhang, Z. Y., Nano Lett. 2007, 7, 2578.
14. Zhong, G., Chan, B., Radom, L., J. Am. Chem. Soc. 2007, 129, 924.
15. Lachawiec, A. J., Qi, G. S., Yang, R. T., Langmuir 2005, 21, 11418.
16. Li, Y. W., Yang, R. T., J. Am. Chem. Soc. 2006, 128, 8136.
17. Li, Y. W., Yang, R. T., J. Am. Chem. Soc. 2006, 128, 726.

18. Chen, L., Cooper, A. C., Pez, G. P., Cheng, H., J. Phys. Chem. C 2007, 111, 18995.
19. Chen, L., Cooper, A. C., Pez, G. P., Cheng, H. S., J. Phys. Chem. C 2007, 111, 5514.
20. Yim, W., Klüner, T., J. Catal. 2008, 254, 349.
21. Woodward, R. B., Hoffmann, R., The conservation of orbital symmetry, Verlag Chemie: Deerfield Beach, FL, 1970.
22. Kresse, G., Hafner, J., Phys. Rev. B 1993, 47, 558.
23. Kresse, G., Hafner, J., Phys. Rev. B 1994, 49, 14251.
24. Kresse, G., Furthmüller, J., Phys. Rev. B 1996, 54, 11169.
25. Kresse, G., Furthmüller, J., Comput. Mat. Sci. 1996, 6, 15.
26. Perdew, J. P., Burke, K., Ernzerhof, M., Phys. Rev. Lett. 1997, 78, 1396.
27. Kresse, G., Joubert, D., Phys. Rev. B 1999, 59, 1758.
28. Monkhorst, H. J., Pack, J. D., Phys. Rev. B 1976, 13, 5188.
29. Henkelman, G., Jónsson, H., J. Chem. Phys. 2000, 113, 9978.
30. Henkelman, G., Uberuaga, B. P., Jónsson, H., J. Chem. Phys. 2000, 113, 9901.
31. Chen, B., Li, B., Chen, L., Appl. Phys. Lett. 2008, 93, 043104.
32. Khabashesku, V. N., Billups, W. E., Margrave, J. L., Acc. Chem. Res. 2002, 35, 1087.
33. Tang, C. C., Bando, Y., Ding, X. X., Qi, S. R., Golberg, D., J. Am. Chem. Soc. 2002, 124, 14550.

Molecular Dynamics Study of Plastic Deformation of Nanocrystalline Palladium

D.V. Bachurin[1,2] and P. Gumbsch[1,3]

[1] Institut für Zuverlässigkeit von Bauteilen und Systemen (IZBS), Universität Karlsruhe, Kaiserstr. 12, 76131 Karlsruhe, Germany
[2] Institute for Metals Superplasticity Problems (IMSP), Russian Academy of Science, 450001 Ufa, Russia
dmitriy.bachurin@izbs.uni-karlsruhe.de
[3] Fraunhofer Institut für Werkstoffmechanik (IWM), 79108 Freiburg, Germany
peter.gumbsch@izbs.uni-karlsruhe.de

Summary. Atomistic simulations of tensile and compressive deformation of three-dimensional nanocrystalline palladium at room temperature and different strain rates were perfomed. Detailed analysis of tensile straining has revealed almost no plasticity and an absence of dislocation activity in the grains right up to the moment of intergranular cracking. During compressive straining the sample exhibits a plastic regime brought about by the motion of extended partial dislocations emitted from the grain boundaries. At higher compressive strains the deformation mechanism changes to one that involves full dislocations and twinning.

1 Introduction

Nanocrystalline metals are polycrystals with a mean grain size below 100 nm. Mechanical properties of nanocrystalline metals are considerably different from those of coarse-grained materials. For example, the yield stress of nanocrystalline metals is up to ten times higher than that of their coarse-grained counterparts [1], similarly the strain-rate sensitivity is much higher for nanocrystalline metals [2]. Their outstanding properties are mostly explained by the peculiarities of the microstructure of nanocrystals: the large number of grain boundaries and the presence of porosity relating to the synthesis technique.

Plastic deformation of coarse-grained materials is mainly carried by dislocations within the individual grains. The decrease of the grain size leads to the limitation of the conventional operation of dislocation sources and deformation of nanocrystalline materials. For very small grain sizes ($d \approx 5\ nm$), plasticity is believed to be carried mostly by the grain boundaries via particular accommodation mechanisms [3, 4]. However, many questions are still under investigation [5]: the transition between different mechanisms of deformation,

W.E. Nagel et al. (eds.), *High Performance Computing in Science and Engineering '09*, DOI 10.1007/978-3-642-04665-0_8,

the influence of the grain boundary structure on the deformation mechanisms and the mechanisms of dislocation nucleation.

Modern experimental visualization techniques do not allow the carrying out of nonintrusive investigations of nanocrystalline materials during deformation without changing their grain boundary structure. Modern computer simulations enable the study and analysis of the processes occuring at the small time scales. Thus large scale molecular dynamics computer simulations help to understand the relationship between structure of grain boundaries and "unusual" mechanical properties of nanocrystals.

Computer simulation of the deformation of three-dimensional nanocrystalline palladium were made earlier [6, 7]. However, the interatomic potential for palladium [8] used in the studies gives an unrealistically low stacking fault energy of 0.008 J/m^2 compared to the experimental value of 0.18 J/m^2 [9]. This is believed to markedly influence the deformation behavior. Moreover, the simulations [6, 7] have been perfomed by applying a constant stress with a view to study grain boundary diffusional creep in nanocrystalline palladium at high temperatures. In this manuscript, the mechanical response of nanocrystalline palladium sample uni-axial tension and compression with different strain rates and at room temperature, is modelled.

2 Molecular Dynamics Method and Simulation Setup

Molecular dynamics (MD) is a form of computer simulation where atoms and molecules interact via known interatomic potential. Because molecular systems generally consist of a vast number of atoms, it is impossible to find the properties of such complex systems analytically. MD simulation circumvents this problem by using numerical methods for solving the equations of motion for the atoms. In the last years MD methods were widely used in solving dislocations dynamics [10, 11, 12, 13], mechanisms of interaction between lattice dislocations and grain boundaries [14, 15, 16, 17], mechanisms of dislocation nucleation and fracture [18, 19, 20, 21], mechanisms of plastic deformation of nanocrystalline materials [4, 7, 22].

MD simulation requires the definition of a potential function, or a description of the terms by which the particles in the simulation interact. In many-body potentials, the potential energy includes the effects of three or more particles interacting with each other and cannot be found by a sum over pairs of atoms alone. One of the widely used methods describing the atomic interactions is the Embedded Atom Method [23, 24].

All simulations were performed using the Molecular Dynamics Program IMD developed at the Institute for Theoretical and Applied Physics in Stuttgart [25]. IMD is a software package for classical molecular dynamics simulations. The parallelization in IMD is done using the standard Message Passing Interface (MPI). Static relaxation was made using the FIRE algorithm [26]. An Embedded Atom Method potential for palladium [27] is used.

The potential is fitted to first-principle local-density-functional calculations [28] and the intrinsic stacking fault energy [29]. This potential was selected since it reproduces unstable and stable stacking fault energies well and is therefore expected to realistically describe the generation of full and partial dislocations. However, the potential gives an unrealistically low surface energy what will lead to premature fracture.

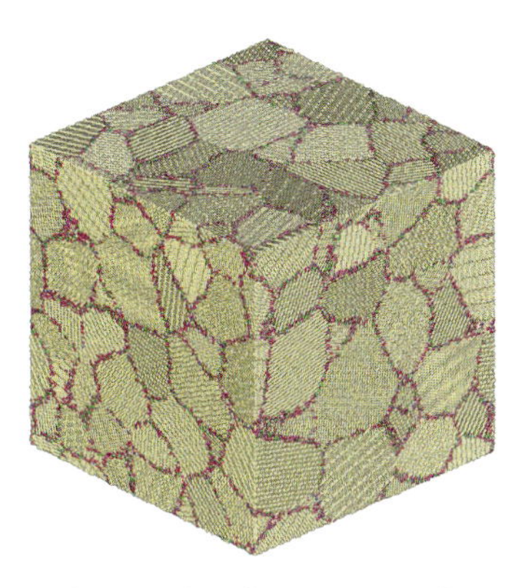

Fig. 1. Computer-simulated sample of nanocrystalline palladium with periodic boundary conditions in all directions. Each side of the box is approximately 41 nm long and the sample contains about 4.6 million atoms. Atoms are colored according to their coordination number. Yellow atoms correspond to the ideal fcc crystal lattice, other colored atoms are grain boundary atoms

The initial three-dimensional nanocrystalline structures were generated using a simple space-filling technique called the Voronoi construction [30]. The sample contains 100 defect-free and randomly oriented grains with mean grain size of 10 nm and approximately 4.6 millions atoms (see Fig. 1). Grain size and grain boundary misorientation distributions of the sample are shown in Fig.2. The deformation behavior of exactly this nanocrystalline structure has been investigated earlier for Al [31]. Initially the sample was rescaled from aluminum to palladium lattice constant. After static relaxation it was annealed for 10 ps at 1360 K (85% of melting temperature of this palladium potential) and then equilibrated at 300 K for 20 ps. The initial density of the sample at 300 K is 96.6% of the perfect crystal value. The percentage of defect atoms (atoms with coordination numbers unequal to 12 calculated with the cutoff radius 3.38Å) is 13.12%. A uni-axial strain with different strain rates $10^7, 10^8, 10^9$ s^{-1} was applied by continuously scaling the atomic coordinates and box-sizes along the x-direction. Along the two other directions, the stresses were kept equal to zero, so the system was allowed to change its sizes freely. MD simulations were made with periodic boundary conditions allowing us to consider the sample as a small part of an infinitely large bulk nanocrystalline sample. All simulations were performed at a constant temperature of 300K.

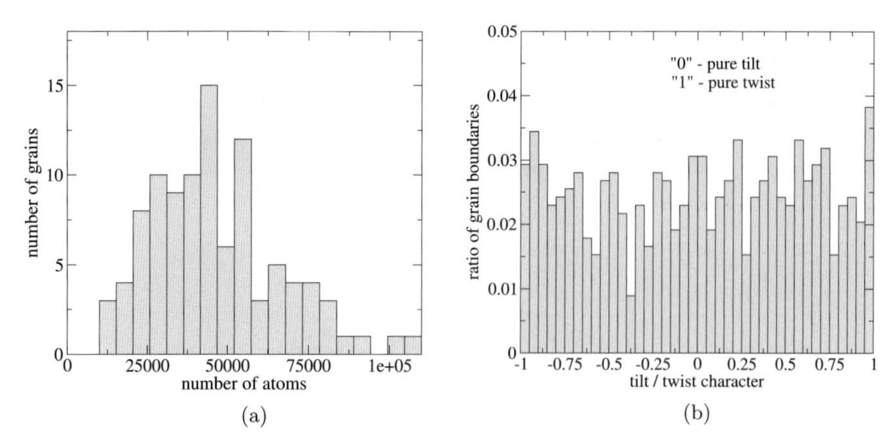

Fig. 2. Grain size (**a**) and grain boundary misorientations (**b**) distributions of the nanocrystalline palladium sample shown in Fig. 1

In order to identify twin planes and stacking faults in the grains, the Honeycutt and Andersen analysis was applied, which allows researchers to classify pairs of atoms according to their local environment [32, 33]. Using this technique four different classes of atoms were defined. Each class of atoms has a separate color: dark-blue represents fcc atoms; light-blue represents first nearest-neighbor hcp coordinated atoms; orange represents other 12 coordinated atoms; dark-red represents all other atoms. The atomic structures are visualized with the use of the program AtomEye [34].

3 Tensile Straining

Figure 3a displays the results of the application of uni-axial tensile deformation made with three different strain rates $10^7, 10^8, 10^9\ s^{-1}$. The curves clearly show the strain rate sensitivity of the nanocrystalline material: the larger the strain rate, the higher the flow stress. The curves reach their maxima at different strains: the larger the strain rate, the larger the strain at which the stress-strain curve has a maximum (see Fig. 3a). The stress-strain curves begin to deviate from the linear slope at a strain of approximately 0.9%. This value is approximately 2 times higher than that of aluminum [35], suggesting that nanocrystalline aluminum is more easily deformable than nanocrystalline palladium. It should be noted that the strain rate of $10^7\ s^{-1}$ is the lowest strain rate to reach a reasonable plastic deformation in an acceptable computer time with the available computational resources. Since all the observed deformation mechanisms generally compete with each other and since this competition may of course be influenced by strain rate and temperature, the present simulations should only be viewed as indications of the relevant mechanisms.

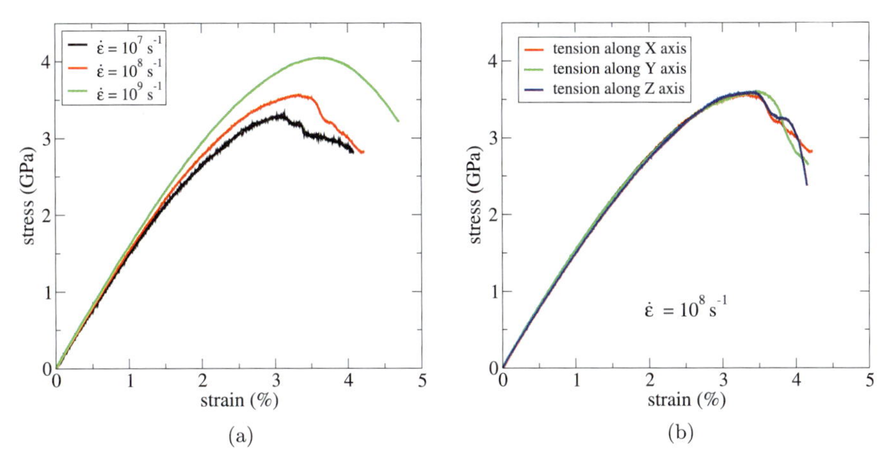

Fig. 3. Stress-strain curves for nanocrystalline palladium at an applied uni-axial tensile strain with 3 different strain rates (**a**) and with 3 different directions of an applied strain (**b**)

Figure 3b shows the results of the application of uni-axial tension along x, y and z directions at a constant strain rate of 10^8 s^{-1}. All three curves are very similar and, as expected, there are no significant differences for these three different uni-axial straining directions of the specimen. It confirms the absence of any anisotrophy or crystallographic texture in the sample.

The strain-stress curves already show that the nanocrystalline palladium sample exhibits almost no plasticity. The elastic regime is followed by a small plastic regime. Thereafter intergranular cracking starts at about 3% strain. The cracks first nucleate at the grain boundaries which are oriented approximately perpendicular to the direction of applied strain and then propagate along other adjacent grain boundaries as shown in Fig. 4. Intergranular cracking occurs at grain boundaries of mixed character (tilt and twist) with misorientation angles between 26° and 45°. No difference in fracture behaviour was found between tilt and twist boundaries. Similar brittle behaviour of nanocrystalline palladium specimens prepared by two different methods – cold rolling and inert-gas condensation – was found in experiments [36]. Although fracturing also appears unavoidable in tensile experiments, the cracking may be of very different origin there. In the simulations presented here the ease of fracture is linked to the unrealistically low surface energy at realistically high stacking fault and grain boundary energies of the potential.

Figure 5a shows the changes in density of the nanocrystalline palladium specimen during tensile deformation. The relative density of the specimen linearly decreases with the applied strain in the elastic and plastic regime up to 3% strain. This linear relation does not hold anymore with further deformation, which indicates the beginning of fracture. The ratio of the defect atoms does not change during elastic deformation but increases significantly in the plastic regime.

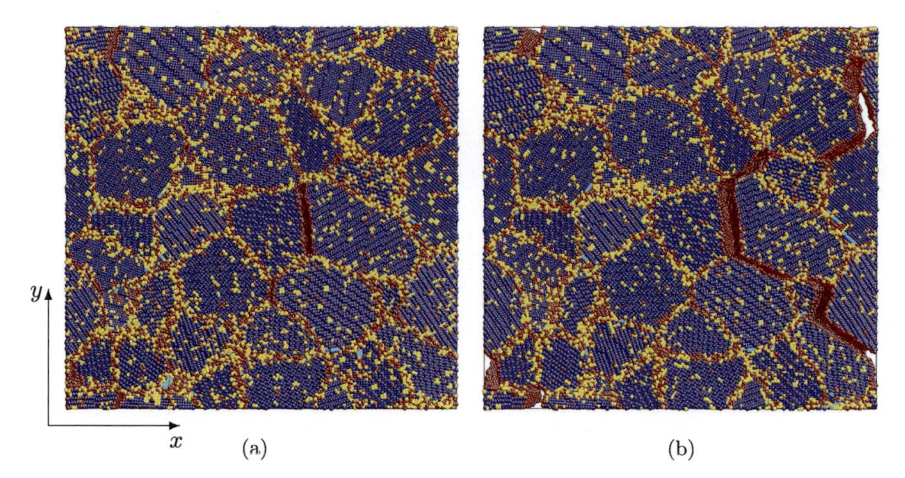

Fig. 4. Results of a MD simulation of nanocrystalline palladium. The sample was deformed along the x-axis in tension with a strain rate of $10^8\ s^{-1}$. Different stages are displayed: (**a**) at 4% strain and (**b**) the evolution at 5% tensile deformation. The cross-sectional view (**a**) clearly shows that the cracks open along grain boundaries, which are oriented perpendicular to the applied strain, and then propagate along adjacent grain boundaries (**b**). Atoms are colored according to their local crystalline structure

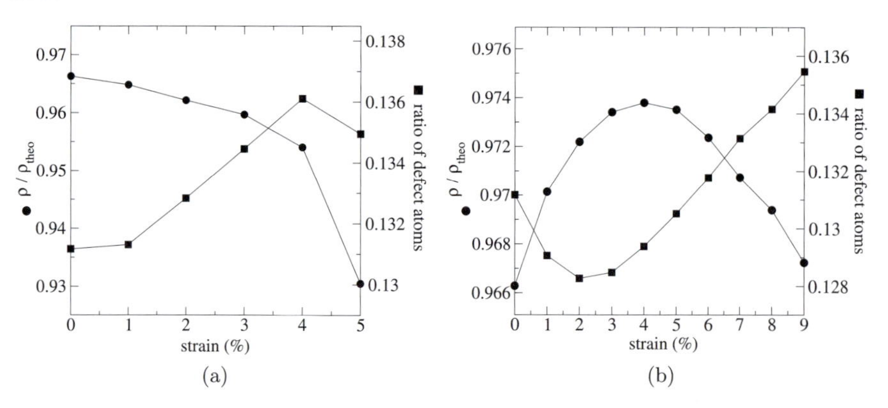

Fig. 5. The dependences of the relative density and the ratio of defect atoms in the nanocrystalline palladium sample versus applied uni-axial tensile (**a**) and compressive (**b**) strains. The data obtained for the constant strain rate of $10^8\ s^{-1}$

Detailed inspection of the atomic configurations shows displacements of the grain boundary atoms suggesting an important role of grain boundary processes such as grain boundary migration, grain boundary sliding, straightening and migration of the triple junctions in nanocrystalline palladium. Extended partial or full dislocations were not found during tensile deformation up to 3% strain. However, several embryos of partial and full dislocation embryos are clearly visible near grain boundaries. The maximum number of dislocation embryos observed in one grain was five. These embryos do not necessary belong to the same glide system. Three possible types of embryos behaviour were recorded: (1) – embryos are generated after some deformation; (2) – dislocation embryos already exist in the initial structure as a result of the relaxation of the grain boundary structures and they may either stay near the grain boundary without any visible change or be partially emitted into the grain; (3) – dislocations embryos that existed in the initial configuration enter back into the grain boundary after deformation. The extent of embryo emission depends on the orientation of the grain and is defined by the resolved shear stress acting on the glide plane. These findings strongly confirm the concept that dislocations can be emitted from stress concentrations at the grain boundaries [37].

Figures 6(a–c) show cross-sections of different grains along a [111] viewing direction at 3% strain. Figure 6c displays the only case of an emission of a full dislocation observed in all simulations. The nucleation of the full dislocation starts by the emission of a leading partial dislocation and afterwards a trailing

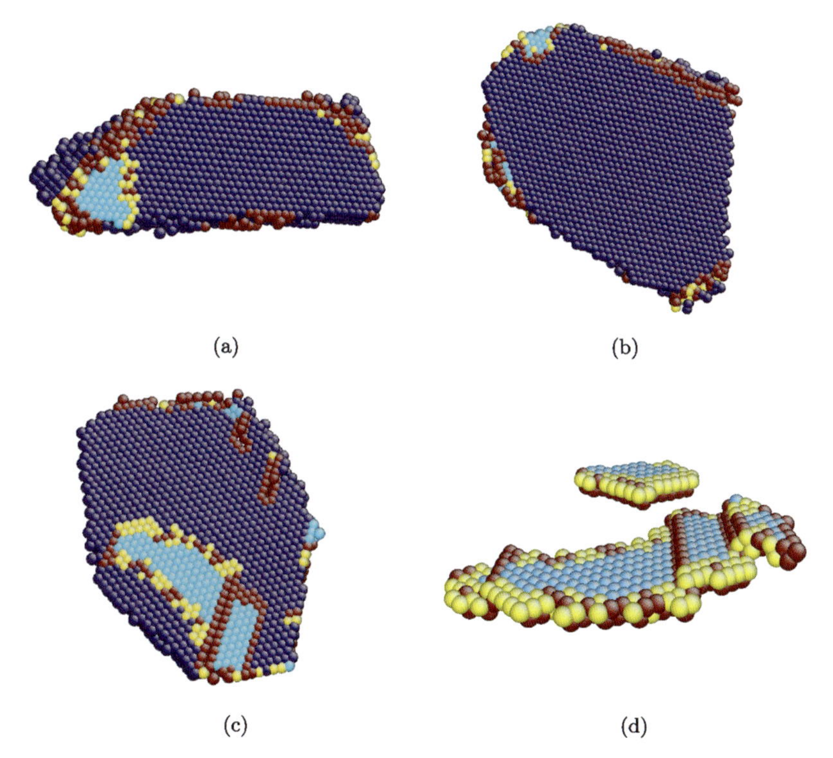

Fig. 6. Dislocation activity in different grains at tensile deformation of 3% strain. Partial dislocation embryos in the grain 11 (**a**) and grain 79 (**b**); partial propagation of the full dislocation embryo into the grain 8 (**c**). The full dislocation (**c**) is jogged (**d**). Atoms are colored according to their local crystalline structure

partial dislocation is emitted from a nearby position at the grain boundary. The stacking fault is clearly seen between the two partial dislocations. The nucleated full dislocation is jogged, i.e. the segments of the dislocation line are not in a specific glide plane, but extend over more than one interplanar spacing as shown in Fig. 6d. The Schmid factor for the active slip system is 0.45 and the resolved shear stress calculated from the applied strain is 1.55 GPa. It is worth noting that the active glide system in this grain is not the one which experiences maximum resolved shear stress. Furthermore, there are even grains with pre-existing dislocation embryos and higher resolved shear stress. This has to do with the presence of the pinning sites at grain boundaries and confirms our results above and the results of [37], which showed that nucleation and propagation of dislocations have to be dealt with as separate processes and that propagation may be more critical than nucleation. For comparison, Fig. 6b shows the leading partial dislocation embryo in grain 79. Even at 3% strain the dislocation can not be emitted although it is almost ideally oriented.

4 Compressive Straining

Figure 7 displays the results of the application of uni-axial compressive deformation at two different strain rates. The stress-strain curves begin to deviate from the linear slope at a strain of approximately 1.5%. This value is about 1.7 times higher than for uni-axial tension obtained above. A finite plastic regime is observed after elastic straining. Like in the case of tensile experiments the higher strain rate leads to a higher flow stress. The plastic regime is followed by the beginning of intergranular cracking after 5% strain. The application of uni-axial compression along x, y and z directions again revealed no significant differences for different directions.

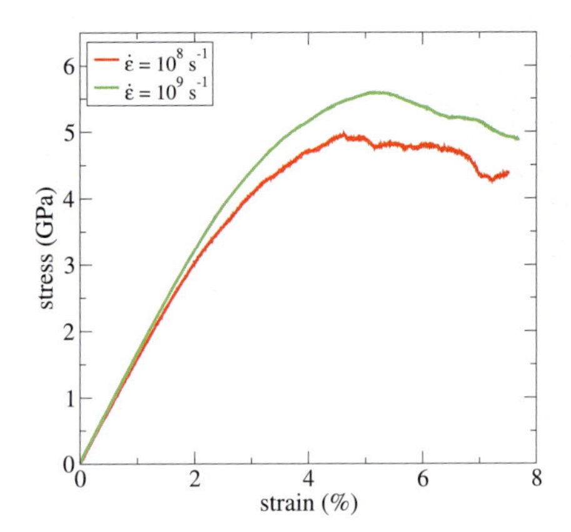

Fig. 7. Stress-strain curves for nanocrystalline palladium for applied uni-axial compressive strain at 2 different strain rates

Figure 5b shows the change of the sample density during compressive straining. The dependence of the relative density on strain is not a linear function as was the case for the tensile simulations. The density increases up to 4% of strain and then decreases. The latter is connected with the beginning of the fracture. It is interesting to note that the ratio of the defect atoms is reduced up to 2% strain and then monotonically increases.

Detailed inspection of the atomic configurations shows no dislocation activity at strains lower than 3%. Emission of partial dislocations was observed after 3% strain. The number of extended partial dislocations is considerably increased with the increase of strain. For example, at 4% strain only two extended partials in two different grains were observed (see Fig. 8a). Eleven extended partial dislocations, two cases of intersection of two partials and three cases of micro-twinning displayed in Fig. 8c and 8d were already registered at 5% strain. The activity of a full dislocation was observed after 6% strain. The latter means that the trailing partials were emitted at the same glide planes as leading partials heretofore. It was detected by calculation of displacement field for every single grain and also by disappearing layers of hcp atoms in the grains where they were revealed at low strains. Thus in-

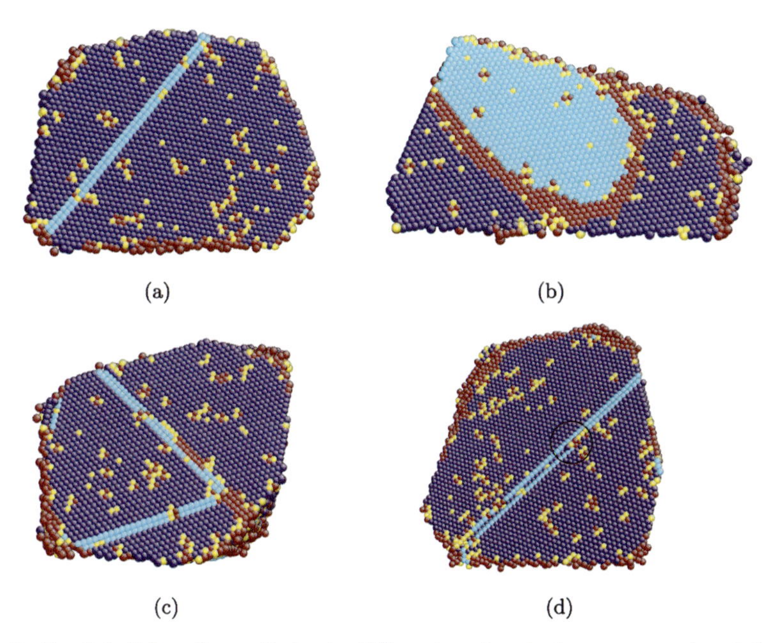

Fig. 8. Partial dislocation activity in different grains during compressive deformation: (**a**) extended partial dislocation at 4% strain; (**b**) travelling through the grain leading partial at 5% strain; (**c**) multislip at 5% strain; (**d**) micro-twinning process at 5% strain. The circle in figure (**d**) shows the partial dislocation emitted on an adjacent slip plane to the initial partial, leaving a micro-twin behind. Atoms are colored according to their local crystalline structure

creased straining may lead to a change in deformation mechanism: at low strains plastic deformation is carried mainly by extended partial dislocations, while at higher strains twinning and then full dislocations activity are added. The latter agrees with earlier MD simulations in nanocrystalline Ni [38].

5 Conclusion

Presented MD simulation results of deformation behavior of three-dimensional nanocrystalline palladium show that it exhibits different defomation mechanisms during uni-axial tensile and compressive straining at room temperature.

Uni-axial tension:

- Nanocrystalline palladium described with the potential used here responds to the loading by fracturing. Cracks are nucleated at high-angle grain boundaries oriented perpendicular to the direction of applied strain. Cracking may, however, be premature due to the low surface energy of the potential.
- Several partial and full dislocation embryos exist near grain boundaries, but no extended partial or full dislocations were found.
- Accommodation processes in grain boundaries (sliding, migration) and rearrangements of triple junctions also contribute to the deformation.

Uni-axial compression:

- Nanocrystalline palladium exhibits plastic deformation. Deformation is governed by extended partial dislocations, at higher strains even by full dislocations and twinning.
- Further increase of straining leads to cracks nucleation at the grain boundaries parallel to the direction of applied strain.

Acknowledgement. Financial support by the DFG (Project Gu 367/18-2) and the DFG-Forschergruppe 714 is gratefully acknowledged.

References

1. Kumar, K.S., Van Swygenhoven, H., Suresh, S.: Mechanical behavior of nanocrystalline metals and alloys. Acta Mater., **51**, 5743–5774 (2003)
2. Schiøtz, J., Di Tolla, F.D., Jakobsen, K.W.: Softening of nanocrystalline metals at very small grain sizes. Nature, **391**, 561–563 (1998)
3. Van Swygenhoven, H., Derlet, P.M.: Grain-boundary sliding in nanocrystalline fcc metals. Phys. Rev. B, **64**, 224105 (2001)
4. Van Swygenhoven, H., Derlet, P.M., Hasnaoui, A.: Atomic mechanism for dislocation emission from nanosized grain boundaries. Phys. Rev. B, **66**, 024101 (2002)

5. Derlet, P.M., Gumbsch, P., Hoagland, R., Li, J., McDowell, D.L., Van Swygenhoven, H., Wang, J.: Atomistic simulations of dislocations in confined volumes. MRS Bulletin, **34**, 184–189 (2009)
6. Yamakov, V., Wolf, D., Phillpot, S.R., Gleiter, H.: Grain-boundary diffusion creep in nanocrystalline palladium by molecular-dynamics simulation. Acta Mater., **50**, 61–73 (2002)
7. Yamakov, V., Wolf, D., Phillpot, S.R., Mukherjee, A.K., Gleiter, H.: Deformation mechanism crossover and mechanical behaviour in nanocrystalline materials. Phil. Mag. Letters, **83**, 385–393 (2003)
8. Foiles, S.M., Adams, J.B.: Thermodynamic properties of fcc transition metals as calculated with the embedded-atom method. Phys. Rev. B, **40**, 5909–5915 (1989)
9. Dillamore, I.I., Smallman, R.E.: The stacking-fault energy of f.c.c. metals. Phil. Mag., **12**, 191–193 (1965)
10. Schroll, R., Vitek, V., Gumbsch, P.: Core properties and motion of dislocation in NiAl. Acta Mater., **46**, 903–918 (1998)
11. Bitzek, E., Gumbsch, P.: Dynamic aspects of dislocation motion: atomistic simulations. Mat. Sci. Eng. A, **400** 40–44 (2005)
12. Jin, Z.H., Gao, H., Gumbsch, P.: Energy radiation and limiting speeds of fast moving edge dislocations in tungsten. Phys. Rev. B, **77**, 094303 (2008)
13. Schroll, R., Gumbsch, P.: Atomistic study of the interaction between dislocations and structural point defects in NiAl. Phys. Stat. Sol. (a) **166**, 475–488 (1998)
14. Jin, Z.H., Gumbsch, P., Ma, E., Albe, K., Lu, K., Hahn, H., Gleiter, H.: The interaction mechanism of screw dislocations with coherent twin boundaries in different face-centred cubic metals. Scripta Mater., **54**, 1163–1168 (2006)
15. Cheng, Y., Mrovec, M., Gumbsch, P.: Atomistic simulations of interactions between the 1/2(111) edge dislocations and symmetric tilt grain boundaries in tungsten. Phil. Mag. **88**, 547–560 (2008)
16. Jin, Z.H., Gumbsch, P., Albe, K., Ma, E., Lu, K., Gleiter, H., Hahn, H.: Interactions between non-screw lattice dislocations and coherent twin boundaries in face-centred cubic metals. Acta Mater., **56**, 1126–1135 (2008)
17. Hashibon, A., Lozovoi, A.Y., Mishin, Y., Elsässer, C., Gumbsch, P.: Interatomic potential for the Cu-Ta system and its application to surface wetting and dewetting. Phys. Rev. B, **77**, 094131 (2008)
18. Gumbsch, P., Beltz, G.E.: On the continuum versus atomistic descriptions of dislocation nucleation versus cleavage in nickel. Modelling Simul. Mater. Sci. Eng., **3**, 597–613 (1995)
19. Ludwig, M., Gumbsch, P.: Cleavage fracture and crack tip dislocation emission in B2 NiAl: an atomistic study. Acta Mater., **46**, 3135–3143 (1998)
20. Gumbsch, P., Zhou, S.J., Holian, B.L.: Molecular dynamics investigation of dynamic crack stability. Phys. Rev. B, **55**, 3445–3455 (1997)
21. Li, J., Ngan, A.H.W., Gumbsch, P.: Atomistic modeling of mechanical behavior. Acta Mater., **51**, 5711–5742 (2003)
22. Van Swygenhoven, H., Spaczer, M., Caro, A., Farkas, D.: Competing plastic deformation mechanisms in nanophase metals. Phys. Rev. B, **60**, 22–25 (1999)
23. Daw, M.S., Baskes, M.I.: Embedded-atom method: Derivation and application to impurities, surfaces, and other defects in metals. Phys. Rev. B, **29**, 6443–6453 (1984)

24. Foiles, S.M., Baskes, M.I., Daw, M.S.: Embedded-atom-method functions for the fcc metals Cu, Ag, Au, Ni, Pd, Pt, and their alloys. Phys. Rev. B, **33**, 7983–7991 (1986)
25. http://www.itap.physik.uni-stuttgart.de/˜imd/index.html
26. Bitzek, E., Koskinen, P., Gähler, F., Moseler, M., Gumbsch, P.: Structural relaxation made simple. Phys. Rev. Lett., **97**, 170201 (2006)
27. von Sydow, B., Hartford, J., Wahnström, G.: Atomistic simulations and Peierls-Nabarro analysis of Schockley partial dislocations in palladium. Comput. Mater. Sci., **15**, 367–379 (1999)
28. Tomanek, D., Sun, Z., Louie, S.G.: Ab initio calculation of chemisorption systems: H on Pd(001) and Pd(110). Phys. Rev. B, **43**, 4699–4713 (1991)
29. Hartford, J., von Sydow, B., Wahnström, G., Lundqvist, B.I.: Peierls barriers and stresses for edge dislocations in Pd and Al calculated from first principles. Phys. Rev. B, **58**, 2487–2496 (1998)
30. Van Swygenhoven, H., Farkas, D., Caro, A.: Grain-boundary structures in polycrystalline metals at the nanoscale. Phys. Rev. B, **62**, 831–838 (2000)
31. Brandstetter, S., Derlet, P.M., Van Petegem, S., Van Swygenhoven, H.: Williamson-Hall anisotropy in nanocrystalline metals: X-ray diffraction experiments and atomistic simulations. Acta Mater., **56**, 165–176 (2008)
32. Honeycutt, J.D., Andersen, H.C.: Molecular dynamics study of melting and freezing of small Lennard-Jones clusters. J. Phys. Chem., **91**, 4950–4963 (1987)
33. Clarke, A.S., Jonsson, H.: Structural changes accompanying densification of random hard-sphere packings. Phys. Rev. E, **47**, 3975–3984 (1993)
34. http://164.107.79.177/Archive/Graphics/A/
35. Bitzek, E., Derlet, P.M., Anderson, P.M., Van Swygenhoven, H.: The stress-strain response of nanocrystalline metals: A statistical analysis of atomistic simulations. Acta Mater., **56**, 4846–4857 (2008)
36. Rösner, H., Boucharat, N., Markmann, J., Padmanabhan, K.A., Wilde, G.: In situ transmission electron microscopic observations of deformation and fracture processes in nanocrystalline palladium and $Pd_{90}Au_{10}$. Accepted in Mat. Sci. Eng. A. (2009)
37. Van Swygenhoven, H., Derlet, P.M., Frøseth, A.G.: Nucleation and propagation of dislocations in nanocrystalline fcc metals. Acta Mater., **54**, 1975–1983 (2006)
38. Van Swygenhoven, H., Derlet, P.M., Frøseth, A.G.: Stacking fault energies and slip in nanocrystalline metals. Nature Mater., **3**, 399–403 (2004)

Conductance of Correlated Nanostructures

Alexander Branschädel[1], Tobias Ulbricht[1], and Peter Schmitteckert[2]

[1] Institut für Theorie der Kondensierten Materie
Wolfgang Gaede Straße 1
Universität Karlsruhe
Karlsruhe Institute of Technology
D-76128 Karlsruhe

[2] Institut für Nanotechnologie
Research Center Karlsruhe
Karlsruhe Institute of Technology
D-76021 Karlsruhe

Transport properties of strongly interacting quantum systems are a major challenge in todays condensed matter theory. In our project we apply the density matrix renormalization group (DMRG) method [2, 3, 4, 5, 6, 7, 8] to study transport properties of quantum devices attached to metallic leads.

The main effort in this project is focused on the calculation of non-equilibrium transport properties. In our last report [8] we reported on first results concerning the finite bias conductance within the interacting resonant level model (IRLM). However, due to the involved complexity of the problem and the corresponding large scale numerics one could still argue, that the results may not be trustworthy. In a cooperation with Edouard Boulat and Hubert Saleur we have been able to show that our approach is in excellent agreement with analytical calculations in the framework of the Bethe ansatz[9]. This agreement is remarkable as the numerics is carried out in a lattice model, while the analytical result is based on field theoretical methods in the continuum. Therefore we have to introduce a scale T_B to compare the field theoretical result to our numerics. Remarkably, at the so called self-dual point the complete regularization can be expressed by a single number, even for arbitrary contact hybridization t'. Most strikingly we proved the existence of a negative differential conductance (NDC) regime even in this simplistic model of a single resonant level with interaction on the contact link.

Concerning the linear transport calculation from resolvents in frequency domain we developed an alternative approach to our Kubo approach [10] by calculating the zero frequency spectral function and applying it to the simplified Meir-Wingreen formula for impurity problems with proportional coupling [11].

W.E. Nagel et al. (eds.), *High Performance Computing in Science and Engineering '09*, DOI 10.1007/978-3-642-04665-0_9,

1 Implementation Details

1.1 Top Level Code

Concerning the implementation of our DMRG code we have completely rewritten the front end code to our DMRG library. This rewrite was driven by the increasing variety of local site types, e.g. spinless-, SU(2)-, and SU(3)- fermions, bosons, hardcore bosons, SU(2)-, SU(3)-, and SU(4)- spins in various representations, and the n-state Potts model, and the demand for more flexibility. The input to our code is now read from std::cin and parsed by our own parser. Within this new scheme we have managed to consolidate our most important codes into a single programme. While this rewrite was a major endeavor we can exploit synergistic effects from combining the different site types with all algorithmic techniques like solving resolvent equations and performing time evolutions. The last missing step is the incorporation of quantum chemistry type four operator terms as used in our fractional hall code [8, 12].

1.2 Library Code

On the library side most of our efforts were spent in designing and implementing a parser which is suited for our needs.

We implemented a new sparse matrix diagonalization scheme where we enhanced the Davidson scheme to stabilize calculations with large (close-) degeneracies. This code is currently tested and results will be reported in the next report. In addition the new version allows us to include complex Hamiltonians, e.g. the inclusion of a magnetic flux via a Peierls phase.

In addition, we have now three cut-off schemes to choose from. For one, we can fix the number of DMRG states for the largest blocks. This mode is important if we are running at the memory limit of the nodes. The discarded entropy (sometimes used as a measure for accuracy) then becomes very small in regions, where the DMRG is presumably very accurate. In these cases, accuracy can be weighted off against computation time and memory consumption by using less DMRG states. Therefore, the other scheme uses a fixed designated discarded entropy and varies the number DMRG states on each DMRG step. Depending on the physical problem, this can reduce computational effort in comparison to the first cut-off scheme. Last, we can also target at a specific Hilbert space size which when used helps to reduce truncation errors in adaptive time evolution schemes.

Within our Posix thread parallelization scheme we tested the pre-sorting of the worker queues in order to perform compute intensive parts first to avoid having to wait for single long lasting operations, e.g. we sort the matrix-matrix multiplications by the size of the involved matrices in a decreasing manner. However, the achieved speed-up was minimal, which shows that such delays are already covered by the high concurrency of our code.

1.3 Benchmark of Different Platforms

In Fig. 1 we compare the timing of distributing matrix-matrix multiplications (BLAS-3 `dgemm`) on different platforms using our master-worker library, for details see [6]. Number of threads is the number of worker threads, and zero worker threads corresponds to a pure serial implementation that serves as a reference calculation without threading overhead. For us the most remarkable

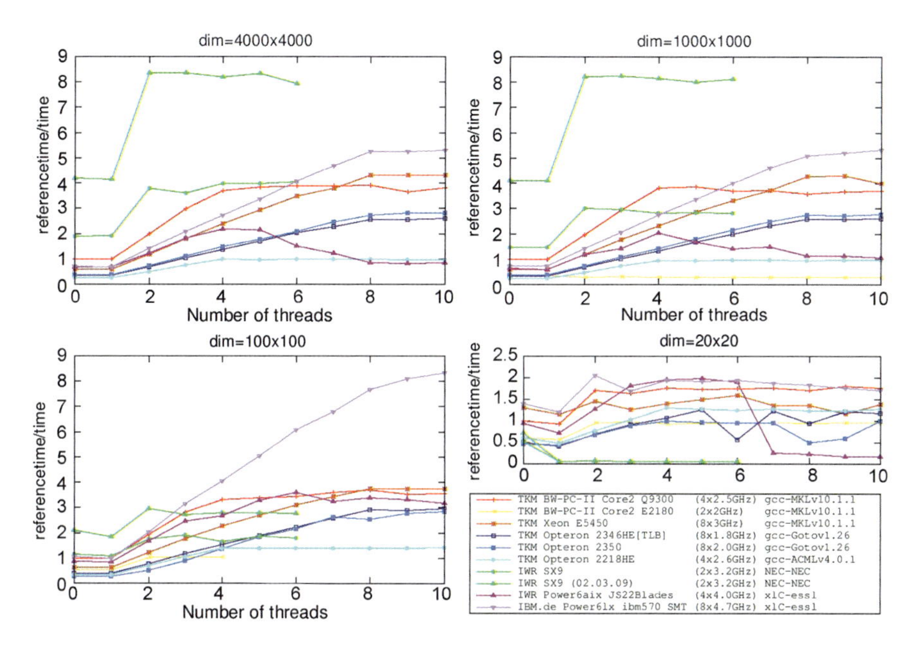

Fig. 1. Comparison of the `dgemm` performance of various architectures within our multi-threaded implementation. All data are in relative speed comparing to the BW-PC of 2008 ('highend edition') [14]. The y-axis denotes the speedup in comparison to a serial run. On the NEC SX-9 we are currently only allowed to use 2 CPUs. The better performing NEC curves corresponds to an update of the BLAS library

result is that finally we can get decent threading performance on the NEC SX architecture, which is in contrast to our previous attempts. Nevertheless the complete DMRG code does not run on the NEC SX-9 due to an unresolved issue in loading our scratch data from the hard disk. We hope to fix this problem soon. Besides this the DMRG code runs fine on all other platforms mentioned in Fig. 1. One should also note, that this threading benchmark's scaling for matrices with size ≥ 100 works equally well on low-end consumer PCs with up to 4 cores, computing nodes with up to 8 cores and high end performance computing machines like SX-9 or Power6.

1.4 Towards Worker Threads on Graphics Cards

In our DMRG code we use Posix threads to parallelize the code which is described in detail in [6]. The advantage of this approach lies in the flexibility of choosing worker threads. Here we present first steps to make use of this flexibility by evaluating the possibility to offload workload to acceleration or graphics cards. Within this parallelization scheme we are not restricted to a single type of worker, but we can schedule different types of workers. Here we report first benchmark results where we use our threading facility to distribute matrix-matrix multiplications over different worker types, namely one worker making use of the GPU on a graphics card to increase throughput.

Single Precision

For first benchmarks, we use an older graphics card with single precision hardware (Radeon HD 2400 XT, GPU RV 610), which we can only feed with `sgemm` calls. For compilation we used gcc 4.3.2 (Debian 4.3.2-1.1). In Fig. 2 we compare ACML (acml-4.2.0), MKL (10.1.1.019/em64t), GotoBLAS [13] (r1.26/penryn) and ACML-GPU (CAL RT & CL v. 1.3.158, acmlg v0.7). Looking at the CPU performance alone one notices that ACML, Goto, and ATLAS give approximately the same performance, with ACML being the fastest of these three libraries. However, MKL gives a factor of two in performance, demonstrating that the choice of BLAS implementation is very important. The platform used was a quad core Intel-based system with 8 GB of RAM [14]. Next, we use one worker thread that calls the ACML-GPU version of `sgemm` while in the remaining threads we employed the ATLAS (gcc-atlas-acmlg) or MKL (gcc-mkl-acmlg). This version of the ACML-GPU library does not allow to take full control over the distribution of `sgemm` calls to CPU or GPU. Thus, we had problems in linking simultaneously against the ACML-GPU and the normal ACML which stems from namespace clashes in these libraries, since they are not supposed to be used at the same time. Also note, that in the ACML-GPU documentation it is stated that the library is not thread safe and must not be used simultaneously by two different programmes or threads. However, since we have only one GPU worker, while the other workers are not accessing the GPU we are in compliance with this restriction albeit running a multi-threaded code. While at first sight these results do not look very encouraging, it is important to note that these benchmarks are only a proof of concept. Note that within our approach we are not restricted to offload work to a single GPU board. We could even utilize a mixture of accelleration boards.

Double Precision

Recently we got access to a machine with a floating point capable graphic card (Radeon HD 4870, GPU RV770). Our benchmark is presented in Fig. 3. With

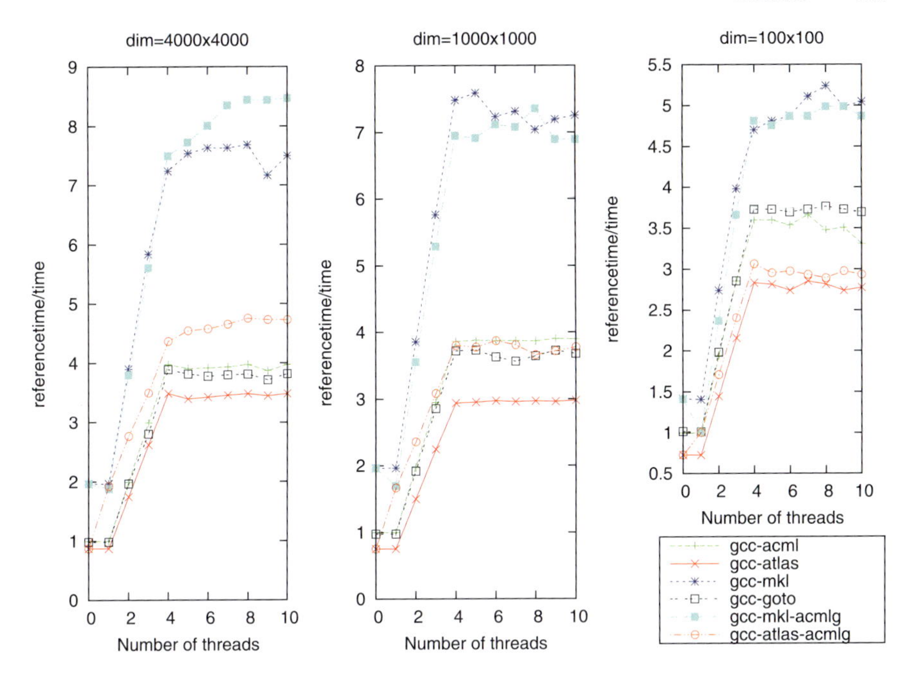

Fig. 2. sgemm performance within our multi-threaded testing code on one platform[14] for different BLAS libraries and in two cases with an additional graphics processor unit running one thread of ACML-GPU sgemm calls. All data are relative to the gcc-acml data serial run at number of threads = 0

this newer hardware the performance boost for the largest matrices due to the graphics card is clearly visible. However, already for the 1000x1000 matrices the quad Intel CPU is of comparable performance. While this benchmark does now look encouraging, the total gain in performance for our DMRG code was marginal. This is to be expected from the benchmark, since the matrices involved were typically smaller than 1000x1000. However, from this benchmark we expect that the use of graphics card is getting interesting for us, if the typical matrix size is at least of the order of 4000x4000. Since we can employ a scheduling depending on the matrix size this will be an interesting option for the future, especially if the BLAS support of the graphics cards improves in the next versions of the ACML. Finally we would like to note that we also tested an NVIDIA based card. However cublas is conflicting with our pthreads leading to spurious deadlocks.

2 Differential Conductance in the Interacting Resonant Level Model

Formally the conductance of a quantum device attached to leads is given by the Meir Wingreen formula [1]. Besides the special case of proportional cou-

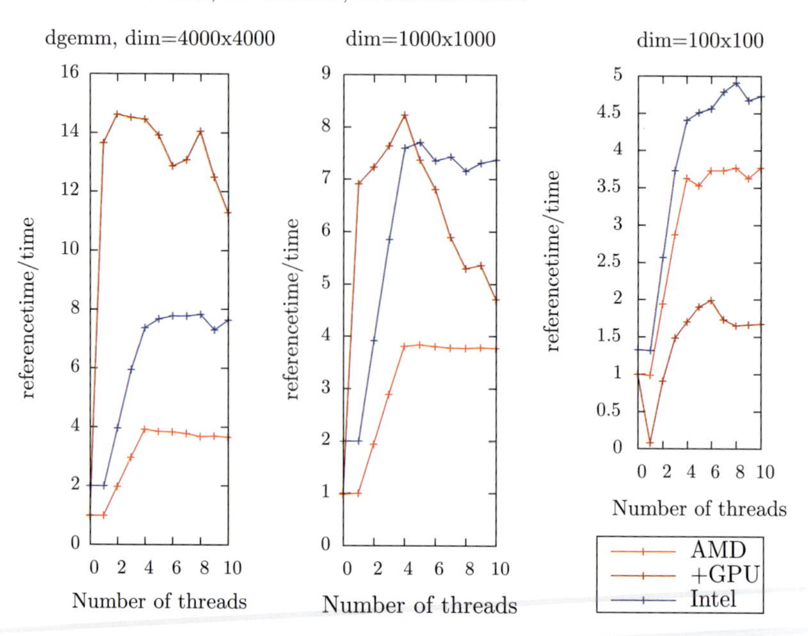

Fig. 3. dgemm performance on an AMD platform using only CPUs (AMD: Phenom II X4 920 Processor, (Quadcore 2.8GHz), BLAS is MKL (10.1.1.019/em64t)), the same platform with the first worker thread scheduled on the GPU (+GPU: AMD + Radeon HD 4870, GPU RV770, gBLAS is CAL RT & CL version: 1.4.227, acmlg v1.0), compared to the gcc-mkl data of Fig. 2 [14] (Intel)

pling, the Meir Wingreen formula can usually only be treated within perturbative approaches. The major problem in non-equilibrium dynamics consists in the fact that the stationary Schrödinger equation is now replaced by the time-dependent Schrödinger equation. Therefore an *eigenvalue problem* is replaced by a *boundary problem* and one has to take care of the initial state. Therefore one has to be very careful by sending all difficulties to time equal minus infinity since at some time t_0 one hits the initial state. In our approach the answer to this problem is to actually start with an initial state and to perform the full time integration of the time dependent Schrödinger equation via a time evolution operator given by the matrix exponential. The details of this approach have been explained in details within this previous report [6, 7, 8] where first results have already been reported in [8]. The main new achievement consists in the fact that we have been able to make quantitative comparison with analytical calculations on a continuum version of the interacting resonant level model (IRLM) in the framework of Bethe ansatz and a dressed Landauer–Büttiker approach and the time dependent simulation on a lattice version of the IRLM.

In Fig. 5 we show the current-voltage characteristic for different interaction values U, which have already been presented in [8]. As one can see, for

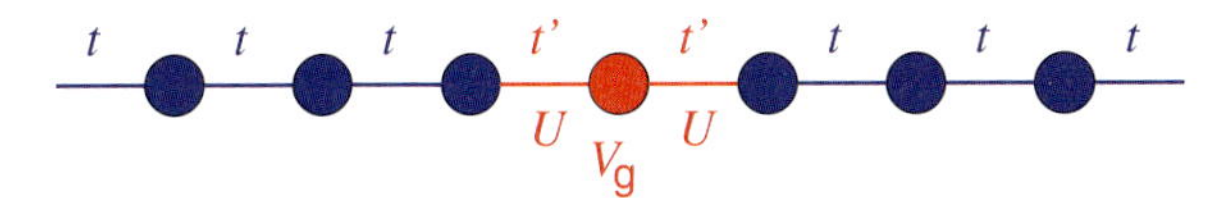

Fig. 4. Sketch of the IRLM model. t' denotes the coupling of the impurity to the left and right lead, $t = 1$ is the hopping element in the leads, U is the interaction on the contact link, and V_g is a gate voltage which is set to zero in this work

$U < 2t$ and not too large voltage we get an enhancement of the resonance width of the differential conductance, while for large interaction we have a shrinking of the resonance width. Most strikingly we find a negative differential conductance regime for large voltages and not too small interaction which is most prominent for $U \sim 2t$ and decays with a power law in the source-drain voltage V_{SD}. In [9] we have shown that at this value the exponent of the decay reaches its maximum while approaching 1/2.

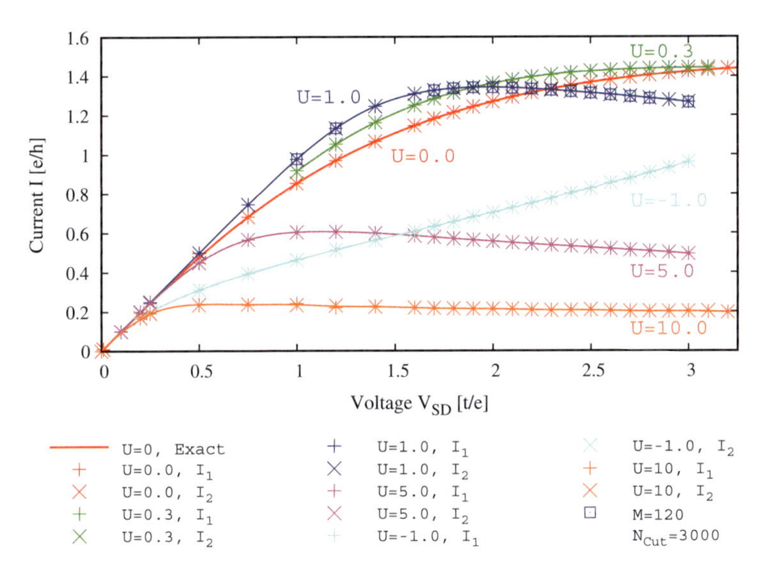

Fig. 5. Current vs. source drain voltage V_{SD} for a hybridization of $t' = 0.5$, and interaction values of $U = -1.0, 0.0, 0.3, 1.0, 5.0$, and 10.0. The subscript '1' ('2') of I refers to data extrapolated from link to the left (right) lead. The lines are guides to the eye, except for $U = 0.0$ where we plotted the exact result for infinite leads. Calculations have been performed with 96 sites, 48 fermions keeping at least 2000 states per DMRG block. In addition data for $U = 1.0$, $M = 120$ sites and $N_{\mathrm{Cut}} = 3000$ states per block is shown

Using a field theoretical treatment of the IRLM based on the thermodynamic Bethe ansatz [9] one can map the problem of interacting fermions on noninteracting Bethe particles which scatter only by a phase shift. Therefore one can employ a Landauer-Büttiker like treatment of the transport problem.

However, in the mapping of the original fermions to the Bethe particles one mixes in general particles from the left and right lead and it is in no way clear how the charge operator $e(N_L - N_R)$ translates into the new basis. Interestingly, the continuum version of IRLM displays a so called self dual point where a mapping between small and large U maps on the same model. Exactly at this point of interaction the most complicated terms in the mapping of the charge operator cancel and the procedure can actually be carried out analytically [9]. Since in the field theory the leads are taken to be continuous with linear dispersion the model is scale free and can not be compared directly to the lattice version. For this one has to regularize the field theoretical result by introducing scales for the observables, which is for example well known in the context of Kondo physics, where one has to introduce a scale T_K. Also here, since at small voltages we have a linear conductance of one, a single scale for the current and voltage axis is sufficient, which we denote by T_B.

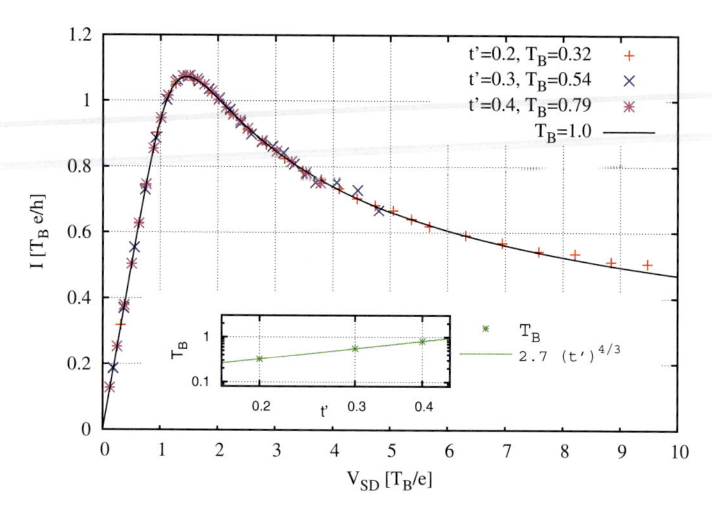

Fig. 6. Comparison between analytical and DMRG results at the self dual point. For each t' the numerical data have been fitted using a single parameter T_B to the analytical result. Finally T_B was fitted against t' leading to the proposed $T_B = c(t')^{4/3}$, $c \approx 2.7$

In Fig. 6 we have rescaled our numerical results for $U = 2.0t$ by fitting the scale T_B for $t' = 0.2, 0.3, 0.4$. Remarkably all the results collapse on the predicted analytical result. It is interesting to note that this scaling works still in the voltage regime where the dispersion of the cosine band of the nearest-neighbour hopping chain in the leads can not be neglected. Even the predicted dependence of T_B on t', $T_\mathrm{B} \sim (t')^{\frac{4}{3}}$ fits very well as shown in the inset with a prefactor of 2.7. Therefore, this prefactor accounts for the complete regularization of the field theory.

In summary we have shown that the field theoretical approach and the results based on simulations within td-DMRG give excellent agreement for an out of equilibrium system with strong interaction. To the best of our knowledge this is the first time that an exact analytical method and a numerical simulation in the field of transport properties of strongly correlated, non-perturbative quantum systems display such an agreement. We therefore hope that this system can serve as a benchmark system for other approaches.

3 Correlations Within the Steady State

We attempt to calculate shot noise within the td-DMRG approach. Shot noise is defined as the zero-temperature contribution to noise in a transport state (compare e.g. [15]). Therefore, the current-current correlations in the time domain

$$S(t,t') = \frac{1}{2}\langle \Delta\hat{I}(t)\Delta\hat{I}(t') + \Delta\hat{I}(t')\Delta\hat{I}(t)\rangle = \Re\langle \Delta\hat{I}(t)\Delta I(t')\rangle \tag{1}$$

have to be calculated in a non-equilibrium zero-temperature state, where $\Delta\hat{I}(t) = \hat{I}(t) - \langle \hat{I}(t)\rangle$. In a steady state the correlation function must fulfil $S(t,t') \equiv S(t-t')$. Then the noise power can be defined as the Fourier transform

$$2\pi\delta(\omega+\omega')S(\omega) = \langle \Delta\hat{I}(\omega)\Delta\hat{I}(\omega') + \Delta\hat{I}(\omega')\Delta\hat{I}(\omega)\rangle \tag{2}$$

where

$$S(\omega) = 2\int_{-\infty}^{\infty} \mathrm{d}t\, \mathrm{e}^{\mathrm{i}\omega t} S(t,t'=0) = 4\Re\int_{0}^{\infty} \mathrm{d}t\, \mathrm{e}^{\mathrm{i}\omega t} S(t,t'=0). \tag{3}$$

The right-hand side of the equation accounts for the symmetry $S(t-t') = S(t'-t)$. In the zero-frequency limit $\omega = 0$ this simplifies to

$$S(\omega=0) = 4\int_{0}^{\infty} \mathrm{d}t\, S(t,t'=0) = 4\int_{0}^{\infty} \mathrm{d}t\, \Re\langle \Delta\hat{I}(t)\Delta\hat{I}(t'=0)\rangle. \tag{4}$$

In a steady state, of course, this expression should be independent of the choice of the time t'

$$S \equiv S(\omega=0) = 4\int_{t'}^{\infty} \mathrm{d}t\, \Re\langle \Delta\hat{I}(t)\Delta\hat{I}(t')\rangle. \tag{5}$$

To calculate the quantity S within the td-DMRG approach, one has to take into account several limitations. First of all, we are in a finite system with M lattice sites and hard walls. Thus a steady state is not well defined. Instead, we make the attempt to calculate the time evolution from an initial non-equilibrium state $|\Psi\rangle$ as described in Section 2. The "switching" of a finite source-drain voltage $V_{\rm SD}$ at initial time causes a ringing of the current [16] which exponentially decays within a settling time $T_{\rm S}$. The current finally enters a plateau regime, where the size of the plateau is given by the transit time $T_{\rm R}$ which is finite due to the finite size of the system. To obtain the quantity S in a situation "close" to a steady state one therefore has to evaluate the integral (5) in a limited time range

$$S_{\rm DMRG} = 4 \int_{T_{\min}}^{T_{\max}} {\rm d}t\, \Re\langle \Delta\hat{I}(t)\Delta\hat{I}(T_{\min})\rangle_{\Psi} \tag{6}$$

where $T_{\min} > T_{\rm S}$ and $T_{\max} < T_{\rm R}$. In a hypothetical situation with a system of infinite size where $T_{\rm R} \to \infty$ the contribution of $\int_{T_{\max}}^{\infty} {\rm d}t\, \Re\langle \Delta\hat{I}(t)\Delta\hat{I}(T_{\min})\rangle_{\Psi}$ can be neglected if $\Re\langle \Delta\hat{I}(t)\Delta\hat{I}(T_{\min})\rangle_{\Psi}$ is small for $t > T_{\max}$ as compared to the mean value in the range $T_{\min} < t < T_{\max}$. One therefore has to choose the size of the system big enough to ensure the correlation function to drop to zero within the transit time. A second weak point of this approach is an effective temperature introduced by finite size effects, which scale with $1/M$. As a result thermal noise is added which can not be neglected especially in the low voltage regime where shot noise is expected to vanish. This shortcoming can be cured by comparing the results for different system sizes M and by an extrapolation $1/M \to 0$.

A good justification for our approach can be given using a non-interacting system with finite size which can be treated by exact diagonalization. The numerical results can be compared to analytical results obtained in the thermodynamic limit. Here, shot noise is calculated for the non-interacting resonant level model with a single impurity coupled to two leads

$$\hat{H} = -t \sum_{m \neq 0} (\hat{c}^{\dagger}_{m+1}\hat{c}_m + {\rm h.c.}) - t'(\hat{c}^{\dagger}_{-1}\hat{c}_0 + \hat{c}^{\dagger}_0\hat{c}_1 + {\rm h.c.}) \tag{7}$$

with spinless electrons at half filling (compare also Fig. 4, where the interaction must be set to $U = 0$ as well as the gate voltage $V_{\rm g} = 0$). The initial state $|\Psi\rangle$ is obtained as the ground state of the system with an applied source drain voltage

$$\hat{H} + \hat{V}_{\rm SD}, \quad \hat{V}_{\rm SD} = \frac{V_{\rm SD}}{2}(\hat{N}_{\rm L} - \hat{N}_{\rm R}) \tag{8}$$

where $\hat{N}_{\rm L/R} = \sum_{\pm m > 0} \hat{c}^{\dagger}_m \hat{c}_m$ is the particle number operator on the left/right lead. The impurity itself is given by a modified hopping matrix element t' which couples the impurity site to the leads. The expression for the analytical result in the thermodynamic limit

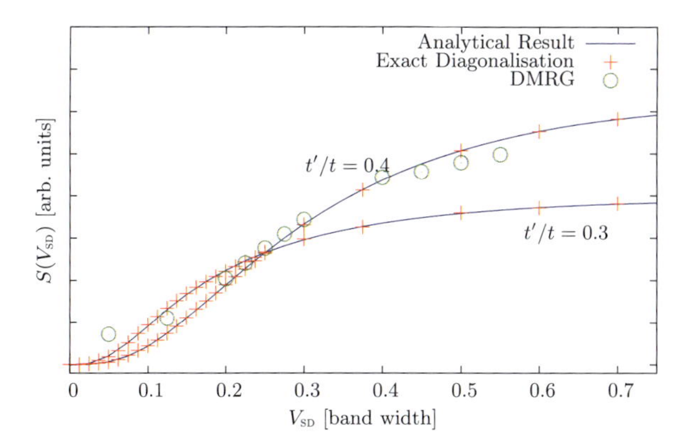

Fig. 7. Shot Noise S as function of the source drain voltage V_{SD} in the non-interacting resonant level model, Eq. (7). The analytical result was obtained using the Landauer–Büttiker theory while the numerical result is computed with an exact diagonalization procedure for a system of different finite sizes of $M = 120 \ldots 180$ lattice sites with a subsequent linear extrapolation of $1/M \to 0$. The two curves correspond to different couplings t' of the impurity to the leads. The preliminary DMRG-result was computed for the case $t'/t = 0.4$ for $M = 96$ lattice sites without extrapolation to the thermodynamic limit. In the low voltage regime, finite size effects play a significant role which add an effective thermal noise. The high voltage regime, however, still suffers from DMRG truncation errors

$$S \propto \int_{-V_{\text{SD}}/2}^{V_{\text{SD}}/2} \mathrm{d}\epsilon \, T(\epsilon)\big(1 - T(\epsilon)\big) \tag{9}$$

can be obtained from Landauer–Büttiker theory. The transmission probability $T(\epsilon)$ for the cosine band can be derived for the non-interacting case using single particle scattering states. The comparison of numerical data from exact diagonalisation and analytical derivation shows very good agreement, Fig. 7.

Also in this figure, we provide first preliminary results for the interacting resonant level model

$$\hat{H} + U \sum_{m=\pm 1} \left(\hat{n}_m - \frac{1}{2}\right)\left(\hat{n}_0 - \frac{1}{2}\right), \quad \hat{n}_m = \hat{c}_m^\dagger \hat{c}_m \tag{10}$$

with $U = 0$, obtained by the td-DMRG method. There is no fundamental problem in applying this method to the interacting case with $U \neq 0$, however, we can not compare the data against analytical results or against exact diagonalization as in the non-interacting case. The DMRG results still suffer from truncation errors in the high voltage regime. This can be overcome by increasing the number of kept states within the DMRG truncation procedure. Furthermore, in the low voltage regime, strong finite size effects introduce additional effective thermal noise. For now we do not have data for different system sizes M which is the reason that we still can not extrapolate to the thermodynamic limit. Results of our ongoing efforts will therefore be reported on later.

4 The Anderson Model with Ferromagnetic Leads

We studied the competition between interference due to multiple single-particle paths and Coulomb interaction in a simple model of an Anderson-like impurity with local magnetic field induced level splitting coupled to ferromagnetic leads (from now on nicknamed the FAB model). The model along with its potential experimental relevance in the field of spintronics serves as a nontrivial benchmark system where various quantum transport approaches can be tested and compared. In [11] we presented results for the linear conductance obtained by a spin-dependent implementation of the density matrix renormalization group scheme which are compared with a mean-field solution as well as a seemingly more advanced Hubbard-I approximation.

In summary it was shown in [11] that the DMRG method was able to provide the linear conductance for arbitrary strength of the on-site Coulomb interaction and arbitrary level splitting. The DMRG data interpolate between the known results of non-equilibrium Green functions (NEGF) for zero interaction and the co-tunneling results for large level splitting. A key result is the strong suppression of conductance with increasing on-site Coulomb interaction if the magnetic field on the dot is opposite to the lead polarization.

The DMRG results also served as a benchmark for different approaches, where we found that both the second order von Neumann (2vN) approach and the NEGF approach with mean-field interaction give reliable results for the conductance. While the 2vN approach also provides plausible results for finite bias, the mean-field NEGF fails due to the wrong treatment of partially occupied states.

5 Further Projects

In a continuation of our Kubo approach in cooperation with Dan Bohr we are currently investigating the linear transport properties of a 6-site ring which is threaded by a phase $\phi = \pi$ by means of Kubo correlation functions and spectral functions. However, it turned out that this system is notoriously hard due to the degeneracies at the Fermi surface. While we already have many results we decided to report on this system in the next report as we are still not convinced whether the results are fully converged.

Acknowledgments. Most of the calculations have been performed at the XC2 of the SCC Karlsruhe under the grant number RT-DMRG. We would like to thank the AMD streamcomputing group for providing us with a pre-release of ACML-GPU library.

References

1. Y. Meir and N. S. Wingreen, Phys. Rev. Lett. **68**, 2512 (1992).
2. S. R. White, Phys. Rev. Lett. **69**, 2863 (1992).
3. S. R. White, Phys. Rev. B **48**, 10345 (1993).
4. Density Matrix Renormalization – A New Numerical Method in Physics, edited by I. Peschel, X. Wang, M.Kaulke, and K. Hallberg (Springer, Berlin, 1999); Reinhard M. Noack and Salvatore R. Manmana, Diagonalization- and Numerical Renormalization-Group-Based Methods for Interacting Quantum Systems, AIP Conf. Proc. **789**, 93–163 (2005)
5. Peter Schmitteckert, Nonequilibrium electron transport using the density matrix renormalization group, Phys. Rev. B **70**, 121302(R) (2004).
6. Peter Schmitteckert and Günter Schneider, Nonequilibrium Transport of Strongly Correlated Systems, p. 113–126, in W. E. Nagel, W. Jäger, and M. Resch (Eds.), High Performance computing in Science and Engineering '06, Springer Verlag Berlin Heidelberg 2006, ISBN 978-3-540-36165-7.
7. Peter Schmitteckert Signal transport in and conductance of correlated nanostructures in W. E. Nagel, W. Jäger, and M. Resch (Eds.), High Performance computing in Science and Engineering '07, Springer Verlag Berlin Heidelberg, ISBN 978-3-540-74738-3 (2007).
8. Tobias Ulbricht and Peter Schmitteckert, Signal transport in and conductance of correlated nanostructures in W. E. Nagel, D. Kröner, and M. Resch (Eds.), High Performance computing in Science and Engineering '08, Springer Verlag Berlin Heidelberg 2008, ISBN 978-3-540-88301-2 (2008).
9. Edouard Boulat, Hubert Saleur, and Peter Schmitteckert, Twofold Advance in the Theoretical Understanding of Far-From-Equilibrium Properties of Interacting Nanostructures, Phys. Rev. Lett. **101**, 140601 (2008).
10. Dan Bohr, Peter Schmitteckert, and Peter Wölfle, DMRG evaluation of the Kubo formula – Conductance of strongly interacting quantum systems Europhys. Lett. **73** (2), 246–252 (2006).
11. Jonas Nyvold Pedersen, Dan Bohr, Andreas Wacker, Tomas Novotny, Peter Schmitteckert, and Karsten Flensberg, Interplay between interference and Coulomb interaction in the ferromagnetic Anderson model with applied magnetic field, Phys. Rev. B **79**, 125403 (2009).
12. Zi-Xiang Hu, Xin Wan, and Peter Schmitteckert, Trapping Abelian anyons in fractional quantum Hall droplets, Phys. Rev. B **77**, 075331 (2008).
13. Kazushige Goto and Robert Van De Geijn, High-performance implementation of the level-3 BLAS, ACM Trans. Math. Softw. **35**, 1 (2008).
14. See the "highend" version with an upgrade to 8 GB RAM and an Intel Core2 Quad Q9300 processor of http://portal.uni-freiburg.de/bw-pc/bwpcII.
15. Ya M. Blanter and M. Büttiker, Shot Noise in Mesoscopic Conductors, Phys. Rep. **336**, 1 (2000).
16. Ned S. Wingreen, Antti-Pekka Jauho, and Yigal Meir, Time-dependent transport through a mesoscopic structure, Phys. Rev. B **48**, 8487 (1993).
17. Dan Bohr and Peter Schmitteckert, Strong enhancement of transport by interaction on contact links, Phys. Rev. B **75** 241103(R) (2007).

Chemistry

Prof. Dr. Christoph van Wüllen

Fachbereich Chemie, Technische Universität Kaiserslautern,
Erwin-Schrödinger-Straße, D-67663 Kaiserslautern, Germany

For a given quantum chemical method, the computational effort grows with the size of the system, so one might think that small computers are good for small molecules, while large computing facilities are used to tackle problems that come with the modeling of large systems. The three contributions in this section show that this is certainly an over-simplification. Depending on what one wants to know and which details one is interested in, even small systems may require access to high-performance computing resources.

The first contribution reports calculations on the most simple systems that arise in atomic physics, namely atomic ions with just a single electron. If the nuclear charge is large, the Schrödinger equation itself cannot describe reality, and one must include relativistic corrections. So-called quasirelativistic Hamiltonians (which are intermediate between the nonrelativistic Schrödinger and the fully relativistic Dirac equation) have attracted much interest because they allow to treat systems with heavy atoms without having to deal with the negative-energy states of the Dirac equation. For the very popular Douglas-Kroll quasirelativistic Hamiltonian, a debate has been around whether the eigenfunctions of this operator behave, close to the nucleus, more or less like Dirac functions, or whether they are much more singular. If this were the case, it would have far-reaching consequences, for example on the rate of convergence if these functions are expanded in a regular basis set.

The problem with the Douglas-Kroll Hamiltonian now is that it is only defined in momentum space. An investigation of the properties of the eigenfunctions should not be based upon the standard way to work around this problem, so a basis set of spherical waves was used. This has the additional advantage that the convergence is slow but systematical, such that the properties of the eigenfunctions close to the nucleus can be extracted from the observed convergence rate. This leads to diagonalizations and multiplications of full matrice of considerable dimension (up to 10^5), such that a parallel calculations with up to 256 CPUs has been set up.

A large part of the work of Brill, Vendrell and Meyer describes investigations on the Zundel cation, $H_5O_2^+$, and its isotopomers. This is a very small molecule by today's standards, but it does not have a molecular structure in the classical sense, because there are large-amplitude motions of the atoms. This makes the vibrational spectra of these molecules different to understand and to assign. A full quantum mechanical treatment of the nuclear motion leads to a problem with 15 degrees of freedom (after separating off the translation and decoupling from the rotations) with a very complicated kinetic energy operator. On a single CPU, even the very efficient multiconfiguration time-dependent Hartree method developed in Heidelberg could not be applied to this problem. Therefore the program was parallelized (using a combination of symmetric multiprocessing and message-passing paradigms). There is still room for improvement, as going beyond 64 processors is not particularly efficient. What counts is, however, that an assignment of the complicated vibrational spectra is not possible, based on the calculations performed in Karlsruhe.

The third contribution, finally, deals with very large systems. Here, however, one looks at the system in a different way: electrons do not exist in the theoretical framework, and the motion of the nuclei is entirely classical. The elementary building blocks are atoms or even atom groups, and the modeling is done via molecular dynamics calculations using force fields. The challenge in this project comes through the presence of charged atoms (ions) in the mixture and the long-range electrostatic forces associated with them. This makes the calculation of seemingly simple quantities like the reduced density of a solution of salt in water extremely difficult. In another set of calculations, the behaviour of a polymer chain in water was studied: depending on the temperature, this chain can either be stretched or collapse into a ball-shaped tangle. The software used in these calculations can use well over 100 processors with high efficiency.

We have thus seen that for systems of any size, there are still open questions which can only be answered by large-scale computations made possible by the high-performance computing centers which offer their resources to scientists. The availability of these computing resources is important in all areas of chemistry, and will most likely become even more important in the future. We see a clear tendency to use more and more CPUs, but many program packages have difficulties to make good use of more than 100 CPUs, and further development in quantum chemistry software is necessary to overcome this.

How do Eigenfunctions of Douglas-Kroll Operators Behave in the Vicinity of Point-like Nuclei?

Christoph van Wüllen

Fachbereich Chemie, Technische Universität Kaiserslautern,
Erwin-Schrödinger-Str., D-67663 Kaiserslautern, Germany

Summary. There is no consensus in the recent literature how singular the eigenfunctions of quasirelativistic operators are close to a point-like nucleus. This question has far-reaching implications, e.g. for the convergence properties when such eigenfunctions are expanded in a basis of regular functions. For this reason the spectrum of a Douglas-Kroll operator in a large basis of spherical waves has been investigated. Such a basis sets shows a very slow but regular convergence pattern from which information on the singularity at the origin can be extracted. This calculation involves multiplications and a diagonalization of very large dense matrices, which has been performed in parallel (with up to 256 CPU cores) using functions from the `ScaLAPACK` library. For first-order Douglas-Kroll operators the eigenfunctions were known to be more singular than for the Dirac operator, and this manifests itself in our results. Second-order Douglas-Kroll and beyond, however, behaves very similar to the Dirac case, and there is ample evidence that the Douglas-Kroll eigenfunctions beyond first-order Douglas-Kroll have the same Singularity close to the nucleus than the Dirac operator. It is thus likely that an expansion of Douglas-Kroll eigenfunctions in basis sets conventionally used in relativistic quantum chemistry will show essentially the same convergence rate as found e.g. for the Dirac operator.

1 Introduction

The spectrum of a relativisitc Dirac operator for a single spin-$\frac{1}{2}$ particle with mass m and charge $-e$ (we have electrons in mind) in the electric field generated by a point-like ("nuclear") charge Ze for not too large Z is composed of two disjoint continua (energies larger than $+mc^2$ and smaller than $-mc^2$, c is the speed of light) as well as discrete energy levels within the gap, most of them slightly below $+mc^2$. The states in the negative continuum are associated with the existence of antiparticles and are important ingredients in a quantum field theory. On the other hand, in a theory with a fixed number of particles they have no (physical) meaning. Since pair creation and annihilation involves energies outside the energy "window" of chemical processes, a Hamiltonian 'for electrons only' is often used in relativistic quantum

W.E. Nagel et al. (eds.), *High Performance Computing in Science and Engineering '09*, DOI 10.1007/978-3-642-04665-0_10,

chemistry. The spectrum of such 'quasirelativistic' Hamiltonians contains the positive-energy continuum and the discrete states of the Dirac operator, while the negative-energy continuum is absent. Such operators can be constructed for a space that contains only two-component wave functions, in contract to the Dirac case where the wave functions have four components. One goal is of course, that the reduced complexity leads to a reduced computational effort, but probably equally important is that quasirelativistic operators can be bounded from below and are then variationally stable. In principle, positive and negative energy parts of the Dirac spectrum can be decoupled, but such an exact Foldy-Wouthuysen transformation is available in closed form only for a free particle ($Z = 0$). It is however possible to generate a series expansion of the exact Foldy-Wouthuysen operator. The original recipe of Foldy and Wouthuysen [1] generates such an expansion in c^{-2} which leads however to a singular quasirelativistic Hamiltonian for any truncation. Douglas and Kroll [2] suggested to use the strength of the external potential (here: Z) as an expansion parameter. This was cast into a working scheme and introduced in relativistic quantum chemistry by Hess [3, 4] using a approximate eigenfunctions of the squared nonrelativistic momentum operator $\hat{\mathbf{p}}^2$ as a basis to construct the Douglas-Kroll operator [5, 6]. For more than a decade, the Douglas-Kroll Hamiltonian has been truncated after the first or second order in Z. Third order Douglas-Kroll has been reported by Nakajima in 2000 [7, 8]. A systematic extension was presentey by Hess and Reiher [9]. Starting with the fifth order, different variants to derive the equation lead to different Douglas-Kroll operators. This puzzling situation was explained and analyzed by the present author [10]. In this work, the working equations for sixth-order Douglas-Kroll were presented for the first time, as well as an algorithm for a computer-based derivation thereof. Such automatic extensions to arbitrary high order were also formulated by Reiher [11]. The Douglas-Kroll method was also formulated as a theory restricted to a finite dimensional space (spanned by the basis function) from the very beginning, and here one gets to 'infinite order' methods in straightforward way (at least numerically) [12, 13, 14, 15]. In either case, one can duplicate the results from the Dirac operator (within the given, finite basis set) to arbitrary accuracy with the Douglas-Kroll method if one goes far enough.

The construction of the exact Foldy-Wouthuysen transformation *within a given finite basis set* is not very difficult, because a diagonalization of the matrix representation of the Dirac operator gives a spectral representation thereof which contains all information needed. Although this 'trick' has been around for some time [16, 17, 18, 19], a concise account has only been given recently [20, 21]. This works claims that any attempt to construct good approximations to the Foldy-Wouthuysen operator itself, rather than to its matrix representation, has thus become essentially obsolete. An important argument put forward in this context is, that the exact Foldy-Wouthuysen operator has very unpleasant properties, for example its eigenfunctions are much more singular than the Dirac functions, and that this makes the use of such operators within a finite basis set representation questionable. The apparent success of the Douglas-Kroll method was then attributed to approximations made in the established computational procedure [22] rather than to the inherent properties of this operators, and it was even claimed that problems with the Douglas-Kroll operator would soon become apparent if one went beyond these approximations. The purpose of the present investigation is to collect numerical evidence that this might not be the case.

Not much is known analytically about the properties of the eigenfunctions of Douglas-Kroll operators. For the first-order Douglas-Kroll operator, which is also known under the name "no pair operator" [23] and "Brown-Ravenhall operator" [24] is is known for a long time, that the eigenfunctions are more singluar than the Dirac eigenfunctions [25], such that normalized bound state solutions only exist for $Z < 124$ in contrast to the Dirac case, where such solutions exist for all $Z < 137$. It might be that this behaviour is the origin of the bad reputation of quasirelativistic Hamiltonians in general. However, for the second-order Douglas-Kroll operator (here given the name "Jansen-Hess operator") it was shown [24] that it spectrum is bounded from below for all $Z < 137$, such that second-order Douglas Kroll behaves like the Dirac operator and is very different from first-order Douglas Kroll. Nothing is known analytically beyond the second order.

2 A Spherical Wave Basis Set

Douglas-Kroll is often called a *momentum space method*. What this means is that most of the basic operators of the theory are only defined through their matrix elements with eigenfunctions of $\hat{\mathbf{p}}^2$. Furthermore, the Douglas-Kroll Hamiltonian finally involves products of such basis operators. For basis functions such as Gaussians that are often used in quantum chemistry, one first builds approximate eigenfunctions of $\hat{\mathbf{p}}^2$ as linear combinations of Gaussians, and computes matrix elements of the basic operators that occur in Douglas-Kroll theory assuming that these linear combinations *are* eigenfunctions of $\hat{\mathbf{p}}^2$, and constructs matrix representations of products of such basis operators through a resolution of the identity, that is, as products of matrix representations of the basis operators. The reader is referred to the original recipe of Hess [5, 6] or to more recent formulations of the theory [8, 9, 10, 26]. The present investigation started with the idea to use spherical waves as basis functions which are indeed exact eigenfunctions of $\hat{\mathbf{p}}^2$ and also make use of atomic symmetry. This eliminates the approximation of the standard procedure whose accuracy is most difficult to control. We restrict our investigation to s type one-electron states, such that we consider s waves of type

$$\Omega(k, \mathbf{r}) = \frac{\sin(kr)}{r} \tag{1}$$

Because spherical waves cannot be normalized, we put the atom into a spherical 'box' with radius R (the nuclear point charge is located in the center of the box), and require that all basis functions vanish at the border of the box. This makes the basis set discrete and the basis functions Ω_i can be normalized

$$\Omega_i(\mathbf{r}) = \frac{1}{\sqrt{2\pi R}} \frac{\sin(k_i r)}{r} \tag{2}$$

with

$$k_i = i\frac{\pi}{R}, \quad i = 1, 2, 3, \ldots \tag{3}$$

It has been verified that choosing $R = 80/Z$ is (by far) large enough for the $1s$ states of hydrogenlike ions. The results obtained in this investigation are therefore fully converged with respect to the size of the 'box'. Higher ionic states have a larger spatial extent and may require a larger R. A *finite* basis set of dimension N is obtained keeping all basis functions from $i = 1$ through $i = N$. The basis functions are eigenfunctions of the squared momentum operator

$$\hat{\mathbf{p}}^2 \Omega_i = k_i^2 \Omega_i, \tag{4}$$

therefore the matrix elements of all operators involved in the free-particle Foldy-Wouthuysen transformation (which is the starting point of the Douglas-Kroll method) are trivial. Furthermore, we need the matrix elements of the nuclear attraction potential $V = -Z/r$ and of $\boldsymbol{\sigma}\mathbf{p}V\boldsymbol{\sigma}\mathbf{p}$ with $\boldsymbol{\sigma} = (\sigma_x, \sigma_y, \sigma_z)$ the vector of the Pauli spin matrices. The matrix element integrals can be solved analytically, and the result is expressed in terms of the cosine integral:

$$-Z < \Omega_i | \frac{1}{r} \Omega_j > = \frac{Z}{R} \left\{ T((i+j)\pi) - T((i-j))\pi) \right\} \tag{5}$$

$$-Z < \Omega_i | \boldsymbol{\sigma}\mathbf{p} \frac{1}{r} \boldsymbol{\sigma}\mathbf{p} \Omega_j > = \frac{\pi^2 Z}{R^3} \Big\{ ij + \frac{i^2 + j^2}{2} \left(T((i+j)\pi) - T((i-j))\pi) \right) \Big\} \tag{6}$$

$$T(x) = \mathrm{Ci}(x) - \ln(x) \tag{7}$$

$$\mathrm{Ci}(x) = -\int_x^\infty \frac{\cos t}{t} \mathrm{d}t \tag{8}$$

Note that $T(-x) = T(x)$, and for $x = 0$, $T(x)$ assumes the finite value $T(0) = \gamma$, the Euler-Macheroni constant $\gamma = 0.577\ldots$. There is a potential problem with numerical accuracy here when taking the difference of the values of the function $T(x)$. Therefore we calculate $T(x)$ in quadruple precision (128-bit accuracy). Fortunately one only needs to evalute this functions only for a limited number of arguments: it is sufficient to evaluate $T(x)$ for integer multiples of π, $x = n\pi$ and $n = 0, 1, 2, \ldots, 2N$. To this end, a table of values for $n = 0$ through $n = 150$ has been evaluated with 50-digit precision and hard-coded into the program, and asymptotic expansions with controlled accuracy are used to calculate $T(n\pi)$ beyond $n = 150$ (for details, see Ref. [27]).

An expansion of atomic wave functions in spherical waves converges quite slowly, because the behaviour of the wave function close to the nucleus is not well represented with this basis set. If one tries to expand the nonrelativistic $1s$ wave function, $\Psi_{nr} \sim \exp(-Zr)$, in our spherical wave basis set,

$$\exp(-Zr) \approx \sum_i c_i \Omega_i(r) \tag{9}$$

one easily evaluates that $c_i \sim i^{-3}$ for large i. This behaviour is dictated by the behaviour of Ψ for small r. While this convergence is actually not fast, it gets much worse if singular relativistic $1s$ functions $\Psi_{rel} \sim r^{\nu} \exp(-Zr)$ ($\nu < 0$) are expanded. Here the coefficients fall off as $c_i \sim i^{-(2+\nu)}$. For Dirac $1s$ states,

$$\nu = \sqrt{1 - \alpha^2 Z^2} - 1, \quad \alpha \approx \frac{1}{137} \tag{10}$$

This makes clear that meaningful results can only be obtained for very large basis sets of spherical waves, up to $N = 10^5$. A diagonalization of a dense matrix of this dimension requires $\sim 10^{16}$ floating point operations and is therefore no longer possible on a single CPU which can perform at most 10^{10} floating point operations per second. In addition, the construction of the matrix elements of Douglas-Kroll operators requires many multiplications of such matrices, each of which involves a computational effort of the same order of magnitude than a diagonalization. We have decided to use the `ScaLAPACK` library [28], using mainly the subroutines `pdsyevd` for matrix diagonalization and `pdgemm` for matrix multiplication. We have used 16, 64 or 256 CPU cores depending on the problem size. The initialization of the program (pre-calculating the $T(x)$ for example) involves negligible computational effort (linear in the problem size) and is done on all CPUs. Each CPU then only calculates the matrix elements which are local to that CPU. Matrix dimensions range from 1000 to 128000.

3 Results

A complete presentation and discussion of the results has been given in Ref. [27]. Here we present results for $Z = 120$ in Table 1. We clearly see convergence to the exact value (-8000) for the nonrelativistic Hamiltonian ("nonrel") with increasing number N of spherical waves in the basis set. The convergence rate r is obtained by fitting the error obtained with a truncated basis set of dimension N to a function of the form AN^r. We see a value close to 3. The last column shows the so-called 'fully relativistic' results from the diagonalization of matrix representation of the Dirac operator (note that this matrix has dimension $2N$ because it still contains the negative-energy eigenstates). The exact Dirac result for $Z = 120$ (rounded to micro-Hartree accuracy) is -9710.783520. From the results for different values of Z (see Ref. [27]) we

Table 1. Nonrelativistic (nonrel), Douglas-Kroll first and fourth order (DK1,DK4) and fully relativistic (Dirac) results (in Hartree) for a hydrogen like ion with $Z = 120$ in a spherical wave basis set. The convergence rate w.r.t. the size of the spherical wave basis is given in the last line

N	nonrel	DK1	DK4	Dirac
1000	−7199.607394	−10685.676109	−9411.765423	−9413.769989
2000	−7199.950212	−11090.064104	−9549.063566	−9553.879425
4000	−7199.993734	−11382.794725	−9621.447056	−9628.877244
8000	−7199.999214	−11594.933420	−9658.979055	−9668.356559
16000	−7199.999902	−11748.847236	−9678.274317	−9688.912065
24000	−7199.999971	−11818.523838	−9684.839175	−9695.960918
32000	−7199.999988	−11860.636133	−9688.162255	−9699.541198
64000	−7199.999999	−11941.903527	−9693.228828	−9705.014575
Conv. rate	−3.08	−0.45	−0.96	−0.95

obtained experimentally that the convergence rate depends on the singularity of the $1s$ function at the origin through

$$r = -(2 + 2\nu), \quad -1 < \nu < 0. \tag{11}$$

The important finding now is that the convergence rate for fourth order Douglas-Kroll (DK4) is similar to that found for the Dirac case, and this also applies to second- and third-order Douglas-Kroll and other values of Z. As expected from the known properties of the first-order Douglas-Kroll (DK1) operator [25], DK1 behaves much different, and the observed convergence rate is indeed very slow. For $Z = 130$, no convergence could be obtained for DK1, while Douglas-Kroll beyond the first order behaves similar to the Dirac case for this high nuclear charge as well.

We cannot go here deeply into the implications of the 'resolution of the identiy' (RI) approximation which is still present in our Douglas-Kroll results beyond DK1. It turns out that this error is smaller than the truncation error, that is introduced by truncating the expansion of the wave function in the spherical wave basis functions. The RI error therefore does not change the overall picture.

Collecting all the numerical findings, there is ample evidence that the eigenfunctions of the Douglas-Kroll operator beyond the first order have a singularity close to a point-like nucleus which is not much different from the singularity of the Dirac functions. Opposite statements in the literature seem to be based on the behaviour of the DK1 eigenfunctions which indeed behave much worse when approaching the nucleus.

4 Hardware and Performance

The calculations were run on the `XC4000` Opteron cluster at the Steinbuch Computing Center in Karlsruhe, Germany. Each node features two dual-core

CPUs (2.6 GHz) and 16GByte memory. We have compiled our own numerical libraries (`ScaLAPACK`, `LAPACK`, `BLACS`, `ATLAS/BLAS`) but use the MPI implementation available in Karlsruhe. For $N = 64000$, a matrix diagonalization on 256 CPU cores takes about 3000 sec wall-clock time. All eigenvectors have been computed because some of them were analyzed. Although we are at most interested in few of them, not much overall saving can be generated because of the large number of matrix multiplications that were required in addition (see below). Note further that the diagonalization step in our program also involves a symmetrization of the matrix, a step which does not do any computation but involves a lot of communication. The wallclock time necessary to construct the second, third and fourth order Douglas-Kroll operators was 1000, 5000 and 20000 seconds. Since DK2, DK3 and DK4 involve 2, 9, and 35 matrix multiplications, this means that a matrix multiplication requires roughly 500 sec. wallclock time, which means that we have achieved a real-world performance of $\approx 10^{12}$ floating point operations per second, which is a very good value for 256 Opteron processors. A complete diagonalization costs as much as six matrix multiplications, which is also a very good value since matrix diagonalizations are more difficult to parallelize than matrix multiplications.

In this fortunate case, nearly all the numerical work is done within two well-define matrix kernels (matrix multiplication and matrix diagonalization). In such a case it is quite straightforward to achieve good parallel performance even when using more than 100 CPU cores. This is still a challenge in many areas of scientific computing.

Acknowledgements. Computer time at the Steinbuch Centre for Computing (SCC) in Karlsruhe (Germany) is gratefully acknowledged. This work has further been supported by the Deutsche Forschungsgemeinschaft (DFG) under grant Wu288/4-1 within the priority programme SPP 1145 *"Modern and universal first-principles methods for many-electron systems in chemistry and physics"*.

References

1. L. L. Foldy and S. A. Wouthuysen, Phys. Rev. **78**, 29 (1950).
2. M. Douglas and N. M. Kroll, Ann. Phys. (N.Y.) **82**, 89 (1974).
3. B. A. Hess, Phys. Rev. A **33**, 3742 (1986).
4. G. Jansen and B. A. Hess, Phys. Rev. A **39**, 6016 (1989).
5. R. J. Buenker, P. Chandra, and B. A. Hess, Chem. Phys. **84**, 1 (1984).
6. B. A. Hess, R. J. Buenker, and P. Chandra, Int. J. Quantum Chem. **29**, 737 (1986).
7. T. Nakajima and K. Hirao, Chem. Phys. Lett. **329**, 511 (2000).
8. T. Nakajima and K. Hirao, J. Chem. Phys. **113**, 7786 (2000).
9. A. Wolf, M. Reiher, and B. A. Hess, J. Chem. Phys. **117**, 9215 (2002).
10. C. van Wüllen, J. Chem. Phys. **120**, 7307 (2004).
11. M. Reiher and A. Wolf, J. Chem. Phys. **121**, 10945 (2004).

12. M. Barysz, A. J. Sadlej, and J. G. Snijders, Int. J. Quantum Chem. **65**, 225 (1997).
13. M. Barysz and A. J. Sadlej, Theochem-J. Mol. Struct. **573**, 181 (2001).
14. M. Barysz and A. J. Sadlej, J. Chem. Phys. **116**, 2696 (2002).
15. D. Kedziera and M. Barysz, J. Chem. Phys. **121**, 6719 (2004).
16. K. G. Dyall, J. Chem. Phys. **106**, 9618 (1997).
17. M. Filatov, J. Chem. Phys. **125**, 107101 (2006).
18. M. Ilias, H. J. A. Jensen, V. Kello, B. O. Roos, and M. Urban, Chem. Phys. Lett. **408**, 210 (2005).
19. M. Ilias and T. Saue, J. Chem. Phys. **126**, 064102 (2007).
20. W. Kutzelnigg and W. J. Liu, J. Chem. Phys. **123**, 241102 (2005).
21. W. J. Liu and W. Kutzelnigg, J. Chem. Phys. **126**, 114107 (2007).
22. W. Kutzelnigg and W. J. Liu, Mol. Phys. **104**, 2225 (2006).
23. G. Hardekopf and J. Sucher, Phys. Rev. A **30**, 703 (1984).
24. R. Brummelhuis, H. Siedentop, and E. Stockmeyer, Doc. Math., J. DMV **7**, 167 (2002).
25. G. Hardekopf and J. Sucher, Phys. Rev. A **31**, 2020 (1985).
26. M. Reiher and A. Wolf, J. Chem. Phys. **121**, 2037 (2004).
27. C. van Wüllen, Chem. Phys. **356**, 199 (2009).
28. ScaLAPACK ist a portable linear algebra library for distributed memory computers. See http://www.netlib.org/scalapack.

Distributed Memory Parallelization of the Multi-Configuration Time-Dependent Hartree Method

Michael Brill, Oriol Vendrell, and Hans-Dieter Meyer

Theoretische Chemie, Physikalisch-Chemisches Institut, Universität Heidelberg,
INF 229, D-69120 Heidelberg, Germany
Hans-Dieter.Meyer@pci.uni-heidelberg.de

1 Introduction

The multiconfiguration time-dependent Hartree (MCTDH) method is an algorithm to efficiently propagate wavepackets. Since its development it has been established as one of the most important methods to study the quantum dynamics, e.g. excitations of vibrations and rotations or scattering events, of molecular systems. The method has been developed in Heidelberg over the last 18 years and is able to treat large systems with $\gtrsim 9$ degrees of freedom in its single processor application. As described in a previous work [1] the code has been parallelized for shared memory hardware and was successfully applied to the investigation of the Zundel cation [2, 3] (15 internal degrees of freedom) using up to 8 processors with speedup factors of >7. However, the Heidelberg MCTDH-code was still lacking the ability to use the more powerful distributed memory hardware of computer clusters. This drawback has been removed recently by applying the Message Passing Interface (MPI) to MCTDH, as described in this report.

The report is structured as follows. The results of the shared memory parallelization of the Heidelberg MCTDH-package are briefly recalled in Section 2. The approach for the distributed memory parallelization of the MCTDH algorithm is discussed in Section 3. The results are shown in Section 4.

2 Equations of Motion and Shared Memory Parallelization of MCTDH

The ansatz for the MCTDH wavefunction reads:

$$\Psi(Q_1,\ldots,Q_f,t) = \sum_{j_1=1}^{n_1} \cdots \sum_{j_f=1}^{n_f} A_{j_1\ldots j_f}(t) \prod_{\kappa=1}^{f} \varphi_{j_\kappa}^{(\kappa)}(Q_\kappa,t) = \sum_J A_J \Phi_J, \quad (1)$$

W.E. Nagel et al. (eds.), *High Performance Computing in Science and Engineering '09*, DOI 10.1007/978-3-642-04665-0_11,

where $\varphi_{j_\kappa}^{(\kappa)}(Q_\kappa, t)$ are the so called single particle functions (SPF) that depend on the κ-th degree of freedom and time. The expansion coefficients $A_{j_1 \dots j_f}(t)$ are time-dependent as well.

In [1] the equations of motion of the MCTDH-algorithm [4–7] have already been introduced:

$$i\dot{\mathbf{A}} = \boldsymbol{\mathcal{K}}\mathbf{A}, \tag{2}$$

$$i\dot{\boldsymbol{\varphi}}^{(\kappa)} = \left(1 - P^{(\kappa)}\right)\left(\boldsymbol{\rho}^{(\kappa)}\right)^{-1}\boldsymbol{\mathcal{H}}^{(\kappa)}\boldsymbol{\varphi}^{(\kappa)}, \tag{3}$$

where $\boldsymbol{\mathcal{K}}$ denotes the matrix representation of the Hamiltonina in the basis of the configurations Φ_J. P is a projector on the space spanned by the single-particle-functions, ϱ is a one-particle density matrix and $\boldsymbol{\mathcal{H}}$ a matrix of mean fields.

These equations can be decoupled in three separate steps if the Constant-Mean-Field-approach, see [8] or Chap. 5.2 in [6], is used:

- Calculation of the Hamiltonian matrix and the mean-fields.
- Propagation of the MCTDH expansion coefficients.
- Propagation of the single particle functions (SPFs).

Since these three tasks represent the mayor part of the runtime of an MCTDH run ($\geq$99%) they have to be considered for the parallelization. Three cases for the runtime distribution among these tasks can be found:

- The balanced one where the runtime is equally distributed.
- The SPF-dominated one where the SPF-propagation needs more then 50% of the runtime.
- The A-vector dominated case where the mean-field computation and the A-vector propagation unify up to 95% or more of the runtime.

Hence for an efficient parallelization of MCTDH for all occurring runtime distributions each part must be parallelized. Unfortunately during the implementation of the shared memory approach it turned out that the SPF-propagation can not be parallelized as efficient as the A-vector propagation. But with growing system size the numerical effort for the SPF-propagation increases slower than the effort for the A-vector propagation and the mean-field computation. This usually results in A-vector dominated cases if the system is large. As the distributed memory parallelization of MCTDH is mainly interesting for large systems it can be restricted to the A-vector propagation and the mean-field computation.

3 Distributed Memory Parallelization of MCTDH

The shared memory parallelization of the A-vector propagation and the mean-field computation has been presented in [1] and will be recalled in brief again.

For the A-vector propagation the time determining step is the evaluation of the right-hand side of the equations of motion:

$$i\dot{A}_J(t) = \sum_L \langle \Phi_J | H | \Phi_L \rangle A_L(t). \tag{4}$$

For an efficient evaluation of the sum MCTDH takes advantage of the product structure of the Hamiltonian. The multidimensional integrals $\langle \Phi_J | H | \Phi_L \rangle$ become a product of one-dimensional integrals. These can be computed separately for each Hamiltonian term. These computations are distributed over the available processors: Different processors compute the integrals for different Hamiltonian terms.

For the mean-field computation the following object must be calculated, the mean field tensor:

$$\mathcal{H}^{(\kappa)}_{rjl} = c_r \sum_J{}^{\kappa} A^*_{J^\kappa_j} \prod_{\lambda=1,\lambda\neq\kappa}^{p} \sum_{l_\lambda} \left\langle \varphi^{(\lambda)}_{j_\lambda} \middle| \hat{h}^{(\lambda)}_r \middle| \varphi^{(\lambda)}_{l_\lambda} \right\rangle A_{L^\kappa_l}. \tag{5}$$

In this case the parallelization can be done similarly to the A-vector propagation: The computations for different Hamiltonian terms, labeled by index r, can be distributed to the available processors.

This parallelization approach has proven its capability to efficiently parallelize A-vector dominated cases (H_2+H_2-scattering [9, 10], $H_5O_2^+$ [2, 3] molecular dynamics). Our *ansatz* for the distributed memory parallelization hence concentrates on this case.

3.1 Parallelization Scheme

For the shared memory parallelization we used POSIX-threads in a scheduler/worker scheme where the main program has the role of a scheduler that does all calculations that are not time consuming and controls the other threads. These other threads are the worker threads that do the time consuming computations in parallel. This scheme has been adopted for the distributed memory parallelization as well. Now a master process controls the slave processes. All non-time consuming computations as well as the SPF-propagation is done by the master process. If the right hand side of equation (4) or the mean-fields (5) have to be calculated the master process distributes the needed data for the computations to the slave processes. Then the work is done in parallel and the results are collected afterwards. This scheme is displayed in Figure 1.

3.2 Combination of the Shared and Distributed Memory Approach

Beside running MCTDH in the pure distributed memory mode it is also possible to combine the distributed and shared memory approach. In this case

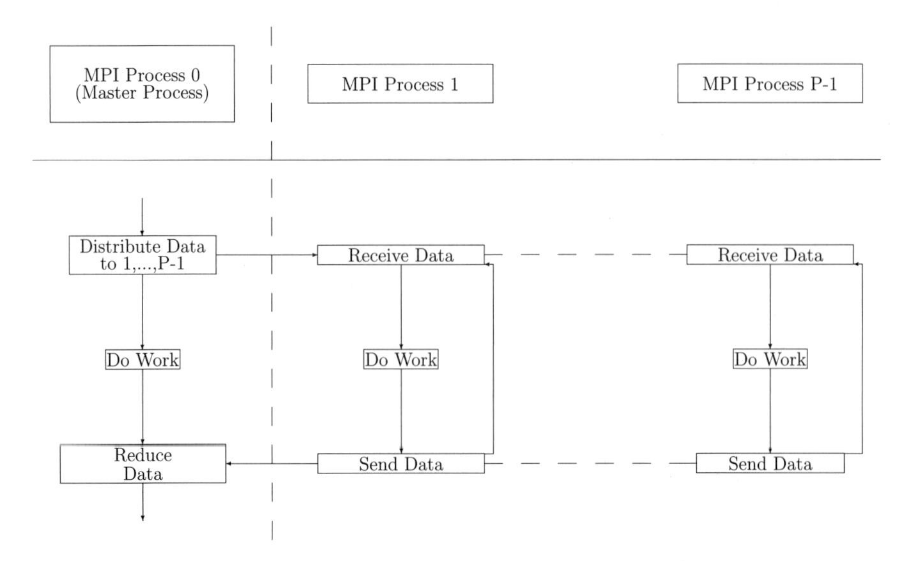

Fig. 1. Organization of the processes in the master/slave approach

each process, the master and the slave processes, additionally use the shared memory parallelization. This means that each MPI-process starts the number of demanded POSIX-threads to further split the computations. This has several advantages. First, the computations that are not considered in the distributed memory parallelization but in the shared memory parallelization are done in parallel in the master process. Second, less memory is needed for the same number of processors used compared to the pure distributed memory approach. This is because in the MPI-parallelization each slave process needs the same amount of memory as the master process, but in the shared memory parallelization each thread only needs a little additionally work space for the storage of intermediate results. Third, the amount of inter process communication is reduced.

3.3 Distributed Memory Parallelization of the Improved Relaxation

A further application of MCTDH beside the wavefunction propagation is the improved relaxation [11, 12]. This variant is used to determine the ground state and some low lying eigenstates and eigenenergies of a quantum system. To do so the Davidson or Block-Davidson algorithm [13] is implemented in MCTDH. This is an iterative method where in each step two new vectors are created that have to be stored in memory. Hence this approach can become unfeasible if the number of vectors needed for convergence is too large to fit in the memory. Thus this part of MCTDH has been parallelized not only to accelerate the computations but also to use the additional memory located on further nodes of the computer cluster. This makes it possible to store more Davidson vectors.

4 Efficiency of the Parallel MCTDH Code

4.1 Benchmark Systems

For the test of the distributed memory parallelization of the MCTDH-code two different A-vector dominated benchmark systems were used. These systems are:

- The H_2+H_2 inelastic scattering [9, 10].
- The quantum molecular dynamics of $H(H_2O)_2^+$ [2, 3].

The size of these two systems differ strongly.

4.2 Propagation Runs

Figure 2 shows the speedup behavior of the MPI-parallel code for the H_2+H_2-scattering calculations measured on the XC2 in Karlsruhe. The first eight data points have been used for a fit to Amdahl's law,

$$S(P) = \frac{1}{q + p/P}, \tag{6}$$

to determine the parallelized part of MCTDH for this system. Here q denotes the strictly serial part of the program and p the parallel one with $p + q = 1$, S is the speedup and P the number of processors. This parallel part is $p = 0.975$ what is in excellent agreement to the part of runtime needed for the A-vector propagation and mean-field computation in a serial run (97.9%). However a further effect is visible if more than 128 processors are used, the speedup is not further increased but decreased. There are two reasons. First, the number of processors is approaching the number of Hamiltonian terms. This means that on each processor only a few steps of the parallelized loop are performed. Hence the workload for each processor might become unbalanced. Second, the computation time drops from 45 minutes on one processor to about one minute on 128 processors. Which means that the time needed for the communication becomes significant for the runtime.

In Figure 3 the speedups for the $H(H_2O)_2^+$-computations are shown. Now not only the pure MPI-parallel version of MCTDH was used but also the hybrid approach combining the shared and distributed memory *ansatz*. The circles show the measured values for the MPI-parallel code, the diamonds and the squares are the values for the combined approach with 2 and 4 POSIX-threads respectively. Each data set was fitted to Amdahl's law, leading to the same parallel part ($p = 0.986$) in spite of their different behavior. The

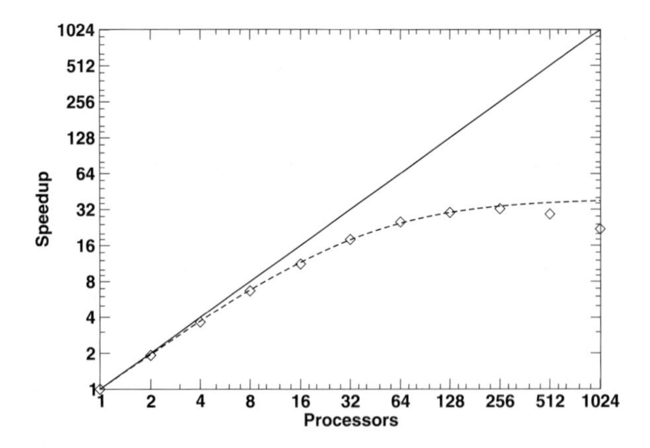

Fig. 2. Speedup of the H_2+H_2-scattering MPI calculations. The diamonds are the measured values, the dashed line is the fit of Amdahl's law to the first eight data points and the solid line is the ideal (linear) speedup

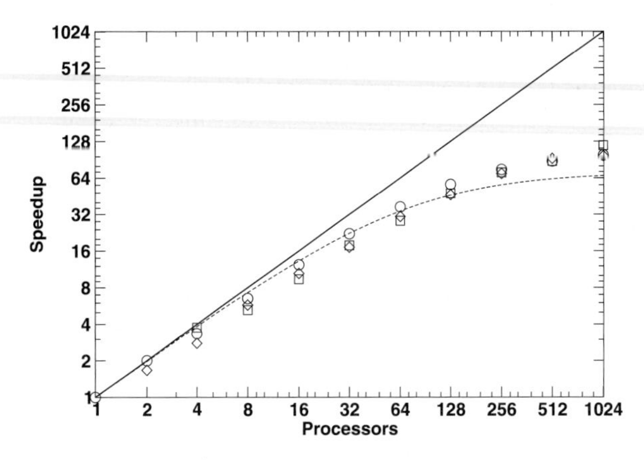

Fig. 3. Speedup of the $H_5O_2^+$ MPI-calculations with 4 POSIX-threads (squares), 2 POSIX-threads (diamonds) and no shared memory parallelization (circles). The circles, diamonds and squares are the measured values, the dashed line is the fit to Amdahl's law (with $p = 0.986$) and the solid line is the ideal (linear) speedup. If Amdahl's law is fitted to the last 3 points only, one determines the parallel part to $p = 0.993$

expected value for the parallel part from the single processor calculation is 99.3% which is still in good agreement with the fitted value. But Amdahl's law does not describe the characteristics of the speedup as well as in the H_2+H_2-scattering case. A comparison between the maximal reached speedup with the single processor calculation suggests a parallel part of $p = 0.993$, the same value as predicted by the one processor test. As one can see up to 128 processors the MPI-parallel version performs better than the hybrid approach.

For higher number of processors the efficiency of the hybrid code catches up and becomes better in some cases, be reminded that the hybrid approach uses considerably less memory.

Beside the advantages of the hybrid approach discussed in Section 3.2 there is unfortunately a drawback. As one can see the speedup of the 2- and 4-thread calculations lies below that one of the pure MPI-computations. The parallelization efficiency is reduced, because – even for the well parallelized A-vector dominated cases – the shared memory parallelization is not as efficient as the distributed memory parallelization. Hence the hybrid parallelization should only be used if a large number of nodes is involved or if memory must be saved. This result is astonishing. We had expected the shared memory approach to be more efficient because of lower communication cost. This was one of the reasons why our first attempt for the parallelization was based on shared memory techniques [1].

4.3 Improved Relaxation Runs

Figure 4 shows the speedups for the $H_5O_2^+$ groundstate relaxation. The circles are the computations where the standard Davidson algorithm is used. All Davidson vectors are stored on one node and just the computation of the preconditioner and the Hamiltonian applied to the A-vector is parallelized. The diamonds show the calculations with the parallel Davidson algorithm where the Davidson vectors are distributed to the nodes to save memory. Both show a parallel part of $p = 0.97$. In the second case more communication is done because each new Davidson vector must be communicated to every

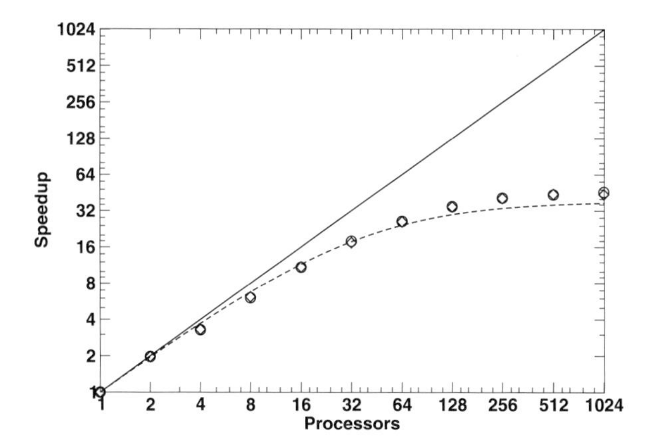

Fig. 4. Speedup of the $H_5O_2^+$ MPI-single-relaxations (no shared memory parallelization). The circles (all Davidson vectors on one node) and diamonds (distributed Davidson vectors) are the measured values, the dashed line is the fit to Amdahl's law (with $p = 0.974$) and the solid line is the ideal (linear) speedup. If Amdahl's law is fitted to the last 3 points only, one determines the parallel part to $p = 0.979$

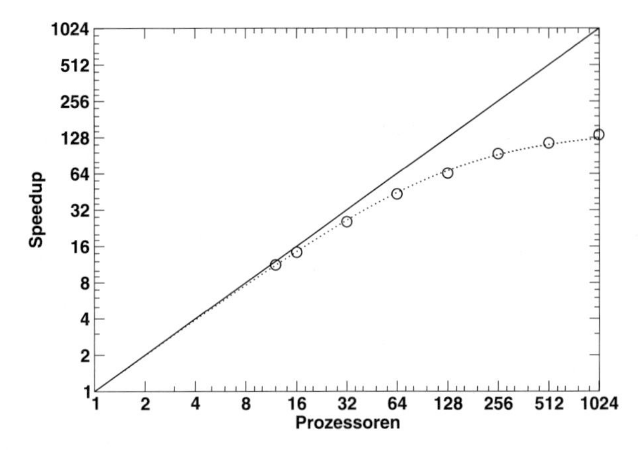

Fig. 5. Speedup of the $H_5O_2^+$ block-relaxations. The diamonds are the measured values for the runtime divided by the fitted value for the serial computation, the dashed line is the fit to Amdahl's law (with $p = 0.993$) and the solid line is the ideal (linear) speedup

node containing old vectors to determine the overlaps. This effect seems to be neglectable.

Further some calculations for the block relaxations were made for the four lowest eigenstates. The same parameters as in the single relaxation case were used. These tests were not feasible with the serial or POSIX-thread parallel code on one node. Hence the MPI-parallel version combined with the shared memory approach was used to have enough memory for the Davidson vectors and to use all available CPUs. The calculation needed at least three nodes to have enough memory for storing the Davidson vectors. To plot the speedup, which is displayed in Figure 5, the data points were fitted to the inverse of Amdahl's law. Now two parameters were sought, the one processor computation time and the parallel part, which was determined to $p = 0.993$. Both relaxation types seem to be well parallelized. For large computations the hybrid approach with distributed Davidson vectors must be used to have enough memory and to exploit all CPUs.

5 Isotopic Effects in the Infrared Spectrum of the Zundel cation

The Zundel cation $H(H_2O)_2^+$ is the smallest protonated water cluster in which an excess proton is shared between two water molecules. Due to the importance of the Zundel structure, and in general of protonated water clusters, in chemistry and biology, this system has been the target of intense investigations during the last years, both from experimental [14–18] and theoretical perspectives [2, 3, 19–22].

5.1 Coordinates and Hamiltonian Operator Setup

The infrared (IR) spectroscopy of the $H(H_2O)_2^+$ form (no deuterated positions) was thoughtfully investigated by full-dimensional quantum dynamics using MCTDH in a series of publications by our group [1–3, 19]. The system was described by a set of 15 internal coordinates based on Jacobi vectors. The derivation of the kinetic energy operator (KEO) in this set of coordinates was discussed elsewhere [2]. To account for the interatomic potential we used the potential energy surface (PES) of Bowman and co-workers, which constitutes the most accurate *ab initio* surface available to date for this system [23]. The potential energy operator (PEO) was constructed as a cut-HDMR [24], also referred to as n-mode representation [25, 26], which was adapted to take advantage of mode combination in the MCTDH wavefunction (*vide infra*). A mode combination based on 5 modes was initially used [2]. We reintroduce here the set of 15 internal coordinates in order to keep the discussion self-contained, which are: the distance between the centers of mass of both water molecules (R). The Cartesian coordinates describing the position of the central proton with respect to the center of mass of the water molecules (x,y,z) where the z coordinate is aligned with the water–water axis and x and y describe the out-of-axis displacements. The Euler angles defining the relative orientation between the two water molecules, waggings: (γ_A, γ_B); rockings: (β_A, β_B); internal relative rotation: (α). The Jacobi coordinates that account for the particular configuration of each water molecule $(R_{1(A,B)}, R_{2(A,B)}, \theta_{(A,B)}))$ where $R_{(A,B)}$ is the distance between the oxygen atom and the center of mass of the corresponding H_2 fragment, $R_{2(A,B)}$ is the H–H distance and $\theta_{(A,B)}$ is the angle between these two vectors.

Several major changes were necessary with respect to the above initial setup in order to deal with deuterated forms of the Zundel cation of general structure E_2O-I-OE_2^+, where E denotes an isotope of H in an external position and I denotes an isotope of H in the internal position. For cations of this general structure the set of internal coordinates introduced above depends on the mass of E, since the position of the center of mass of the water molecules depends on this mass. For increasing mass of E, e.g. deuteration of the external positions, the center of mass of the water molecules is displaced from the vicinity of the O atom towards the H-H axis. This has the undesirable effect of an increased potential coupling between angular (waggings, rockings, internal rotation) displacements of each water molecule, and water-water and central proton motions. This is due to the fact that with increasing mass in the external positions, the aforementioned angular displacements contain a larger component of a change in position of the corresponding oxygen atom in space. This would have no further consequences if it was not for the fact that convergence of a n-mode representation of the potential relies on moderate coupling of the potential in the selected set of coordinates. In the case of the Zundel cation this leads to a complete break-down of the n-mode representation for externally deuterated species. A new set of internal coordinates was therefore introduced, in which the R coordinate now denotes the distance between both oxygen atoms instead of the distance between the centers of mass of both water molecules. The new set of coordinates is based on the mixed Jacobi-valence vectors depicted in Fig. 6. This change alone in the definition

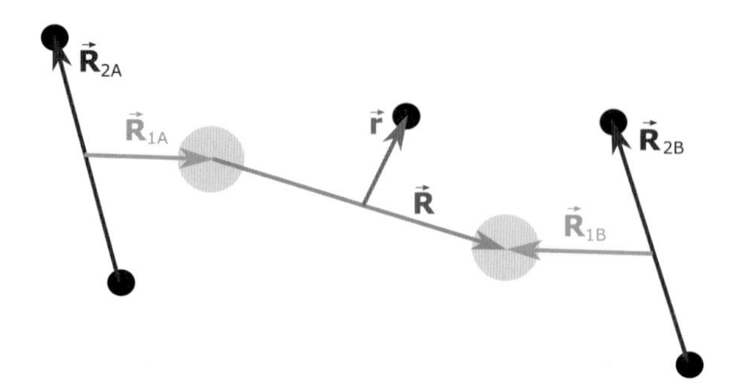

Fig. 6. Mixed Jacobi-valence description of the $H_5O_2^+$ system. The two big circles represent the position of the oxygen atoms, while the small circles represent the positions of the hydrogens

of R introduces new coupling terms to in the KEO, which results in more than doubling its size [27]. The potential coupling is reduced, and a n-mode representation of the PEO provides a good description of the potential again. Additionally, the grids conforming the n-mode representation need to be computed only once and can be used for all isotopologues of the form E_2O-I-OE_2^+, since now the set of polyspherical vectors in Fig. 6 becomes independent of the atomic masses. A final *a posteriori* change is introduced to the system of coordinates, namely the position of the central proton along the O–O axis, z, is normalized with respect to R as

$$z' = \frac{z}{R - 2d_0} \tag{7}$$

where d_0 is the shortest allowed distance between the central proton and one of the oxygen atoms. The new z' coordinate spans the $[-0.5, 0.5]$ range and has no units. This redefinition avoids having in the primitive grid configurations in which the central proton is unphysically close to one of the oxygen atoms. These very high energy configurations occur when z is at the edge of its grid and at the same time the system is in a configuration with short O–O distance. The presence of such configurations in the primitive grid causes problems to both the representation of the potential, due to large correlations, and to its subsequent use in dynamical calculations. The necessary modifications to the KEO in order to account for the change of coordinates in Eq. 7 are discussed in Ref. [27]. From now on we refer to the newly defined z' as z for notational simplicity.

The number of configurations of the MCTDH expansion is drastically reduced by combining physical coordinates into logical coordinates, a method also referred to as mode combination. The 15 degrees of freedom are combined into six logical coordinates Q_i according to the following combination scheme: $Q_1 = [z, R]$, $Q_2 = [\alpha, x, y]$, $Q_3 = [\gamma_A, \gamma_B]$, $Q_4 = [\beta_A, \beta_B]$, $Q_5 = [R_{1A}, R_{2A}, \theta_{1A}]$ and $Q_6 = [R_{1B}, R_{2B}, \theta_{1B}]$. The MCTDH wavefunction thus reads:

$$\Psi(Q_1, \cdots, Q_6, t) = \sum_{j_1}^{n_1} \cdots \sum_{j_6}^{n_6} A_{j_1,\cdots,j_6}(t) \prod_{\kappa=1}^{6} \varphi_{j_\kappa}^{(\kappa)}(Q_\kappa, t). \tag{8}$$

Coordinates which are strongly coupled to each other are good candidates to be combined since their correlation is then taken into account at the level of the single particle functions (SPF) $\varphi_{j_\kappa}^{(\kappa)}$. The correlation between the combined modes is addressed by the multiconfigurational expansion. Mode Q_1 basically accounts for the configuration of the hydrogen bond, modes Q_2, Q_3 and Q_4 contain all angles describing the relative orientation of both water monomers A and B, while Q_5 and Q_6 contain their internal coordinates, respectively. This mode combination scheme differs from the 5-mode scheme used previously [2], in which z was combined together with (α, x, y) while R was combined with (β_A, β_B). Since z and R are strongly correlated, their correlation is better accounted for at the level of SPF dynamics, instead of having to be accounted for by the multiconfigurational expansion. With six combined modes instead of five the typical number of configurations increases by about one order of magnitude. However, the modes are smaller and easier to propagate. Additionally, this scheme is favored by the parallel version of the MCTDH code presented here, since propagation of the vector of coefficients scales better with the number of processors than propagation of the SPFs [1].

The combined coordinates, $(Q_1, \ldots, Q_6)$, used to define the PES representation have been already introduced as they are the same ones used to define the wavefunction. This makes the evaluation of matrix elements of the potential very efficient and enormously speeds up the subsequent dynamical calculations. Details about the numerical scaling of this particular variant of the n-mode approach are found in Ref. [2]. The PES representation used in this work includes all clusters up to second order in the combined coordinates plus the third order clusters correlating Q_1, which defines the hydrogen bond, and Q_2, Q_3 and Q_4, which contain all angular coordinates and the out-of-axis displacements of the central proton. The PES representation thus reads

$$\begin{aligned} V(Q_1, \ldots, Q_6) = &\sum_{i=1}^{6} v_i^{(1)}(Q_i) + \sum_{i<j} v_{ij}^{(2)}(Q_i, Q_j) \\ &+ v_{123}^{(3)}(Q_1, Q_2, Q_3) + v_{134}^{(3)}(Q_1, Q_3, Q_4) \\ &+ v_{124}^{(3)}(Q_1, Q_2, Q_4) + v_{234}^{(3)}(Q_2, Q_3, Q_4). \end{aligned} \tag{9}$$

As detailed elsewhere [2], the clusters are constructed by averaging over various reference geometries to guarantee that the cluster expression has the full symmetry. A detailed investigation of the error introduced by this representation of the PES with respect to the reference one has been carried on and is detailed in Ref. [27]. For the sake of completeness we mention that 2nd and 3rd order clusters in Eq. 9, which correspond to 4-, 5- , 6- and 7-dimensional grids in the set of 15 internal coordinates, are transformed to product form using the potfit algorithm [28–30]. This results in a much more compact form of the PEO and makes MCTDH calculations much more efficient [2, 27].

5.2 IR Spectrum of $D(D_2O)_2^+$, $H(D_2O)_2^+$ and $D(H_2O)_2^+$ Isotopologues

The interpretation of the IR spectra requires definite assignments of the spectral lines and an understanding of their origin. Zeroth-order states are used as a tool to perform such assignments. They correspond to well defined local excitations of the system, e.g. the bending mode of the water molecules or the one-quantum excitation of the proton-transfer mode, and they are constructed as products of eigenfunctions of low dimensional Hamiltonians. A more specific definition and procedures to obtain them in the context of MCTDH was presented elsewhere [3]. In the following, $|\Phi_l\rangle$ refers to the vibrational wavefunction of a zeroth-order state while $|\Psi_m\rangle$ corresponds to a vibrational eigenstate. The quantities used for assignments are the $|\langle\Phi_l|\Psi_m\rangle|^2$ products, which tell to which extent a particular and well defined zeroth-order vibration participates in a certain spectral line. Even though each line contains contributions from all or some of the considered zeroth-order states (non-vanishing $|\langle\Phi_l|\Psi_m\rangle|^2$ elements), there is usually a zeroth-order state that contributes to a specific transition appreciably more than the others. Thus, when we refer to a certain spectral line as the (lab) transition or to the corresponding eigenstate as $|\Psi_{lab}\rangle$, it is because it is possible to identify the zeroth-order state $|\Phi_{lab}\rangle$ as the *leading* contribution to $|\Psi_{lab}\rangle$. In the case of very large coupling it may not be possible to cleanly disentangle the spectrum into one-to-one assignments of spectral peaks to zeroth-order states, since a given transition may present similar contributions from two or more zeroth-order states. In such a case the matrix containing the $|\langle\Phi_l|\Psi_i\rangle|^2$ loses its diagonal dominance and non-diagonal elements denoting the mixture of various zeroth-order states in the description of eigenstates become non-negligible. The most important $|\langle\Phi_l|\Psi_i\rangle|^2$ elements for the four considered isotopologues are given elsewhere [21].

Fig. 7 presents the IR spectra of $H(H_2O)_2^+$, $D(D_2O)_2^+$, $H(D_2O)_2^+$ and $D(H_2O)_2^+$. The IR spectrum of the $H(H_2O)_2^+$ isotopologue was extensively discussed in References [19] and [3]. It is briefly reintroduced here since it serves as a reference to understand the spectra of the other isotopologues. The lowest frequency parts of the four spectra are composed of two lines related to the 1-quantum wagging motions and its combination with the internal rotation motion of one of the monomers with respect to the other. The highest

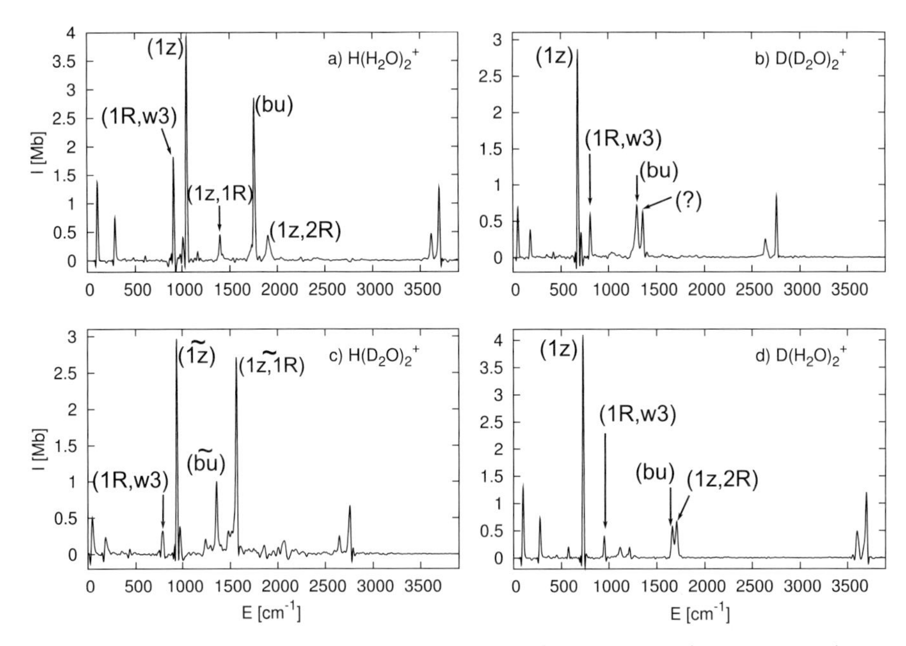

Fig. 7. Computed IR spectra of the a) $H(H_2O)_2^+$, b) $D(D_2O)_2^+$, c) $H(D_2O)_2^+$ and d) $D(H_2O)_2^+$ isotopologues of the Zundel cation with the corresponding assignments for the most intense absorptions

frequency parts of the spectra are composed of two bright lines related to the terminal O–H vibrations of the water molecules. Neither the assignment of the just discussed peaks in the lowest and highest energy domain, nor the relative positions between each other, change after deuteration. Only the expected red-shifts take place for $D(D_2O)_2^+$ and $H(D_2O)_2^+$, i.e. when the terminal hydrogens are substituted by deuterium atoms.

The situation turns out to be much more complex in the spectral region between 600 and 2000 cm^{-1}. The middle range spectrum of the $H(H_2O)_2^+$ cation in Figure 7a features five clearly visible absorptions in the range between 900 and 1900 cm^{-1}. They could be assigned and explained as arising from a set of five coupled zeroth-order states composed of: [3, 19] $|\Phi_{1R,w_3}\rangle$, a combination of two modes, a two-quanta asymmetric wagging (w_3) mode the one-quantum ($1R$) mode, where R is the O–O stretching coordinate; $|\Phi_{1z}\rangle$, the one-quantum asymmetric proton stretch along the central O–O axis (z refers the proton position along the O–O axis); $|\Phi_{1z,1R}\rangle$, the combination of the $1z$ and $1R$ excitations; $|\Phi_{1z,2R}\rangle$, the combination of the $1z$ and two-quanta O–O excitations; $|\Phi_{bu}\rangle$, the *ungerade* water-bending mode.

The most intense line of the $H(H_2O)_2^+$ spectrum centered at 1040 cm^{-1} is related to the ($1z$) transition since the displacement of the central proton along the O–O axis causes the largest variation of the dipole moment. Thus the $|\Phi_{1z}\rangle$ zeroth-order state has a large contribution to this eigenstate, but the second most important contribution to this line arises from the $|\Phi_{1R,w_3}\rangle$ zeroth-order state. The situation is reversed for the transition centered at 915 cm^{-1}, whose leading contribution is$|\Phi_{1R,w_3}\rangle$ and the second most important one is $|\Phi_{1z}\rangle$. Therefore the doublet of peaks centered at about 1000 cm^{-1} in the $H(H_2O)_2^+$ arises from a Fermi resonance between the strongly coupled, zeroth-order states $|\Phi_{1z}\rangle$ and $|\Phi_{1R,w_3}\rangle$ [3, 19, 20]. The next three lines of the spectrum correspond to the 1415 cm^{-1} $(1z, 1R)$, 1750 cm^{-1} (bu) and 1905 cm^{-1} $(1z, 2R)$ transitions, respectively. All three transitions, and specially (bu), have a non-negligible contribution from the $|\Phi_{1z}\rangle$ zeroth-order state, from which they obtain a large part of their spectral intensity [3]. Moreover, the strong coupling between the $|\Phi_{1z}\rangle$ and $|\Phi_{bu}\rangle$ modes is responsible for shifting the $(1z)$ and (bu) lines about 150 cm^{-1} down and up, respectively, with respect to their estimated uncoupled positions [3, 20].

The IR spectrum of $D(D_2O)_2^+$ is shown in Figure 7b. The $(1z)$ peak is found here at 678 cm^{-1} and is, as in $H(H_2O)_2^+$, the most intense IR absorption. The $(1R, w_3)$ peak is found at 807 cm^{-1}. Therefore the characteristic doublet at about 1000 cm^{-1} in $H(H_2O)_2^+$ is also found in $D(D_2O)_2^+$, but with its constituent peaks in reverse order. The (bu) peak is found at 1298 cm^{-1}, about 450 cm^{-1} below its position in $H(H_2O)_2^+$. Neither $(1z, 1R)$ nor $(1z, 2R)$ peaks appear in the spectrum of $D(D_2O)_2^+$. The position of these two eigenstates has been computed to be 1150 and 1600 cm^{-1}, respectively. Therefore they are located far from absorptions from which they could borrow intensity. Moreover, after deuteration the coupling between z (proton position) and R (O–O distance) is reduced since the system remains in deeper, less anharmonic regions of the potential, thus reducing even more the possibility for direct absorption of the $(1z, 1R)$ and $(1z, 2R)$ combinations. Close to the (bu) peak a line is found, which is still unassigned. This unknown state borrows some intensity from (bu) in order to become bright in $D(D_2O)_2^+$. This absorption is also seen as a shoulder to the (bu) peak in experimental spectra in this region. [18] Its exact nature will be addressed in a forthcoming publication.

The most complex of all considered spectra is $H(D_2O)_2^+$, shown in Figure 7c. Here the deuteration of the external positions brings the position of the zeroth-order state $|\Phi_{bu}\rangle$ down to lower frequencies by about 300 cm^{-1}, while the zeroth-order states $|\Phi_{1z}\rangle$ and $|\Phi_{1z,1R}\rangle$ remain almost unaffected. This results in a situation in which the zeroth-order $|\Phi_{bu}\rangle$ is found between $|\Phi_{1z}\rangle$ and $|\Phi_{1z,1R}\rangle$. These three zeroth-order states strongly couple to each other and are responsible for the triplet absorption with peaks at 938, 1355 and 1564 cm^{-1}. The peak at 938 cm^{-1} has almost equal relative contribution from $|\Phi_{1z}\rangle$ and $|\Phi_{bu}\rangle$. The central peak at 1355 cm^{-1} has almost equal participation from $|\Phi_{1z}\rangle$, $|\Phi_{bu}\rangle$ and $|\Phi_{1z,1R}\rangle$, while the peak at 1564 cm^{-1} is a mixture of $|\Phi_{bu}\rangle$ and $|\Phi_{1z,1R}\rangle$ with a slightly larger participation of the latest. The strong couplings shaping the middle region of the spectrum are reflected in the loss of diagonal dominance of the matrix composed of the $|\langle\Phi_l|\Psi_m\rangle|^2$ elements (see Table in supporting information) for $H(D_2O)_2^+$. The use of a tilde in these three assignments in Figure 7c indicates that the tag assignments are a bit arbitrary because of the strong mixing of underlying zeroth-order states, in contrast to other assignments in which one zeroth-order state is mainly responsible for a given peak. The $(1R, w_3)$ peak is of reduced intensity due to its red shift and consequent decoupling from the zeroth-order $(1z)$. The $(1z, 2R)$ state is located far from peaks from which it can borrow intensity and shows no IR absorption.

In contrast to $H(D_2O)_2^+$, deuteration of the central position alone in $D(H_2O)_2^+$ leads to the simplest and most diagonally dominant IR spectrum of this series. Here the zeroth-order $(1z)$ shifts to lower frequencies and decouples from $(1R, w_3)$. The peak at 785 cm^{-1} is cleanly assigned to $(1z)$. $(1z)$ decouples as well from (bu). This brings the position of the (bu) peak down to 1662 cm^{-1}, closer to the bending frequency of an isolated water molecule and explains the reduced intensity. Note that the position of (bu) at 1758 cm^{-1} in $H(H_2O)_2^+$ was due to strong coupling with the central proton $(1z)$ mode. Due to the isotopic substitution the $(1z, 2R)$ peak is shifted down, ending relatively close to (bu). The doublet formed by (bu) and $(1z, 2R)$ is the only structure related to a resonance in $D(H_2O)_2^+$. However, both lines can be cleanly assigned, as seen by inspecting the contribution of the corresponding zeroth-order states.

6 Conclusions

The Heidelberg MCTDH-code has been parallelized using the Message Passing Interface to make distributed memory hardware accessible. The already known scheduler/worker shared memory parallelization scheme has been adapted in a master/slave approach for the MPI-parallelization. Two of the three main tasks of the MCTDH propagation, the propagation of the A-vector and the computation of the mean fields, are parallel now. Hence the so called A-vector dominated computations can be done in parallel on distributed memory hardware. The resulting code was tested on benchmark systems to determine the efficiency of the parallelization. It turned out, that the code is well parallelized for A-vector dominated systems involving a large number of Hamiltonian-terms. The degree of parallelization lies close to $p = 0.99$ leading to speedups of up to 120 when running on 1024 processors or 55 on 128 processors. The distributed memory parallelization of MCTDH can be used in two different modes. In the first case only the MPI-parallelization is used in the second case the MPI- and the POSIX-thread-parallelization are combined. The second variant can be used to save memory on a node - shared memory parallelization needs less memory than distributed memory parallelization - and to save communication costs especially for applications involving a large number of processors.

Further, not only the computations were made faster by the MPI-parallelization but also the memory of all used nodes was made accessible for the improved Relaxation runs. Now the iteration order can be significantly increased for single- and block-relaxation runs. Because the number of Davidson vectors needed for convergence increases with excitation energy, more eigenstates may now be computed. This holds in particular for block-improved-relaxation runs.

The latest developments in the parallel MCTDH code described here have been applied to the dynamics and IR spectroscopy of isotopically substituted forms of the Zundel cation. We present assignments for the most important peaks of the IR spectra of $D(D_2O)_2^+$, $H(D_2O)_2^+$ and $D(H_2O)_2^+$ isotopologues in the spectral range 0–4000 cm^{-1}. Dramatic differences are found between the peaks of the various isotopologues, which are a consequence of the soft, anharmonic, and coupled interatomic potential of the Zundel cation. Precise assignments can be reached for these clusters. The calculations present important fundamental information on the dynamics and spectroscopy of protonated water clusters and represent a step forward in our understanding of these complex systems.

The successful application of the shared- and distributed-memory implementations of the MCTDH algorithm to the complicated dynamics of the Zundel cation, which is accurately solved in its full dimensionality, demonstrates the tremendous potential of these new tools for dealing with a large variety of complex physical and chemical problems from a full quantum perspective. The parallelization of MCTDH and its application to the dynamics of the Zundel cation took a comparable manpower cost. This stresses the importance of development, implementation and improvement of computing algorithms to make them efficient for the most modern hardware, as an integral part of the scientific task, rather than a simple byproduct of it.

Acknowledgments. The authors thank Prof. J. Bowman for providing the potential-energy routine, F. Gatti for deriving the kinetic energy operator, D. Lauvergnat for help in testing the kinetic energy operator with the TNUM program and the Scientific Supercomputing Center Karlsruhe for generously providing computer time. O.V. is grateful to European Commission for financial support through the Marie Curie (FP6) program.

References

1. M. Brill, O. Vendrell, F. Gatti, and D. Meyer, in *High Performance Computing in Science and Engineering 07*, edited by W. E. Nagel, D. B. Kröner, and M. Resch (Springer, Heidelberg, 2008), pp. 141–156.
2. O. Vendrell, F. Gatti, D. Lauvergnat, and H.-D. Meyer, J. Chem. Phys. **127**, 184302 (2007).
3. O. Vendrell, F. Gatti, and H.-D. Meyer, J. Chem. Phys. **127**, 184303 (2007).
4. H.-D. Meyer, U. Manthe, and L. S. Cederbaum, Chem. Phys. Lett. **165**, 73 (1990).
5. U. Manthe, H.-D. Meyer, and L. S. Cederbaum, J. Chem. Phys. **97**, 3199 (1992).
6. M. H. Beck, A. Jäckle, G. A. Worth, and H.-D. Meyer, Phys. Rep **324**, 1 (2000).

7. H.-D. Meyer and G. A. Worth, Theor. Chem. Acc. **109**, 251 (2003).
8. M. H. Beck and H.-D. Meyer, Z. Phys. D **42**, 113 (1997).
9. F. Otto, F. Gatti, and H.-D. Meyer, J. Chem. Phys. **128**, 064305 (2008).
10. A. N. Panda, F. Otto, F. Gatti, and H.-D. Meyer, J. Chem. Phys. **127**, 114310 (2007).
11. H.-D. Meyer, F. Le Quéré, C. Léonard, and F. Gatti, Chem. Phys. **329**, 179 (2006).
12. L. J. Doriol, F. Gatti, C. Iung, and H.-D. Meyer, J. Chem. Phys. **129**, 224109 (2008).
13. E. Davidson, J. Comp. Phys. **17**, 87 (1975).
14. K. R. Asmis *et al.*, Science **299**, 1375 (2003).
15. T. D. Fridgen *et al.*, J. Phys. Chem. A **108**, 9008 (2004).
16. J. M. Headrick, J. C. Bopp, and M. A. Johnson, J. Chem. Phys. **121**, 11523 (2004).
17. N. I. Hammer *et al.*, J. Chem. Phys. **122**, 244301 (2005).
18. L. McCunn, J. Roscioli, M. Johnson, and A. McCoy, J. Phys. Chem. B **112**, 321 (2008).
19. O. Vendrell, F. Gatti, and H.-D. Meyer, Angew. Chem. Int. Ed. **46**, 6918 (2007).
20. O. Vendrell and H.-D. Meyer, Phys. Chem. Chem. Phys. **10**, 4692 (2008).
21. O. Vendrell, F. Gatti, and H.-D. Meyer, Angew. Chem. Int. Ed. **48**, 352 (2009).
22. A. B. McCoy *et al.*, J. Chem. Phys. **122**, 061101 (2005).
23. X. Huang, B. J. Braams, and J. M. Bowman, J. Chem. Phys. **122**, 044308 (2005).
24. O. F. Alis and H. Rabitz, J. Math. Chem. **25**, 197 (1999).
25. S. Carter, S. J. Culik, and J. M. Bowman, J. Chem. Phys. **107**, 10458 (1997).
26. J. M. Bowman, S. Carter, and X. Huang, Int. Rev. Phys. Chem. **22**, 533 (2003).
27. O. Vendrell *et al.*, J. Chem. Phys. **submitted**, (2009).
28. A. Jäckle and H.-D. Meyer, J. Chem. Phys. **104**, 7974 (1996).
29. A. Jäckle and H.-D. Meyer, J. Chem. Phys. **109**, 3772 (1998).
30. J. M. Bowman, T. Carrington Jr., and H.-D. Meyer, Mol. Phys. **106**, 2145 (2008).

Development of Models for Large Molecules and Electrolytes in Solution for Process Engineering

Jonathan Walter[1], Stephan Deublein[1], Jadran Vrabec[2], and Hans Hasse[1]

[1] Lehrstuhl für Thermodynamik, Technische Universität Kaiserslautern, Erwin-Schrödinger-Str. 44, 67663 Kaiserslautern, Germany
jonathan.walter@mv.uni-kl.de
[2] Thermodynamik und Energietechnik, Universität Paderborn, Warburger Straße 100, 33098 Paderborn, Germany

1 Introduction

Two of the most challenging tasks in molecular dynamics simulation are the simulation of long-range interactions like in electrolyte solutions and the internal degrees of freedom like in hydrogels. Both tasks lead to time consuming simulations with big systems. Therefore, massively parallel high performance computers are needed for both tasks. The first step of the present work was to test and validate different force fields for electrolyte solutions and hydrogels.

2 Simulation of Electrolyte Solutions

The characterization of electrolyte systems is computationally very expensive. The ionic influence, noticeable even at long distances away from the charge, requires large simulation systems and the use of special, time consuming algorithms that allow for a truncation of the ionic interactions at length scales accessible in molecular simulation. Examples of such algorithms are the classical Ewald-Summation [9] and its derivates like Particle-Mesh-Ewald-Summation [7] or Particle-Particle Particle-Mesh method [8].

In the early stages of research on electrolyte systems, the effort was mainly directed to the development of ion models, which were capable of reproducing static structural properties of the ionic solutions [27, 10, 6]. Recently, however, such models have been proven to be too inaccurate for the prediction of basic thermodynamic properties of the electrolyte solution [12]. Since the models were parametrized solely at short distances, the long distance effects of the ions are underestimated. These effects can be seen in particular in the density and the activity of the salt mixture.

W.E. Nagel et al. (eds.), *High Performance Computing in Science and Engineering '09*, DOI 10.1007/978-3-642-04665-0_12,

Our recent research effort has evaluated numerous single charge ion models regarding their ability to predict the density of ionic solutions. Currently, it is focused on the development of molecular ion models that are capable to reproduce and predict thermodynamic properties of ionic systems at various temperatures and ion concentrations.

2.1 Methods

The first investigated system consists of sodium and chloride ions as electrolytes and explicit water as solvent. The ions were modelled as Lennard-Jones spheres with a central charge. Water models were taken from literature. We determine the density of the ionic solution in isobaric-isothermal (NPT) Monte-Carlo simulations with 1000 particles at ambient temperature and pressure. The long-range charge-charge interactions were calculated using Ewald summation with a cutoff radius of half the box length L and the Ewald parameter κ of $5.6/L$. The simulation program employed was *ms*2, which is developed by our group.

2.2 Results

The results for the density of electrolyte systems are highly dependent on the uncertainties of density of the pure solvent itself. To diminish these influences and for a better comparison between different solvent models we show all results in reduced form $\widetilde{\rho}$, normalizing the density of the electrolyte system ρ_e by the density of pure solvent ρ_s.

$$\widetilde{\rho} = \frac{\rho_e}{\rho_s} \tag{1}$$

Influence of Water Model

At first, the influence of the molecular model for the solvent on the density of the ionic system was investigated. This study is limited to the three most established water models, namely SPC [5], SPC/E [2], and TIP4P [14], and is performed at ambient temperature of 25 °C, covering a wide range of salt concentrations in solution. The sodium chloride ions were described by the force field of Balbuena et al. [3], which is widely in use. Their parameters, shown in Table 1, were kept constant for all simulations.

The calculations of the electrolyte solutions at various ion concentrations show that their density in reduced form is independent of the solvent model, cf. Figure 1. It was confirmed that this behavior is not only valid for the ion model of Balbuena et al., but for other models of sodium chloride and for other ions as well (not shown here).

The independence of $\widetilde{\rho}$ from the solvent model facilitates the evaluation of literature models for ions as well as the development of own ion models significantly. In future studies, only one solvent description has to be considered for the characterization of an electrolyte system.

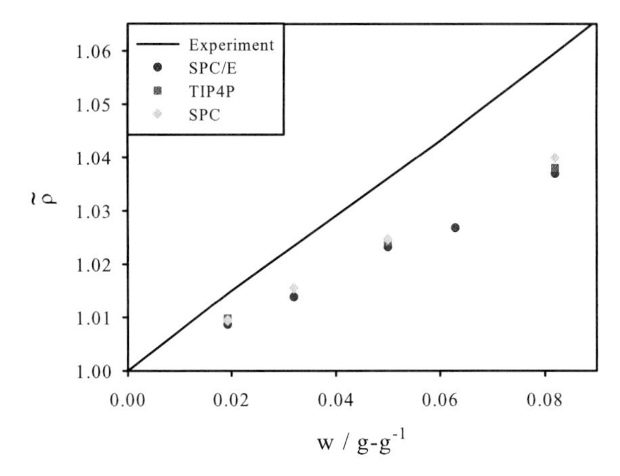

Fig. 1. Plot of the reduced density $\widetilde{\rho}$ of an aequous sodium chloride solution as a function of the salt concentration, given in weight fractions w. Different symbols indicate different solvent models

Model Evaluation and Development

In a subsequent study, different ion models were investigated regarding their ability to predict densities of electrolyte solutions over a wide range of ion concentrations. A total of four different models, listed in Table 1, were analyzed. The ions were completely dissolved in SPC/E-water.

Table 1. Different ion models for sodium chloride from literature

Model	Ion	σ [Å]	ϵ/k_B [K]	q
Balbuena et al. [3]	Na^+	2.69	61.87	+ 1
	Cl^-	3.79	68.07	− 1
Lenart et al. [20]	Na^+	2.443	14.329	+ 1
	Cl^-	3.487	117.76	− 1
Wheeler et al. [34]	Na^+	2.35	55.30	+ 1
	Cl^-	4.42	54.2	− 1
Straatsma et al. [28]	Na^+	2.85	24.12	+ 1
	Cl^-	3.75	64.69	− 1
Jensen et al. [15]	Na^+	4.07	0.06	+ 1
	Cl^-	4.02	85.383	− 1

All ion models in the study capture the characteristic linear increase of density with increasing sodium chloride concentration in solution, as it is clearly visible in Figures 1 and 2. However, the simulation results deviate from experimental values considerably for the normalized density. A possible reason for this abberation are the problems in describing the solvent "water"

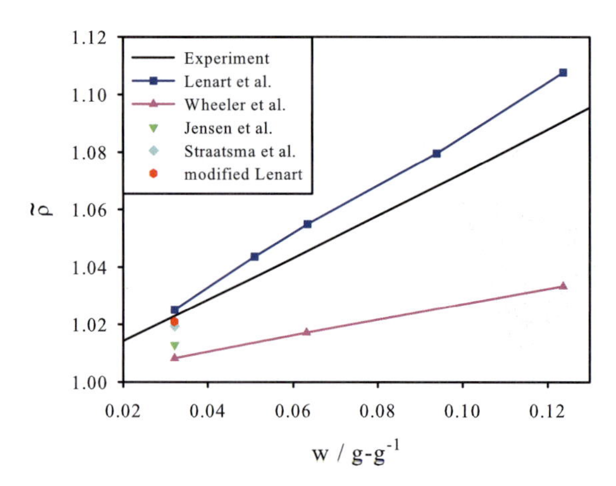

Fig. 2. Plot of the normalized density $\tilde{\rho}$ of an aequous sodium chlorid solution as a function of the salt concentration, given in mass fractions w. Different symbols indicate different ion models

accurately. The errors of its model influence the parametrization of the ion, which causes the deviation visible in our analysis.

First results indicate that it is possible to improve the results by fitting the ion force fields to the density data, also shown in Figure 2. This is currently explored in ongoing work.

2.3 Conclusions

Current models of simple single charged ions are not yet capable of reproducing basic thermodynamic properties quantitatively. Further research effort on the model development is needed to gain accurate predictions for thermodynamic properties of electrolyte solutions. Our approach of a quantitative characterization of electrolyte systems with normalized densities facilitates and accelerates this process. The amount of single simulations necessary for the model development is decreased significantly without loss of accuracy.

2.4 Computational Demands

All simulations presented in Section 2.2 were carried out with the MPI based molecular simulation program *ms*2 developed in our group. Typical simulation runs employ 8 CPUs running for 96 hours. For model optimization, a large number of independent simulations are necessary and can be performed in parallel. For the prediction of other thermodynamic data like activity coefficients, up to 32 CPUs running for 96 hours or more depending on the system are required.

3 Simulation of Hydrogels

Hydrogels have been a research topic of growing interest in recent years. Some of the applications for hydrogels are absorbers, drug delivery [26], microvalves, microactuators [30] or immobilised catalysts. There is substantial research in the field of hydrogels, however, these promissing materials are not yet fully understood.

In this work, the behavior of the Poly(n-isopropylacrylamide) (PNiPAM) single chain is examined by molecular dynamics simulation on the atomistic level. PNiPAM is used here as it is one of the better known hydrogels and shows a significant temperature dependence with respect to swelling. Early experimental investigations on PNiPAM polymers in aqueous solutions [11] were done in 1968. E.g. the lower critical solution temperature (LCST) of PNiPAM in water of about 305 K was found.

With molecular dynamics it is possible to predict the behavior of hydrogels. In the literature the hydrogels are simulated at two different levels of detail, coarse-grain models and atomistic models. With the coarse-grain models the influence of solvent [18] or ions in the solvent on the hydrogel [22, 19] was investigated. The results of these simulations are different states of conformation for the hydrogel and the transition between them.

Atomistic models need more computer resources but the full detail of the simulated hydrogel can be analysed. The advantage of atomistic models is that the dynamic properties like diffusion coefficient the complete structure of the hydrogel like hydration shell or hydrogen bonds can be obtained. For polymers, atomistic simulations with e.g. PAA [24], Polyvinylalcohol (PVA) [25, 23], PNiPAM [21, 29] can be found in literature. In these investigations mainly the properties of water like first hydration shell, diffusion coefficient, orientation or hydrogen bonds were evaluated.

These investigations were mostly done with small systems of up to 1000 solute molecules. Therefore, it was not possible to simulate polymers and hydrogels of realistic size. There also seem to be no validations of different force fields for hydrogels in the literature. The second problem is that in many cases the comparison with experiments is difficult or not attempted at all. The approach of this work is to find a suitable force field for PNiPAM and to compare the simulation results to experimental data. For this approach different force fields from the literature were tested with single PNiPAM chains in water at different temperatures.

3.1 Methods and Molecular Models

The simulations for the PNiPAM single chains were all performed with version 4.0.3 of the GROMACS simulation package [31, 13]. The adventage of GROMACS is the optimized code for single processors and also massively parallel simulation. Three force fields for the polymer PNiPAM Gromos-87 [32], Gromos-96 53a6 (G53a6) [33] and OPLS-AA [17, 16] and two models for

Table 2. Forcefield parameters for PNiPAM

Atom	Gromos-87			G53a6			opls-aa		
	σ / nm	ϵ / J	charge	σ / nm	ϵ / J	charge	σ / nm	ϵ / J	charge
CH2	0.3965	0.5856	0.00	0.4070	0.4105	0.00	0.3500	0.2761	-0.12
CH	0.6211	0.0544	0.00	0.5019	0.0949	0.00	0.3500	0.2761	-0.06
C	0.3361	0.4058	0.38	0.3581	0.2774	0.45	0.3750	0.4393	0.50
O	0.2626	1.7245	-0.38	0.2760	1.2791	-0.45	0.2960	0.8786	-0.50
N	0.2976	0.8767	-0.28	0.3136	0.6398	-0.31	0.3250	0.7113	-0.50
H	0.0000	0.0000	0.28	0.0000	0.0000	0.31	0.0000	0.0000	0.30
CH	0.6211	0.0544	0.00	0.5019	0.0949	0.00	0.3500	0.2761	-0.14
CH3	0.3786	0.7533	0.00	0.3748	0.8672	0.00	0.3500	0.2761	-0.18
H in CH_x	-	-	-	-	-	-	0.0000	0.0000	0.06

water SPC/E [2] and TIP4P [14] for water were used in this work. Gromos96 and G53a6 are united-atom force fields, OPLS-AA is an all-atom force field. The parameters of the PNiPAM force fields are listed in Table 2.

For all force fields a mixing rule with the geometrical mean for both σ and ϵ was used.

$$\sigma_{ij} = \sqrt{\sigma_i \cdot \sigma_j} \tag{2}$$

$$\epsilon_{ij} = \sqrt{\epsilon_i \cdot \epsilon_j} \tag{3}$$

To test the suitability of the force fields two issues were tackled here. The force field should be able to show the temperature dependence on the conformation and both the collapse and the stretching of single PNiPAM chain. Simulations with all six combinations of the polymer/water force fields were performed at the temperatures 280 and 360 K. These temperatures are well below and well above the LCST. At the lower temperature, the single chain should be in a stretched conformation, at the higher temperature it should be in a collapsed conformation. The simulations were started both from stretched and from collapsed initial conformations.

For equilibration, the single chain in water was simulated in the NPT ensemble over 100 000 timesteps. The Berendsen barostat [4] and the velocity-rescale thermostat [1] were used. The pressure was 1 bar and the timestep length 1 fs for all simulations. After equilibration, 1 to $3 \cdot 10^7$ production time steps were performed. The initial configurations for different degrees of collapse were taken from different temporal stages of the simulation at 360 K.

To analyze the results, the radius of gyration was calculated in all simulations. The radius of gyration was defined as

$$R_g = \left(\frac{\Sigma_i ||r_i||^2 m_i}{\Sigma_i m_i} \right)^{1/2} \tag{4}$$

and characterizes the degree of stretching of polymer chains.

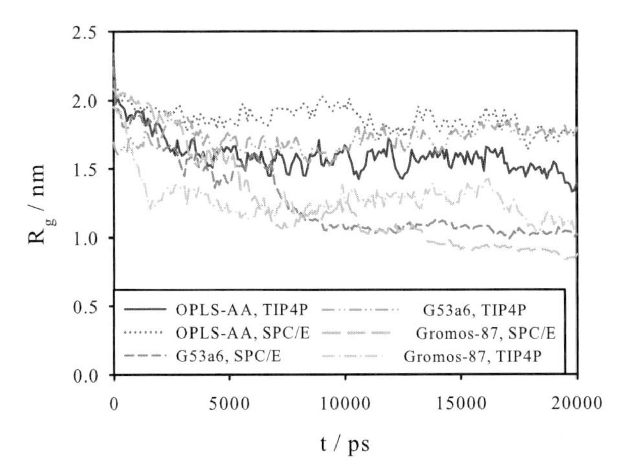

Fig. 3. Radius of gyration for different force field combinations at 280 K over simulation time. The simulations were performed with a chain of 30 monomers of PNiPAM and 14 482 water molecules

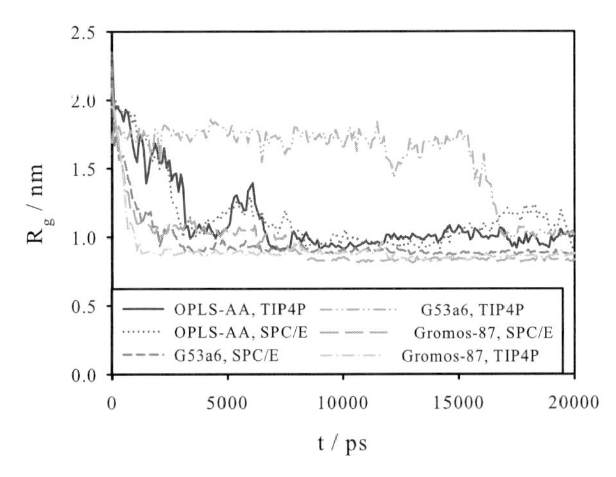

Fig. 4. Radius of gyration for different force field combinations at 360 K over simulation time. The simulations were performed with a chain of 30 monomers of PNiPAM and 14 482 water molecules

3.2 Results

Figures 3 and 4 show the simulation results for PNiPAM with the different force fields at 280 and 360 K, being below and above the LCST. In Figure 3 the simulation results at 280 K are presented. At this temperature the single chain in equilibrium should be in a stretched conformation. Only two force field combinations, i.e. OPLS-AA with SPC/E and G53a6 with TIP4P, are able to produce the stretched configuration at 280 K.

Figure 4 shows the results for the same type of simulations at 360 K. At this temperature, the single chain should be in a collapsed configuration. All force field combinations produce the collapsed conformation for this temperature. The only noticeable difference in these simulations is that in the simulation with the force fields G53a6 and TIP4P the conformation transition occurs later in the simulation. 360 K is near the LCST that this force field shows for PNiPAM (results are not shown here). That is the reason for the late conformation transition in the simulation. The lowest line for 280 K shows that once completely collapsed the chain will not unfold again in reasonable time.

This comparison of the different force fields shows two results: The two force field combinations that are able two yield both the stretched and the collapsed configuration of the single chain are OPLS-AA with SPC/E and G53a6 with TIP4P. Furthermore, the force field combinations that predict qualitatively correct conformation changes are not suggested by the authors. Note that the OPLS-AA force field was developed for TIP4P water and G53a6 for SPC/E water.

Secondly, it was shown that the force fields are able to show conformation transitions in both directions. For this test, only the force fields OPLS-AA with SPC/E and G53a6 with TIP4P were used. The results of these simulations can be seen in Figures 5 and 6. Figure 5 shows simulation results of single PNiPAM chain with OPLS-AA and SPC/E water. The two lines at the top and at the bottom are the simulations starting in the stretched configuration. In the center are simulations starting with configurations in between the stretched and the collapsed single chain. These simulations were performed at 280 K in order to test the conformation transition from the collapsed to the stretched single chain. The result is that this conformation transition can be simulated but is slower than the conformation transition in the other direction.

The same simulations for G53a6 with TIP4P water can be seen in Figure 6. For this force field combination, the transition of the conformation is fast in both directions.

The comparison of the two force field combinations OPLS-AA with SPC/E and G53a6 with TIP4P shows that both force fields are able to predict the conformation transition in both directions. And for both force field combinations the radius of gyration in equilibrium for 280 K is about 1.8 nm and for 360 K about 0.8 nm.

3.3 Conclusion

The present results show two things about the usability of protein force fields from the literature for PNiPAM simulations. The first is that force fields that were fitted to thermophysical data like OPLS-AA show better results than force fields that were only fitted to structural data. The second is that it is possible to simulate PNiPAM in water with these force fields and the results

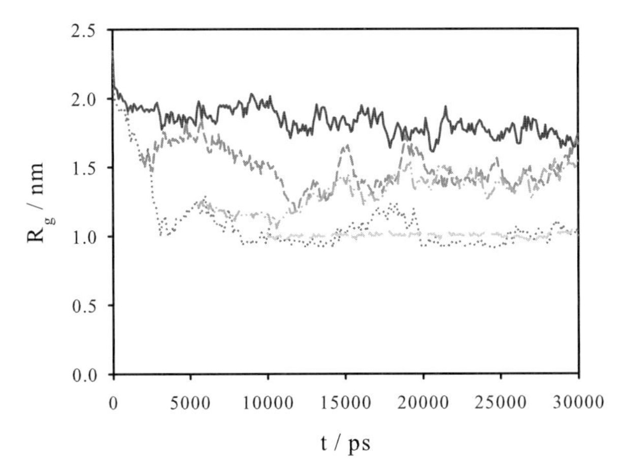

Fig. 5. Radius of gyration at 360 (dotted line) and 280 K (all other lines) over simulation time with different initial configurations for OPLS-AA with SPC/E water. The simulations were performed with a chain of 30 monomers of PNiPAM and 14 482 water molecules

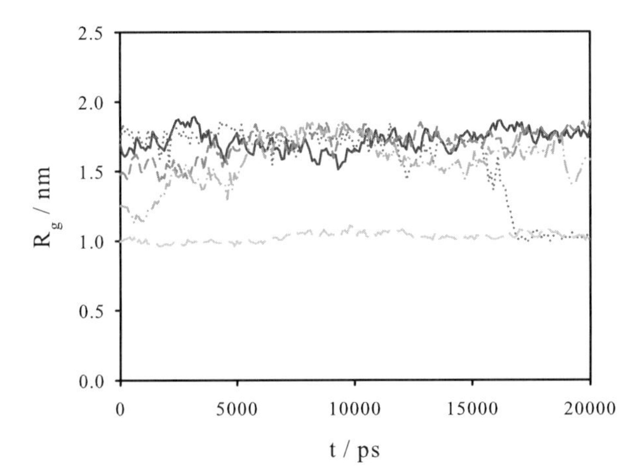

Fig. 6. Radius of gyration at 360 (dotted line) and 280 K (all other lines) over simulation time with different initial configurations for G53a6 with TIP4P water. The simulations were performed with a chain of 30 monomers of PNiPAM and 14 482 water molecules

also show a qualitatively correct temperature dependence of the radius of gyration for these polymer chains.

3.4 Computational Demands

All presented simulations in Section 3.2 were carried out with the MPI based molecular simulation program GROMACS. The parallelization of the molec-

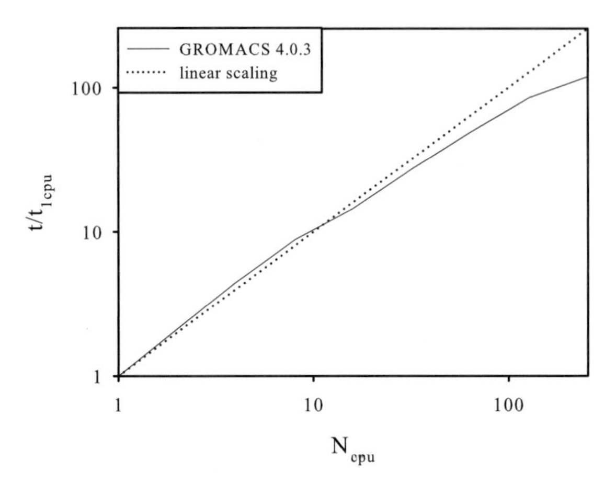

Fig. 7. Scaling of the massively parallel program GROMACS 4.0.3 on HP XC4000, simulation time reduced with the simulation time on one processor over number of processors. Simulated was a PNiPAM chain with 30 monomeres in 14 482 water molecules

ular dynamics part of GROMACS is based on the eighth shell domain decomposition method [13]. With GROMACS, typical simulation runs to determine the radius of gyration in equilibrium employ 64–128 CPUs running for 1–3 days. For these simulations very large systems must be considered comprising typically about 58 800 interaction sites. Figure 7 demonstrates the good scaling of the program on the HP XC4000 cluster at the Rechenzentrum of the Universität Karlsruhe (TH). For these simulations a maximum memory of 284 MB and a maximum virtual memory of 739 MB was used.

References

1. G. Bussi, D. Donadio, and M. Parrinello. Canonical sampling through velocity rescaling. *Journal of Chemical Physics*, 126:014101, 2007.
2. H. J. C. Berendsen, J. R. Grigera, and T. P. Straatsma. The missing term in effective pair potentials. *Journal Of Physical Chemistry*, 91:6269–6271, 1987.
3. P. B. Balbuena, K. P. Johnston, and P. J. Rossky. Molecular dynamics simulation of electrolyte solutions in ambient and supercritical water. 1. Ion solvation. *Journal of Physical Chemistry*, 100:2706–2715, 1996.
4. H. J. C. Berendsen, J. P. M. Postma, W. F. van Gunsteren, A. DiNola, and J. R. Haak. Molecular dynamics with coupling to an external bath. *Journal of Chemical Physics*, 81:3684–3690, 1984.
5. H. J. C. Berendsen, J. P. M. Postma, W. F. van Gunsteren, and J. Hermans. *Intermolecular Forces*. D. Reidel Publishing Company, 1981.
6. L. X. Dang and T. M. Chang. Molecular mechanism of ion binding to the liquid/vapor interface of water. *Journal of Physical Chemistry B*, 106:235–238, 2002.

7. T. Darden, D. York, and L. Pedersen. Particle mesh ewald - an n.log(n) method for ewald sums in large systems. *Journal of Chemical Physics*, 98:10089–10092, 1993.
8. J. W. Eastwood, R. W. Hockney, and D. N. Lawrence. P3m3dp - the 3-dimensional periodic particle-particle-particle-mesh program. *Computer Physics Communications*, 19:215–261, 1980.
9. P. P. Ewald. The calculation of optical and electrostatic grid potential. *Annalen der Physik*, 64:253–287, 1921.
10. F. G. Fumi and M. P. Tosi. Ionic sizes + born repulsive parameters in nacl-type alkali halides. I. Huggins-Mayer + Pauling forms. *Journal of Physics and Chemistry of Solids*, 25:31–44, 1964.
11. M. Heskins and J. E. Guillet. Solution properties of poly(n-isopropylacrylamide). *Journal of Macromolecular Science, Part A*, 8:1441–1455, 1968.
12. B. Hess, C. Holm, and N. van der Vegt. Osmotic coefficients of atomistic nacl (aq) force fields. *Journal of Chemical Physics*, 124:164509, 2006.
13. B. Hess, C. Kutzner, D. van der Spoel, and E. Lindahl. Gromacs 4: Algorithms for highly efficient, load-balanced, and scalable molecular simulation. *Journal of Chemical Theory and Computation*, 4:435–447, 2008.
14. W. L. Jorgensen, J. Chandrasekhar, J. D. Madura, R. W. Impey, and M. L. Klein. Comparison of simple potential functions for simulating liquid water. *Journal of Chemical Physics*, 79:926–935, 1983.
15. K. P. Jensen and W. L. Jorgensen. Halide, ammonium, and alkali metal ion parameters for modeling aqueous solutions. *Journal of Chemical Theory and Computation*, 2:1499–1509, 2006.
16. W. L. Jorgensen, D. S. Maxwell, and J. Tirado-Rives. Development and testing of the opls all-atom force field on conformational energetics and properties of organic liquids. *Journal of American Chemical Society*, 118:11225–11236, 1996.
17. W. L. Jorgensen and J. Tirado-Rives. The opls potential functions for proteins, energy minimizations for crystals of cyclic peptides and crambin. *Journal of the American Chemical Society*, 110:1657–1666, 1988.
18. A. V. Lyulin, B. Dünweg, O. V. Borisov, and A. A. Darinskii. Computer simulation studies of a single polyelectrolyte chain in poor solvent. *Macromolecules*, 32:3264–3278, 1999.
19. H. J. Limbach and C. Holm. Single-chain properties of polyelectrolytes in poor solvent. *Journal of Physical Chemistry B*, 107:8041–8055, 2003.
20. P. J. Lenart, A. Jusufi, and A. Z. Panagiotopoulos. Effective potentials for 1 : 1 electrolyte solutions incorporating dielectric saturation and repulsive hydration. *Journal of Chemical Physics*, 126:044509, 2007.
21. G. Longhi, F. Lebon, S. Abbate, and S. L. Fornili. Molecular dynamics simulation of a model oligomer for poly(n-isopropylamide) in water. *Chemical Physics Letters*, 386:123–127, 2004.
22. B. A. Mann, K. Kremer, and C. Holm. The swelling behavior of charged hydrogels. *Macromolecular Symposium*, 237:90–107, 2006.
23. F. Müller-Plathe and W. F. van Gunsteren. Solvation of poly(vinylalcohol) in water, ethanol and an equimolar water-ethanol mixture: structure and dynamics studied by molecular dynamics simulation. *Polymer*, 38/9:2259–2268, 1997.
24. P. A. Netz and T. Dorfmüller. Computer simulation studies on the polymer-induced modification of water properities in polyacrylamide hydrogels. *Journal of Physical Chemistry B*, 102:4875–4886, 1998.

25. B. Nick and U. W. Suter. Solubility of water in polymers - atomistic simulations. *Computational and Theoretical Polymer Science*, 11:49–55, 2001.
26. N. A. Peppas, P. Bures, W. Leobandung, and H. Ichikawa. Hydrogels in pharmaceutical formulations. *European Journal of Pharmaceutics and Biopharmaceutics*, 50:27–46, 2000.
27. B. G. Rao and U. C. Singh. A free-energy perturbation study of solvation in methanol and dimethyl-sulfoxide. *Journal of the American Chemical Society*, 112:3803–3811, 1990.
28. T. P. Straatsma and H. J. C. Berendsen. Free-energy of ionic hydration - analysis of a thermodynamic integration technique to evaluate free-energy differences by molecular-dynamics simulations. *Journal of Chemical Physics*, 89:5876–5886, 1988.
29. Y. Tamai, H. Tanaka, and K. Nakanishi. Molecular dynamics study of water in hydrogels. *Molecular Simulation*, 16:359–374 1996.
30. H. van der Linden, W. Olthuis, and P. Bergveld. An efficient method for the fabrication of temperature-sensitive hydrogel microactuators. *Lab on a Chip - Miniaturisation for Chemistry and Biology*, 4:619–624, 2004.
31. D. van der Spoel, E. Lindahl, B. Hess, G. Groenhof, A. E. Mark, and H. J. C. Berendsen. Gromacs: Fast, flexible, and free. *Journal of Computational Chemistry*, 26:1701–1718, 2005.
32. W. F. van Gunsteren and H. J. C. Berendsen. *Gromos-87 manual*. Biomos BV Nijenborgh 4, 9747 AG Groningen, The Netherlands, 1987.
33. W. F. van Gunsteren, S. R. Billeter, A. A. Eising, P. H. Hünenberger, P. Krüger, A. E. Mark, W. R. P. Scott, and I. G. Tironi. *Biomolecular Simulation: The Gromos 96 Manual and User Guide*. vdf Hochschulverlag AG an der ETH Zürich, Zürich, Switzerland, 1996.
34. D. R. Wheeler and J. Newman. Molecular dynamics simulations of multicomponent diffusion. 1. Equilibrium method. *Journal of Physical Chemistry B*, 108:18353–18361, 2004.

Reacting Flows

Prof. Dr. Dietmar Kröner

Abteilung für Angewandte Mathematik, Universität Freiburg,
Hermann-Herder-Str. 10, 79104 Freiburg, Germany

Although there is a strong increase of the use of alternative energies, combustion will be most important for the production of energy during the next years. The articles in this chapter mainly concern the development of better combustions concepts with respect to higher efficiency and lower pollutant emission. Special combustors are analyzed numerically and the results can be used for the constructions of better combustors. Although the necessary CPU-times on the NEC SX-8 or the HP XC4000 systems for the corresponding simulations are still of the order of weeks, the results can have a strong impact on the most important problems in culture and society. This emphasizes the importance of High Performance Computing. On the other hand the huge CPU-times show that we still need more powerful systems to solve the big challenges of the future.

In the following we will briefly describe the main topics of the articles in this chapter.

In the paper of Widenhorn et al. the reacting turbulent flow field of a swirl-stabilized gas turbine combustor is simulated numerically using the commercial CFD software package ANSYS CFX-11.0. The simulation is based on the mathematical model consisting of equations for conservation of mass, momentum and energy, an additional turbulence model and species equations. A new contribution in this work is the application of the scale adaptive simulation turbulence model. This model allows the dynamical adjustment of the turbulence model to the length scale of the resolved structures. This is used in combination with the combined eddy dissipation and finite rate chemistry model. The implicit solver is based on a finite volume scheme for structured and unstructured grids (hexaeder) consisting of approximately 2 million grid points. The discretization in space and time is of second order and a multi-grid technique is used for solving the coupled systems. The parallelization is organized by MPI utilities. The calculations have been performed on the HPXC 4000 system of the High Performance Computing Center in Karlsruhe. The numerical results are in very good agreement with experimental measurements.

The problem of NO_X-emissions of aircrafts and in particular of scram jets have been considered in the paper of Kindler et al. The underlying mathematical model consists as before of the compressible Navier-Stokes equation and additional equation for the species and the turbulent transport. For the simulations 13 species and 32 reaction steps have been taken into account. The unsteady system is discretized by a finite volume scheme and the chemistry is treated fully coupled with a fluid motion. The discrete system is solved by an implicit Gauss-Seidel iteration. These are the main ingredients of the inhouse code TASCOM3D. The underlying grid consists of about 800 000 volumes in 3D which are strongly refined at all near wall regions as well as in the main combustion zone. The numerical simulations are used to study the temperature, the fraction of NO and NO_2, the geometry of the flame as a consequence of the air-inlet temperature. These calculations have been performed on the NEC SX-8 using 8 CPUs. These simulations demonstrate a strong dependency of the emission indices on the air inlet temperature. The average CPU-time for one simulation was about 400 hours.

In order to decrease the pollution and to use the resources more efficiently better combustion concepts have to be developed. Therefore burners with strong swirl and additional pilot jet are often used. In the paper of Wang et al. the occurrence of central recirculation zones and the processing vortex core for two variants of the burner as well as the flame response to a variation of the mass flow rate in the pilot jet are investigated. The numerical scheme is based on a low Mach-number version of the compressible Navier-Stokes equations and additional transport equations which take into account six species and two reactions. A turbulence model is included. The numerical results show good agreement with experimental measurements.

Numerical Characterization of a Gas Turbine Model Combustor

A. Widenhorn, B. Noll, and M. Aigner

Institut für Verbrennungstechnik der Luft- und Raumfahrt, Universität Stuttgart, Pfaffenwaldring 38-40, 70569 Stuttgart, Germany
axel.widenhorn@dlr.de

Summary. In this contribution the three-dimensional reacting turbulent flow field of a swirl-stabilized gas turbine model combustor is analyzed numerically. The investigated partially premixed and lifted CH_4/air flame has a thermal power load of P_{th}=35kW and a global equivalence ratio of φ=0.65. To study the reacting flow field the Scale Adaptive Simulation (SAS) turbulence model in combination with the Eddy Dissipation/Finite Rate Chemistry combustion model was applied. The simulations were performed using the commercial CFD software package ANSYS CFX-11.0. The numerically achieved time-averaged values of the velocity components and their appropriate turbulent fluctuations (RMS) are in very good agreement with the experimental values (LDA). The same excellent results were found for other flow quantities like temperature and mixture fraction. Here, the corresponding time-averaged and the appropriate RMS profiles are compared to Raman measurements. Furthermore, instantaneous flow features are discussed. The simulations have been performed on the HP XC4000 system of the High Performance Computing Centre Karlsruhe.

1 Introduction

In order to achieve low levels of pollutants modern gas turbine combustion systems are operating in swirl-stabilized lean premixed mode. The central recirculation zone, which arises due to the swirl of the incoming flow, serves to stabilize the flame within the combustion zone. However, especially under lean premixed conditions severe self-excited combustion oscillations may arise. These unwanted instabilities are in conjunction with high amplitude pressure oscillations, which can decrease the lifetime and availability of the gas turbine. Depending on the swirl number swirling flows can exhibit different topologies [1, 2, 3, 4]. A typical flow instability at high swirl numbers is the precessing vortex core (PVC). This flow phenomenon can be detected at the outlet of injector systems. The PVC exhibits a rotation around the swirl flow axis at a certain frequency. Further typical flow instabilities, which may have a

W.E. Nagel et al. (eds.), *High Performance Computing in Science and Engineering '09*, DOI 10.1007/978-3-642-04665-0_13,

significant impact on combustion dynamics, can arise due to an unsteady vortex breakdown.

In the future the design process of modern gas turbine combustion systems will rely more and more on numerical simulation. In order to allow a reliable design the CFD methods have to predict accurately the aerodynamics and the combustion driven dynamics. However, there is still a large need for improvements in the field of turbulence and combustion modeling as well as in the definition of appropriate boundary conditions [5, 6, 7, 8]. To achieve this aim unsteady three-dimensional CFD simulations are mandatory. Nowadays different approaches are available to capture unsteady flow fields. These are Unsteady Reynolds Averaged Navier-Stokes Simulation (URANS), Large Eddy Simulation (LES) and hybrid RANS/LES methods. The URANS approach, which is commonly used in practical applications, uses complete statistical averaging [9]. This allows the prediction of the ensemble-averaged quantities for velocity, temperature and species distribution of non-reacting and reacting flow fields. However, experience shows that the URANS approach can lead to an excess of turbulent dissipation. Hence, there is the risk that important flow structures are dissipated. LES methods allow resolving the large turbulent structures [10, 11, 12]. For this reason only the smaller structures have to be modeled. However, the LES turbulence models often require prohibitive high computation times for the simulation of gas turbine combustion systems. To get high quality results in combination with an acceptable computational effort hybrid RANS/LES turbulence models like the Scale Adaptive Simulation (SAS) or the Detached Eddy Simulation (DES) approach can be used. These turbulence models combine the strength of LES and RANS. The potential of hybrid RANS/LES models for the simulation of non-reacting and reacting flow fields were demonstrated by Widenhorn et al. [13, 14, 15, 16, 17]. Compared to LES, which has a similar capability the computational effort is reduced drastically.

For the simulation of reacting flows the complex combustion processes have to be modeled accurately. This is necessary since for example in a partially premixed flame the changes in the flow and the fuel to air ratio have a direct influence on the rate of combustion. The change in the heat release causes pressure waves, which under certain conditions may lead to combustion oscillations. Furthermore, due to chemical-kinetic effects the heat release rate and the location of the flame can be influenced. This has an effect on the pressure field and time lag between the flame front and the air and fuel inlets thus influencing the stability map of the flame. Therefore, the accurate prediction of heat release rate and flame location by the combustion model is important. Up to now it is not finally clear which combustion model can be used in conjunction with SAS or DES turbulence models.

In the present work the reacting dynamic turbulent flow field of a model gas turbine combustor was investigated using the SAS turbulence model in combination with the EDM/FRC combustion model by comparing the simulation data against the experimental data set. Furthermore, the detailed numerical analysis was used to get a deeper understanding of the mixing and stabilization processes, the shape of the recirculation and reaction zones and of the unsteady flow structures. In addition the required computational resources were assessed. For the numerical simulations the commercial CFD package ANSYS CFX-11.0 was used.

2 Physical Model

2.1 Conservation Equations

The initial set for the numerical simulation of reacting flows includes the continuity, momentum, energy, turbulence and species equations. In this paper the compressible formulation is used. The equations are given by:

$$\frac{\partial Q}{\partial t} + \frac{\partial (F - F_v)}{\partial x} + \frac{\partial (G - G_v)}{\partial y} + \frac{\partial (H - H_v)}{\partial z} = S \tag{1}$$

The conservative variable vector Q consists of the density, the velocity components, the total specific energy, the turbulent kinetic energy, the specific dissipation rate and the species mass fractions and is defined as:

$$Q = \left[\bar{\rho}, \bar{\rho}\tilde{u}, \bar{\rho}\tilde{v}, \bar{\rho}\tilde{w}, \bar{\rho}\tilde{E}, \bar{\rho}k, \bar{\rho}\omega, \bar{\rho}\tilde{Y}_i\right]^T \qquad i = 1, 2, \ldots, N_{k-1} \tag{2}$$

Here, Favre-averaged quantities are used.

F, G and H are the inviscid and F_v, G_v and H_v are the viscous fluxes in x-, y- and z-direction, correspondingly. The vector S in eq. (1) contains the source terms and is defined as:

$$S = [0, S_u, S_v, S_w, S_E, S_k, S_\omega, S_{Y_i}]^T \qquad i = 1, 2, \ldots, N_{k-1} \tag{3}$$

2.2 Turbulence Modelling

For the closure of the above system of partial differential equations for turbulent flows the Boussinesq hypothesis is used. The required values for the eddy viscosity can be obtained by appropriate turbulence models.

Scale Adaptive Simulation Model (SAS)

In the present work the Scale Adaptive Simulation (SAS) turbulence model was applied. In contradiction to other turbulence models SAS adjusts dynamically to the length scale of the resolved structures. The length scale of the resolved eddies is taken into account by the introduction of the von Karman length scale into the turbulence scale equation. This information allows the SAS model to operate in LES or in RANS like mode. In attached boundary layers a RANS model is in effect. The SAS model can be based on the k-kL formulation given in Menter and Egorov [18]. In this work the model of Menter, which is based on the SST turbulence model is applied [19, 20].

$$\begin{aligned} \frac{\partial(\rho\omega)}{\partial t} + \frac{\partial\left(\rho\bar{U}_j\omega\right)}{\partial x_j} &= \tilde{\alpha}\rho S^2 - \tilde{\beta}\rho\omega^2 + \frac{\partial}{\partial x_j}\left[\frac{\mu_t}{\tilde{\sigma}_\omega}\frac{\partial\omega}{\partial x_j}\right] \\ &+ \frac{2\rho}{\sigma_\Phi}\frac{1}{\omega}\frac{\partial k}{\partial x_j}\frac{\partial\omega}{\partial x_j} + F_{SAS-SST} \end{aligned} \tag{4}$$

The additional term is given by eq. (5):

$$\begin{aligned} F_{SST-SAS} = \rho F_{SAS}\ \max &\left[\bar{\zeta}_2\kappa S^2\frac{L}{L_{\nu k}}\right. \\ &\left.-\frac{2}{\sigma_\Phi}k\ \max\left(\frac{1}{\omega^2}\frac{\partial\omega}{\partial x_j}\frac{\partial\omega}{\partial x_j}, \frac{1}{k^2}\frac{\partial k}{\partial x_j}\frac{\partial k}{\partial x_j}\right); 0\right] \end{aligned} \tag{5}$$

Since the grid spacing is not an explicit term in (4) the SAS model can operate in scale resolving mode without explicit grid information. The issue of grid induced separation of the flow in the boundary layer as it can appear in DES model is eliminated. The model constants are given in [21].

2.3 Combustion Modelling

To take combustion processes into account the respective chemical production rate of all species has to be modelled. The required values for this quantity can be obtained by appropriate combustion models. In the present work the combined Eddy Dissipation/Finite Rate Chemistry combustion model is used applying a one step global reaction mechanism for methane combustion.

Combined Eddy Dissipation/Finite Rate Chemistry Model

The Eddy Dissipation combustion model concept, which was introduced by Magnusson [22] is based on the hypothesis that the chemical reaction is fast in relation to the transport processes of the flow. The reaction rate is assumed to be proportional to a mixing time defined by the turbulent kinetic energy k and the specific dissipation rate ϵ. The reaction rate is not kinetically controlled, which may result in poor predicts for processes where the chemical kinetics limit the reaction rate. Thus, the source term of the species conservation equation of a species i is calculated by eq. (6).

$$S_i = M_i \sum_{r=1}^{Nr}\left(\nu''_{ir} - \nu'_{ir}\right) R_i \tag{6}$$

where R_r defines the reaction rate and ν' and ν'' are the stoichiometric coefficients. The constants for the reaction and product limiter of the EDM combustion model implemented in ANSYS CFX-11.0 are set to A=4.0 and B=0.5. To prevent unphysical behavior the mixing rate in the EDM combustion model is limited by a value of 2,500s^{-1}. Further information about the combustion models and the constants can be found in [22, 23].

The Finite Rate Chemistry combustion model is based on the assumption that the mixing is much faster than the kinetically controlled processes. Here, the chemical production rate of the species i is defined by

$$S_i = M_i \sum_{r=1}^{Nr} \left[(\nu''_{ir} - \nu'_{ir}) \cdot \left(k_{fr} \prod c_\beta^{\eta'_{\beta r}} - k_{br} \prod c_\beta^{\eta''_{\beta r}} \right) \right] \tag{7}$$

In eq. (7) k_{fr} and k_{br} are the forward and backward reaction rates of the reaction r. The reaction can be described by the Arrhenius function:

$$k_r = A_r T^{n_r} exp \left(\frac{-E_r}{R_m T} \right) \tag{8}$$

A detailed explanation is given in [23].

In the combined model both reaction rates are computed first independently from each other. Then, for the calculation of the effective chemical production rate the minimum value of both models is used. Thus, the chemical production rate is either limited by the chemical kinetics or by the turbulent mixing. Therefore, this model is potentially applicable for a wide range of turbulent reacting flows from low to high Damköhler numbers. The chemical reaction is represented by a global one step reaction mechanism for CH_4/air combustion.

$$CH_4 + 2O_2 \rightarrow CO_2 + H_2O \tag{9}$$

3 Numerical Method

The simulations were performed applying the commercial software package ANSYS CFX-11.0. The fully implicit solver is based on a finite volume formulation for structured and unstructured grids. A mulitgrid strategy is applied to solve the linear set of coupled equations. For the spatial discretization a second order scheme is used except for the species and energy equations. For these equations a high order resolution scheme, which is essentially second order accurate and bounded is exploited. For the time discretization an implicit second order time differencing scheme is used.

The parallelization in CFX is based on the Single Program Multiple Data (SPMD) concept. The numerical domain is decomposed into separate tasks, which can be executed separately. The communication between the processes is realized using the Message Passing Interface (MPI) utility. The partitioning process is fully automated and the memory usage is equally distributed among all processors.

4 Results and Discussion

The aim of the present work was to elaborate the strengths and weaknesses of different turbulence and combustion models for the numerical prediction of gas turbine combustor flows. To this end numerical simulation runs and comparisons of calculated and measured values were done for an aero engine model combustor. Within the present project numerical calculations were set up and non-reacting and reacting flow simulations were performed. Exemplarily the results of a reacting flow simulation will be reported in the subsequent chapters.

4.1 Test Case

The gas turbine model combustor is schematically illustrated in Fig. 1. The burner is a modified version of an aero engine gas turbine combustor [25, 26, 27]. Dry air at room temperature and atmospheric pressure is supplied from a common plenum and admitted through a central (d_i=15mm) and an annular nozzle (d_i=17mm, d_o=25mm contoured to an outer diameter of d_o=40mm) to the flame. The plenum has an inner diameter of d_i=79mm and a height of h=65mm. The inner swirler consists of 8 channels for the central nozzle and the outer swirler of 12 channels for the annular nozzle. Non-swirling gaseous fuel is injected between the two co-rotating air flows. The annular fuel injection slot, which forms a ring between the air nozzles is divided into 72 sections. The area of each fuel supply segment equals to A=0.5x0.5mm^2. The exit planes of the fuel and the center air nozzle are lo-

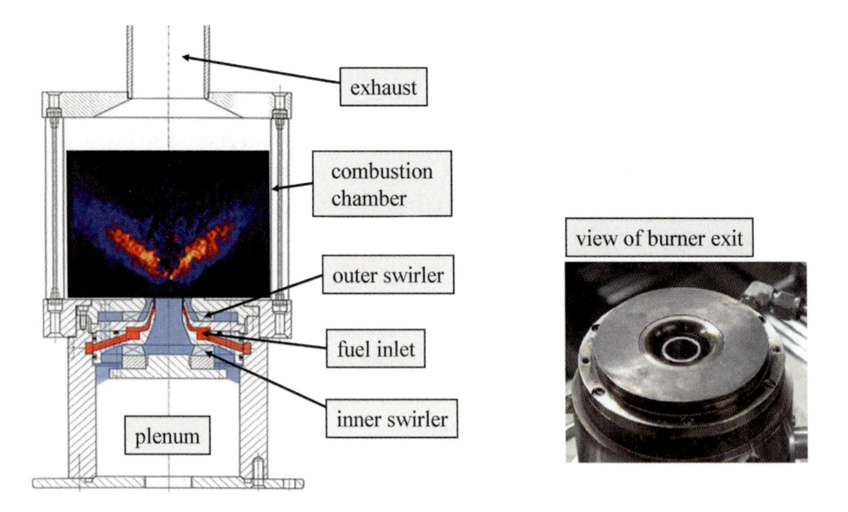

Fig. 1. Schematic of the gas turbine model combustor

cated 4.5mm below the exit plane of the annular air nozzle, which is taken as the reference height h=0mm. The combustion chamber, which permits a good optical access consists of 4 quartz plates held by four steel posts (d=10mm) in the corners. The square cross section of the chamber is A=85x85mm^2 and the height is h=114mm. The combustion chamber is connected via a conical top plate to a central exhaust pipe with an inner diameter of d_i=40mm and a height of h=50mm. Microphones are installed in the plenum and the combustion chamber to detect the pressure fluctuations. Depending on the load conditions the combustion is stable or strongly oscillating. In this work the stable CH_4/air flame operating at a thermal power load of P_{th}=35kW and a global equivalence ratio of φ=0.65 is investigated.

4.2 Numerical Setup

The computational grid consists of 1.91 million grid points. For the injector system and the combustion chamber an unstructured hexaeder grid with 1.6 million grid points was created. In the regions of potential turbulence generation and large velocity gradients a fine mesh was used in order to fulfill the LES requirements. Furthermore, the growth of the adjacent cells was limited to 10% in these zones. For the plenum and the air support to the swirlers an unstructured tetrahedral mesh was applied. It consists of 1.79 million tetrahedral elements and 0.31 million grid points. Figure 3 shows the computational domain and the location of the numerical boundary conditions. At the air inflow boundary condition a mass flow of 0.01762kg/s is set. The air temperature is set to 330K. The turbulent quantities are defined by using the medium

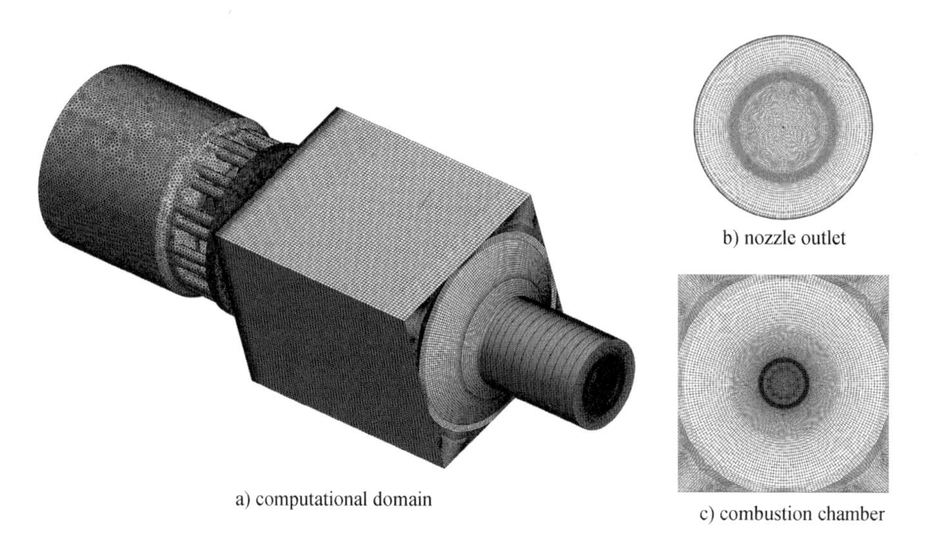

Fig. 2. Computational grid of the gas turbine model combustor

intensity option of 5% in ANSYS CFX. The boundary condition at the fuel inlet specifies a pure methane mass flow of 0.000696kg/s at 330K. Here, the turbulent intensity is denoted by 15% and the eddy length scale by 0.0005m. The wall of the plenum and the injector system is assumed to be adiabatic. The wall temperature of the combustion chamber is set to 1050K and the one of the combustion chamber bottom to 600K. For the outlet a static pressure boundary condition is used. The relative static pressure is set to 0Pa. The reference pressure of the computational domain is defined by 101,325Pa. Furthermore, Fig. 3 shows the locations where a comparison of measured and calculated averaged velocity profiles will be presented.

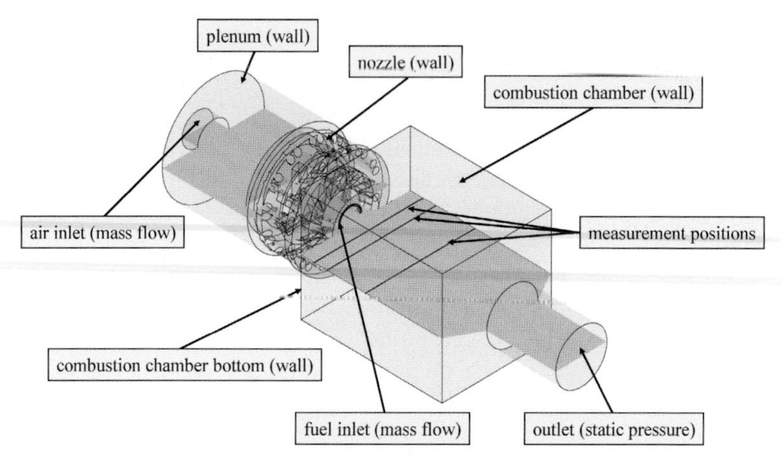

Fig. 3. Computational domain of the gas turbine model combustor including numerical boundary conditions and measurement locations

4.3 Time-Averaged Results

Velocity

In Fig. 4 a simulated two-dimensional streak line plot is used to visualize the time-averaged velocity field within the combustion chamber. The plot shows a flow field, which is typical for enclosed swirl stabilized burners with a concentrically shaped inflow. As expected the mean flow field shows an inner recirculation zone (IRZ) as well as an outer recirculation zone (ORZ) near the walls of the combustion chamber. The visual appearance exhibits despite the square combustion chamber an axial symmetrical flow field. The deviations in axial symmetry are observable in both the experimental and numerical data. The swirl number, which was evaluated from the velocity profiles at the nozzle exit is approximately 0.9.

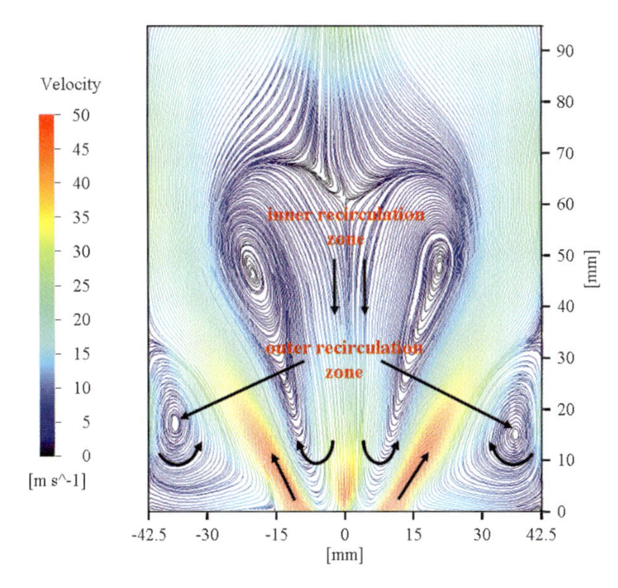

Fig. 4. Time-averaged axial velocity plot on a cutting plane

Figure 5 and Fig. 6 shows the comparison of the numerically obtained time-averaged axial, radial and tangential velocity profiles and their turbulent fluctuations (RMS) with appropriate LDA measurements. At h=5mm the positive peaks in the axial velocity profile reflects the incoming stream of fresh gas. The simulated maximum peak time-averaged axial velocity is around u_{mean}=42m/s. In the radial velocity component the ORZ becomes apparent by the inward-directed radial velocity in the region of y>16mm. The maximum radial velocity is roughly the half of the one in the axial direction. In the ORZ the tangential velocity is rather constant. In the region of the incoming stream maximums in the tangential velocity reflecting the flows from the central and annular nozzle are visible. The minimum between the two peak values (5mm$\leq$y$\leq$15mm) in the experimental data is caused by the fuel nozzle and its wake. This phenomenon can not be covered by the numerical simulation. The axial velocity at the exit of the fuel nozzle is around 54m/s. The turbulent fluctuations of the axial velocity at h=5mm have a pronounced maximum of u_{rms}=28m/s in the shear layers between the incoming stream of fresh gas and the IRZ. The radial velocity exhibits further peak values in the shear layers between the incoming stream and the ORZ. The high RMS level in the shear layers indicates the strong turbulent fluctuations in the region of the flame. Therefore, intensive mixing of the incoming cold fresh gas with the hot burned gas coming from the IRZ and the ORZ can be expected. Moreover, the broad peaks of the turbulent fluctuations in the axial direction around y=6mm points out that the shear layers are not locally stable. Therefore, the IRZ should not be regarded as a stable structure, which hardly varies its position.

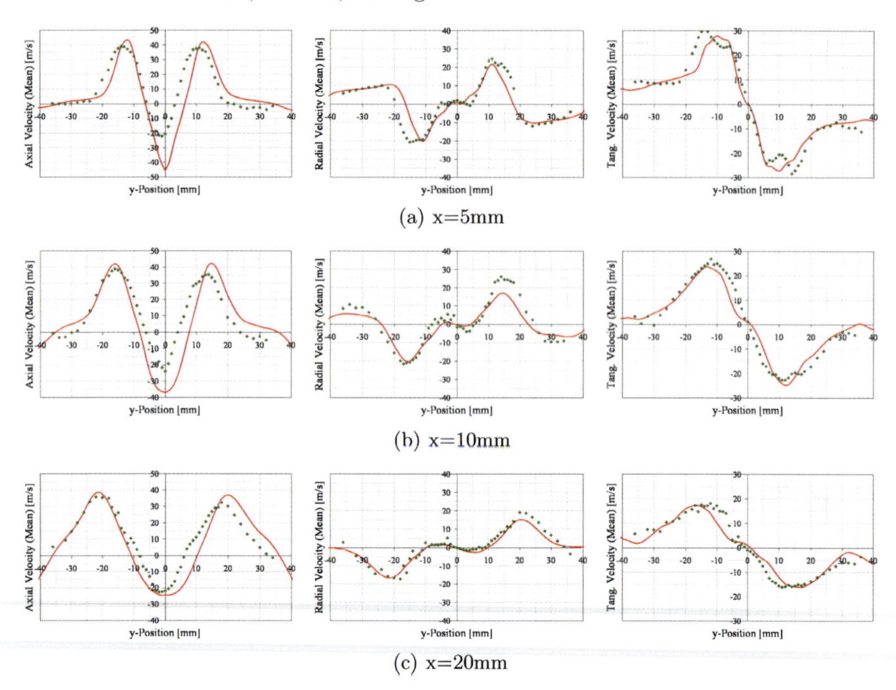

(a) x=5mm

(b) x=10mm

(c) x=20mm

Fig. 5. Time-averaged axial velocity profiles (left), radial velocity profiles (middle) and tangential velocity profiles (right); green diamonds (LDA), red line (simulation)

The IRZ rather fluctuates mainly in the axial direction whereas the ORZ alters in the radial direction. In the central region the level of turbulence is much smaller for the axial and radial velocity component. At a high of h=10mm the basic flow features are not changed compared to h=5mm. At h=20mm, the velocity profiles have broadened and the ORZ has reduced in size. The RMS values of the tangential velocity component are quite constant over the radius whereas the turbulent fluctuations of the axial and radial velocity component exhibit broad maxima. With increasing downstream position h the profiles smooth out and the ORZ and IRZ vanish. At all positions the simulated time-averaged radial and tangential velocity profiles agree very well with the experiment. A deviation occurs at h=5mm and h=10mm which is close to the burner mouth for the time-averaged axial velocity profile. Here, the positive peak values which belong to the incoming stream of fresh gas are predicted well whereas the strength of the recirculation zone is too high. Nevertheless at all other positions the time-averaged axial velocity profiles are predicted well. The calculation slightly over-predicts the peak values of the turbulent fluctuations of the axial velocity component at h=5mm. At h=5mm and h=10mm the strength of the turbulent fluctuations close to the centerline are faintly under-predicted for this quantity. The RMS values of the radial velocity components at h=5mm and h=10mm are predicted well close to the centerline but the peak

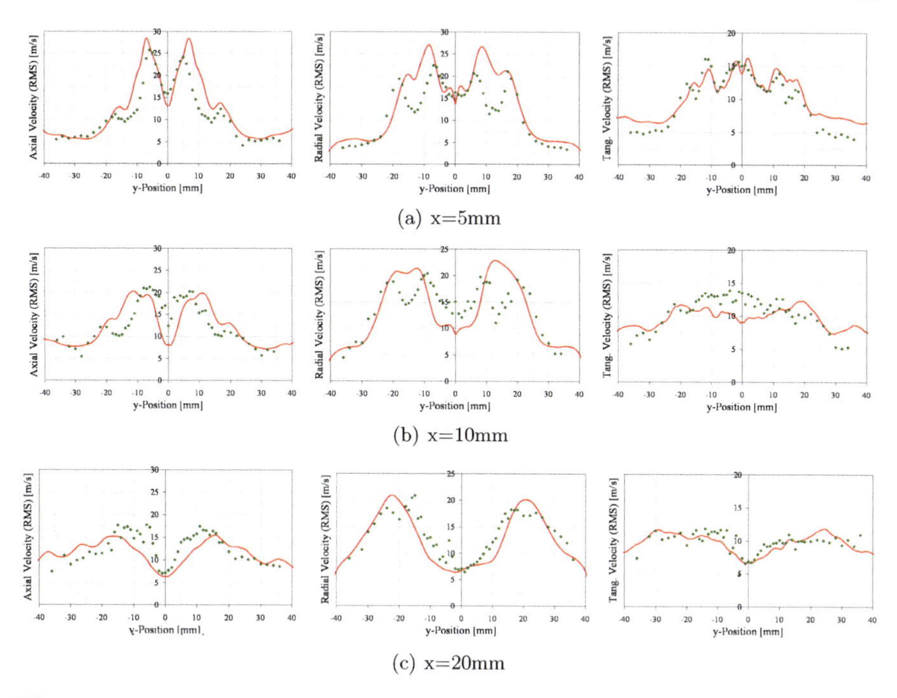

Fig. 6. Time-averaged turbulent axial velocity fluctuations (left), radial velocity fluctuations (middle) and tangential velocity fluctuations (right); green diamonds (LDA), red line (simulation)

values differs compared to the experiment. Nevertheless the values at h=20mm show an excellent agreement. For the tangential velocity components there is an excellent agreement at all measuring positions.

Time-Averaged Temperature Profiles

Figure 7 shows the comparison of the numerical time-averaged temperature profiles and the associated turbulent fluctuations (RMS) together with appropriate Raman measurements. The distribution of the mean temperatures reflects the shape of the flame. At h=5mm the flame exhibits a numerically obtained temperature of T=1600K at the centerline. The low temperature regions at $5mm \leq y \leq 15mm$ reflect the incoming stream of the fresh gas. Here, the lowest simulated mean temperature is around 500K. The temperature increase for $y > 10mm$ is mainly due to the mixing of the hot exhaust gas from the ORZ with the unburned gas. With increasing height the time-averaged temperature values rise and reach a maximum between h=10mm and h=20mm. Afterwards the temperature decreases slowly. The highest mean temperature obtained within the IRZ equals to T=1868K, which is close to the experimental value of T=1827K. In this region the time-averaged value of the mixture

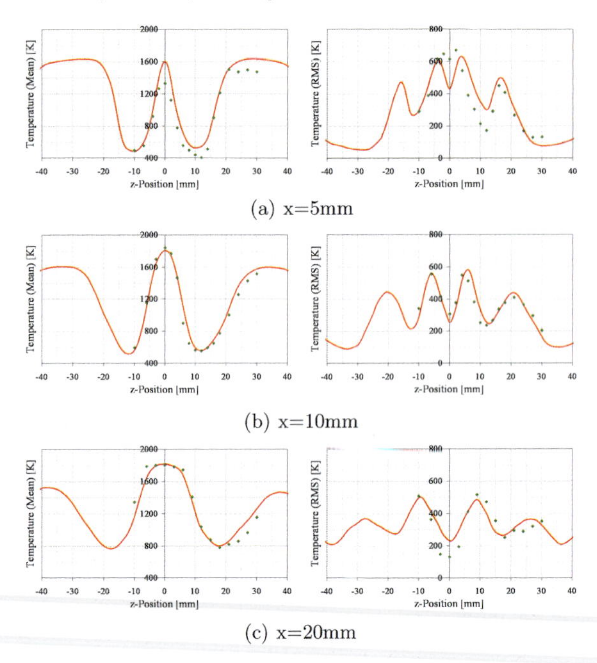

Fig. 7. Time-averaged temperature (left) and their appropriate turbulent fluctuations (right); green diamonds (Raman), red line (simulation)

fraction f lies above the global one. With increasing downstream position h the profiles smooth out. For h>10mm the temperature level in the ORZ is lower than in the IRZ. This effect can be explained by the leaner mixture and the heat losses to the wall in the ORZ. The temperature fluctuations reach a level of 500-600K close to the nozzle and decrease to 40-70K at larger heights. Considering the temperature fluctuations especially within the lower part of the IRZ evidences that the IRZ is a highly unsteady flow region. At all positions the simulated time-averaged temperature profiles agree excellent with the experimental values. The only nameable deviation occurs at h=5mm where the temperature around the centerline is over-predicted. The comparison of the simulated turbulent temperature fluctuation profiles shows a very good agreement with the experiment at most positions. At h=5mm the calculation under-predicts the turbulent intensities of the temperature at the centerline. At h=20mm the same quantity is slightly over-predicted at the central axis. Nevertheless the peak values and the shape agree very well with the measured data.

Time-Averaged Mixture Fraction Profiles

Figure 8 compares the numerically obtained time-averaged mixture fraction profiles with appropriate Raman measurements. The highest mixture frac-

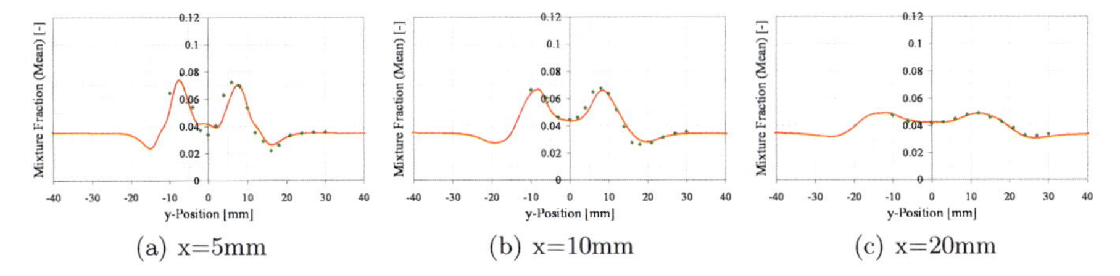

(a) x=5mm (b) x=10mm (c) x=20mm

Fig. 8. Time-averaged mixture fraction; green diamonds (Raman), red line (simulation)

tion value appears above the fuel nozzle exit. As a result of the fast mixing generated by this nozzle configuration the mean mixture fraction is already quite small at h=5mm. The simulated peak value is f_{max}=0.072. Near stoichiometric mixtures occur at y=5mm and y=10mm. Taking into account the low temperatures in this regions this is a strong indication that the flame is far away from equilibrium state. Furthermore, it is clearly visible that in the IRZ the mixture fraction is higher than the global one (f_{global}=0.036), whereas in the ORZ the mixture fraction equals f_{global}. The higher mixture fraction in the IRZ has a positive effect on the ignition and flame stabilization since temperature levels above the global adiabatic flame temperature are enabled. At h=20mm the variation of the time-averaged mixture fraction along the y-axis is reduced. The broadened maximum of the mixture fraction reveals the fast mixing of the fuel stream. In the time-averaged plots the mixing is completed at h=40mm. Generally it can be stated that the IRZ exhibits a high temperature level and a mixture fraction larger than the global value of f_{global}=0.036. Furthermore, the profiles of the time-averaged temperature and mixture fraction do not well correlate in the lower part of the flame, which indicates non-equilibrium effects.

Instantaneous Heat Release

The reaction takes place mainly in the inner and outer shear layers where the hot burned gas form the IRZ and ORZ is mixed with the unburned incoming gas. The reaction zones are highly unsteady and vary in time and space. The instantaneous plots show that the root of the heat release can penetrate into the central air nozzle. In the time-averaged data set the heat release starts at a height of approximately h=5mm and extends up to about h=70mm. Instantaneous pictures show that in seldom cases a reacting mixture is transported further downstream. Here, reaction takes place even at h>70mm as indicated in Fig. 9. The flame fronts are strongly wrinkled thus reflecting the strong local turbulence present in the flow field. Here, the existence of the small vortical structures plays a vital role in the formation and destruction of the reaction zone structures which can extend over several centimeters.

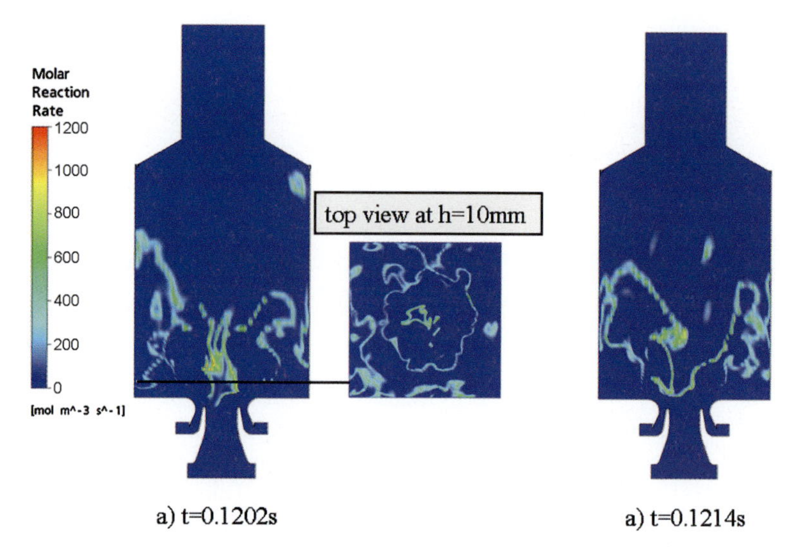

Fig. 9. Spatial position and structure of the reaction zone deduced from the methane molar reaction rate on the cutting plane

5 Computational Resources

The simulations have been performed on the HP XC4000 system. According to our experience to obtain a statistically converged solution a total integration time of at least 4 combustor residence times is required. This corresponds for the applied test case to a simulated real time of 0.12s. Additionally for the start-up phase 2 combustor residence times are needed. Therefore, the overall simulated real time is 0.18s. Since the SAS turbulence model is applied in combination with a combustion model a relatively fine mesh has to be used to achieve low CFL numbers. Hence, the numerical effort is very large. A typical grid, which is applied for the simulation has 1.91million grid points. In order to perform the calculations within adequate turnaround times the CPU numbers varies between 20 and 24. The typical total CPU time required for such runs is about 60 days and the total RAM requirements are 1040MB. In all simulations one time step consists of four inner iteration loops. The time step is set to $1e^{-5}$s. This leads to 18000 time steps and 72000 inner iteration loops per run.

6 Conclusions

The potential of the SAS turbulence models in combination with the EDM/FRC combustion model for the simulation of the reacting flow in gas turbine combustion chambers has been worked out. In general the results show

Table 1. CPU requirements

Task	CPU Requirements
Momentum and Mass	40.4 %
Heat Transfer	9.5 %
Turbulence	13.0 %
Mass Fraction	19.9 %
Variable Update	14.9 %
File Writing	0.2 %
Miscellaneous	2.1 %

a remarkable predictive capacity of these methods. Compared to the classical LES approach the SAS models can save one order of magnitude of computing power, since the large portions of the computational domain can be treated by a RANS method thus partly enabling larger grid sizes. Another advantage is the stationary RANS boundary conditions formulation, which can be applied more easily than unsteady LES boundary conditions. Nevertheless, high performance computing is necessary to perform the calculations within adequate turnaround times. The numerically obtained time-averaged velocity components as well as the time-averaged temperature profiles match very well the LDA and Raman measurements. Likewise the turbulent velocity and temperature fluctuations agree very well with the experimental data.

Acknowledgement. The authors would like thanks the High-Performance Computing-Centre Karlsruhe for the always helpful support and the computation time on the high performance computers.

References

1. Liang, H., Maxworthy, T.: An Experimental Investigation of Swirling Jets. J. Fluid Mech., **525**, pp. 115–159 (2005)
2. Fernandez, E.C., Heitor, M.V., Shtork, S.I.: An Analysis of Unsteady Highly Turbulent Swirling Flow in a Model Vortex Combustor. Exp. Fluids, **40**, pp. 177–187 (2006)
3. Midgley, K., Spencer, A., McGuirk, J.J.: Unsteady Flow Structures in Radial Swirler Fed Fuel Injectors. ASME Turbo Expo 2004, GT2004-53608
4. Thompson, K.W.: Time Dependent Boundary Conditions for Hyperbolic Systems. J. of Comput. Phys., **68**, pp. 1–24 (1987)
5. Poinsot, T., Lele, S.: Boundary Conditions for Direct Simulations of Compressible Viscous Flows. J. of Comput. Phys., **101**, pp. 104–129 (1992)
6. Widenhorn, A., Noll, B., Aigner, M.: Accurate Boundary Conditions for the Numerical Simulation of Thermoacoustic Phenomena in Gas-Turbine Combustion Chambers. ASME Turbo Expo 2006, GT2006-90441

7. Widenhorn, A., Noll, B., Aigner, M.: Impedance Boundary Conditions for the Numerical Simulation of Gas-Turbine Combustion Systems. ASME Turbo Expo 2008, GT2008-50445
8. Noll, B.: Numerische Strömungsmechanik. Springer Verlag (1993)
9. Dunham, D., Spencer, A., McGuirk, J.J., Dianat, M.: Comparison of URANS and LES CFD Methodologies for Air Swirl Fuel Injectors. ASME Turbo Expo 2008, GT2008-50278
10. Garcia-Villalba, M., Fröhlich, J.: LES of a free annular swirling jet; Dependence of coherent structures on a pilot jet and the level of swirl. Int. J. of Heat and Fluid Flow, **27**, pp. 911–923 (2006)
11. Schlüter, J., Schönfeld, T., Poinsot, T., Krebs, W., Hoffmann, S.: Characterization of Confined Swirl Flows Using Large Eddy Simulation. ASME Turbo Expo 2001, GT2001-0060
12. Widenhorn, A., Noll, B., Stöhr, M., Aigner, M.: Numerical Investigation of a Laboratory Combustor Appling Hybrid RANS-LES Methods. DESider 2007, Second Symposium on Hybrid RANS-LES Methods, Corfu, Greece
13. Widenhorn, A., Noll, B., Stöhr, M., Aigner: Numerical Characterization of the Non-Reacting Flow in a Swirled Gasturbine Model Combustor. The 10th Results and Review Workshop of the HLRS
14. Widenhorn, A., Noll, B., Aigner, M.: Numerical Characterization of the Reacting Flow in a Swirled Gasturbine Model Combustor. The 11th Results and Review Workshop of the HLRS
15. Widenhorn, A., Noll, B., Aigner, M.: Numerical Study of a Non-Reacting Turbulent Flow in a Gas Turbine Model Combustor. AIAA-Paper, 2009-647
16. Widenhorn, A., Noll, B., Aigner, M.: Numerical Characterization of a Gas Turbine Model Combustor applying Scale-Adaptive Simulation. ASME Turbo Expo 2009, GT2009-59038
17. Menter, F.R., Egorov, Y.: Re-visting the Turbulent Scale Equation. IUTAM Symposium (2004)
18. Menter, F.R., Egorov, Y.: A Scale-Adaptive Simulation Model Using Two-Equation Models. AIAA-Paper, 2005-1095
19. Menter, F.R.: Two Equation Eddy Viscosity Turbulence Models for Engineering Applications. AIAA Journal **32(8)**, pp. 269–289 (1995)
20. Menter, F.R., Kuntz, M., Bender, R.: A Scale Adaptive Simulation Model for Turbulent Flow Prediction. AIAA Paper, 2003-0767
21. ANSYS CFX-11.0 User Guide
22. Hjertager, B.H., Magnusson, B.F.: On Mathematical Modelling of Turbulent Combustion with Special Emphasis on Soot Formation and combustion. Sixteenth Symposium on Combustion, 1976
23. Gerlinger, P.: Numerische Verbrennungssimulation, Springer Verlag, 2005
24. Giezendanner, R., Keck, O., Weigand, P., Meier, W., Meier, U., Stricker, W., Aigner, M.: Periodic Combustion Instabilities in a Swirl Burner Studied by Phase-Locked Planar Laser-Induced Fluorescence. Combust. Sci. Technol., **175**, pp. 721–741 (2003)
25. Weigand, P., Meier, W., Duan, X.R., Stricker, W., Aigner, M.: Investigation of swirl flames in a gas turbine model combustor; I. Flow field, structures, temperatures and species distributions, Combustion and Flame, **144**, pp. 205–224 (2006)

26. Weigand, P., Meier, W., Duan, X.R., Stricker, W., Aigner, M.: Investigation of swirl flames in a gas turbine model combustor; II. Turbulence-chemsitry interactions, Combustion and Flame, **144**, pp. 225–236 (2006)
27. Cao, M., Eickhoff, H., Joss, F., Simon, B.: ASME Propulsion and Energetic, 70th Symposium, AGARD Conf. Proc. 422

Numerical Investigations of NO_X-Formation in Scramjet Combustors

Markus Kindler, Benjamin Rust, Peter Gerlinger, and Manfred Aigner

Institut für Verbrennungstechnik der Luft- und Raumfahrt, Universität Stuttgart, Pfaffenwaldring 38-40, 70569 Stuttgart, Germany

Summary. In the present paper the impact of NO_X-emissions of aircrafts on ozone is highlighted. The chemical processes of NO_X in the troposphere and stratosphere are explained briefly and the formation of NO_X under the specific conditions of a scramjet combustor is discussed. Finally a model scramjet combustor with hydrogen injection is investigated numerically using the scientific in-house code TASCOM3D. The results of the simulations demonstrate the formation of NO_X in scramjet combustors and are discussed in detail. With regard to the development of NO_X reduction strategies further investigations are performed in terms of the variation of the inlet temperature. Variations of other relevant parameters concerning NO_X-formation like the equivalence ratio, Mach number at combustor entrance, types of fuel injection, etc. are part of ongoing investigations but will not be discussed in this paper. Finally the performance of the scientific code TASCOM3D on the NEC SX-8 is analyzed.

1 Introduction

Research on hypersonic flight has been performed since more than fifty years [1]. Within these studies the development of air-breathing engines (i.e. ramjets for flight Mach numbers 2-7, scramjets for flight Mach numbers 5-15) has always been of great interest since the payload of rockets is very small compared to the total mass and hence the costs for transportations are very high. Air-breathing engines could advance the payload/cost-ratio and make hypersonic transportation accessable for new markets. In the recent years ambitious projects have started like the EC-funded LAPCAT [2] or the research training group GRK 1095/1 [3] to attain the goal of an efficient and safe hypersonic propulsion system for civil aircrafts in the future.

Beside the proof of profitability and safeness such a propulsion system can only be established if the ecological impact could be minimized. Relating to environmental protection in civil aviation the emission of NO_X is a demanding issue that has to be regarded seriously. The Intergovernmental Panel on Climate Change (IPCC) [4] highlighted NO_X-emissions of aircrafts in general

W.E. Nagel et al. (eds.), *High Performance Computing in Science and Engineering '09*, DOI 10.1007/978-3-642-04665-0_14,

as one major influence on atmospheric ozone chemistry. Despite the relatively small percentage of 3% of all anthropogenic NO_X-emissions that are caused by aviation these emissions modify the atmospheric NO_X background concentration at cruise altitudes [5]. Due to the low mixing rate at higher altitudes the residence time of those emissions is relatively long so that NO_X molecules participate multiple times in the ozone production (troposphere) or destruction (stratosphere) chain. Additionally due to the low temperature at cruise altitudes the effect of greenhouse gases is enhanced so that few amounts of those gases have a much greater effect compared to sea level. The role of NO_X in the chemistry of the troposphere and stratosphere is a very complex phenomena which leads beyond the scope of this paper. Thererfore the interested reader is refered to [6]. As a concluding remark it is mentioned that further studies and future scenarios [7, 8] expect that especially the appearence of a supersonic aircraft fleet leeds to a depletion of the ozone layer in the stratosphere due to the higher cruise altidude of supersonic aircrafts.

2 NO_X-Formation in Scramjet Combustors

Related to the operating conditions of scramjet combustors using hydrogen two main procedures of NO_X-formation are identified. First there is the so called "Zeldovich-NO" or "Thermal-NO" which is built according to

$$O + N_2 \longleftrightarrow NO + N \tag{1}$$

$$N + O_2 \longleftrightarrow NO + O \tag{2}$$

$$N + OH \longleftrightarrow NO + H \; . \tag{3}$$

Due to the strong bonding of N_2 reaction (1) has a high activation energy and a relatively low reaction rate and is therefore the rate limiting step in the mechanism [9]. Hence "Zeldovich-NO" strongly depends on the temperature that is needed to reach the high activation energy. Another procedure that occurs in scramjet combustors is the formation of nitrogen dioxid according to

$$O_2 + NO \longleftrightarrow NO_2 + O \tag{4}$$

$$NO + O \longleftrightarrow NO_2 \tag{5}$$

or

$$NO + HO_2 \longleftrightarrow NO_2 + OH \; . \tag{6}$$

Reaction steps (4) and (5) describe a rapid conversion of NO in NO_2 mainly at the flame edges where sufficient NO and O molecules exist [10]. Reaction (6) takes place at regions with lower temperature where the concentrations of HO_2 and OH are high and NO from high temperature regions has been transported to [11].

3 Governing Equations and Numerical Scheme

The investigations of NO_X-formation in model scramjet combustors presented in this paper are performed using the scientific in-house code TASCOM3D (Turbulent All Speed Combustion Multigrid). It describes reacting flows by solving the full compressible Navier-Stokes, species and turbulence transport equations. Additionally an assumed PDF (probability density function) approach is used to take turbulence chemistry interaction into consideration. Therefore two additional equations (for the variance of the temperature and the variance of the sum of species mass fractions) have to be solved. Thus the described set of averaged equations in three-dimensional conservative form is given by

$$\frac{\partial \mathbf{Q}}{\partial t} + \frac{\partial(\mathbf{F} - \mathbf{F}_\nu)}{\partial x} + \frac{\partial(\mathbf{G} - \mathbf{G}_\nu)}{\partial y} + \frac{\partial(\mathbf{H} - \mathbf{H}_\nu)}{\partial z} = \mathbf{S}\,, \tag{7}$$

where

$$\mathbf{Q} = \left[\bar{\rho}, \bar{\rho}\tilde{u}, \bar{\rho}\tilde{v}, \bar{\rho}\tilde{w}, \bar{\rho}\tilde{E}, \bar{\rho}q, \bar{\rho}\omega, \bar{\rho}\sigma_T, \bar{\rho}\sigma_Y, \bar{\rho}\tilde{Y}_i\right]^T, \quad i = 1, 2, \ldots, N_k - 1\,. \tag{8}$$

The variables in the conservative variable vector $\mathbf{Q}$ are the density $\bar{\rho}$ (averaged), the velocity components (Favre averaged) $\tilde{u}$, $\tilde{v}$ and $\tilde{w}$, the total specific energy $\tilde{E}$, the turbulence variables $q = \sqrt{k}$ and $\omega = \epsilon/k$ (where k is the kinetic energy and ϵ the dissipation rate of k), the variance of the temperature σ_T and the variance of the sum of the species mass fractions σ_Y and finally the species mass fractions $\tilde{Y}_i$ ($i = 1, 2, \ldots, N_k - 1$). Thereby N_k describes the total number of species that are used for the description of the gas composition. The vectors $\mathbf{F}$, $\mathbf{G}$ and $\mathbf{H}$ specify the inviscid fluxes in x-, y- and z- direction, $\mathbf{F}_\nu$, $\mathbf{G}_\nu$ and $\mathbf{H}_\nu$ the viscous fluxes, respectively. The source vector $\mathbf{S}$ in (7) results from turbulence and chemistry and is given by

$$\mathbf{S} = \left[0, 0, 0, 0, 0, \bar{S}_q, \bar{S}_\omega, \bar{S}_{\sigma_T}, \bar{S}_{\sigma_Y}, \bar{S}_{Y_i}\right]^T, \quad i = 1, 2, \ldots, N_k - 1\,, \tag{9}$$

where $\bar{S}_q$ and $\bar{S}_\omega$ are the averaged source terms of the turbulence variables, $\bar{S}_{\sigma_T}$ and $\bar{S}_{\sigma_Y}$ the source terms of the variance variables (σ_T and σ_Y) and $\bar{S}_{Y_i}$ the source terms of the species mass fractions. For turbulence closure a two-equation low-Reynolds-number q-ω turbulence-model is applied [12]. The momentary chemical production rate of species i in (7) is defined by

$$S_{Y_i} = M_i \sum_{r=1}^{N_r} \left[\left(\nu_{i,r}^{''} - \nu_{i,r}^{'}\right) \left(k_{fr} \prod_{l=1}^{N_k} c_l^{\nu_{l,r}^{'}} - k_{br} \prod_{l=1}^{N_k} c_l^{\nu_{l,r}^{''}}\right)\right], \tag{10}$$

where k_{f_r} and k_{b_r} are the forward and backward rate constants of reaction r (defined by the Arrhenius function), the molecular weight of a species M_i, the species concentration $c_i = \rho\tilde{Y}_i/M_i$ and the stoichiometric coefficients $\nu_{i,r}^{'}$ and $\nu_{i,r}^{''}$ of species i in reaction r. The averaged chemical production rate for a species i due to the use of an assumed PDF approach is described in detail in [13, 14]. In the present paper all simulations have been performed using the modified Jachimowski hydrogen/air reaction mechanism (13 species/ 32 steps) [15] including the species and reaction steps concering the formation of NO_X. The unsteady set of equations (7) is solved using an implicit Lower-Upper-Symmetric Gauss-Seidel (LU-SGS) [16, 17, 18, 19] finite-volume algorithm, where the finite-rate chemistry is treated fully coupled with the fluid motion. More details concerning TASCOM3D may be found in [14, 18, 19, 20, 21].

4 Geometry and Boundary Conditions

The model scramjet combustor with strut injection of the present investigations has been subject for several studies in the past [22, 23, 24, 25]. Fig. 1 shows a sketch of the simulated combustor. Upstream of the strut injector

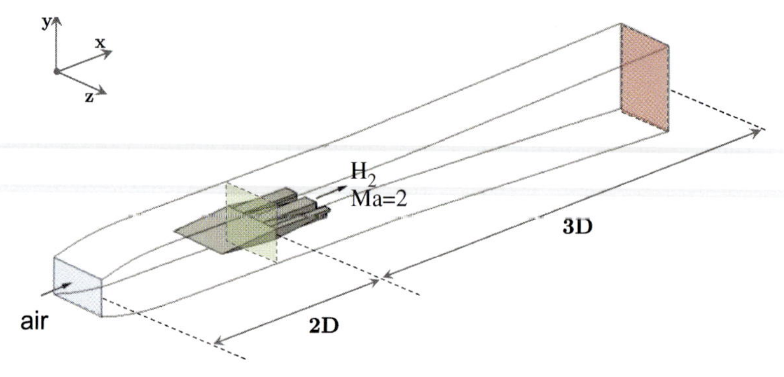

Fig. 1. Sketch of the investigated model scramjet combustor with lobed strut injector

the flow is expanded in the laval nozzle. After the laval nozzle the channel has a constant cross section (40mm x 38mm, length = 100mm). Afterwards the diverging part of the combustion chamber (expansion angle = 3°, length of diverging part = 390mm) begins. The constant section of the combustion chamber secures that the temperature is high enough to ensure auto-ignition. On the other hand the diverging part of the channel is needed to avoid thermal choking and to compensate effects from heat release due to combustion. Hydrogen is injected through the blunt of a lobed strut (80mm x 40mm x 6mm) directly into the core of the air flow. The simulations have been performed in two-steps: two-dimensional simulations from the laval nozzle up to the middle of the strut and three-dimensional simulations from the middle of the strut to the outlet of the combustion chamber (where the results of the two-dimensional simulations are taken as inlet conditions). Thereby the grid of the two-dimensional simulations has about 8000 volumes, the one of the three-dimensional simulations about 800000 volumes. Both grids are strongly refined at all near wall regions as well as in the main combustion zone. Tab. 1 summarizes the boundary conditions of the simulations concerning the investigations of the impact of different air inlet temperature on NO_X-formation. Thereby the inlet temperature varies from 1070K - 1500K whereas the boundary conditions for the hydrogen injection are identical for all simulations (equivalence ratio $\Phi = 0.2$). The wall temperature of the combustor and the strut injector is constant ($T_{\text{wall}} = 420$K) in all investigations.

Table 1. Inflow conditions for the simulation of the combustion chamber with lobed strut injector for $T_{in} = 1070$K, $T_{in} = 1300$K, $T_{in} = 1500$K, respectively

		air		**strut**
T_{in} [K]	1070	1300	1500	
pressure p [Pa]	80450	97742	112780	38005
inlet temperature T_{in} [K]	1070	1300	1500	164
wall temperature T_{wall} [K]	420	420	420	420
velocity u $[\frac{m}{s}]$	1824	1824	1824	1954
mass fraction Y_{H_2} [-]	0	0	0	1
mass fraction Y_{O_2} [-]	0.23	0.23	0.23	0
mass fraction Y_{N_2} [-]	0.77	0.77	0.77	0

5 Results and Discussion

Fig. 2 shows the calculated distribution of temperature, NO molar fraction and NO_2 molar fraction of the model scramjet combustor in case of an air inlet temperature of $T_{in} = 1500$K, respectively. For all investigated inlet conditions an detached flame is observed (in case of $T_{in} = 1500$K the flame stabilizes at about 25mm downstream of the strut). Shortly after ignition the flame covers only a small area which increases over the channel length. The vorticities enhanced by the lobed strut injector spread the flame with increasing distance from injection and form a W-shaped flame at the end of the combustion chamber. The distribution of the molar fraction of NO and NO_2 demonstrates clearly the characteristics of the NO_X-formation described in sec. 2. Formation of NO occurs at the high temperature regions due to the high activation energy described by the "Zeldovich"-mechanism so that the distribution of NO looks similar to the temperature distribution. NO_2-formation occurs in regions of low temperature and at the flame edges. Hence the distribution of NO_2 is concentrated at the flow surrounding the flame. Fig. 3 shows the calculated distribution of temperature, NO molar fraction

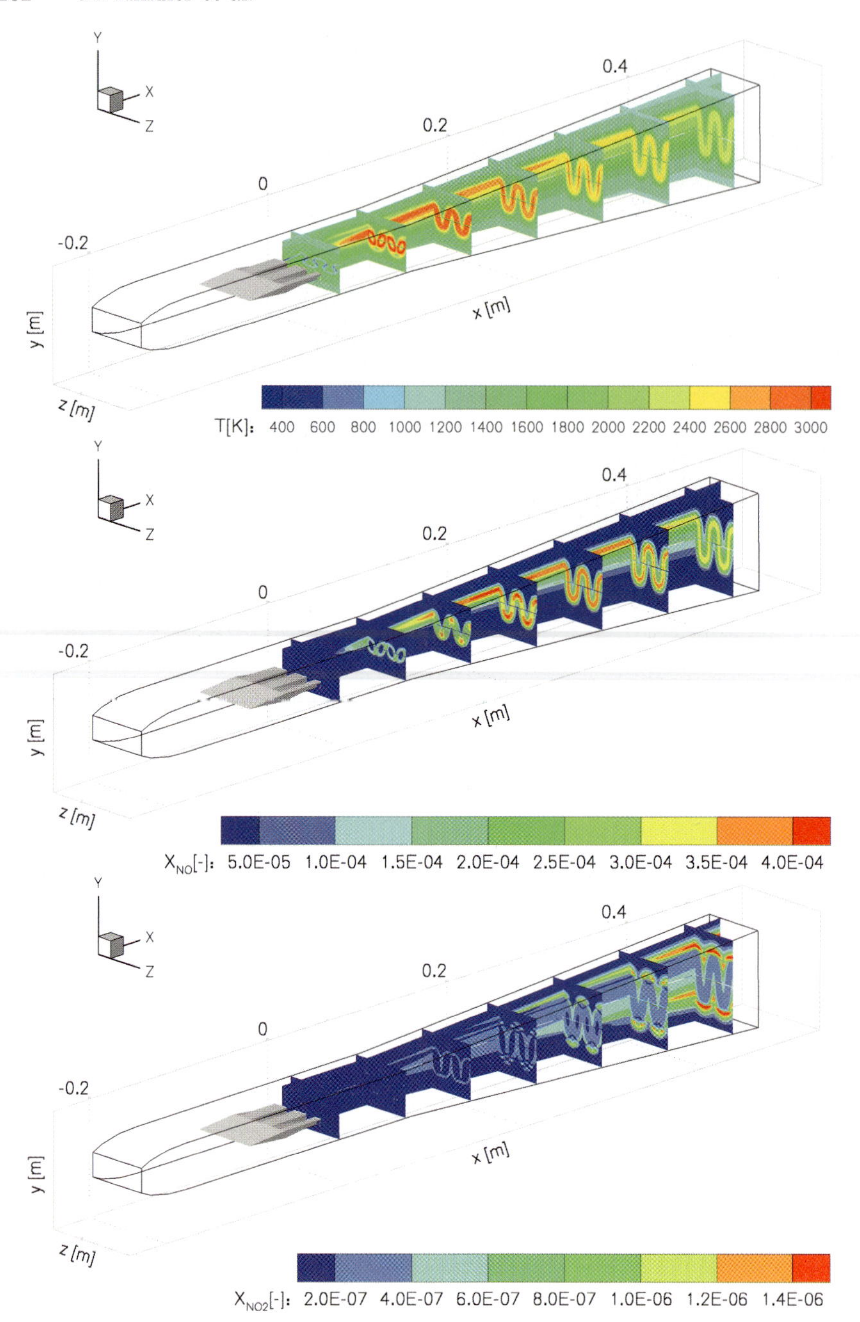

Fig. 2. Calculated distribution of temperature, NO molar fraction and NO_2 molar fraction (from top to bottom) of the model scramjet combustor, respectively (T_{in} = 1500 K)

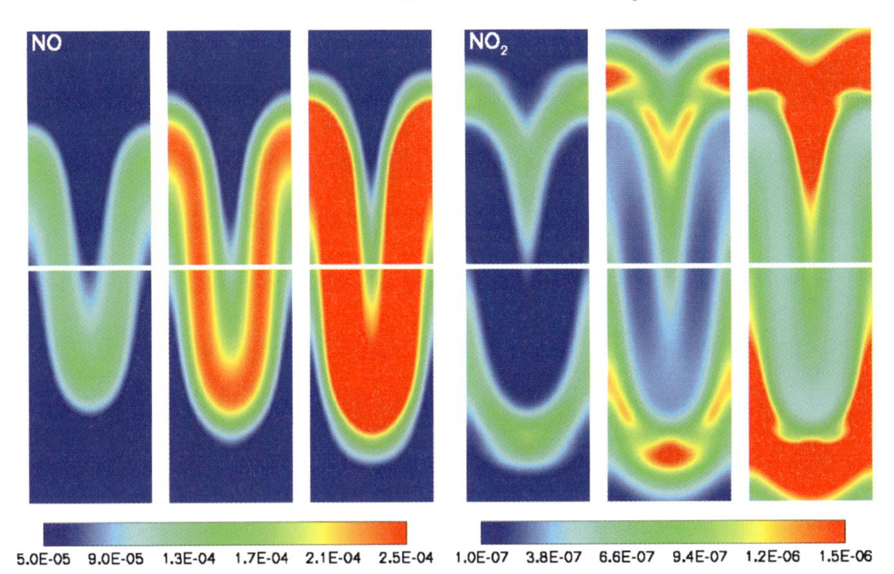

Fig. 3. Calculated distribution of NO molar fraction (left) and NO_2 molar fraction (right) at combustor exit in case of T_{in} = 1070K, 1300K and 1500K (from left to right), respectively

and NO_2 molar fraction at the combustor exit in case of the inlet temperature T_{in} = 1070K, 1300K and 1500K, respectively. Again the characteristics of the NO and NO_2 formation can be observed independently from the investigated inlet temperature. Furthermore the distribution of NO shows a strong dependence of the inlet temperature and the amount of NO that increases with higher temperatures. Consequently also the formation of NO_2 is enhanced by higher temperatures because of the necessity of NO that could be converted. Fig. 4 demonstrates the averaged temperature, burning efficiency and NO and NO_2 emission indices ($EI_{NO_X} = \dot{m}_{NO_X}(x)/\dot{m}_{H_2}$ in $\left[\frac{g}{s}\right] / \left[\frac{kg}{s}\right]$) in case of different air inlet temperatures over the channel length, respectively. Thereby the averaged temperature shows that the ignition delay is decreased significantly with increasing temperature (about 35mm in case of T_{in} = 1500K, about 90mm in case of T_{in} = 1070K). The averaged temperature maxima vary from about 1500K to 1888K (not to be confused with the local peak temperatures that vary from 2877K to 3122K). At the diverging part of the channel the temperature is decreased due to the acceleration of the flow after a distance of 100mm downstream of the strut. The burning efficiency shows a complete consumption of the fuel in case of T_{in} = 1500K and 1300K and only few unburned hydrogen at the channel exit in case of the lowest temperature (η_V = 0.98). The ignition delay caused by the different inlet temperatures can be observed clearly, too. The averaged emission indices for NO and NO_2 both show a significant temperature dependency. The values of the emis-

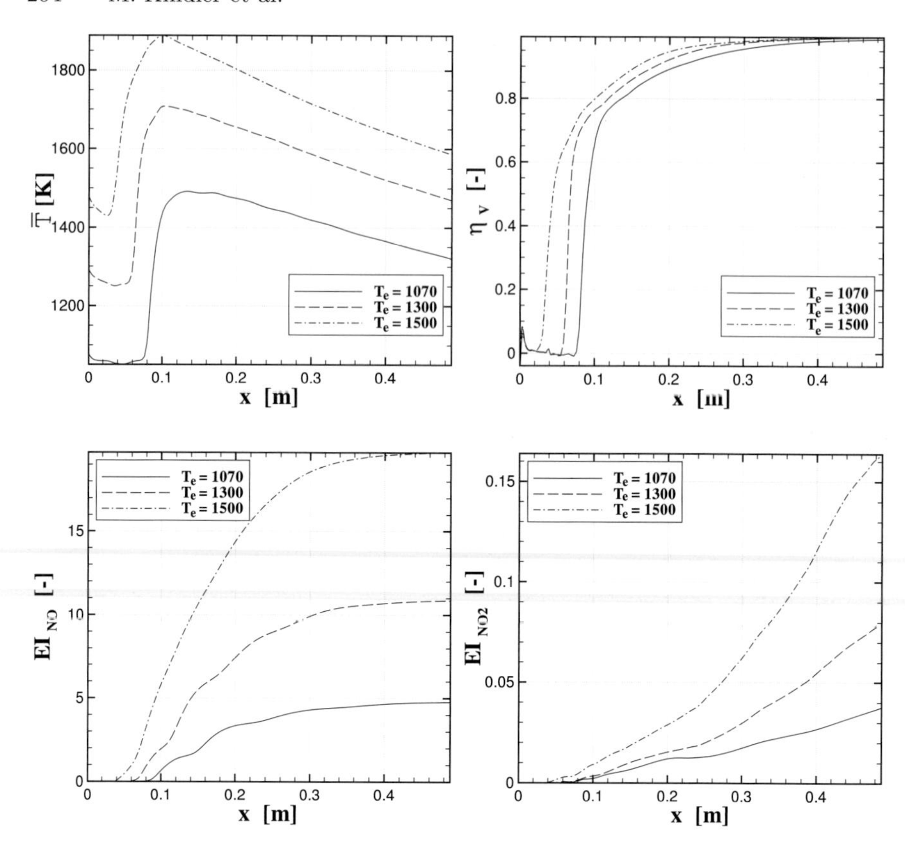

Fig. 4. Averaged temperature (top left), burning efficiency (top right), NO emission index (bottom left) and NO_2 emission index (bottom right) in case of T_{in} = 1070 K, 1300 K and 1500 K over the channel length, respectively

sions indices of NO and NO_2 are about four times as high in case of T_{in} = 1500K compared to T_{in} = 1070K ($EI_{NO,1500K} = 19.71$, $EI_{NO,1070K} = 4.76$; $EI_{NO_2,1500K} = 0.16$, $EI_{NO_2,1070K} = 0.039$). Furthermore the emission indices show that the formation of NO_2 is marginal compared to the formation of NO in the model scramjet combustor investigated. The investigations of the NO_X-formation show that under certain circumstances the emissions of an scramjet should not be underestimated (modern aircrafts' EI_{NO_X} = 10-15). Additionally it has to be considered that the equivalence ratio of $\Phi = 0.2$ is relatively low and an increase of the equivalence ratio probably enhances the NO_X-emissions. Hence the development of low NO_X-scramjet combustors will be necessary to establish such an aircraft in the future. Thereby different injection techniques and sophisticated combustor geometries could be helpful to reach a better mixing and to avoid temperature peaks in the combustion chamber.

6 Performance

The investigations have been performed on the NEC SX-8 using 8 CPUs. The averaged cpu-time for one simulation is about 300h - 400h. About 49% of the computational time is used in three subroutines: one for the calculation of the chemical source terms (19.5%), another for the calculation of the gas properties (15.7%) and finally one for the solution of the imlicit part of the unsteady set of equation (13.9%). Thereby the calculation of the chemical source terms and the gas properties are local phenomena so that no dependency of the neighbour cells occurs and one dimensional arrays can be used without difficulty. This results in a good performance (vector operation ration: 99.4% - 99.46%, averaged vector length: 255.2 - 255.7, 5.7 - 7.3 GFLOPS). To resolve the data dependency for the solution of the implicit part the code has to be vectorized along diagonal lines (2D) and planes (3D), respectively. In case of two-dimensional simulations this could be a strong limitation due to relatively short vector lengths [26]. In the present three-dimensional simulations the solution of the implicit part is less critical because the amount of cell volumes corresponding to data-independent planes is besides of few planes at block corners relatively high. Hence the implicit solver reaches an vector operation ratio of 99.11%, an averaged vector length of 210.8 and about 2.9 GFLOPS. Nevertheless there is a potential for further optimizations of the implicit part by a sophisticated arrangement of each block in the computational domain (number of blocks as little as possible, amount of cell volumes in each block as many as possible) which has to be regarded in every simulation. The performance analysis for the whole scientific code TASCOM3D shows a vector operation ratio of 98.8%, an averaged vector length of 144 and a performance of 4.72 GFLOPS.

7 Conclusion

The problem of NO_X-emissions in aviation in general and especially in scramjets has been discussed and the formation of NO_X in scramjet combustors illustrated. The in-house code TASCOM3D has been used to investigate a model scramjet combustor with hydrogen injection using a lobed strut injector. Thereby the inlet temperature of the incoming airflow has been varied (T_{in} = 1070K, 1300K and 1500K). The simulations demonstrate a strong dependency of air inlet temperature and NO_X-formation so that the emission indices in case of T_{in} = 1500K are four times as high as corresponding indices in case of T_{in} = 1070K. Furthermore the investigations have shown that under certain circumstances the emissions of NO_X are too high and probably raise even more with increasing equivalence ratios (equivalence ratio for present investigations = 0.2). Additionally there are some uncertainties concerning the chosen reaction mechanism. Previous studies have shown that the use of different reaction mechanisms results in varying predictions of NO_X-emissions. In order to achieve a better understanding of the matter further investigations on NO_X-formation in scramjet combustors concerning the reaction mechanisms and varying flight conditions have to be performed. Furthermore techniques for the minimization of the NO_X formation (e.g. better mixing of fuel and air, avoiding temperature peaks, etc.) have to be developed to limitate the environmental pollution due to future aircrafts. The performance analysis of the presented simulations demonstrates that TASCOM3D reaches quite sufficient vector operation ratios and vector lengths. Even the implicit solver shows a good performance but still has some potential for optimization, e.g. by an improved arrangement of the computational domains.

Acknowledgements. This work was performed within the 'Long-Term Advanced Propulsion Concepts and Technologies II' project investigating high-speed airbreathing propulsion. LAPCAT, coordinated by ESA-ESTEC, is supported by the EU within the 7th Framework Programme, Transport, Contract no.:ACP7-GA-2008-21 1485. Further information on LAPCAT II can be found on http://www.esa.int/ techresources/lapcat_II.

We wish to thank the *Deutsche Forschungsgemeinschaft (DFG)* for financial support within the GRK 1095/1 at the University of Stuttgart.

The simulations were performed on the national super computer NEC SX-8 at the High Performance Computing Center Stuttgart (HLRS) under the grant number scrcomb.

References

1. Bertin, J.J., Cummings, R.M., Fifty years of hypersonics: where we've been, where we're going. Prog Aerospace Science **39**, pp. 511–536, 2003.
2. Steelant, J., LAPCAT: A Technical Feasibility Study on Sustained Hypersonic Flight, ISABE-2007-1205, 2007.
3. Gaisbauer, U., Weigand, B., Reinartz, B., Research Training Group GRK 1095/1: Aero-Thermodynamic Design of a Scramjet Propulsion System, ISABE-2007-1131, 2007.
4. Penner, J.E., Lister, D.H., Griggs, D.J., Dokken, D.J., McFarland, M., Aviation and the Global Atmosphere, Intergovernmental Panel on Climate Change, Cambridge University Press, UK. pp. 373, 1999.
5. Köhler, I., Sausen, R., Reinberger, R., Contributions of aircraft emissions to the atmospheric NOx content, Atmospheric Environment, Volume 31, Issue 12, June 1997, pp. 1801–1818.
6. Crutzen, P.J., The influence of nitrogen oxides on the atmospheric ozone content. Quarterly Journal of Royal Meteorological Society 96, pp. 320–325, 1970.
7. Grooss, J.U., Bruehl, C., Peter, T., Impact of aircraft emissions on tropospheric and stratospheric ozone. Part I: Chemistry and 2-D model results, Atmos Environ 32 (18), pp. 3173–3184, 1998.
8. Dameris, M., Grewe, V., Koehler, I., Sausen, R., Bruehl, C., Grooss, J.U., Steil, B., Impact of aircraft emissions on tropospheric and stratospheric ozone. Part II: 3-D model results, Atmos Environ 32 (18), pp. 3185–3199, 1998.

9. Warnatz, J., Maas, U., Dibble, R.W., Combustion : physical and chemical fundamentals, modeling and simulation, experiments, pollutant formation, Springer, Berlin-Heidelberg-New York, 2006.
10. Slack, M., Grillo, A., Investigation of Hydrogen-Air Ignition Sensitized by Nitric Oxide and by Nitrogen Dioxide, NASA CR-2896, 1977.
11. Kuo, K.K., Principles of combustion, John Wiley & Sons, ISBN 0-471-04689-2, Hoboken, New Jersey, 2005.
12. Coakley, T.J., Huang, P.G., Turbulence Modeling for High Speed Flows, AIAA paper 92-0436, 1992.
13. Gerlinger, P., Numerische Verbrennungssimulation, Springer, ISBN 3-540-23337-7, Berlin-Heidelberg, 2005.
14. Gerlinger, P., Investigations of an Assumed PDF Approach for Finite-Rate-Chemistry, Combustion Science and Technology, **175**, pp. 841–872, 2003.
15. Wilson, G.J., MacCormack, R.W., Modeling Supersonic Combustion Using a fully Implicit Numerical Method, AIAA Journal Vol. 30, No. 4, April 1992.
16. Shuen, J.S., Upwind Differencing and LU Factorization for Chemical Non-Equilibrium Navier-Stokes Equations, Journal of Computational Physics, **99**, pp. 233–250, 1992.
17. Jameson, A., Yoon, S., Lower-Upper Implicit Scheme with Multiple Grids for the Euler Equations, AIAA Journal, **25**, pp. 929–937, 1987.
18. Gerlinger, P., Brüggemann, D., An Implicit Multigrid Scheme for the Compressible Navier-Stokes Equations with Low-Reynolds-Number Turbulence Closure, Journal of Fluids Engineering, **120**, pp. 257–262, 1998.
19. Gerlinger, P., Möbus, H., Brüggemann, D., An Implicit Multigrid Method for Turbulent Combustion, Journal of Computational Physics, **167**, pp. 247–276, 2001.
20. Stoll, P., Gerlinger, P., Brüggemann, D., Domain Decomposition for an Impicit LU-SGS Scheme Using Overlapping Grids, AIAA-paper 97-1896, 1997.
21. Stoll, P., Gerlinger, P., Brüggemann, D., Implicit Preconditioning Method for Turbulent Reacting Flows, Proceedings of the 4th ECCOMAS Conference, **1**, pp. 205–212, John Wiley & Sons, 1998.
22. Gerlinger, P., Brüggemann, D., Numerical Investigation of Hydrogen Strut Injections into Supersonic Air Flows, Journal of Propulsion and Power, **16**, pp. 22–28, 2000.
23. Gerlinger, P., Stoll, P., Kindler, M., Schneider, F. and Aigner, M., Numerical Investigation of Mixing and Combustion Enhancement in Supersonic Combustors by Strut Induced Streamwise Vorticity, Aerospace Science and Technology, **12**, pp. 159–168, 2008.
24. Kindler, M., Gerlinger, P. and Aigner, M., Numerical Investigations of Mixing Enhancement by Lobed Strut Injectors in Turbulent Reactive Supersonic Flows, ISABE-2007-1314, 2007.
25. Kindler, M., Blacha, T., Lempke, M., Gerlinger, P., Aigner, M., Numerical Investigations of Model Scramjet Combustors. In: Nagel, Wolfgang E.; Kröner, Dietmar B.; Resch, Michael [Hrsg.]: High Performance Computing in Science and Engineering '08, Transactions of the High Performance Computing Center, Stuttgart (HLRS), S. 153 - 166, ISBN 978-3-540-88301-2, 2009.
26. Kindler, M., Gerlinger, P. and Aigner, M., Assumed PDF Modeling of Turbulence Chemistry Interaction in Scramjet Combustors, High Performance Computing in Science and Engineering '07, pp. 203–213, 2008.

Large-Eddy Simulation of Lean Premixed Flames in a Model Swirl Burner

Ping Wang[1], Jochen Fröhlich[2], and Ulrich Maas[1]

[1] Institut für Technische Thermodynamik, Universität Karlsruhe, Kaiserstr. 12, D-76131 Karlsruhe
ping.wang@mailbox.tu-dresden.de
[2] Institut für Strömungsmechanik, Technische Universität Dresden, George-Bähr-Str. 3c, D-01069 Dresden
jochen.froehlich@tu-dresden.de

Summary. Lean premixed flames in an un-confined model swirl burner are studied numerically by means of large eddy simulation. In the burner considered, swirling flow and a richer pilot flame are used to stabilize the flame. The paper investigates the occurrence of the central recirculation zone (CRZ) and the precessing vortex core (PVC) for two variants of the burner, as well as the flame response to a sinusoidal variation of the mass flow rate in the pilot jet. The corresponding isothermal swirling flows are also simulated for providing reference. Detailed comparisons with experimental data are performed for velocity statistics and generally good agreement is achieved. Additionally, coherent structures are analyzed and local power spectra computed. It is found that a very strong PVC is observed in numerical results and experiment for the non-reactive case with retracted pilot jet. In the reactive cases where the whole CRZ is enclosed in the high-temperature post-flame region, the PVC is almost completely suppressed.

1 Introduction

Despite tremendous efforts to increase the use of alternative energy, combustion will still be the main source for thermal energy in the next decades. Due to increasing concerns about the impact on the environment, new and better combustion concepts have to be developed which maximize the efficiency and at the same time minimize the pollutant formation. Lean premixed (LP) combustion is one of these concepts. The principle of LP combustors is to operate under lean and highly homogeneous conditions, to keep the local and global combustion temperature in a desired narrow window, which is good for lower pollutant emission. However, this can lead to undesired effects such as lean blow-off and thermo-acoustic resonance. To deal with these, burners with strong swirl and an additional pilot jet with richer mixture are often utilized

W.E. Nagel et al. (eds.), *High Performance Computing in Science and Engineering '09*, DOI 10.1007/978-3-642-04665-0_15,

for flame stabilization. In the resulting complex flow system, flow instabilities are observed in experiments [1] and simulations [2].

A typical flow phenomenon in swirling flows is the so-called central recirculation zone (CRZ). In the reactive swirling flow the CRZ recirculates hot gas to the root of the flame. Furthermore, the flow speed is reduced locally by the recirculation, such that points exist where the flow speed matches the turbulent flame velocity. The ability to predict the size, shape and the position of the CRZ therefore is essential to the prediction of this type flow. Generally, the CRZ is not stable but precesses around the axis, hence termed precessing vortex core (PVC). The occurrence of a PVC in an isothermal swirling flow depends on several factors such as swirl number and geometrical setup, e.g. Under combustion conditions its behavior is complicated further by equivalence ratio, mode of fuel entry, flow rate etc. [3]. However, the strong PVC occurring in an isothermal swirling flow is often damped or even suppressed in the corresponding reactive flow, depending on the considered configuration [4, 5]. Numerous studies have been reported on the influence of different parameters such as the level of swirl, the inflow profile, the flow configuration, Reynolds number, and heat release etc., as reviewed in [3, 6].

Another practically important issue is combustion-driven instability. If the Rayleigh criterion is fulfilled oscillations of heat release are amplified via inducing oscillations in static pressure which trigger oscillations in mass flow rate and hence in heat release. A substantial part of that loop was studied experimentally by measuring the flame response to a sinousoidally oscillation of mass flow rate in an axial swirl burner issuing into the ambient, using a dedicated device [7, 8].

The main objectives of the present work are to investigate the impact of heat release and the location of the pilot jet on the swirling flow structures in an unconfined co-annular model burner, and to investigate the occurrence of CRZ and PVC in this burner and the flame response to a sinusoidal variation of the mass flow rate in the pilot jet, with the aid of LES.

2 Numerical Method

In the present work, the so-called thickened-flame (TF) model [9] is employed to simulate the turbulent premixed combustion. Although it is comparatively simple it has shown to perform well for the type of flow under consideration [4]. In the TF model, the pre-exponential constants in the Arrhenium form reaction rates and the transport coefficients are both modified by a thickening factor F to yield a thicker reaction zone which can then be resolved with LES mesh. To account for the subgrid-scale (SGS) wrinkling, an efficiency function E related to the local subgrid-turbulent velocity and the filter width is employed. The equation for the mass fraction of the k-th species is thus modified to read

$$\frac{\partial \rho Y_k^{th}}{\partial t} + \frac{\partial \rho u_j Y_k^{th}}{\partial x_j} = \frac{\partial}{\partial x_j}(\rho D_k EF \frac{\partial Y_k^{th}}{\partial x_j}) + \frac{E}{F}\dot{\omega}_k(Y_j^{th}, T^{th}) \quad (1)$$

where the superscript '*th*' represents thickened quantities. D, $\dot{\omega}$, T are the molecular diffusivity, reaction rate and temperature, respectively. The efficiency function E is estimated as proposed by Colin et al. [9]. The thickening factor F is a local quantity and generally estimated as

$$F(n) = n\Delta/\delta_L^0 \quad (2)$$

where Δ is the characteristic grid cell length scale, and δ_L^0 is the laminar flame thickness. Hence, the thickened flame is resolved with n grid cells. An optimal value for the parameter n generally is in the range 5-10 (T. Poinsot, private communication). In the present work, the effect of n is studied, demonstrating that n=5 is a good choice for the flow cases under consideration (see section 4.1). The dynamic procedure of [10] was used to determine F. It reduces F to unity remote from the flame and hence provides realistic transport coefficients ther

To simulate the lean premixed methane/air combustion of the experiment, the two-step chemical scheme 2sCM2 [4] is used which takes into account six species (CH4, O2, CO2, CO, H2O and N2) and two reactions:

(a) $CH_4 + 1.5O_2 \rightarrow CO + 2H_2O$

(b) $CO + 0.5O_2 \leftrightarrow CO_2$

The numerical method employed is based on the so-called low-Mach number version of the compressible Navier-Stokes equations. With this approach, the pressure P is decomposed into a spatially constant component $P^{(0)}$, interpreted as the thermodynamic pressure, and a variable component $P^{(1)}$, interpreted as the dynamic pressure. $P^{(0)}$ is connected to temperature and density, while $P^{(1)}$ is related to the velocity field only and does not influence the density. Due to this decomposition, sonic waves are eliminated from the flow, so that the time step is not restricted by the speed of sound.

Applying large eddy filtering to the low-Mach number equations, the corresponding filtered LES equations are obtained. The unclosed terms in these equations have to be determined by a SGS model. The variable-density dynamic Smagorinsky model by Moin et al. [11] is used to determine the SGS eddy viscosity in the momentum equations. The SGS scalar flux is modeled by the gradient diffusion model.

These models were then implemented in the Finite-Volume code LESOCC2C [12, 13, 14], which is a compressible version of LESOCC2 [15]. LESOCC2 is highly vectorized, and parallization is accomplished by domain decomposition and explicit message passing via MPI. It solves the incompressible Navier-Stokes equations on body-fitted curvilinear block-structured grids employing second-order central schemes for the spatial discretization and a 3-step Runge-Kutta method for the temporal discretization. The convection

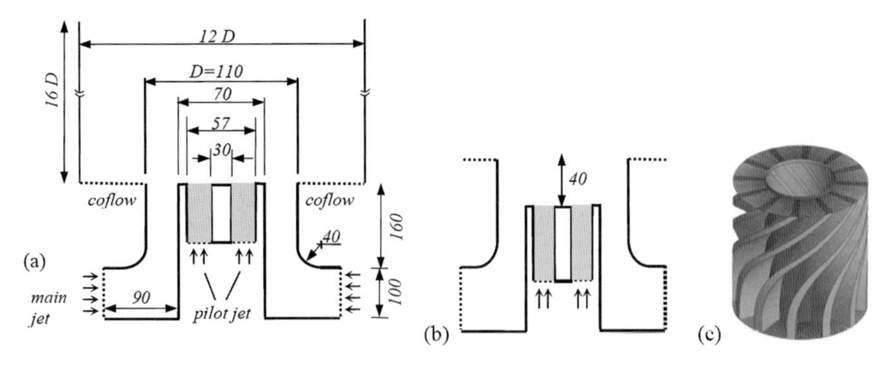

Fig. 1. (a) Sketch of the computational domain, pilot jet flush with main jet, units is mm. (b) Pilot jet retracted by 40 mm. (c) 3D view of the pilot swirler located at the position indicated by gray color in part (a) and (b)

term of the species equation is discretized with the HLPA scheme [16]. In order to simulate reactive flows, LESOCC2C is designed to solve the above low-Mach-number version of the compressible Navier-Stokes equations, by means of a pressure-based method. Apart from the evaluation of additional non-linear terms, this approach amounts to solving a Poisson-type equation in each time step, similarly to the approach for constant density flows in LESOCC2, so that the algorithm of both versions is very much the same.

3 Computation Set up and Computational Effort

The configuration of the model burner investigated is depicted in Fig. 1a, together with the computational domain (not to scale). It consists of two co-annular swirling jets discharging into the ambient air which is at rest. The tube in the centre of this figure (indicated by gray color) is the pilot lance. Two geometrical setups are studied: pilot jet flush with outlet (Fig. 1a) and pilot jet retracted by 40 mm (Fig. 1b). The Reynolds number of the flow based on the bulk velocity U_b = 11.5 m/s, and the radius of outer jet nozzle R=55 mm, is Re=35,000. The mass flow rates are 180 m3/h and 20 m3/h at normal condition, and the theoretical swirl numbers are 0.9 and 0.79 for the main and the pilot jet, respectively. Swirl is co-rotating. A slip condition was used on the outer boundary and a convective outflow condition was imposed at the exit. The Werner-Wengle wall function was employed at all solid walls. A very small uniform co-flow was imposed remote from the jets as validated in [17].

The blades of the axial swirl generator for the pilot jet used in the experiment were represented geometrically using 12 inclined blades as seen in Fig. 1c. For the inflow of the main jet, 12 gaps where positioned regularly along the circumference of the cylindrical surface, with radial and tangential

Table 1. Overview of the principal cases simulated

Cases	Reactive	Retracted	ϕ_{main}	ϕ_{pilot}	Pulsating	n	$Grid/10^6$	$\Delta t/\mu s$
Cold00	no	no	0.0	0.0	no	/	8.45	6.69
Cold40	no	yes	0.0	0.0	no	/	8.88	6.69
Hot00	yes	no	0.667	0.833	no	5	4.34	2.84
Hot40a	yes	yes	0.667	0.833	no	5	4.56	2.52
Hot40b	yes	yes	0.667	0.833	no	10	4.56	1.25
Hot40c	yes	yes	0.667	0.935	no	5	4.56	2.52
Pulse1	yes	yes	0.667	0.935	100Hz	5	4.56	2.52
Pulse2	yes	yes	0.667	0.935	200Hz	5	4.56	2.52

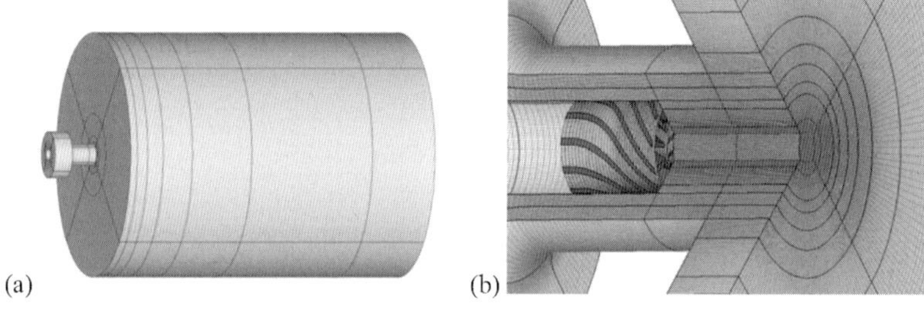

Fig. 2. (a) 3D view of the computational domain. Dark lines indicated block boundaries of the computational grid. (b) Zoom view of the grid around the jet nozzle

velocity component being imposed. Different swirl levels are achieved by adjusting the tangential velocity component and maintaining the radial velocity component.

In order to study the impact of reaction and pilot jet location on the flow, and to investigate the flame response to a sinusoidal variation of the mass flow rate in the pilot jet, numerous simulations have been performed. Table 1 provides an overview of the principal flow cases simulated in this work. For cases with steady inflow in the pilot jet, velocity statistics at several cross sections are provided for comparison. Pulse1 and Pulse2 are two flames excited by periodic oscillations of the mass flow rate in the pilot jet at 100Hz and 200Hz, respectively. Fig. 2 shows the computational grid used for the retracted setup.

4 Results and Discussion

4.1 Effect of Thickening-Factor

The results obtained with n=5 and n=10 in Eqn. (2), i.e. from case Hot40a and Hot40b was found to be very much the same, as seen in Fig. 5 below. Since n=5 allows to use a larger time step in the present explicit method in time, this value was used in the sequel [14].

4.2 Behaviour of Recirculation Zone

In order to investigate the behaviour of the recirculation zone under the different conditions (reactive vs. non-reactive, retracted vs. non-retracted) instantaneous flowfields obtained from four flow cases are compared in Fig. 3, where the recirculation zone is denoted by the zero streamwise velocity contour lines. The CRZ of the two reactive cases (Fig. 3e and 3f) is entirely located in the high temperature post-flame region. Hence the local molecular viscosity is substantially higher than in the two non-reactive cases (Fig. 3a and 3b). This

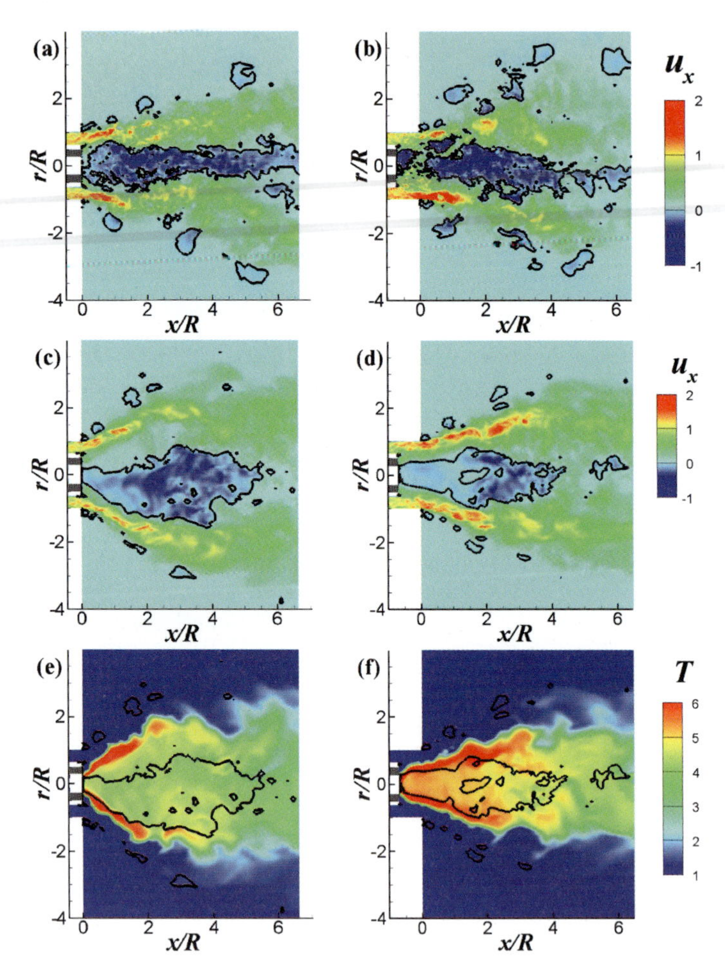

Fig. 3. (a)-(d) instantaneous contours of streamwise velocity, (e) and (f) instantaneous contours of temperature. Solid lines, u_x=0. (a)-(d) are for case Cold00, Cold40, Hot00, Hot40a, respectively. (e) and (f) depict the temperature field corresponding to (c) and (d), respectively. Note: u_x and T are normalized by the reference values, 11.5m/s and 298.15K, respectively

explains why the CRZ in the two reactive cases is much smoother than that in the non-reactive ones. The CRZ of cases Cold00, Hot00 and Hot40a (shown in Fig. 3a, 3c and 3d, respectively) is stable and does not precess around the axis of symmetry. In contrast, the CRZ of case Cold40 is not stationary but precesses regularly around the geometrical axis. Hence, the instantaneous streamwise velocity at the vicinity of the main jet nozzle oscillates accordingly. This is seen in Fig. 3b, in which higher values of u_x are seen in the lower side of the jet nozzle. Three dimensional animations have been created backing this statement.

The structures at the outer border of the jets in Fig. 3 show different spreading angles. This can be quantified with the help of the profiles of the mean velocity (two cases of them are shown in Fig. 4 and 5). Taking the profiles at x/R=2 and $r_{1/2}$, the outer point where $\langle u_x \rangle = 0.5 \langle u_x \rangle_{max}$, provides a corresponding measure. The behaviour is as follows: when the pilot jet nozzle is retracted in the iso-thermal case, the spreading angle increases somewhat, whereas it decreases in the reactive case.

4.3 Statistics of Velocity

Radial profiles of streamwise and angular velocity statistics for the retracted setup are shown in Fig. 4 and 5. For the non-reactive cases, the overall agreement between the LES result and the experimental data [8] is very good. It is seen that the negative axial velocity region (CRZ) is extended to $x/R > 3$ for Cold00 (not shown here), while for case Cold40, it is slightly shorter (Fig. 6a). The profiles of the mean flow show the CRZ, and at the same time, large tangential velocity exists in this region. Downstream turbulence becomes more and more isotropic. The rms fluctuations of angular velocity and streamwise velocity reach the same level, and their profiles become more and more flat.

For the reactive cases Hot40a and Hot40b shown in Fig. 5, the overall agreement is not as satisfactory as for the non-reactive cases. In the vicinity of the main jet nozzle, x/R=0.1, the agreement for the axial velocity is good. The profile of the tangential velocity of case Hot40a however differs from the experimental data in the centre region, $r/R < 0.5$. The rms-fluctuations at this station are also under-predicted. Numerous efforts were made to improve the result but failed because of the relatively low grid resolution between the blades of the pilot jet. In the reactive cases, only approximately 12×5 grid cells were used for resolving a single inter-blade channel to keep the total number of points in tangential direction reasonable. For the cases Cold00 and Cold40, more grid cells were used. Another source for disagreement is the bias of the LDA data at the outer edge of the jet towards higher velocities since only the jet and not the co-flow was seeded. Finally, a small systematic error occurs in some experimental data reflected by a difference between data from the two sides of the flame, plotted together in these figures.

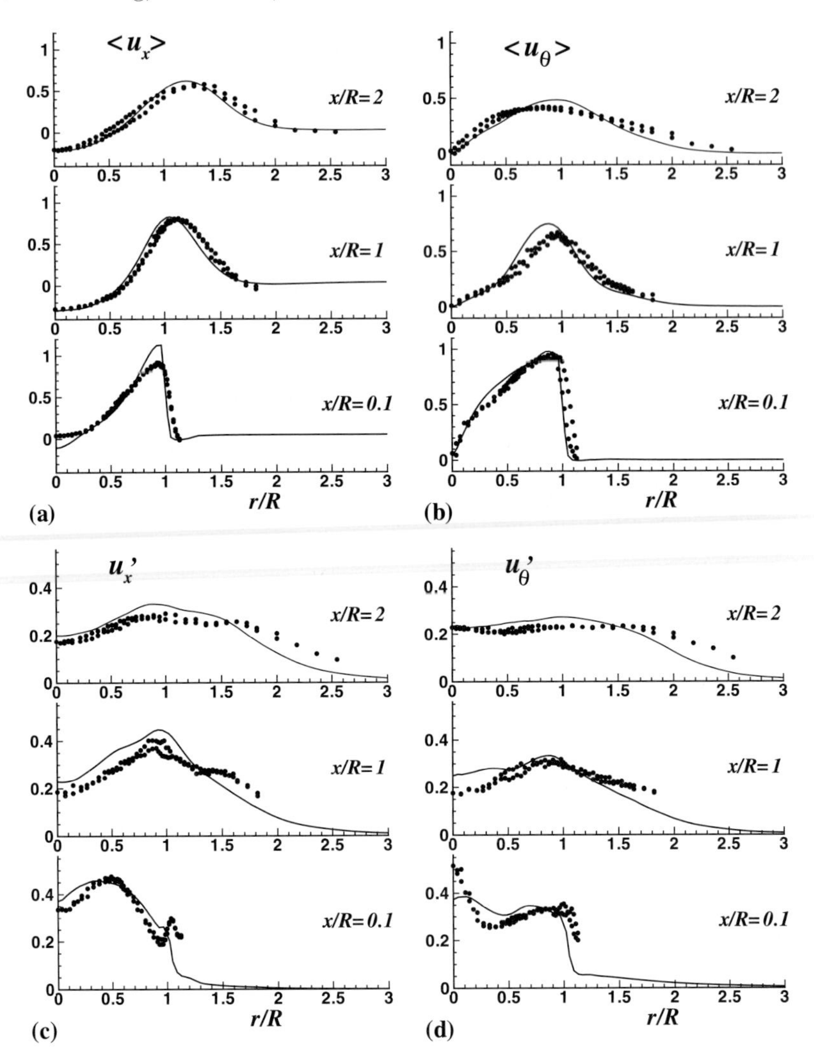

Fig. 4. Radial profiles of velocity statistics for case Cold40. (a) mean streamwise velocity, (b) mean tangential velocity, (c) rms-fluctuations of streamwise velocity, (d) rms-fluctuations for tangential velocity. Symbols: experiment; solid lines: LES results

4.4 Coherent Structures

In order to investigate the formation and evolution of coherent structures, iso-surfaces of the pressure fluctuation $p - \langle p \rangle$ are employed, filtered with a 3D box filter of twice the step size of the grid in a post-processing step [12]. In Fig. 6b, the PVC in the center region and an induced spiral structure in the outer region are observed together with further small vortex features,

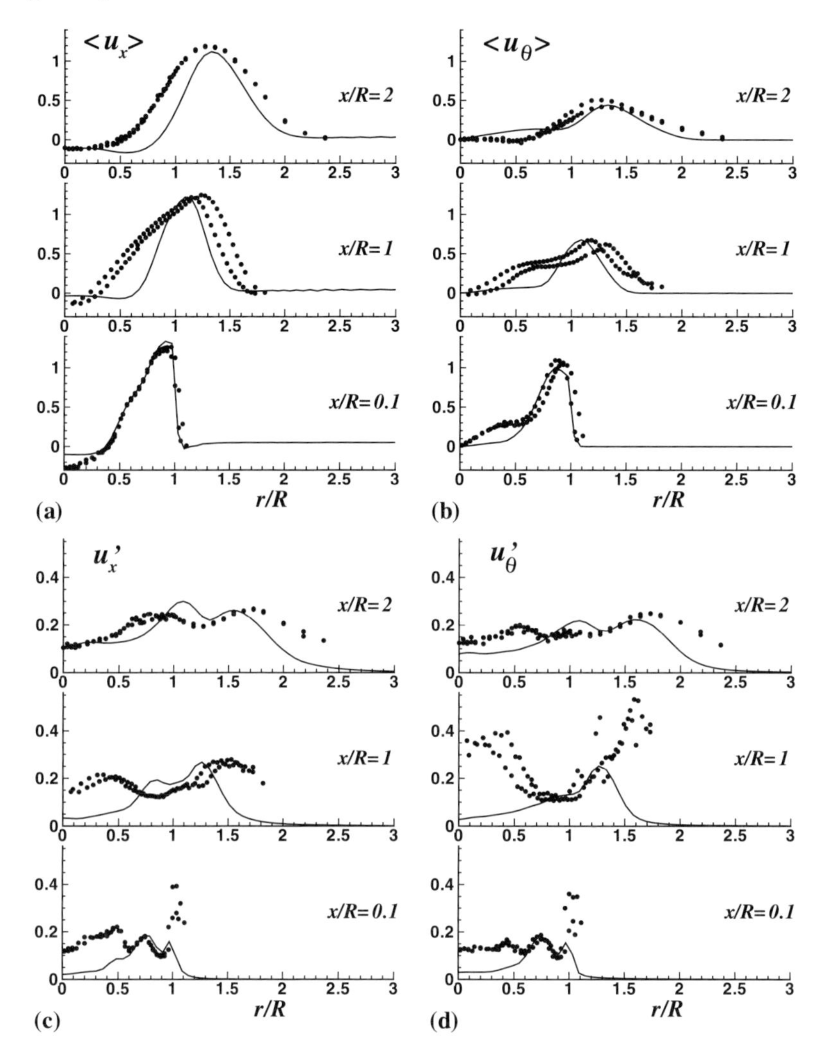

Fig. 5. As Fig. 4, but for reactive cases. Symbols: experiment; solid lines: Hot40a; dashed lines: Hot40b

rotating around the jet center regularly in the same rotating direction as the PVC. The agreement with the earlier simulations of [2] is excellent. This PVC is not observed in case Cold00, where a big, almost axisymmetric structure is found (Fig. 6a).

The PVC has important impact on combustion instabilities [3]. In order to investigate the occurrence of a PVC in the reactive cases, several visualization technologies were employed. Fig. 7 shows vortex structures obtained from the two reactive cases as well as the instantaneous flame front, represented as an iso-surface of the temperature. The resolved flame front is quite wrinkled,

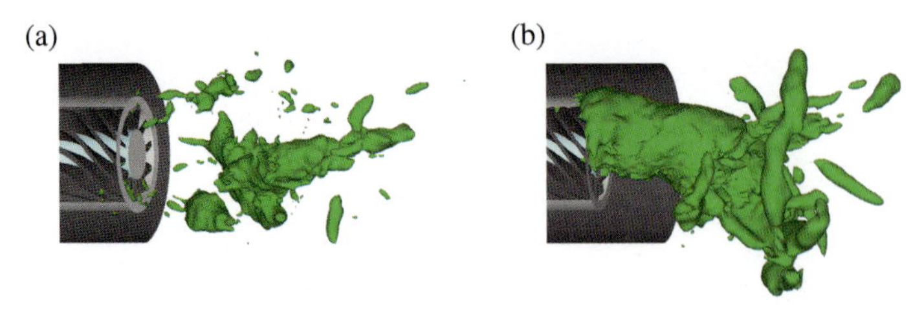

Fig. 6. Smoothed iso-surface of pressure fluctuation, $p - \langle p \rangle = -0.4$. (a) case Cold00, (b) case Cold40

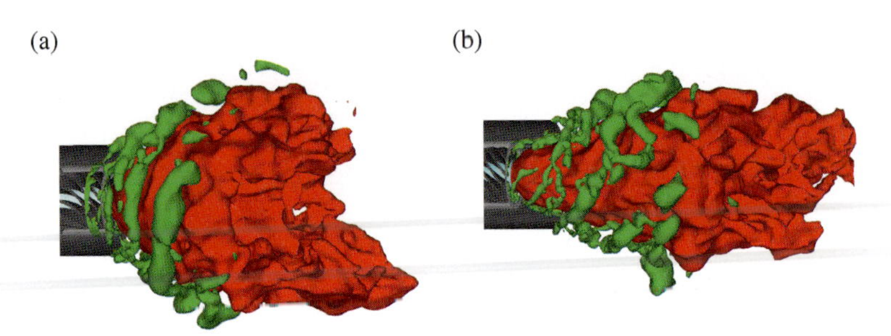

Fig. 7. Smoothed iso-surface of pressure fluctuation in brighter color, $p - \langle p \rangle = -0.1$, and iso-surface of temperature in dark color, T=1340K. (a) case Hot00, (b) case Hot40a

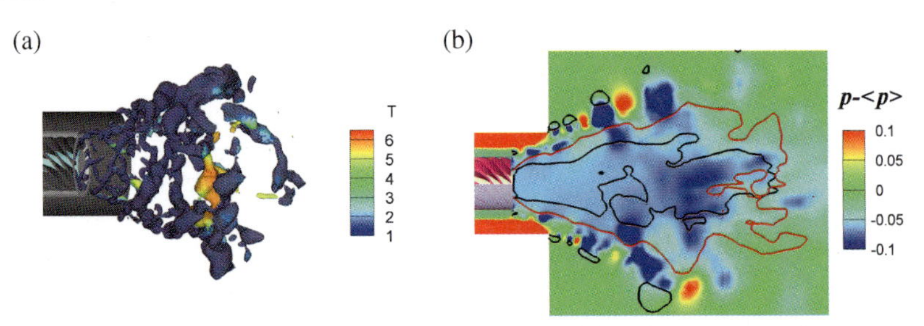

Fig. 8. Coherent structures for reactive case Hot40a. (a) Same iso-surface of pressure fluctuation as in Fig. 7(b), but colored with T. (b) contour for pressure fluctuation in the centre plane; red solid line is for T=1340K; black solid lines are for u_x=0.0

reflecting the large-scale interaction between turbulence and flame, while the unresolved wrinkling of the flame is modeled via the factor E.

In Fig. 7, the spiral-type vortices are located outside the flame front. For a better view of the centre structures, the same iso-surface of pressure fluctuation in Fig. 7b is shown separately in Fig. 8a, but colored with temperature. Fig. 7 and Fig. 8 and numerous other visualizations reveal that large structures with high temperature are only seen far downstream of the nozzle, i.e. for x/R >2.5. No big regular vortex structure was found close to the axis near the nozzle, i.e. no obvious PVC exists in the reactive flow. This is presumably

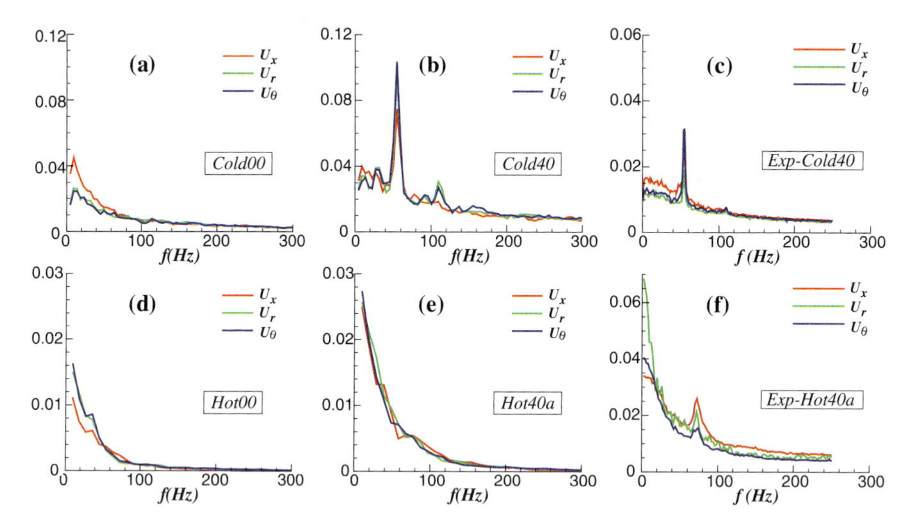

Fig. 9. Power spectrum of velocity components at location (x/R=1, r/R=0.18) for four cases. Experimental data from [8]

due to the fact that the CRZ is fully enclosed in the high-temperature post-flame region [3]. Fig. 8b shows the corresponding flowfield in the centre plane for case Hot40a. Clearly, the whole central reverse flow is of high temperature. In Fig. 8b the black circles located at the outer shear layer are related to the spiralling vortices seen in Fig. 7b.

The suppression of the PVC when turning from the cold flow to the hot flow is supposed to result from the substantially higher viscosity of the recirculating flow, which "produces conditions not favorable to PVC formation" [3]. The profiles of tangential velocity in Fig. 4b and 5b, for x/R=0.1 reflect this, in line with the results of [5]. In Fig. 4b at axial station x/R=0.1, a high gradient in the tangential velocity profile is found very close to the centre line. Although high gradients in the tangential velocity are also found in other cases (e.g. Fig. 5b), these are located somewhat more outward.

4.5 Power Spectrum Analysis

Power spectra of velocity components at location (x/R=1.0, r/R=0.18) are shown in Fig. 9, for four cases. For the non-reactive retracted case (Fig. 9b), a dominant peak at f=54Hz, is clearly seen. This frequency is the so-called PVC frequency and coincides very well with the experimental result. For the other three cases, no such dominant peak is found. In these spectra, larger amplitudes generally occur at lower frequency, f <30Hz.

Quantitative comparison between the LES result and the experimental data is only possible at few points, for instance at position (x/R=1, r/R=0.18), and only for the retracted geometry [8]. Fig. 9f is the experimental result for case Hot40a at this point, where a small peak at frequency

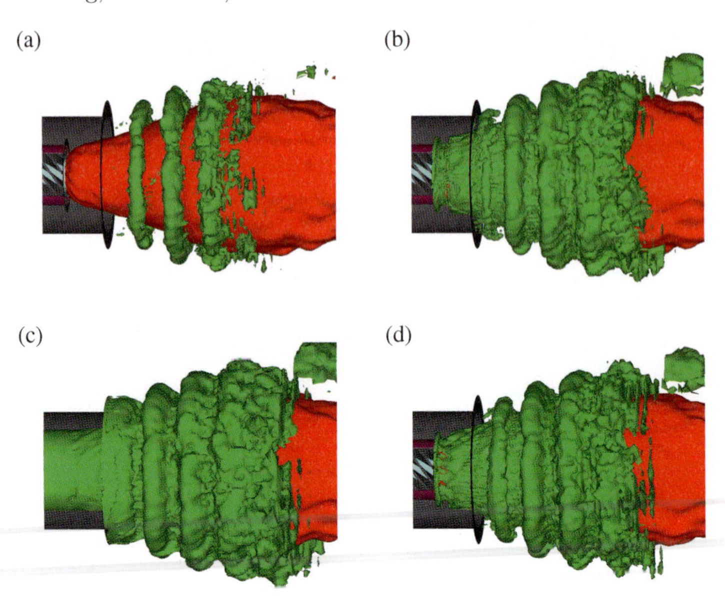

Fig. 10. Flame response to the pulsating pilot mass flow rate, case Pulse2. Red: iso-surface of T=1340K; blue: iso-surface of pressure p= -0.005. The phase angles are 0°, 90°, 180° and 270° for a, b, c, d, respectively

f=71Hz is observed. This peak is missing in the LES, but the overall tendency of the spectra is captured (Fig. 9e).

4.6 Flame Response to Pulsating Flow Rate of Pilot Jet

In the pulsating cases, the mass flow rate of the pilot jet varies according to

$$u(t) = U\left\{1 + \sqrt{2} \times 0.25 \times sin(2\pi t/T_0)\right\} \tag{3}$$

where U is the mean bulk velocity of the pilot jet, so that eqn. (3) produces 25% rms fluctuation of the mass flow rate in the pilot jet. The flow rate in the main jet was constant.

Flow structures of case Pulse2 are shown in Fig. 10. Unlike Fig. 6 and 7, iso-surface of the pressure itself is employed here instead of pressure fluctuation, since the time averaged pressure $\langle p \rangle$ in pulsating flows is not as meaningful as in non-pulsating flows. Fig. 10 shows a phase average over 30 periods. The phase angles are defined according to Eqn. (3). The large structures in Fig. 10 are ring vortices compared to the spiral type in the non-pulsating case Hot40a. During phase 0°–90° ring vortices form at the outer edge of the main jet and travel downstream. In [7], it was shown that reaction can occur in the vortex rings structure and hence produce a fluctuation of the heat release thus exciting combustion-driven oscillations. The geometry studied in [7] however

Table 2. Normalized integrated heat release rates

	Hot40c	Pulse1	Pulse2
mean	8.701	8.506	8.709
rms	0.172	0.185	0.155

is different from the one considered in the present work. It exhibits only one jet, and the whole jet is forced periodically so that reaction occurs in the vortex rings while in Fig. 10 the vortex rings are located outside the flame front.

Similar vortex rings are obtained in case Pulse1 (not depicted here). They are larger so that three rings are shown simultaneously compared to four in Fig. 10 when using the same visualization technique. In Pulse1 the vortex rings occasionally are slightly inclined with respect to the x-axis.

The integrated heat release rate was also determined and compared with to the case with steady inflow, Hot40c. The results are shown in Table 2. Here, all heat release rates are normalized with the same reference heat release rate. The fluctuation level of Pulse1 is the highest, whereas its mean heat release is the lowest. The fluctuation level of Hot40c is also high, although it is solely due to the natural turbulent flow. If it is desired to suppress oscillations of the heat release rate and to increase the overall heat release, the high frequency (200Hz) controlled pulsating seems preferable.

In addition, an attempt to quantitatively compare the flame response function to the experimental data from [8] was undertaken: the amplitude of the heat release rate oscillation and the phase angle between the heat release rate and the mass flow rate was investigated [13]. The overall agreement was found unsatisfactory. A better way for quantitatively compare the heat release rates obtained from the LES and the experiment is needed and subject of current work.

5 Conclusions and Outlook

In this work, the impact of heat release and location of a pilot jet on the swirling flow structures in a co-annular swirl burner is studied with LES. The agreement between the LES results and the corresponding experiment is good bearing in mind the uncertainties in boundary conditions and experimental data. A PVC was found only in the non-reactive retracted configuration but not in the other setups, although a CRZ occurs in all configurations. It is also seen that the dominating vortex structures in the combustion cases are generally found outside the flame front. They exhibit a spiral shape with steady inflow and ring shape with pulsating inflow. The method prescribed here is currently applied to a closed burner. Furthermore, different variants of the flame surface density model are applied and compared to the results with the thickened-flame model applied in the present study.

Acknowledgements. The authors gratefully acknowledge the support of the German Research Foundation (DFG) through the Collaborative Research Center SFB606 'Unsteady Combustion'. The computations were performed on both the HP-XC clusters of SSCK Karlsruhe and the SGI Altix 4700 cluster of ZIH Dresden. The authors thank W. Rodi for inspiring discussions on LES and H. Büchner and his co-workers (Karlsruhe University) for providing their experimental data in electronic form.

References

1. P. Habisreuther, C. Bender, O. Petsch, H. Büchner, and H. Bockhorn. Prediction of pressure oscillations in a premixed swirl combustor flow and comparison to measurements. *Flow, Turbulence and Combustion*, 77:147–160, 2006.
2. J. Fröhlich and D. A. von Terzi. Hybrid LES/RANS methods for the simulation of turbulent flows. *Prog. Aerospace Sci.*, 44:349–377, 2008.
3. N. Syred. A review of oscillation mechanisms and the role of the precessing vortex core (pvc) in swirl combustion systems. *Prog. Energy Combustion Sci.*, 32:93–161, 2006.
4. L. Selle, G. Lartigue, T. Poinsot, R. Koch, K.-U. Schildmacher, W. Krebs, B. Prade, P. Kaufmann, and Veynante D. Compressible large eddy simulation of turbulent combustion in complex geometry on unstructured meshes. *Combustion and Flame*, 137:489–505, 2004.
5. S. Roux, G. Lartigue, T. Poinsot, U. Meier, and C. Bérat. Studies of mean and unsteady flow in swirled combustor using experiments, acoustic analysis, and large eddy simulations. *Combustion and Flame*, 141:40–54, 2005.
6. O. Lucca-Negro and T. O'Doherty. Vortex breakdown: a review. *Prog. Energy Combustion Sci.*, 27:431–481, 2001.
7. C. Külsheimer and H. Büchner. Combustion dynamics of turbulent swirling flames. *Combustion and Flame*, 131:70–84, 2002.
8. H. Büchner. Teilprojekt c1: Experimentelle untersuchungen zur schwingungsneigung pilotierter vormischflammen. In *Forschungsbericht SFB 606*. University of Karlsruhe, 2004.
9. O. Colin, F. Ducros, D. Veynante, and T. Poinsot. A thickened flame model for large eddy simulations of turbulent premixed combustion. *Phys. Fluids*, 12:1843–1863, 2000.
10. J.P. Legier, T. Poinsot, and D. Veynante. Dynamically thickened flame les model for premixed and non-premixed turbulent combustion. In *Proc. summer program – 2000*. Center for Turbulence Research, 2000.
11. P. Moin, K. Squires, W. Cabot, and S. Lee. A dynamic subgrid-scale model for compressible turbulence and salar transport. *Phys. Fluids*, pages 2746–2757, 1991.
12. P. Wang, J. Fröhlich, V. Michelassi, and W. Rodi. Large eddy simulation of variable-density turbulent axisymmertric jets. *Int. J. Heat and Fluid Flow*, 29:654–664, 2008.

13. P. Wang, S.K.M. Ayyash, and J. Fröhlich. Flame response to a pulsating pilot jet in an unconfined double-concentric swirl burner. In *4th European Combustion Meeting*, page P811389, 2009.
14. P. Wang and J. Fröhlich. Impact of reaction and location of a pilot jet on the flow structures in a co-annular swirl burner. In *6th Symp. on Turbulence Shear Flows Phenomena*, pages 327–332, 2009.
15. C. Hinterberger. *Dreidimensionale und tiefengemittelte Large–Eddy–Simulation von Flachwasserströmungen.* PhD thesis, Institute for Hydromechanics, University of Karlsruhe, 2004.
16. J. Zhu. A low diffusive and oscillation-free convection scheme. *Comm. Appl. Num. Meth.*, 7:225–232, 1991.
17. M. Garcia-Villalba, J. Fröhlich, and W. Rodi. Identification and analysis of coherent structures in the near field of a turbulent unconfined annular swirling jet using large eddy simulation. *Phys. Fluids*, 18(5):055103, 2006.

Computational Fluid Dynamics

Prof. Dr.-Ing. Siegfried Wagner

Institut für Aerodynamik und Gasdynamik, Universität Stuttgart, Pfaffenwaldring 21, 70550 Stuttgart

A large number of highly qualified papers report upon the research work that was performed during the last 12 months using high performance computers (HPC) of HLRS Stuttgart and SCCK (Steinbuch Centre for Computing Karlsruhe). Only approximately 50 per cent of these papers could be selected for inclusion in the present book because of restricted space. Again these papers had to be preferred that demonstrated the unalterable usage of HPC for the solution of the problems besides high scientific standard.

Direct Numerical Simulation (DNS) and Large Eddy Simulation (LES) or hybrid methods, i.e. a combination of DNS or LES with RANS (Reynolds Averaged Navier-Stokes) methods require HPC with big capability both with respect to storage and performance. Thus, when the NEC SX-8 and NEC SX-9 at HLRS as well as the HP XC4000 at SCCK became available the capability of these computing centres grew enormously. In addition, further algorithms were developed simultaneously in order to improve the performance.

As a first example of progress in algorithms Dedner et al. explain the development of an interface library to solve partial differential equations, especially non-linear evolution equations. They demonstrate the versatility of the software packages DUNE and DUNE-FEM by several examples to be used for efficient simulations on the parallel computer HP XC4000 of SCCK including a recently developed stabilization technique using local grid adaptivity in combination with dynamic load balancing. They show also the advantage of higher order methods like DG (discontinuous Galerkin) that allow much coarser grids and reach the same accuracy as fine grids and lower order methods. They performed calculations with up to $1.34 \cdot 10^8$ elements and $1.35 \cdot 10^8$ unknowns using 512 processors on the HP XC4000. In this case they could reach an efficiency of over 0.96. So far the computations were restricted to relatively simple geometric configurations. It will be interesting to apply DUNE to more complex configurations in engineering applications.

The next six papers deal with DNS. Selent and Rist show by DNS that jet vortex generators can positively control boundary layer flows. Although this was already known from experiments DNS allows a lot new insights into the mechanisms of flow. However, it is very demanding since it exhibits multiple scale behaviour with respect to both geometry and flow physics. Because of the singular behaviour of the partial differential equations special care was applied to grid generation. Numerical instabilities had to be avoided by a special filtering system. One node with 16 CPUs on the NEC SX-9 and NEC microtasking shared-memory parallelization were applied to achieve a performance of 230 GFlops and a vector operation ratio of 99.67%. A total of 24,576 GB memory was accumulated. It would be interesting to find whether an application of two or more nodes would be possible and what order the performance would be then.

Wissink and Rodi use DNS to study the flow over and heat transfer to the stagnation region of a circular cylinder affected by free-stream fluctuations. They get the impressive results through computations on the NEC SX-8 using 192 processors on 24 nodes and 746.5 million grid points. The vector operation ratio was 98.8% with an average vector length of 146. The 3D simulation reached an average speed of 3.3 GFlops per processor which amounts to almost 634 GFlops in total. Each simulation takes 240,000 time steps using chain jobs with duration of up to 23.5 hours. To finish the 3D problems 258.5 wall-clock hours were required in total corresponding to 49632 CPU hours.

Wang studies turbulent flows with DNS. Starting from sizable capacity data for homogeneous shear turbulence he investigates the conditional statistics along gradient trajectories and gets new results different from the classical Kolmogorov scaling. The remarkable results are obtained on the NEC SX-8 using 16 nodes and MPI. The average performance was 1.5 GFlops per processor on 128 processors or 192 GFlops totally. The average vector length is 146. The total memory used is 1086 GB or 8.6 GB per process.

Weking et al. treat the rise behaviour of single air bubbles of different size in viscous liquids with DNS. The authors use their own code FS3D that solves the incompressible Navier-Stokes equations using a Volume-of-Fluid (VOF) technique. To minimize the computational effort, a moving frame of reference technique is used. The simulations demonstrate a good agreement of the rise velocity and deformation of the bubbles with experiments. When using an average vector length of 208 vector operation ratios of about 98.1% are achieved. The performance reached 6.2 GFlops per processor on 1 CPU and 3.5 GFlops per processor on 8 CPUs or 28 GFlops totally.

Panara and Noll study the wall heat load in unsteady and pulsating combustor flow within a channel. Their goal is to improve the validity and limitations of turbulence models in URANS (Unsteady Reynolds Averaged Navier-Stokes) or hybrid LES/RANS equations to be able to more accurately study heat load and high-cycle thermal fatigue of combustion chambers in non-isothermal unsteady flow conditions. The computer code used was developed using OpenFaom Toolbox version 1.3. Computations for CFD code validation and optimization were conducted on 2 and 4 nodes of the HP XC4000 at SCCK. The grids used contained up to 1,600,000 points requiring up to 8 GB. The typical total CPU time for 1 run required 5 to 6 days.

Devesa et al. study passive-scalar transport by extending the Adaptive Local Devolution Method to LES. They show that the resulting adaptive advection algorithm enables the reliable prediction of the turbulent transport of passive scalars in isotropic turbulence and in turbulent channel flow for a wide range of Schmidt numbers. Simulations of the flow in a confined rectangular-jet reactor agree well with experiments. A simulation involves 13.65 million finite volumes or a grid size of 14.4 million points. The computation on a NEC SX-8 of HLRS requires around 15 GB RAM and for one run 4600 CPU hours.

Klumpp et al. analyze the fundamental physical mechanism that determines the damping effect of a riblet surface on 3D spatial transition in a flat plate zero-pressure gradient boundary layer using LES. They investigate two types of transition scenarios and the effects on transition delay. The simulations are run on the NEC SX-8 with 5 nodes or 40 CPUs of HLRS and require 21 million grid points with a total memory of 20.1 GB. They achieve a vector operations ratio of maximal 99.6% and a performance of 158.3 GFlops. They could even increase the performance on a NEC SX-9 of HLRS to 483.2 GFlops using 2 nodes or 32 CPUs when performing an LES around a DRA2303 airfoil.

Uddin et al. investigate the flow field and heat transfer of turbulent pulsating and swirling jet impingements by LES using the CFD Code FASTEST. The computations are performed on the CRAY Opteron Cluster using 20 CPUs and on the NEC SX-8 using MPI. The simulations of a pulsating jet requires 10000 CPU hours whereas the swirling jet needs 15000 CPU hours in average on the CRAY Opteron Cluster. The authors report that the computations on the NEC SX-8 consume only one third to one fourth of the computation time compared to the CRAY Opteron platform.

Pritz et al. calculate the resonance characteristics of a combustion system by LES that consists of Helmholtz-resonance-type components. The goal is to predict the resonance characteristics already during the design phase. They apply their in-house flow solver SPARC in combination with MPI on the HP XC4000 of SCCK using 108 Opteron processors. Since communication between CPUs of the HP XC4000 are handled by the InfiniBand 4X DDR Interconnect Pritz et al. reach a parallel efficiency of close to 98%.

Von Terzi et al. investigate the flow in asymmetric diffusers with complex 3D separation employing LES and RANS simultaneously. Their objective is to obtain reference data for assessing the performance of recently developed hybrid LES/RANS codes. They use the code LESOCC and study five different grid sizes between 1.6 and 7.4 million grid cells. Different geometries, turbulence models and grids are investigated. The most time consuming run required 120,000 CPU hours. The code was tested on various platforms including the NEC SX-8, the SGI Altix 4700 and Linux clusters. Scaling of the code LESOCC was tested on the HP XC4000 for up to 512 processor cores achieving a speed-up of approximately 80%. Findings in the literature are confirmed that these complex flows pose a severe challenge to RANS. Accurate and reliable results cannot be expected from such simulations and LES has to be applied which underlines the necessity of high performance computers.

Beronov and Özylmaz use the Lattice Boltzmann (LBM) simulation tool for the investigation of grid generated turbulence in order to provide qualitatively new data for practically relevant modelling of inhomogeneous turbulence. They apply their in-house LBM-solver BEST that is MPI parallelized and has versions optimized for various cache and vector-based platforms, e.g. NEC SX-8 and SGI Altix 4700. While using these high performance computers the authors show that decaying grid-generated turbulence is self-similar sufficiently far from the grid. The computations were performed on the NEC SX-8 of HLRS using 8 nodes. An excellent scaling of the LBM code was achieved and allowed to perform a series of DNS runs and thus a parametric study of several effects as well as a good convergence of all velocity statistics of interest.

Adolph et al. perform numerical simulations of journal bearings that include Reynolds differential equations under consideration of inertia terms which are often neglected. Using the finite difference method for the numerical solution of the governing equations the authors show that the FDEM code is able to give global error estimates. The computations were carried out on the HP XC4000 of SCCK including AMD Opteron processors with 5.2 GFlops peak performance. The authors report results for 8 different Reynolds numbers on an 11x21 grid. For the maximum investigated Reynolds number the authors achieve a relative maximum estimated error as low as $0.852 \cdot 10^{-3}$ and a CPU time per computation step of 0.090 seconds. When Adolph et al. refined the grid to 21x41 nodes the maximum error is reduced to $0.230 \cdot 10^{-3}$.

Krause and Reinartz perform numerical simulations of the performance of two possible intake concepts of a scramjet demonstrator model at a flight Mach number of 8. The computations are performed with the RANS solvers FLOWER and QUADFLOW. The authors investigate 2D intakes with non-converging sidewalls and 3D intakes with converging sidewalls. The computations were performed on 16 processors of the NEC SX-8 of HLRS. The number of grid cells was between 3 and 4 millions. The memory in the 3D cases amounts to around 80 GB. The authors reach a vectorization level of 99.7% using the FLOWer code. The answer time lies between five to seven days. A single batch job performs approximately 30000 time steps and requires 10 hours of CPU-time per node. The application of the NEC SX-8 seems to be inevitable to achieve reliable design results in reasonable time.

Chen et al. present numerical investigations within the frame of a new project entitled "Aero-Structural Dynamics Methods for Airplane Design". The goal is to study the dynamic vibration decay behaviour of the HIRENASD model. The Computational Aero-Structural Dynamics code SOFIA of RWTH Aachen is used for the numerical simulations. 1.6 million grid points are necessary for the numerical simulations that were performed on 16 XEON processors of the computing centre of RWTH Aachen and on 16 processors of the NEC SX-8 at HLRS. The NEC SX-8 was five times faster than the XEON in case of dynamic simulations. The code performance on the NEC SX-8 reached an average of 2480 MFlops and a vector operation ratio of 97%.

Higher Order Adaptive and Parallel Simulations Including Dynamic Load Balancing with the Software Package DUNE

Andreas Dedner[1], Robert Klöfkorn[2], and Dietmar Kröner[3]

[1] Abteilung für Angewandte Mathematik, Universität Freiburg (funded by the Landesstiftung Baden–Württemberg)
dedner@mathematik.uni-freiburg.de
[2] Abteilung für Angewandte Mathematik, Universität Freiburg (funded by the BMBF and DFG)
robertk@mathematik.uni-freiburg.de
[3] Abteilung für Angewandte Mathematik, Universität Freiburg
dietmar@mathematik.uni-freiburg.de

Summary. In this paper we describe the recent development of the interface library DUNE and in particular the module DUNE-FEM. We demonstrate that the development of the software package DUNE and in particular the module DUNE-FEM has reached a state where DUNE and DUNE-FEM can be used for efficient parallel simulations including higher order discretizations and techniques like local grid adaptivity in combination with dynamic load balancing. This is shown by test problems reaching from academical up to more sophisticated problems.

1 Introduction

The main aspect of this project is the development of an interface library for solving partial differential equations, especially non-linear systems of evolution equations. These systems are used in many different fields of applications from astrophysics to industrial problems, to model the evolution of physical quantities U, e.g., density, momentum, and energy of a fluid. A very general form of these equations is

$$\partial_t U(x,t) + \nabla \cdot (F(U(x,t),x,t) - a(U,x,t)\nabla U(x,t)) = \\ S(U(x,t),x,t) + \mathcal{T}[U(\cdot,t)](x) \; .$$

This is an advection-diffusion-reaction equation including a non-local term $\mathcal{T}$; this term is used to model for example self-gravitation or radiation in astrophysics or the pressure distribution in porous media flow problems. Many applications studied in different projects at our institute can be modeled by this general system of evolution equations:

W.E. Nagel et al. (eds.), *High Performance Computing in Science and Engineering '09*, DOI 10.1007/978-3-642-04665-0_16,

1. simulation of geophysical flows (funded by the Landesstiftung Baden-Württemberg)
2. simulation of fuel cells (funded by the BMBF)
3. simulation of phase transitions (funded by the DFG)
4. weather forecast simulations (funded by the DFG)
5. coupled hydrological processes (funded by the BMBF)
6. evolution on and of surfaces (funded by the DGF)
7. scavenging of a two-stroke engine with detailed combustion (funded by the Landesstiftung Baden–Württemberg)

Our goal is to develop a general framework for solving this type of equations based on modern software design techniques. Hereby efficiency and the reuse of code is a major aspect in the design process. Since an important aspect of our work is the design and the improvement of numerical methods, it has to be easy to apply the code to different test cases, model equations in one or two space dimensions, and must allow an easy comparison of new methods with established ones. Also, due to the complexity of the applications mentioned above, parallelization combined with local grid adaptation is a central ingredient for the development of efficient numerical schemes. To be able to concentrate on the underlying applications and on the design of new numerical schemes, aspects like load-balancing in a distributed computational environment should not be a concern of the researcher working on the project; the software library we are developing in this project should handle all these aspects in a transparent manner.

2 The Dune Interface Library – Recent Development

All the design restrictions mentioned in the introduction require the use of modern software design techniques. These strategies have already been successfully applied during the development of the DUNE grid interface [1, 2, 3]. The grid interface library allows the generic access to the grid structure which forms the basis of most numerical schemes for solving partial differential equations. Lately, version 1.2 of DUNE (consisting of the the DUNE core modules: DUNE-COMMON, DUNE-GRID, DUNE-ISTL) has been released showing that the development on the grid interface and the iterative solver template library has reached a stable state. So far the following grid implementations are available through the grid interface:

- `ALUGrid`: simplex grid in 2d/3d and cube grid in 3d with non-conform grid adaption, parallelization and dynamic load balancing
- `AlbertaGrid`: simplex grid in 2d/3d with conform grid adaption (bisection)
- `NetworkGrid`: grid for 1D networks
- `PrismGrid`: tensor product prismatic grid ($\Omega \times [0, h]$)
- `PSGrid`: parallel simplex grid also on manifolds

- `UGGrid`: hybrid grid with non-conform adaption and with red-green closure
- `YaspGrid`: parallel cartesian grid

Based on the DUNE core modules we are developing the package DUNE-FEM of which version 1.0 is about to be released [5]. In this package interfaces for discrete functions and operators are defined and a wide range of examples are implemented. During the design process the efficient and easy use of parallel computers has always been a central goal. The structure of the interfaces closely follows the mathematical formalism used to define grid based numerical schemes, e.g., the finite-element framework found in [4]. The package thus includes interfaces for

- Function spaces and functions
- Discrete function spaces (combining function space and vector valued finite base function set), DG and Lagrange higher order implemented
- Discrete functions (with element wise representation, dof handling)
- Discrete spatial operators (with efficient operations, e.g., $+$ and $\circ$) DG for elliptic and hyperbolic operators, CG for elliptic operators
- Inverse operators (Newton, Krylov methods...)
- IMEX Runge-Kutta methods for time dependent problems
- Automatic handling of grid adaptation, parallelization, and load balancing

In the following we present some test cases showing in particular the parallel efficiency of the developed code.

3 Applications

In this section we present several numerical examples using the DUNE-FEM module developed at our institute. The first example is a rather well known test example, which has been added here to show the efficiency of the parallelization of the linear solvers implemented in DUNE-FEM (in particular the communication during each iteration of the linear solver). The second example is a cutting edge example; a well known benchmark problem for the Euler equations is solved with a third order method (including a newly developed stabilization technique [6]) using local grid adaptivity in a parallel simulation including dynamic load balancing. The third and last example considers shallow water flow. The development of the scheme and its implementation was performed during a diploma thesis [8] and is based on a 3D hydrostatic free surface Navier-Stokes model.

3.1 Poisson's Equation

In this section we consider a benchmark problem for the Poisson equation

$$\begin{aligned} -\Delta u &= f \quad \text{in } \Omega \subset \mathbb{R}^d, \quad d \in \{1,2,3\} \\ u &= g \quad \text{on } \partial\Omega \end{aligned} \tag{1}$$

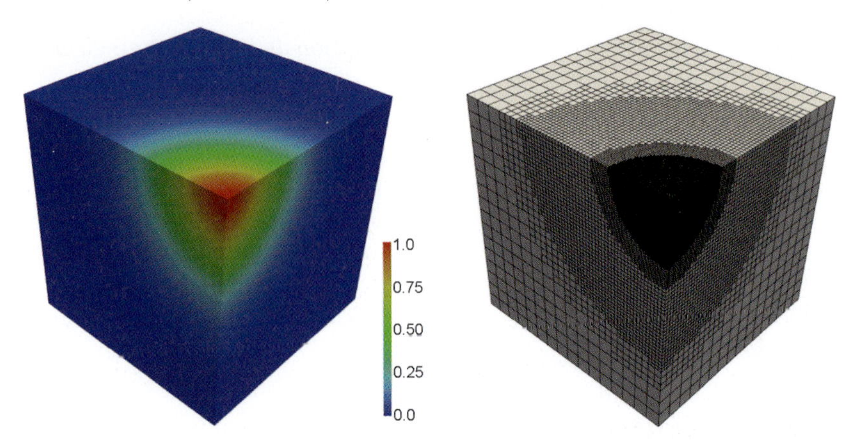

Fig. 1. Solution of the Poisson equation on a non-conform refined hexahedral grid

on the domain $\Omega = [0,1]^d$. The data is chosen to yield an exact solution

$$u(x) = e^{-10|x|^2}.$$

Therefore, the right hand side of (1) is given by

$$f(x) = -\Delta u(x) = \big(20d - 400|x|^2\big)e^{-10|x|^2}$$

and the the boundary data is simply $g = u_{|\partial\Omega}$.

Discretization

This problem is solved using a standard conforming finite elements approach.

Numerical Results

In Figure 1 the approximative solution of equation (1) on a non-conforming adaptively refined hexahedral grid is shown. The simulation used `ALUCubeGrid` (providing unstructured hexahedral grids) and P_1 Lagrange elements. In the left part of Figure 2 the solution of (1) on a non-conforming adaptively refined triangular grid is shown. The simulation used `ALUSimplexGrid` and P_1 Lagrange elements. The local grid adaptivity uses a residual based error estimator for this problem (see for example [12]). In the right part of Figure 2 the solution of (1) on a Cartesian grid using P_6 Lagrange elements is shown.

In Figure 3 the run times for solving this problem on a very fine uniformly refined grid containing about $1.34 \cdot 10^8$ hexahedrons using different grid implementations and different numbers of processors is shown. Each computation was done twice using different methods of communication. One communication method uses the data communication provided by the DUNE grid interface. The other method builds a cache holding information of all DoFs that

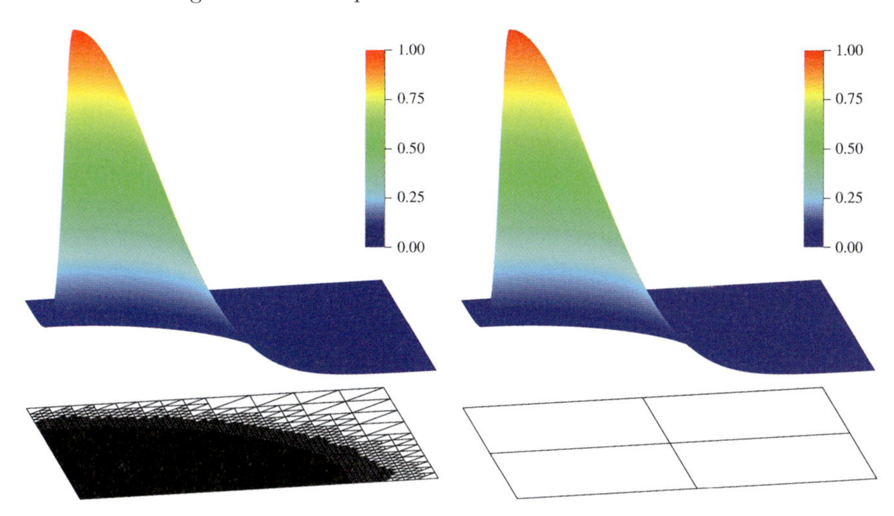

Fig. 2. Solution of the Poisson equation: on a non-conform refined triangular grid using P_1 finite elements (left) and on a rectangular mesh with 4 elements and P_6 finite elements (right). For both solutions the L^2-error is approximately $2.85 \cdot 10^{-5}$. For the P_1 solution 105338 grid cells were necessary while for the P_6 solution only 4 cells

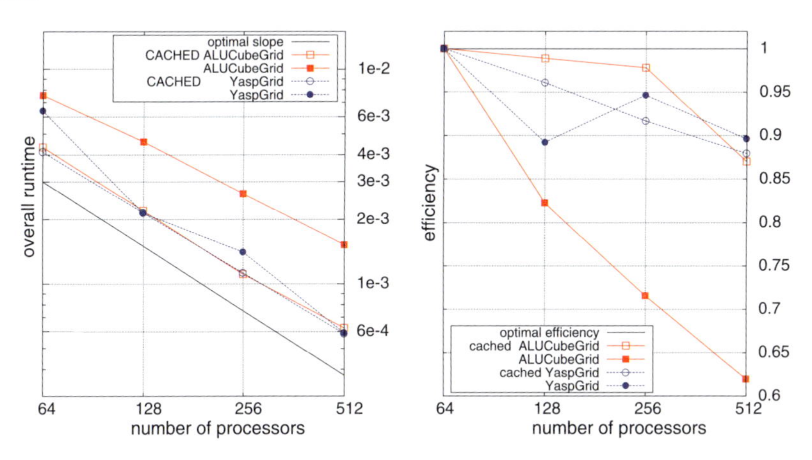

Fig. 3. Run times for a parallel computation (left) and efficiency of the parallel code (right) for the Poisson equation. The plots show the results for two different grid implementation (`ALUCubeGrid` and `YaspGrid`) on a hexahedral grid containing about $1.34 \cdot 10^8$ elements and the problem size is about $1.35 \cdot 10^8$ unknowns. The computations have been done on the parallel super computer XC4000 of the SCC Karlsruhe

need to be exchanged during a communication procedure and thus a grid traversal is not necessary and the message buffers can be allocated at once since the size is already known. The computations using cached communication are tagged with **cached** in both plots of Figure 3. In the left part of Figure 3 one can see that the computation using `ALUCubeGrid` and the cached communication is more than twice as fast as the run using the non-cached communication. With a higher number of processors the gap becomes even larger. For `YaspGrid` (a Cartesian grid) the cached communication is only slightly faster since the traversal of grid cells is already very cheap. Also, the results for the computation using the cached communication are more uniformly decreasing. `YaspGrid` was used with no overlap and for `ALUCubeGrid` all vertices with partition type `ghost` are neglected as unknowns for the solution, right hand side, and the system matrix. Comparing both grids, we can see from Figure 3 (left) that they show almost identical run times. We state that the use of the cached communication is essential for an efficient parallel communication using a large number of communication operations during execution of the numerical solution algorithm such as a linear solver or also an ODE solver does.

3.2 The Euler Equations

For a compressible inviscid fluid the Euler equations of gas dynamics have the following form:

$$\partial_t \boldsymbol{u} + \sum_{j=1}^{d} \partial_{x_j} \boldsymbol{f}_j(\boldsymbol{u}) = 0, \qquad \text{in } ([0,T) \times \Omega \subset \mathbb{R}^d), \quad d \in \{1,2,3\} \tag{2}$$

where the vector of the conservative variables is

$$\boldsymbol{u} = (\rho, \rho\boldsymbol{v}, e)^T, \quad \rho\boldsymbol{v} = (\rho v_1, \ldots, \rho v_d)^T, \quad e = \rho E.$$

We assume that solutions $\boldsymbol{u}$ of (2) take it's values in the set of states

$$\Psi := \big\{(\rho, \rho\boldsymbol{v}, e) \mid \rho > 0, \boldsymbol{v} \in \mathbb{R}^d, e - \frac{\rho}{2}|\boldsymbol{v}|^2 > 0\big\},$$

and the convective flux functions for $i = 1, ..., d$:

$$\boldsymbol{f}_i(\boldsymbol{u}) := \begin{pmatrix} \boldsymbol{u}_{i+1} \\ \boldsymbol{u}_{i+1}\boldsymbol{u}_2/\boldsymbol{u}_1 + \delta_{i,1}\, P(\boldsymbol{u}) \\ \vdots \\ \boldsymbol{u}_{i+1}\boldsymbol{u}_{d+1}/\boldsymbol{u}_1 + \delta_{i,d}\, P(\boldsymbol{u}) \\ (\boldsymbol{u}_{d+2} + P(\boldsymbol{u}))\, \boldsymbol{u}_{i+1}/\boldsymbol{u}_1 \end{pmatrix},$$

where $\delta_{i,d}$ is the Kronecker delta. The system is closed by the equation of state for an ideal gas where the pressure is given by

$$P(\boldsymbol{u}) = (\gamma - 1)\Big[\boldsymbol{u}_{d+2} - \frac{\boldsymbol{u}_1}{2} \sum_{i=1}^{d} (\boldsymbol{u}_{i+1}/\boldsymbol{u}_1)^2\Big] = (\gamma - 1)\Big[e - \frac{\rho}{2}|\boldsymbol{v}|^2\Big],$$

where γ is the adiabatic constant (see for example [9]).

Discretization

The system is discretized using a Runge-Kutta Discontinuous Galerkin method. As basis functions we use quadratic monomial basis functions that are orthonormal with respect to the L^2 scalar product. As numerical flux function the HLL flux function was used which can be found in standard textbooks on the subject (see for example [9, 10]). The discretization includes a limiter based stabilization technique which is described in detail in [6].

Numerical Results

As benchmark problem we consider the *Forward Facing Step* (see [6] for details) for $d = 3$.

In Figure 4 the density distribution including the adaptively refined grid can be found. One can also see the underlying grid partitioning. The simulation has been done on the parallel super computer XC4000 of the SCC Karlsruhe using 512 processors. Quadratic polynomial basis functions were used. The final grid contains about 4.5 million grid cells which leads for this example to about $2.25 \cdot 10^8$ unknowns. Grid adaptation is performed in each time step. If the local grid adaptation leads to an unbalance of work load then a **dynamic load balancing** is performed such that the work is again equally distributed between the processors[1] again. A detailed EOC analysis of this discretization as well as other benchmark problems for this problem can be found in [6].

In Figure 5 the parallel performance of the code is presented. In the left part the run times for one time step as well as only for the ODE solver are plotted for the runs on 128, 256, and 512 processors. In the right part the efficiency due to this run times is shown. One can see that the overall efficiency is above 0.93 and the efficiency of the serial part (ODE solver) of the code is above 0.96 which is very close to the optimal value of 1. This indicates that the parallelization of the code is very efficient for this problem.

3.3 Free Surface Shallow Water Flow

Problem Formulation

We solve the 3D free surface Navier-Stokes equations with a hydrostatic pressure assumption, a model suitable for shallow flows. Taking into account the orography b and denoting with h the free surface the computational domain is

$$\Omega(t) = \left\{ (\boldsymbol{x}, z)^T \in \mathbb{R}^d \, : \, \boldsymbol{x} \in \Omega_{\boldsymbol{x}}, \, b(\boldsymbol{x}) < z < b(\boldsymbol{x}) + h(\boldsymbol{x}, t) \right\}.$$

[1] The load balancing considers the number of grid elements considered in the numerical algorithm and is based on the graph partitioning algorithm provided by METIS [11].

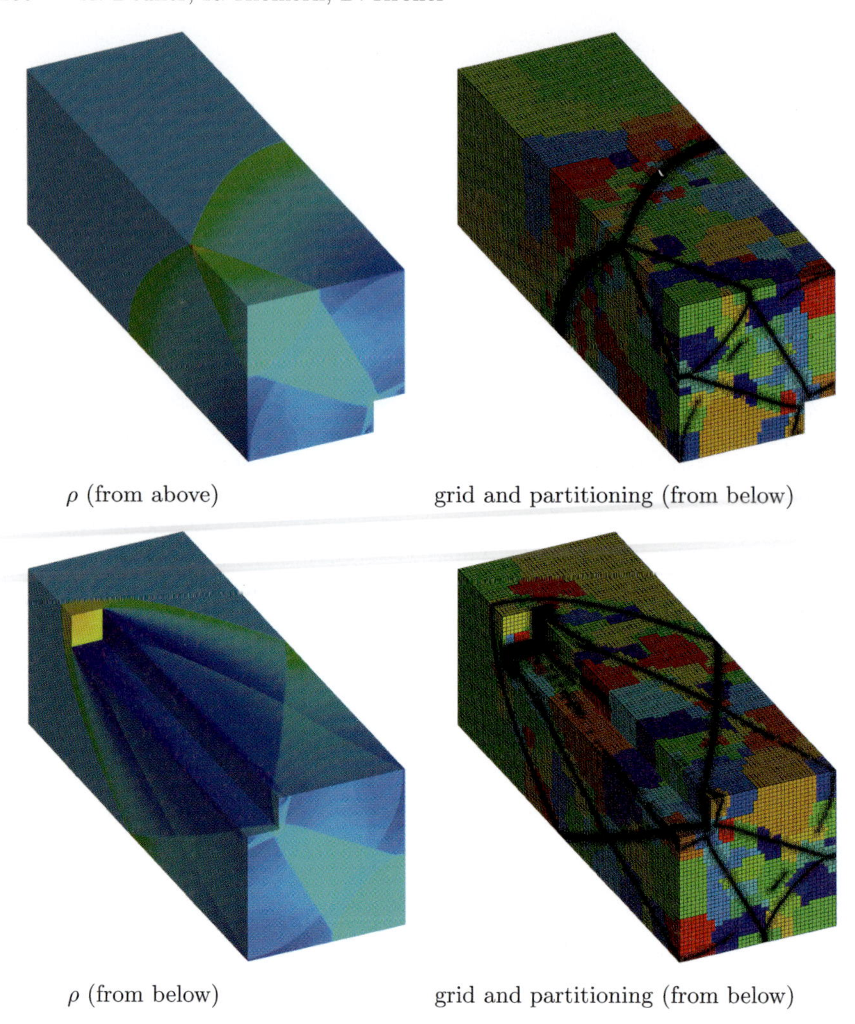

Fig. 4. Density distribution, adapted grid and partitioning of the grid at $T = 2$. The calculation used `ALUCubeGrid` and 512 processors. Quadratic basis functions were used

where $\Omega_{\boldsymbol{x}} \subset \mathbb{R}^{d-1}$ is a fixed domain over which b and $h(\cdot, t)$ can be represented as functions. We will denote the d-dimensional velocity field by the following system for the 3d velocity field $\mathbf{u} = (\mathbf{u}_{\boldsymbol{x}}, w)^T$ results from a *Shallow Water Scaling* of the incompressible Navier-Stokes equations as it is presented in [7]:

$$\begin{aligned} \partial_t h + \nabla_{\boldsymbol{x}} \cdot \left(\int_b^{b+h} \mathbf{u}_{\boldsymbol{x}} dz \right) &= 0 && \text{in } \Omega_{\boldsymbol{x}}, \\ \partial_t \mathbf{u}_{\boldsymbol{x}} + (\mathbf{u} \cdot \nabla) \mathbf{u}_{\boldsymbol{x}} + g \nabla_{\boldsymbol{x}} h &= -g \nabla_{\boldsymbol{x}} b + \partial_z \left(\mu \partial_z \mathbf{u}_{\boldsymbol{x}} \right) && \text{in } \Omega(t), \\ \nabla \cdot \mathbf{u} &= 0 && \text{in } \Omega(t), \end{aligned} \tag{3}$$

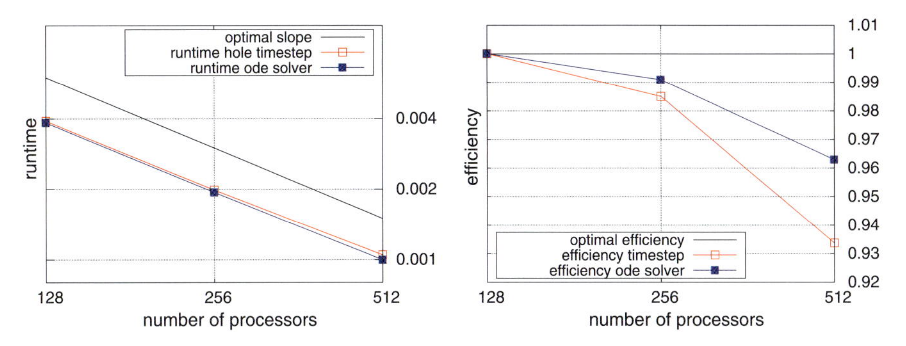

Fig. 5. Run times for a parallel computation (left) and efficiency of the parallel code (right) for the Euler equations using the third order stabilized DG discretization using hexahedral elements (`ALUCubeGrid`). On the left plot the runtime for one complete timestep (all) and the runtime of one timestep taken by the ODE solver (ode) are shown. In the right plot the efficiency of the code is shown. The simulations have been done on the parallel super computer XC4000 of the SCC Karlsruhe

where $g > 0$ is the gravitational constant. For the full set of boundary conditions associated with this system we refer to [8].

Discretization

The system is solved using a Local Discontinuous Galerkin approach. Using a sigma transformation the problem is represented on a fixed domain $\omega \times [0,1]$ where $\omega \subset \mathbb{R}^2$. As basis function linear basis functions that are orthonormal with respect to the L^2 scalar product are used.

Implementation Details

The method uses the concept of combined operators employing three steps: $(h, \mathbf{u}_{\boldsymbol{x}})$:

1. compute the integrals of the horizontal velocities $\int_b^{b+h} \mathbf{u}_{\boldsymbol{x}} dz$
2. compute the vertical velocity based on the divergence constraint: $\partial_z w = -\nabla_{\boldsymbol{x}} \cdot \mathbf{u}_{\boldsymbol{x}}$
3. compute the advection-diffusion terms in (3)

This spatial discretization is combined with an implicit-explicit Runge-Kutta solver treating the diffusion terms implicitly to increase the stability of the method.

To increase the efficiency of the first two steps of the algorithm a special semi-discrete prisma grid (Figure 6) is used, the grid is structured in the z direction so that the computation of the vertical integral and transport is easy to compute.

Fig. 6. (*left*) partitioning of a 3d semi-structured prism grid, (*middle*) 3d representation of initial conditions, and (*right*) solution to a latter time

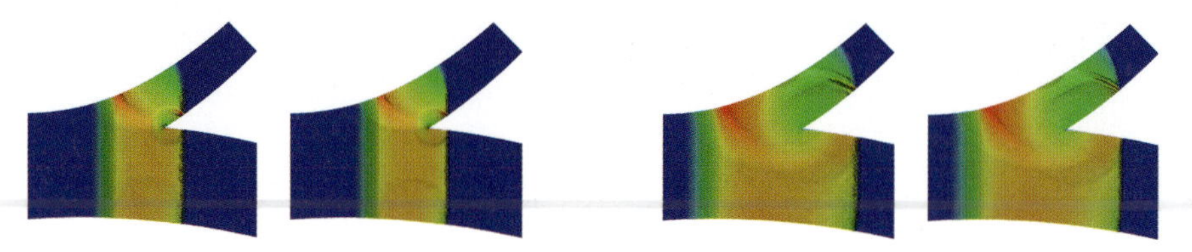

Fig. 7. Two points in time of a simulation using piecewise linear basis functions on coarse grid (917760 elements) (left) and on finer grid (3671040 elements) (right)

Numerical Results

Figure 7 shows an example computation of a wave moving from left to right hitting a fork in the river. The convergence of the scheme is demonstrated showing results on two different grids for two time steps.

4 Conclusion

We could show that the development of the software package DUNE and in particular the module DUNE-FEM has reached a state where from academical test problems up to more sophisticated problems DUNE and DUNE-FEM can be used for efficient parallel simulations including higher order discretizations and techniques like local grid adaptivity in combination with dynamic load balancing.

References

1. P. Bastian, M. Blatt, A. Dedner, C. Engwer, R. Klöfkorn, R. Kornhuber, M. Ohlberger, and O. Sander. A generic grid interface for parallel and adaptive scientific computing. II: Implementation and tests in DUNE. *Computing*, 82(2-3):121–138, 2008.

2. P. Bastian, M. Blatt, A. Dedner, C. Engwer, R. Klöfkorn, M. Ohlberger, and O. Sander. A generic grid interface for parallel and adaptive scientific computing. I: Abstract framework. *Computing*, 82(2-3):103–119, 2008.
3. P. Bastian, M. Droske, C. Engwer, R. Klöfkorn, T. Neubauer, M. Ohlberger, and M. Rumpf, *Towards a Unified Framework for Scientific Computing.* In Proc. of the 15th International Conference on Domain Decomposition Method, 2004.
4. P.G. (ed.) Ciarlet and J.L. (ed.); Lions. *Handbook of numerical analysis. Volume II: Finite element methods (Part 1).* Amsterdam etc.: North-Holland. ix, 928 p. Dfl. 275.00, 1991.
5. A. Dedner, A. Klöfkorn, M. Nolte, and M. Ohlberger. DUNE-FEM – The FEM Module. URL http://dune.mathematik.uni-freiburg.de/, Applied Mathematics, University Freiburg, 2007.
6. A. Dedner and R. Klöfkorn. A generic stabilization approach for higher order Discontinuous Galerkin methods for convection dominated problems. Preprint no. 8 (submitted to SIAM Sci. Comp.), Mathematisches Institut, Unversität Freiburg, 2008.
7. J.-F. Gerbeau and B. Perthame. Derivation of viscous Saint-Venant system for laminar shallow water; numerical validation. *Discrete Contin. Dyn. Syst., Ser. B*, 1(1):89–102, 2001.
8. C. Gersbacher. Local Discontiunous Galerkin Verfahren zur Simulation flacher dreidimensionaler Strömungen mit freier Oberfläche. Diplomarbeit, Abteilung für Angewandte Mathematik, Albert-Ludwigs-Universität Freiburg, Dezember 2008.
9. D. Kröner. *Numerical Schemes for Conservation Laws.* Verlag Wiley & Teubner, Stuttgart, 1997.
10. Randall J. Leveque, *Finite volume methods for hyperbolic problems.Cambridge Texts in Applied Mathematics.* Cambridge: Cambridge University Press., 2002.
11. METIS-homepage. URL http://glaros.dtc.umn.edu/gkhome/views/metis.
12. A. Schmidt and K. G. Siebert, *Design of Adaptive Finite Element Software – The Finite Element Toolbox ALBERTA.* Springer, 2005.

Direct Numerical Simulation of Jet in Crossflow Actuators

B. Selent[1] and U. Rist[1]

Institut für Aerodynamik und Gasdynamik, Universität Stuttgart, Pfaffenwaldring 21, D-70550 Stuttgart, Germany
lastname@iag.uni-stuttgart.de

Summary. Jet vortex generators have been proven to provide a mechanism to positively control boundary layer flows. The present paper illustrates a method to perform direct numerical simulations (DNS) of a jet actuator flow inside a laminar boundary layer. A structured finite difference method is used for the simulations. The numerical scheme was adapted to account for the large scale differences both in geometric and fluid dynamic aspects. Analytical mesh transformations have been implemented to resolve the jet orifice. Suitable boundary conditions are established to model the jet flow. Arising numerical instabilities have been suppressed by implementing a compact filter scheme. Test simulations are done for jet actuator configurations in laminar baseflow with jet to freestream velocity ratios of up to $R = 3.0$. The computational effort on a NEC SX-8/9 is also investigated.

1 Introduction

Jet actuators or jet vortex generators have been proven to provide a mechanism to positively control boundary layer (BL) flows. Experimental work by Johnston et al. [7] has shown the general ability to suppress separation in flows with adverse pressure gradients. The effect stems from the fact that longitudinal vortices are established inside the boundary layer and a mixing of the BL's faster layers with low-speed layers closer to the wall takes place. The mixing in turn leads to increased skin friction thus enabling the flow to overcome larger pressure gradients in downstream direction. This is a very similar effect observed from passive vortex generators [4]. The advantage of jet vortex generator systems over existing solid generators lies in their flexibility to be applied only when necessary and thus to avoid any parasitic drag. An exhaustive parameter study was undertaken by Godard et al. [6] covering many aspects of jet actuators such as velocity ratio R, skew angle β and pitch angle α, hole geometry and direction of rotation.

Albeit the outcomes of these experiments yield a very good general idea of the mechanisms of active flow control devices there are still a number of

W.E. Nagel et al. (eds.), *High Performance Computing in Science and Engineering '09*, DOI 10.1007/978-3-642-04665-0_17,

open questions involved as no detailed picture of the formation of the vortex and its interaction with the boundary layer could be gained from experiment yet. Therefore, any design suggestions for actuators rely heavily on empirical data and are difficult to transpose to different configurations. Within the present research numerical simulations of jet actuators are to be performed by means of the DNS technique. The DNS approach is used for its lack of any model assumptions. Therefore, it is well suited to provide a reference solution for coarser or "more approxiamtive" numerical schemes. Furthermore, DNS allows for a computation of the unsteady flow formation especially in the beginning of the vortex generation and detailed analysis of the fluid dynamics involved.

2 Numerical Method

2.1 General Setup

The numerical simulations are based on the complete Navier-Stokes equations for fully compressible, transient and three-dimensional flow. The computational domain consists of a flat plate and is depicted in figure 1.

Simulations start with initial conditions for a laminar baseflow and allow for the introduction of harmonic sinusoidal perturbations through a turbulator strip as well as discrete jet-disturbances through a round hole or slit. The jets can be skewed to the freestream by an arbitrary angle α and inclined by an angle β relative to the flat plate. The outflow contains a sponge zone where all perturbations are ramped to zero values. The code *NS3D* [1] solves the NS equations written in conservative variables [10]

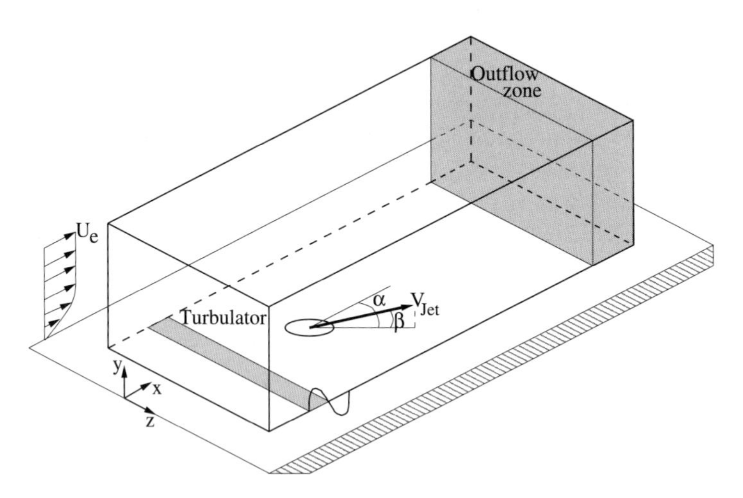

Fig. 1. Computational domain

$$\mathbf{Q} = [\rho, \rho u, \rho v, \rho w, E]^T.$$

In terms of the solution vector **Q** and the according flux vectors **F**, **G**, **H** the basic equation to be solved becomes

$$\frac{\partial \mathbf{Q}}{\partial t} + \frac{\partial \mathbf{F}}{\partial x} + \frac{\partial \mathbf{G}}{\partial y} + \frac{\partial \mathbf{H}}{\partial z} = 0.$$

Time integration of the base equation is carried out by a standard four-step Runge-Kutta method of order Δt^4. In order to stabilize the simulation the stencil used to compute the spatial derivatives can be shifted forward and backward at each sub step of the Runge-Kutta integration in order to add increased numerical viscosity.

The spatial derivatives in downstream and wall-normal direction (x,y) are approximated by compact finite differences with spectral-like resolution of order h^6 [9]. The resulting tridiagonal linear systems of equations are solved by the Thomas algorithm. In spanwise direction z a Fourier spectral method is implemented to approximate spatial derivatives. (Inverse)FFT routines are used for transformation from physical to spectral space and vice versa. Nonlinear terms are computed in a pseudo-spectral manner, i. e. only the derivatives of the primitive variables $q = [\rho, u, v, w, p, T]^T$ are computed in Fourier space and afterwards multiplied accordingly in physical space.

Inflow boundary conditions consist of characteristic boundary conditions for subsonic flow. Additionally, at the inflow input of periodic disturbances is possible providing means for turbulent inflow conditions and/or harmonic disturbance waves. At the wall boundary all velocities are set to 0 (no-slip) and the wall is chosen to be isothermal (cooled wall). Pressure on the wall is recovered by setting the wall-normal derivative equal to zero. Inhomogeneous boundary conditions on the wall are used to introduce perturbations through both periodic and steady suction and blowing. The nozzle flow of the jet actuators is modelled by a polynomial $\rho v(r) = R(1 - 6r^5 + 15r^4 - 10r^3)$ of $O5$ which guarantees smooth functions for the derivatives. Furthermore the jet exit flow can be skewed and pitched by arbitrary angles. This polynomial approximation of the jet flow has proven sufficiently accurate albeit small-scale interactions close to the nozzle edge as observed in experiments are not simulated. Figure 2(a) depicts the jet exit velocity profile approximation and the resulting azimuthal vorticity $\omega_\Theta = -\frac{\partial v(r)}{\partial r}$ in comparison to a circular pipe flow.

At the freestream boundary an exponential decay condition is used to eliminate perturbations and provide potential flow conditions.

Outflow boundary conditions are realized through a sponge zone containing a ramping function (relaminarization). Alternatively the grid can be stretched and a filter applied. In either case all deviations from initial conditions are fully damped before the domain exit is reached.

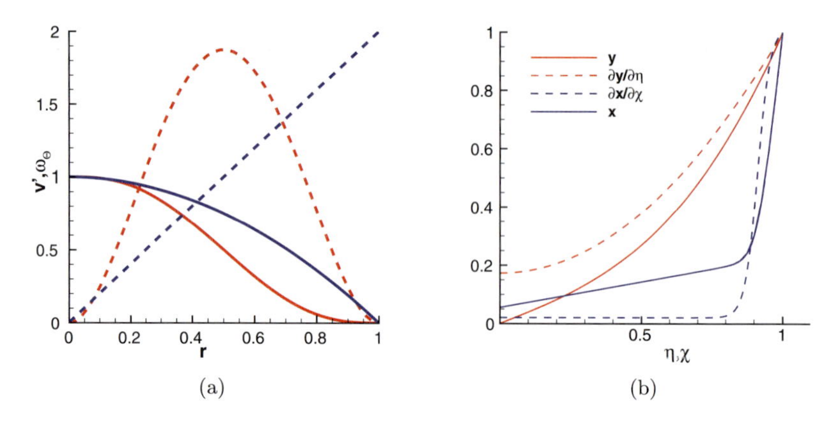

Fig. 2. (a) Jet approximation: Solid lines are velocity profiles, dashed lines indicate vorticity. Red: Polynomial representation, blue: Hagen-Poiseuille profile. (b) Grid transformations

2.2 Problem-Specific Extensions

The regarded Jet-in-Crossflow computation constitutes a severe test to any numerical method as it exhibits multiple scale behaviour both in geometry and flow physics. Firstly, the jet orifices are usually comparatively small with respect to the domain dimensions influenced by the jet. Secondly, the jet-to-freestream velocity ratio is quite large depending on the objectives to be reached. For separation control, values between $R = 3 - 6$ are commonly used. Furthermore, flow control is used in low speed flight as encountered during take off and landing. Thus the freestream Mach numbers are in the range usually modelled by incompressible formulations of the Navier-Stokes equations whereas the high-speed jet flows exhibit compressibility effects. The compressible equations show a decisively singular behaviour with decreasing Ma which can only be counteracted by deploying computationally expensive time-integration methods, i.e. implicit methods or small time-step size.

In order to overcome the aforementioned difficulties, grid transformation techniques have been incorporated allowing for grid compression and/or stretching in x and y direction. With respect to the present problem grid stretching in both x and y direction is employed. Step sizes and the rate of change of the step sizes are depicted in figure 2(b) for a typical mesh. This allows to resolve the jet exit with a sufficient number of grid points (6 per radius) in spite of its small extension $d \sim 1\delta^*$ where δ^* is the displacement thickness.

Two methods to include artificial viscosity are implemented in the scheme. Either the spatial derivative stencil can be shifted alternately backwards and forwards in between time integration sub steps or a filter can be applied after each full Runge-Kutta step. Numerical tests have shown that large gradients at the jet-exit boundary lead to numerical instabilities which can not be

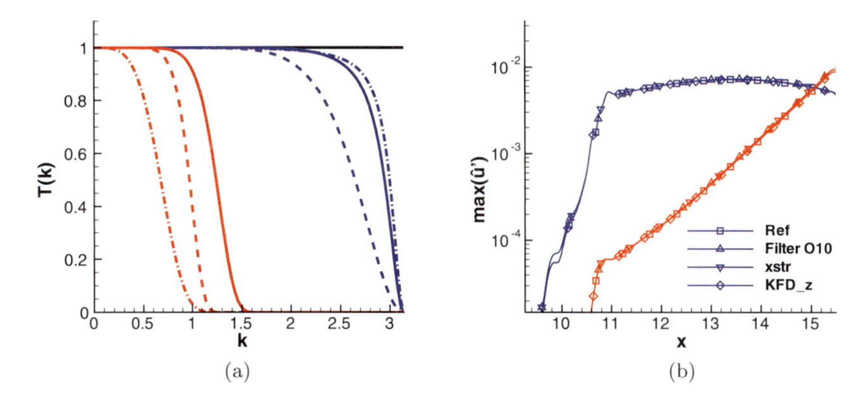

(a) (b)

Fig. 3. (a) Filter transfer function: Blue lines depict analytic transfer functions, red lines are exponential transfer functions. Solid: $h^{10}, \alpha = 0.49$, dashed: $h^{10}, \alpha = 0.4$, dash-dotted: $h^4, \alpha = 0.495$. (b) Test case for numerical scheme, blue lines represent 2D perturbations, red lines are 3D perturbations

suppressed by use of alternating stencils. Thus, it is necessary to add a filter scheme in x- and y-directions. As no subgrid scale model is used for unresolved scales of the flow, the filter has to be tailored suitably to allow all scales to pass which can be represented on the mesh otherwise the solution might be smoothed non-physically. Various compact filter schemes [9] have been examined and the respective transfer functions are compared in figure 3(a). It can be seen that the analytical transfer functions depend largely on the filter constant α rather than on the order of the filter. Nevertheless, higher-order schemes are by far superior in terms of keeping the order of the numerical approximation in space and time. After 10000 time steps both h^{10} order filters keep scales resolved on more than four mesh points almost unfiltered whereas the h^4 order filter exhibits a strong damping effect on the amplitudes of scales as large as eight mesh points. These wavenumbers are usually part of the physical solution rather than being numerical artefacts and numerical damping is not desired. Therefore, a filter of order h^{10} finally has been implemented in combination with one-sided filters for near-wall points [5].

A test case scenario has been computed in order to test and verify the implemented extensions to the numerical scheme. Therefore, subharmonic laminar-turbulent transition on a flat plate with zero pressure gradient is simulated in accordance to Thumm [10]: one 2D and one 3D disturbance are introduced via a disturbance strip. The maximum amplitudes of the perturbed streamwise velocity is analysed in Fourier space in order to compare the outcome of the simulations for each wave. Results are shown in figure 3(b). The reference is taken from Thumm [10] and it can be seen that simulations with a stretched grid as well as simulations with a filter applied yield identical results to the reference. One more computation has been performed replacing the spectral discretization in spanwise direction by compact finite differences. Again agreement of the solution to the reference is excellent.

3 Results

Numerical simulations have been performed for two distinct Jet-in-Crossflow configurations which differ by the jet angles. The first configuration was chosen because it has been investigated both experimentally [8] and numerically [2]. The second setup was chosen with a configuration representative for separation control devices [3]. Freestream conditions are identical in both cases with inflow $Re = 165$ based on freestream velocity U_∞, kinematic viscosity ν and displacement thickness δ^*. Freestream $Ma = 0.25$ which was chosen in order to avoid transonic effects. The mesh used in both cases consists of $800 \times 180 \times 128$ nodes in x, y and z direction, respectively. Grid stretching is applied in x and y direction and a filter is employed for the inclined jet. Initial step sizes are $\Delta x = 0.22$, $\Delta y = 0.1$ and $\Delta z = 0.2$ and the time step is $\Delta t = 0.0065$. Initial baseflow conditions represent a laminar boundary-layer flow on a flat plate with zero pressure gradient. The inhomogeneous boundary conditions mimic round hole jet nozzles with a diameter of $\delta^*/D = 1/3$. The freestream-to-jet-exit-velocity ratio is $R = 3$ in all cases.

3.1 Vertical Jet

The vertical Jet-in-Crossflow simulation serves as test and reference case for succeeding simulations of skewed and inclined jet actuators. Computations have been carried out for 31 time units L_x/U_∞. Albeit this does not suffice to provide data for statistical analysis, it yields a good picture of the evolution of the perturbed flow field. The instantaneous flow is visualised in figure 4 by use of the λ_2 vortex detection method and streamwise velocity contour surfaces. Even if the jet itself is only modelled the typical structures of a jet in crossflow are simulated realistically. Such structures include the bending of the jet due to the oncoming freestream, formation of half-ring vortices in the strong shear layer on the top side of the jet as well as a horseshoe vortex which wraps around the nozzle and induces a near-wall high-speed streak in the wake behind the nozzle.

Even though both the jet and the laminar boundary layer are initially steady the resulting flow regime becomes highly unsteady with the jet exhibiting unstable modes leading to the formation of periodic vortex rings on its top outside of the boundary layer. These rings merge and dissipate while convecting downstream. Close to the wall a different flow pattern can be observed. At the downstream side of the jet exit, fluid is drawn upstream and upwards rolling into two longitudinal vortices. As these vortices convect downstream they disperse and form a widening gap where high speed fluid is transported downwards. Again periodic instabilities are visible.

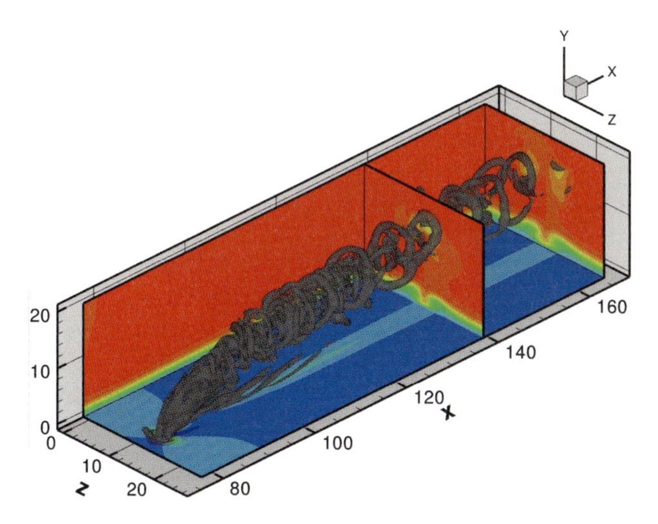

Fig. 4. Vertical Jet-in-Crossflow: Isosurfaces at $\lambda_2 = -0.01$, contours represent u

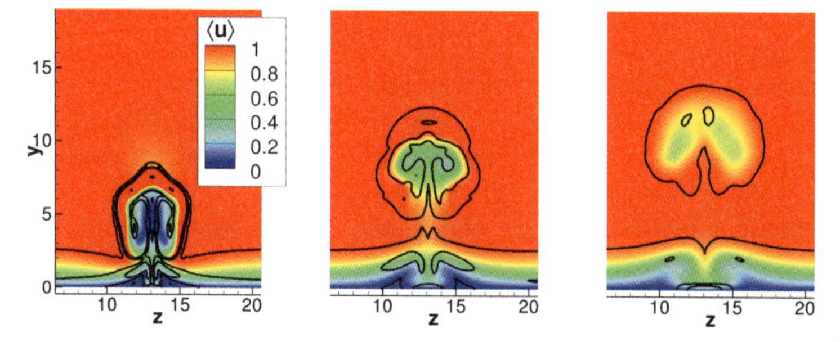

Fig. 5. Mean u contour levels and contour lines for mean ω at positions $\Delta x = 1.5D$, $\Delta x = 5D$ and $\Delta x = 9D$ (from left to right)

Figure 5 shows downstream time averaged velocity and vorticity magnitude contours at stations $\Delta x = 1.5D$, $\Delta x = 5D$ and $\Delta x = 9D$ and demonstrates how the boundary layer low-speed layers roll up and a high-speed streak forms in the centreline behind the jet. Along the trajectory of the jet a vortex pair can be observed which merges into a half ring and is dissipated after a short distance downstream.

3.2 Inclined Jet

Further simulations have been set up in accordance to an active flow-control-device configuration. For that the jet is inclined by $\alpha = 30°$ and rotated against the freestream by $\beta = 30°$. The peak jet-exit velocity magnitude is kept at $R = 3$. Simulations have been carried out on the same mesh for the same time period as the vertical jet simulations. An instantaneous snapshot

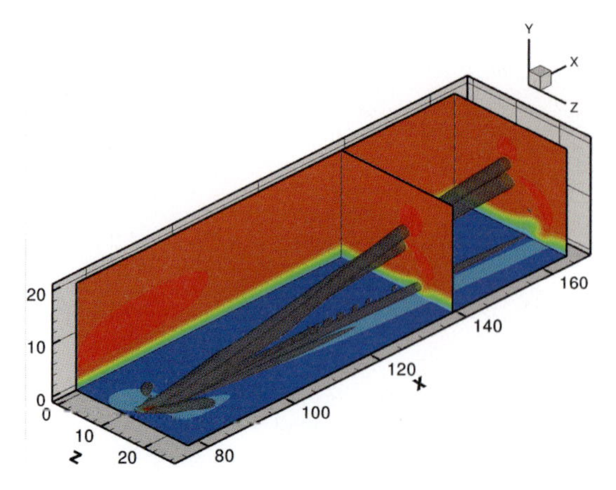

Fig. 6. Vertical Jet-in-Crossflow: Isosurfaces at $\lambda_2 = -0.001$, contours represent u

of the vorticity system is shown in figure 6 by means of the λ_2 method and streamwise velocity contours. The very different flow pattern results from the skewing of the jet. Two longitudinal vortices establish in the flow and wrap around each other due to the induced velocities in the transversal plane. The development of a high-speed streak close to the wall takes place with an inclination to the centreline of the flow. The jet is bound closer to the wall because of the smaller vertical velocity component.

The two longitudinal vortices seem not subject to instabilities at that point in time. They stretch rather well-defined downstream and upwards along the plate. By choice of a smaller λ_2 magnitude a third region of vorticity becomes visible which is bound to the wall. This wake region does exhibit instabilities as observed in the previous simulation. Figure 7 depicts the time averaged vorticity magnitude and streamwise velocity at downstream positions $\Delta x = 5D$, $\Delta x = 9D$ and $\Delta x = 23D$, respectively. The oncoming freestream deflects the jet in both spanwise and wall-normal direction. A strong shear layer develops on the top side of the jet. The sweep of the jet over the boundary layer results in the near-wall fluid layers being entrained upwards into the trajectory of the jet. The entrainment in combination with the deflection of the jet leads to the formation of a near-wall vortex-pair surrounded be the shear layer. This pair moves away from the wall while travelling downstream and it can be seen that the two vortices identified by the λ_2 method are the remaining cores of the vortex-pair after the rotational region has largely dissipated. The exchange of high- and low-speed layers inside the boundary layer is not as strong as in the case of the vertical jet.

3.3 Boundary Layer Control

As already mentioned in section 1, Jet-in-Crossflow configurations are investigated as a means to suppress boundary layer separation. This is to be achieved

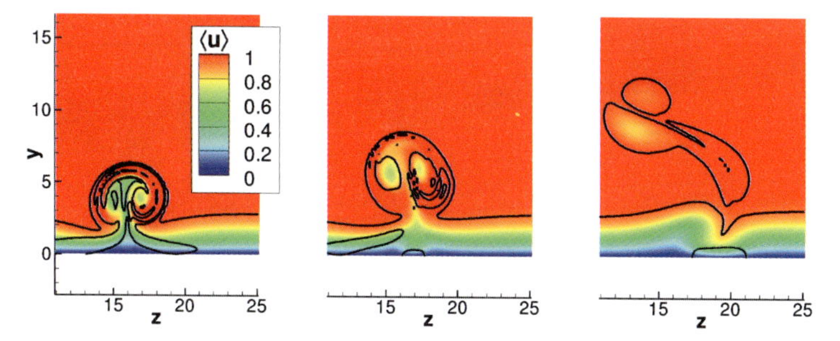

Fig. 7. Mean u contour levels and contour lines for mean ω at positions $\Delta x = 5D$, $\Delta x = 9D$ and $\Delta x = 23D$ (from left to right)

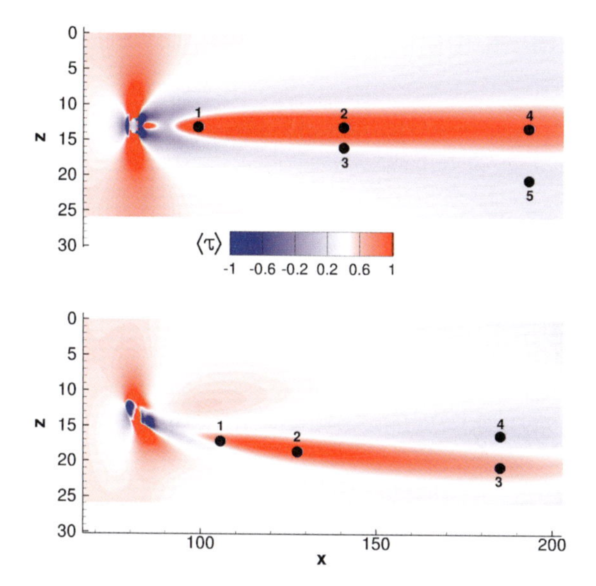

Fig. 8. Comparison of mean τ distribution. Top: vertical jet, bottom: inclined jet. Circles indicate probe position

by increasing the wall friction through moving faster fluid layers closer to the wall. Figure 8 shows a comparison of the time-averaged wall-shear-stress distribution for the simulated cases. Both configurations lead to a net increase of wall shear stress in a confined stripe behind the jet. This region extends further in both downstream and spanwise direction when a vertical jet is used. Also the magnitudes of τ are larger for this case. Corresponding to the circles in figure 8, figures 9 and 10 depict the time-averaged u velocity profiles. These profiles are compared to the profiles of the unperturbed steady laminar flow.

The velocity profiles show the mean-flow deformation in the wake. The wall gradients of the profiles along the wake centreline (numbers 1,3,5) are steeper and are similar to turbulent mean profiles. Outside the boundary layer a defect

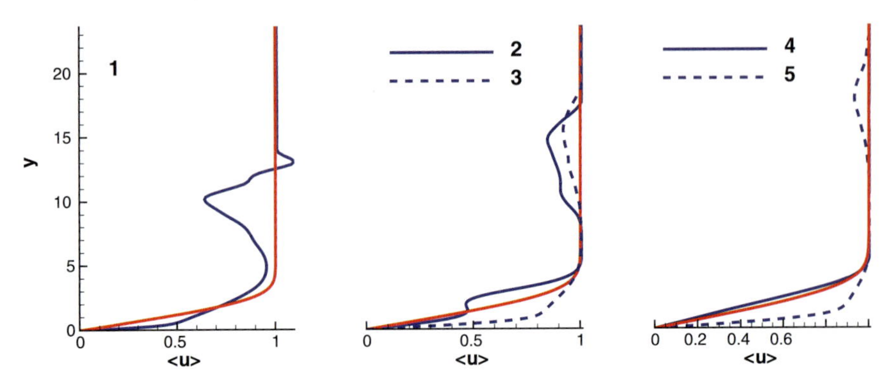

Fig. 9. Mean downstream velocity u profiles. Numbers correspond to the probe positions in fig. 8

is visible which stems from the periodic sweeps of the vortex rings. Profiles measured on the edge or outside of the wake are not as strongly affected. The probe at position 2 can still see the sweep of the jet in the freestream whereas at station 5 no effects on the mean flow are recorded at all.

Time-averaged streamwise velocity profiles for the inclined jet are illustrated in figure 10. The wall gradients do not increase as much as in the previous case at any measured station and the gain in momentum thus is not as large. At position 1 the profile exhibits layers of increased and decreased velocity corresponding to the deflection shear and the vortex-core position in wall-normal direction. At point 2 the momentum gain at the wall reaches a maximum and the streak just outside the boundary layer increases the streamwise speed. Further downstream the profile positioned along the wake centreline (number 3) still possesses a steeper gradient albeit already at a smaller inclination than point 2. Furthermore, the streak on top of the boundary layer widens and losses on its maximum speed are due to dissipation. The probe at position 4 on the other hand measures a profile containing an inflection point inside the boundary layer and a velocity-magnitude defect. The boundary layer is lifted upwards.

Regarding the findings of the simulations some comments on the jet actuator's feasibility to suppress separation are to be made. Firstly, the simulations where done in a laminar boundary layer regime. Thus, the increase in momentum on the wall especially in case of a vertical jet stems mostly from introducing instabilities which the boundary layer is receptible to. These instabilities get amplified and the boundary layer undergoes a transition from laminar to turbulent state. The jet mainly acts as obstacle and source for three-dimensional perturbations similar to a solid surface roughness element. The idea of the inclined jet is of a different nature. The jet not only serves as turbulator but is supposed to generate a very distinct longitudinal vortical motion in order to enhance the mixing of the fluid layers. Albeit the vortex is formed it does not lead to a larger increase of wall friction in the

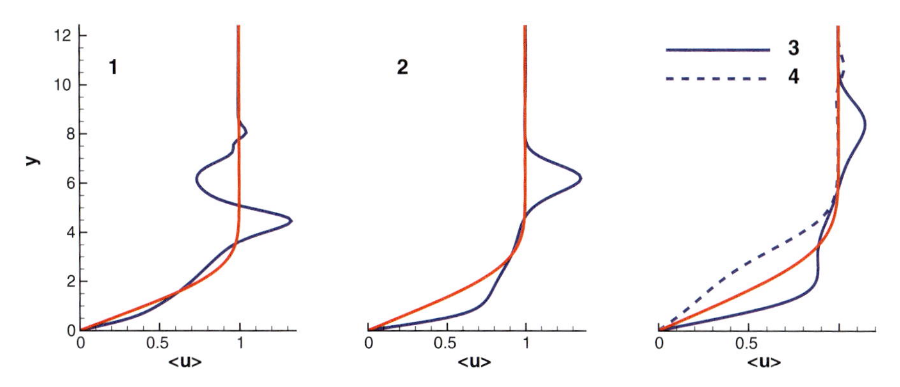

Fig. 10. Mean downstream velocity u profiles. Numbers correspond to the probe positions in fig. 8

Table 1. Computational resources for JIC simulations

Case	Wall time [sec]	CPU time [sec]	time/TS/GP [μsec]	GFLOPS
VJ	33488	532228	0.60	230
INCJ	34294	545444	0.62	230
VJ – spec	33091	524314	0.59	231

boundary layer than the vertical jet. The inclined jet might nontheless be the better option in case of a turbulent boundary layer regime where the blockage of a vertical jet does not deform the mean flow to the same amount as in the laminar case. Secondly, the parameters of the inclined jet may be altered to enhance the desired effect, either by increasing the velocity ratio and/or changing the pitch and skew angles.

3.4 Computational Aspects

All simulations have been carried out on the NEC SX-9 supercomputer at HLRS Stuttgart. One node with 16 CPUs has been employed for all computations. Each computation consisted of one MPI process which was parallelized in z-direction employing NEC Microtasking shared-memory parallelization. Details of the algorithm can be found in [1]. All computations have employed a fully compact finite differencing scheme. Preliminary computations using a spectral ansatz in spanwise direction have shown to introduce unphysical wiggles due to the global representation of the jet. The fully compact discretization has additionally been found to benefit from an increased time step limit. An overview of the main characteristics for the presented computations is shown in Table 1.

Memory usage accumulated to 24.576 GB for all cases and the vector operation ratio was at 99.67%. The slightly longer computational time in case of the inclined jet can be traced to the filter scheme. Due to the compact formulation one additional tridiagonal linear system of equations has to be solved for each spatial variable that is filtered. Filtering acquires about 1.5% of computational time for each variable. Thus an overall penalty of 3% has to be taken into account. The spanwise discretization does not show a siginificant difference between the spectral and compact schemes. The CPU time per time step and gridpoint averages to 0.60μsec which is about three times faster than computations on the NEC SX-8.

4 Conclusions

Simulations of two different Jet-in-Crossflow configurations have been carried out in order to investigate the effect on a laminar boundary layer at $Ma = 0.25$. The main characteristics, i.e. inflow Reynolds number Re and velocity ratio R have been kept constant. A reference case has been established by simulating a vertical jet geometry. In a subsequent simulation the jet has been skewed and inclined by values typically found in an active-flow-control device setup. The simulations gave insight into the generation of a jet-vortex system and wake perturbations. In case of a vertical jet a mean-flow deformation is achieved by instabilities in the wake leading to the onset of laminar-turbulent transition. An inclination and skew of the jet leads to the generation of a longitudinal vortex-pair which form the core of a region with increased vorticity. Instablities in the wake are not as dominant in this case. Thus, fluid layers in the boundary layer are mixed by an up- and downwash motion and only to a lesser extend by amplification of the unstable modes. The jets have been compared by means of measuring the mean wall-shear stress. A larger increase has been found for the vertical jet in the investigated scenario. Further simulations may include a variation of the pitch and skew angles of the jet as well as the velocity ratio. Of interest is also a variation of initial conditions towards a fully developed turbulent boundary layer. This can be achieved by exploiting the findings of the present work by using an array of vertical jets to perturb the flow sufficently and simulate the boundary layer laminar-turbulent transition process.

Acknowledgement. We thank the Höchstleistungsrechenzentrum Stuttgart (HLRS) for provision of supercomputing time and technical support within the project "LAMTUR."

References

1. A. Babucke. Direct numerical simulation of a serrated nozzle end for jet-noise reduction. In M. Resch, editor, *High Performance Computing in Science and Engineering 2007*. Springer.

2. S. Bagheri, P. Schlatter, P. J. Schmid, and D. Henningson. Global stability of a jet in crossflow. *J. Fluid Mech.*, 624:33–44, 2009.
3. M. Casper, C. J. Kähler, and R. Radespiel. Fundamentals of boundary layer control with vortex generator jet arrays. In *4th Flow Control Conference*, number AIAA 2008-3995.
4. D. A. Compton and J. P. Johnston. Streamwise vortex production by pitched and skewed jets in a turbulent boundary layer. *AIAA Journal*, 30(3):640–647, 1992.
5. D. V. Gaitonde and M. R. Visbal. Further Development of a Navier-Stokes Solution Procedure Based on Higher-Order Formulas. In *37th AIAA Aerospace Science Meeting & Exhibit*, number AIAA 99-0557.
6. G. Godard and M. Stanislas. Control of a decelerating boundary layer. Part 3: Optimization of round jets vortex generators. *Aerospace Science and Technology*, 10:455–464, 2006.
7. J. P. Johnston and M. Nishi. Vortex generator jets – a means for flow separation control. *AIAA Journal*, 28(6):989–994, 1990.
8. R. M. Kelso, T. T. Tim, and A. E. Perry. An experimental study of round jets in cross-flow. *J. Fluid Mech*, 306:111–144, 1996.
9. S. K. Lele. Compact finite difference schemes with spectral-like resolution. *Journal of Comp. Physics*, 103:16–42, 1992.
10. A. Thumm. *Numerische Untersuchung zum laminar-turbulenten Strömungsumschlag in transsonischen Grenzschichtströmungen.* PhD thesis, Universität Stuttgart, 1991.

Laminar Heat Transfer From the Stagnation Region of a Circular Cylinder at Re=140 000

Jan G. Wissink and Wolfgang Rodi

Institute for Hydromechanics, University of Karlsruhe
Kaiserstr. 12, D-76128 Karlsruhe
rodi@uka.de

Summary. Direct Numerical Simulations (DNS) of flow over and heat transfer to the stagnation region of a circular cylinder have been performed at a Reynolds number of $Re_D = 140\,000$ - based on the far field velocity and the diameter of the cylinder D -. To avoid the formation of a vortex street, behind the cylinder a splitter plate was introduced. The computations were carried out on the NEC SX-8 using up to 192 processors and 746.5 million grid points. Two simulations were performed: A two-dimensional, fully laminar simulation to obtain a base-line solution and a fully three-dimensional simulation to address the influence of oncoming free-stream fluctuations on the heat transfer from the stagnation region of the heated cylinder. The incoming free-stream turbulence stems from a separate simulation of isotropic turbulence in a box. Compared to the fully laminar simulation, the addition of free-stream fluctuations at the inflow plane was found to lead to an increase in heat transfer in the stagnation region of the cylinder by 66%.

1 Introduction

The temperature at which a turbine can be operated is limited by the material used for the manufacturing of the blades and the effectiveness of the blade-cooling strategy. A peak in the heat load on turbine blades is reached in the stagnation region, where - due to the strongly accelerating external flow - the bounday layer is laminar. External fluctuations, originating from the upstream row of blades, that impinge on the stagnation-region-boundary-layer will lead to an increase in heat transfer. The latter is referred to as "laminar heat transfer".

The Reynolds number of this flow problem, $Re = 140\,000$ is based on the free-stream velocity U_0 and the diameter of the cylinder D. It is in the higher subcritical range, which means that without a splitter plate and incoming free-stream fluctuations we would obtain a laminar separation of the boundary layers on the upper and lower surface of the cylinder. The two free-shear layers would subsequently undergo transition to turbulence behind the cylinder. The main purpose of the splitter plate is to hinder the formation of a von Karman

W.E. Nagel et al. (eds.), *High Performance Computing in Science and Engineering '09*, DOI 10.1007/978-3-642-04665-0_18,

vortex street that could lead to a streamwise periodic oscillation in the location of the stagnation line.

With the high-quality simulations we present in this paper, we aim to produce good quality data 1) to elucidate the physical mechanisms that play a role in laminar heat transfer and 2) to serve as reference data with which to improve transition models. By employing such improved models in industrial codes, a more accurate prediction of the heat load in the stagnation region of turbine blades can be achieved and more efficient blades can be designed.

Several experiments have been performed to study the influence of free-stream turbulence on "laminar heat transfer" that is: the heat transfer in regions where the boundary layer is laminar - and on boundary layer transition. In their experimental studies of laminar heat transfer along a flat plate affected by free-stream fluctuations, Kestin et al. [4] and Junkhan and Serovy [3] discovered that in order for these fluctuations to be able to increase heat transfer, the affected laminar boundary layer flow needs to be accelerating. These findings were confirmed by the experiments of Schulz [9], who measured the heat transfer distributions around a typical fore-loaded airfoil for several free-stream turbulence levels. Along the stagnation region at the leading edge, where the acceleration is very strong, the highest heat transfer was found. On the suction surface, where a laminar to turbulent transition occurs, the main effect was to cause an earlier onset of transition inducing a subsequent increase in heat transfer. Along the pressure surface, where the highly accelerated boundary layer remains laminar, the main effect of the free stream turbulence was to cause a large increase in laminar heat transfer. Liu and Rodi [5, 6] performed heat transfer measurements of flow in a turbine cascade with incoming wakes. With increasing wake-frequency, the heat transfer from the accelerating laminar flow portions of both the pressure and suction side to the free-stream, was found to increase. Recently, Choi et al. [2], studied the effect of free-stream turbulence on turbine blade heat transfer. The grid-generated turbulence was shown to affect the heat transfer along both areas of the turbine blade with flow acceleration (i.e. the entire pressure side and approximately the upstream half of the suction side) and along the areas where the boundary-layer flow was exhibiting transition to turbulence: Increasing the free-stream turbulence level - while keeping the Reynolds number fixed - was observed to lead to an increase in heat transfer.

The first DNS involving laminar heat transfer from a heated turbine blade was performed by Wissink and Rodi [10]. The setup of the computational domain was largely taken in accordance with the setup of the experiments of Liu and Rodi [5, 6]. The DNS showed a slight increase in heat transfer in the laminar portion of the boundary layer that could be attributed to the presence of free-stream fluctuations.

The stagnation region of a circular cylinder is very similar to the stagnation region of a turbine blade. Hence, the findings of a study of the increase in laminar heat transfer in this region by external fluctuations are directly transferable to the fluctuation-induced increase in heat transfer to the stagnation region of turbine blades. The present paper reports the results of a numerical study carried out at a relatively high Reynolds number. While the laminar base line simulation is effectively two-dimensional, in the second simulation - which is fully thee-dimensional - fluctuations were added at the inflow plane. To be able to perform a detailed study of the effect of these fluctuations on the heat transfer from the stagnation region, we were given the opportunity to perform direct numerical simulations of this computationally expensive flow problem on the NEC SX-8 at HLRS in Stuttgart.

2 Computational Details

2.1 Numerical Aspects

The three-dimensional, incompressible Navier-Stokes equations were discretised using a collocated, curvilinear finite-volume code that combined a second-order central discretization in space with a three-stage Runge-Kutta method for the time-integration. To avoid a decoupling of the velocity field and the pressure field, the momentum interpolation technique of Rhie and Chow [7] was employed. The Poisson equation for the pressure was iteratively solved using the SIP-solver of Stone [8]. The code was parallellized using the standard Message Passing Interface (MPI) protocol. For a more detailed desciption of the code see [1].

The computational mesh was subdivided into 192, partially overlapping, submeshes (blocks) of equal size. Each block was assigned to a unique processor to ensure a near-optimal load balancing. Because of the usage of an implicit Poison solver for the pressure, a frequent exchange of data between neighbouring blocks needed to be carried out. Hence, the simulation speed benefits greatly when running on a computational platform with powerful processors and very fast intra and internodal communication between processors. Hence, both simulations were performed on the NEC SX-8. In the major three-dimensional (3D) simulation $2406 \times 606 \times 512$ grid points were used in the circumferential, radial, and spanwise direction, respectively. To perform this simulation 192 processors on 24 nodes were used. The vector operation ratio was 98.8% with an average vector length of 146. The 3D simulation reached an average speed of $3.3GFlop/s$ per processor. In total, 240 000 time steps were performed using chain jobs with a duration of up to 23.5 hours. To finish the 3D job in total 258.5 clock hours were required corresponding to $258.5 \times 192 = 49\,632$ CPU hours.

2.2 Setup of the Computations

The (spanwise cross-section of the) computational geometry that was employed in both simulations is shown in Figure 1 (left). At the inflow plane a flow field $(u, v, w)^t = (1, 0, 0)^t U_0 + (u', v', w')^t$ was prescribed, in which the free-stream fluctuations $(u', v', w')^t$ originated from a separate large-eddy simulation of isotropic turbulence in a box. In the two dimensional laminar simulation the turbulence intensity $Tu = \sqrt{\frac{1}{3}\frac{\overline{u'u'}+\overline{v'v'}+\overline{w'w'}}{U_0^2}} \times 100\%$ at the inflow plane was $Tu_{in} = 0\%$, while in the 3D simulation $Tu_{in} = 30\%$. The spanwise size of the 3D simulation was $l_z/D = 3.0$ and the integral length-scale of the free-stream fluctuations was approximately $\frac{1}{10}l_z$. As a dimensionless representation of the heat-flux we use the local Nusselt number Nu, which - using D as length-scale - is defined by

$$Nu = \frac{hD}{k} = \frac{-1}{1-\alpha}\frac{\partial\left(\frac{T}{T_0}\right)}{\partial\left(\frac{r}{D}\right)},$$

where $\alpha = 0.7$ is the ratio between the oncoming flow $T = \alpha T_0$ and the wall-temperature of the cylinder $T = T_0$ and r is the radial coordinate with respect to the cylinder axis. At the upper and lower boundary free-slip bound-

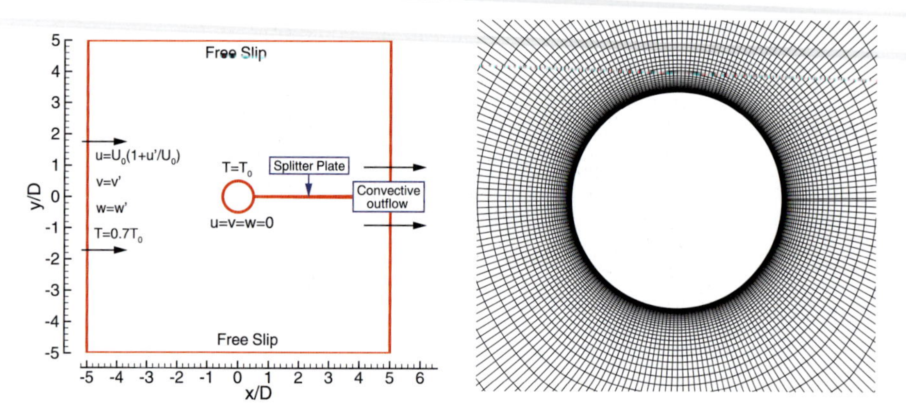

Fig. 1. left: Computational geometry, Right: O-mesh in the vicinity of the cylinder showing every 12th grid line from the simulations

ary conditions were used while at the surface of the cylinder a no-slip condition was employed. At the outflow plane a convective outflow condition was used and in the spanwise direction of the 3D simulation periodic boundary conditions were employed. The splitter plate (with free-slip boundary conditions) mounted behind the cylinder was used to counteract the formation of a von Karman vortex street behind the cylinder that might cause a quasi-periodic motion of the stagnation line at the front of the cylinder. In Figure 1 (right) a detail of the O-mesh in the (x, y) plane employed in Simulations D and E is shown. Because of the dense grid, only every twelfth grid-line is displayed. It can be seen that the grid near the cylinder is only stretched in the radial direction. This was done to avoid the occurence of numerical inaccuracies due to unfavourable cell aspect ratios. Towards the boundary of the computational domain the circular shape of the circumferential direction of the O-mesh is smoothly changed into a rectangular shape. As a result, in almost the entire computatinal domain the shape of the grid volumes is near-optimal, while only the volumes in the outer corners - which are far enough away to not have any effect on the flow around the cylinder - have a less favourable shape. The time step employed in the simulations is $\delta t = 5 \times 10^{-5} D/U_0$.

3 Results

3.1 Study of the Instantaneous Flow

In Figure 2, contours of the instantaneous Nusselt number, Nu along the surface of the cylinder are shown. The stagnation line is located at an angle of $\alpha = 0^o$. In the region around the stagnation line (the stagnation region), a

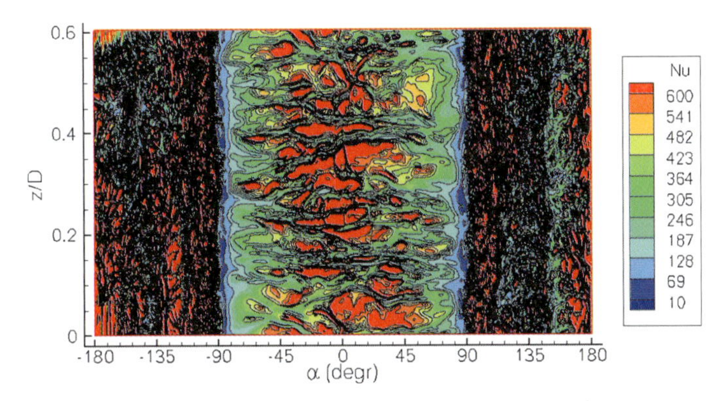

Fig. 2. Instantaneous Nusselt number in the stagnation area of Simulation 2

relatively coarse pattern consisting of streamwise elongated strips of high Nu can be seen. The strips are the footprints of impinging free-stream vortical structures that are elongated by the strongly accelerating flow around the front of the cylinder. Downstream of the location of separation at $\alpha = \pm 89^o$ - where Nu reaches a minimum - a fine-grained pattern in the Nu-contours is obtained. This pattern reflects the presence of turbulent re-circulating flow along the entire downstream half of the circular cylinder.

A study of the origin of the coarse pattern in the Nu-contours that was observed in the stagnation region (see Figure 2) is presented in Figure 3. Here, a sequence of six radial cross-sections through the boundary layer at $\alpha = 0^o, 10^o, \ldots, 50^o$ - showing fluctuating velocity vectors and contours of the instantaneous temperature T - is shown. The approximate loaction of the edge of the boundary layer is identified by δ. The actual location of the zoomed snapshots is identified in the schematic at the top of the figure. As

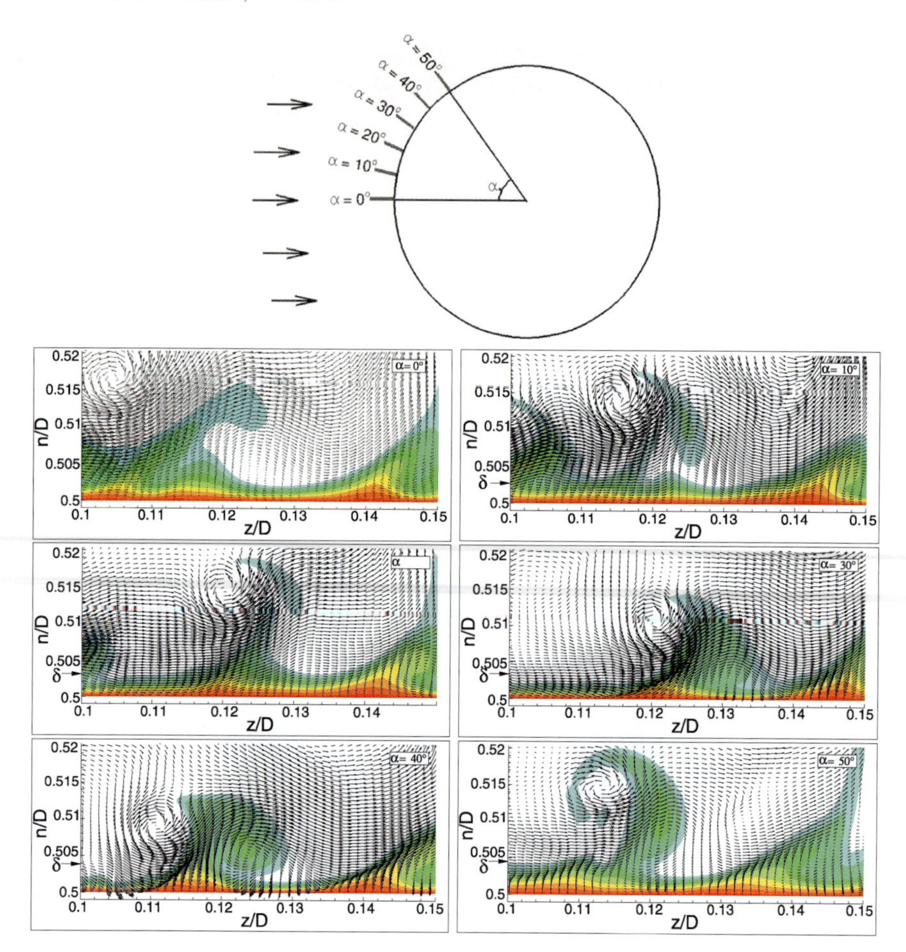

Fig. 3. A sequence of six orthogonal cross-sections at $\alpha = 0^o, 10^o, 20^o, 30^o, 40^o, 50^o$ (zoomed view) originating from a single snapshot showing the fluctuating velocity vectors and contours of the instantaneous temperature. The pane at the top identifies the location of the cross-sections

the flow moves along the surface of the cylinder, the boundary layer thickness can be seen to increase from $0.0027D$ at $\alpha = 10^o$ to $0.0045D$ at $\alpha = 20^o$. The rotating structure that is observed near $(z/D, r/D) = (0.115, 0.515)$ in the snapshots at $\alpha = 10^o, \ldots, 50^o$ corresponds to a single, slightly meandering streamwise vortical structure that is wrapped around the stagnation region of the circular cylinder by the accelerating wall-parallel streamwise flow. At the left, the counter-clockwise rotation forces cold fluid from the free-stream towards the cylinder's surface, while on the right hot fluid that originates from the cylinder's surface is swept upwards. It is this hot fluid being lifted upward that is reflected by the formation of the distinctive coarse pattern in Figure 2.

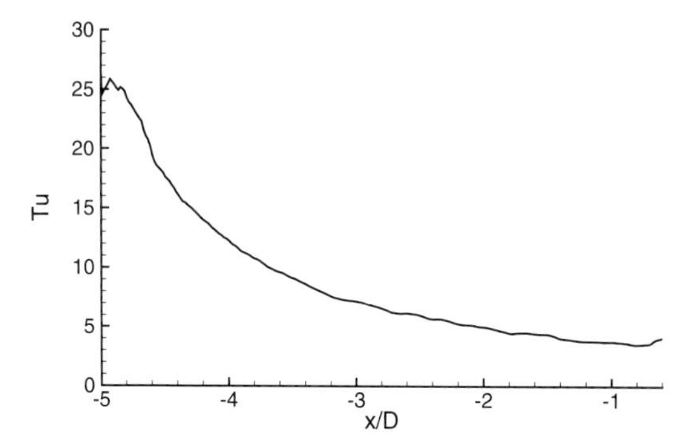

Fig. 4. Turbulence intensity in the horizontal centre plane ($y/D = 0$) of the inflow region

In Figure 4, the decay of the isotropic turbulence that is introduced at the inflow plane is studied along $y/D = 0$. Immediately downstream of the inflow plane a sharp drop (caused by interpolation errors) in the Tu-level from $Tu_{in} = 30\%$ to $Tu = 25\%$ can be observed. After a transient behaviour between $x/D = -5$ and $x/D = -4.7$, Tu is observed to approximately undergo an exponential decay until it reaches a minimum value of $Tu = 3.5\%$ at $x/D = -0.75$.

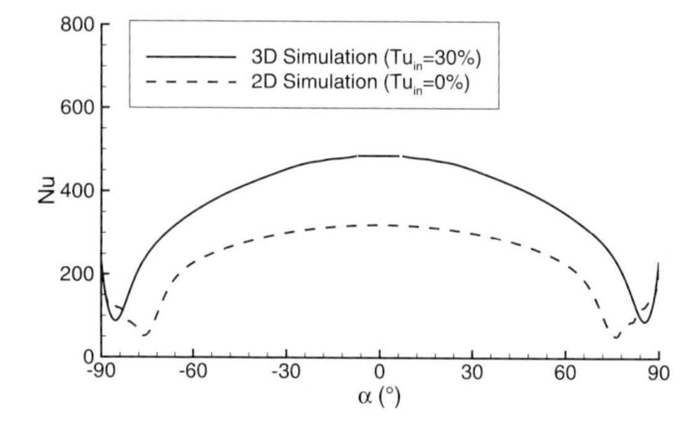

Fig. 5. Time-averaged local Nusselt number in the stagnation region of the cylinder

Figure 5 shows the mean local Nusselt number in the stagnation region of the cylinder. For both the laminar 2D simulation and the 3D simulation with oncoming free-stream turbulence the maximum Nu is reached at the stagnation point, where the boundary layer is thinnest. Farther downstream, Nu gradually decreases until slighly downstream of the location where the boundary layer separates. Behind the location of separation, at the back of the cylinder, turbulent flow re-ciculates (evidenced by the fine-grained instantaneous Nu-contours in Figure 2) causing a significant increase in Nu. Even though the actual turbulence level has dropped from $Tu = 30\%$ at the inflow

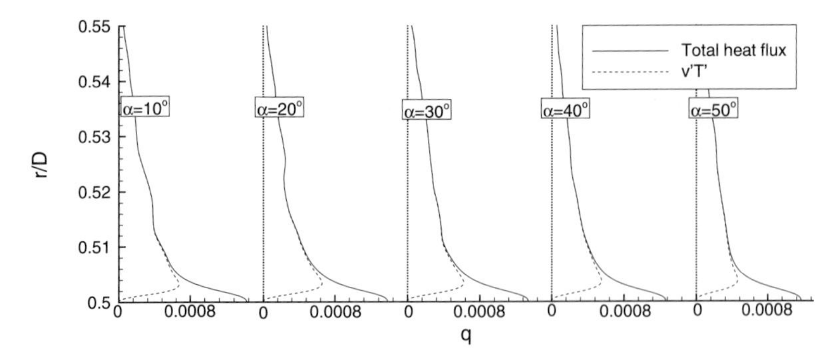

Fig. 6. Total heat flux and fluctuating heat flux at $\alpha = 10^o, 20^o, \ldots, 50^o$ in the stagnation region of the cylinder plotted against the radial distance from the centre of the cylinder

plane to $Tu = 3.5\%$ slightly upstream of the cylinder (see Figure 4), comparing the turbulent 3D to the laminar 2D simulations a significant increase in Nusselt number (of almost 66% at the stagnation point) can be seen in the entire stagnation region. Also, the minimum Nu (which approximately corresponds to the location of separation) in the 3D simulation is reached approximately 10° farther downstream as compared to the laminar simulation. This can be partially explained by the free-stream turbulence in the 3D simulation that energizes the boundary layer and thereby delays transition.

The fluctuating normal heat flux, $\overline{v'T'}$, and the total heat flux, $q_{total} = \overline{v'T'} - \frac{1}{Pr\,Re}\frac{\partial T}{\partial r}$ are shown in Figure 6 at various locations, $\alpha = 10^o, 20^o, \ldots, 50^o$, in the stagnation region of the cylinder. At all locations, very close to the cylinder the laminar heat flux is the sole contributor to q_{total} as the presence of the wall completely damp the wall-normal fluctuations such that $\overline{v'T'} \approx 0$. Farther away from the cylinder's surface, the fluctuating normal heat flux quickly becomes the main contributor to the total heat flux. Moving downstream from the leading edge, the viscous layer adjacent to the cylinder (identified by the lack of fluctuating heat transfer) can be seen to slowly become thicker, which causes the local Nusselt number to decrease. Another remarkable result is how far heat is transported away from the cylinder by the fluctuating normal heat flux: In the radial direction, $\overline{v'T'}$ only very gradually declines and heat is transported up to 0.05D away from the cylinder. From this we can draw the conclusion that turbulent diffusion of heat is a very important transport mechanism that significantly contributes to the observed increase in Nusselt number by 66% as compared to the base-line simulation.

4 Conclusions

The possibility to perform Direct Numerical Simulations (DNS) of flow around and heat transfer from a heated circular cylinder with and without incoming

free-stream turbulence on the NEC SX-8 supercomputer in Stuttgart, allowed a comprehensive and accurate investigation of the effects that impinging fluctuations have on the heat transfer in the stagnation region of a circular cylinder at a Reynolds number in the higher subcritical range.
From the simulations, we reach the following conclusions:

- The strongly accelerating wall-parallel flow stretches vortical structures in the inflow plane of the fully 3D simulation and orientates them in the direction of flow as the structures approach and finally impinge on the stagnation region of the cylinder.
- Upon impingement, the elongated vortical structures promote the transport of cold fluid from the free-stream towards the cylinder, while at the same time hot fluid near the cylinder is swept up towards the free-stream, resulting in a coarse, elongated pattern of high values of the instantaneous Nusselt number.
- In the viscous sublayer of the boundary layer thermal diffusion is found to be fully laminar. Higher up in the boundary layer, the fluctuating wall-normal transport of heat (turbulent diffusion of heat) takes over and is found to be able to transport heat well away from the surface of the cylinder (up to $0.05D$ in the radial direction) even against the main direction of flow.
- Compared to the laminar base line simulation, in the three-dimensional simulation the observed level of fluctuations of $Tu = 3.5\%$ - reached immediately upstream of the cylinder (down from $Tu = 30\%$ at the inflow plane) - were found to lead to an increase in Nusselt number of 66% in the stagnation region.

Acknowledgements. The authors would like to thank the German Research Foundation (DFG) for funding this project and the Steering Committee for the Supercomputing Facilities in Stuttgart for granting computing time on the NEC SX-8.

References

1. M. Breuer and W. Rodi: Large eddy simulation for complex flow of practical interest. In: Flow Simulation with High-Performance Computers II, Notes on Num. Fluid Mech., **52**, Vieweg Verlag, Braunschweig (1996).
2. J. Choi, S. Teng, J. Han, F. Ladeinde: Effect of free-stream turbulence on turbine blade heat transfer and pressure coefficients in low Reynolds number flows, Int. J. Heat Mass Transfer, **47**, 3441–3452 (2004).
3. G.H. Junkhan and G.K. Serovy: Effects of free-stream turbulence and pressure-gradient on flat plate boundary layer velocity profiles and on heat transfer. ASME J. Heat Transfer, **89**, 169–176 (1967).
4. J. Kestin, P.F. Maeder, H.E. Wang: Influence of turbulence on heat transfer of heat from plates with and without a pressure gradient. Int. J. Heat Mass Transfer, **3**, 133–154 (1961).

5. X. Liu, W. Rodi: Velocity measurements in wake-induced unsteady flow in a linear turbine cascade. Exp. in Fluids, **17**, 45–58 (1994).
6. X. Liu, W. Rodi: Surface pressure and heat transfer measurements in a turbine cascade with unsteady oncoming wakes. Exp. in Fluids, **17**, 171–178 (1994).
7. C.M. Rhie and W.L. Chow: Numerical study of the turbulent flow past an airfoil with trailing edge separation. AIAA J., **21**, 1525–1532 (1983).
8. H.L. Stone: Iterative Solution of Implicit Approximations of Multidimensional Partial Differential Equations. SIAM J. Numerical Analysis, **5**, 530–558 (1968)
9. A. Schulz: Zum einfluß höher Einstromturbulenz, intensiver Kühlung und einer Nachlaufströmung auf den Aüßeren Wärmeübergang einer konvektiv gekühlten Gasturbinenschaufel, Ph.D. Thesis, U. Karlsruhe, Karlsruhe, Germany, 1986 (also S. Wittig, A. Schulz, H.J. Bauer, and K.H. Sill, 1985, Effects of Wakes on Heat Transfer in Gas Turbine Cascades, AGARD CP 390, pp. 6-1–6-13.
10. J.G. Wissink and W. Rodi: DNS of passive heat transfer in a turbine cascade with incoming wakes. J. Fluid Mech., **569**, 209–247 (2006).

Conditional Statistics Along Gradient Trajectories in Fluid Turbulence

Lipo Wang

Institut für Technische Verbrennung
RWTH-Aachen, Aachen
Templergraben 64, 52056, Aachen, Germany
wang@itv.rwth-aachen.de

Summary. To investigate turbulent flows, direct numerical simulations (DNS) is playing more and more important roles benefitting from the modern computing technologies. Once the DNS data are obtained, according to different theory and purposes, the data analysis will be the core of the work at next stages. From the sizable capacity data for homogeneous shear turbulence, the conditional statistics along gradient trajectories have been investigated. It has been derived and also proved numerically that the two-point velocity difference structure functions along the same gradient trajectories have a linear scaling with respect to the arclength between the two points, different from the classical Kolmogorov scaling. In addition, the performance of the OpenMP parallelized code is satisfactory.

1 Introduction and Previous Achievements

Because of the apparent frustrations from the mixture of chaos and order and the multiscale behavior, exact theoretical solution of turbulence does not exist, even for the simplest cases. Experimentation, as a commonly used method for data collecting, has many remarkable advantages. However, due to some inevitable pitfalls in experimentation, such as the response lag, perturbation and poorly changeability of turning parameters etc, direct numerical simulation (DNS) plays a very important and indispensable role in the research of turbulence. Even for theoretical research, DNS becomes possible only in recent years, benefitting from the modern computer technology.

Different from other numerical works, DNS, on the one hand, pursues the Reynolds number as high as possible; one the other hand, must be constringed by the resolution requirements, which make DNS quite hard to reach the applicable level for engineering application. Particularly for our concerns of the fine topological structure in turbulent flows, the requirement of DNS is even harsher. In this report, after a brief introduction of the main results, which have been introduced already in previous works about the performance

W.E. Nagel et al. (eds.), *High Performance Computing in Science and Engineering '09*, DOI 10.1007/978-3-642-04665-0_19,

and algorithms of DNS [1], we will concentrate on the conditional statistics along gradient trajectories of scalar field variables.

1.1 Dissipation Elements and Gradient Trajectories

Because of the prohibitive difficulties in solving turbulence mathematically from the Navier-Stokes equations, alternatively a better understanding of turbulence may rely more on physics than mathematics. There are many efforts for the analysis of turbulence from geometrical aspects. The advantage and importance of the decomposition of flow field into geometrical elements in physical space has been realized and attacked for quite a long time. However, there was no success to get a determined and complete partition of the total turbulent field into many units. One needs to construct a suitable method which can identify specific geometrical elements in the turbulent flow. Which method one should choose is by no means evident.

The application of dissipation element analysis in turbulence is a new attempt to decompose the entire flow fields into relative simple small geometrical units from some pre-defined scalar field variables [2]. Starting from any material point, its gradient trajectory will inevitably reach a maximal and minimal point along the descending and ascending directions, respectively. The ensemble of material points from which the trajectories can share the same pair of minimum and maximum points define a spatial region called a dissipation element. By definition, dissipation elements are space-filling and non-arbitrarily defined geometrical objects, which is uniquely different from other methods. From DNS data, some examples of dissipation elements, interacting together with vortex tubes, are shown in Fig. 1, where the color shows the value of the scalar along trajectories. It can be shown that generally dissipation elements incline to be perpendicular to vortex tubes, which can be explained by the compressive strain exerted by vortex tubes on the passive scalar. Further discussions on the physical properties of dissipation elements can be refereed to [2, 3].

1.2 DNS for Homogeneous Shear Turbulence

In the previous HRC-DNS (High Resolution Calculation of DNS) project, homogenous shear turbulence simulations have been performed. The schematic configuration can be explained in Fig. 2. The time-evolving flow field and passive scalar field are calculated simultaneously. Turbulent motion can be sustained because of the presence of the external shear forcing, and similarly the random fluctuation of passive scalar is maintained by the mean scalar gradient. The mean velocity gradient S and passive scalar gradient K, two free tuning parameters for different simulations, are exerted in the same direction.

Spectrum method is specially advantageous in DNS because of the high numerical fidelity to solve the structure of small scales. FFT (Fast Fourier

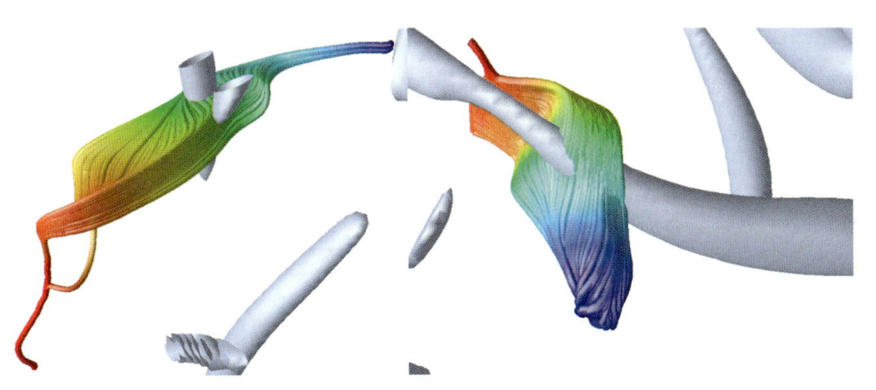

Fig. 1. Examples of the interaction of dissipation elements with vortex tubes

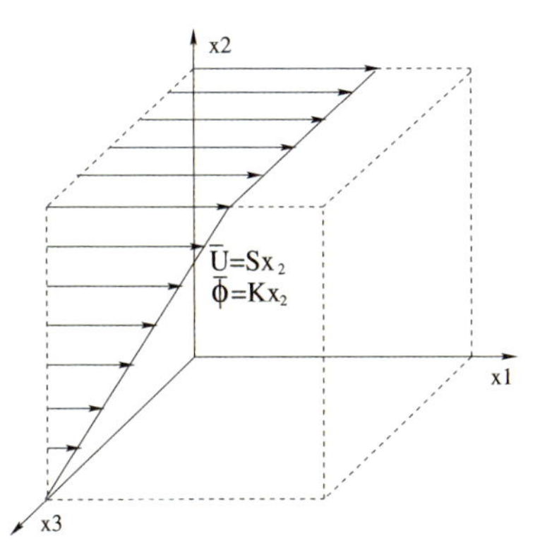

Fig. 2. The physical configuration of homogenous shear turbulence

Transform) is a favorite spectrum method on the condition that the boundary conditions (B.C.) of the problems are periodic. To meet this periodicity requirement, a moving coordinate attached with the mean flow has been adopted and, correspondingly remeshing (regriding) are needed to adjust the deformation of total calculation domain.

To simulate turbulent flows, the most strict requirement is to resolve the motion at different scales and, at the same, to control the pollution from Boundary Conditions(B.C.). The largest and smallest scales in turbulence are regarded as the integral scale L and the Kolmogorov dissipative scale η, respectively. For homogeneous shear flows, mainly we need to consider the resolution of η. Roughly speaking, the grid number N in each direction need to be

$$N \approx Re^{\frac{3}{4}} \approx Re_{\lambda}^{\frac{3}{2}}, \tag{1}$$

where Re and Re_{λ} are the Reynolds number based on the integral scale and the Taylor scale, respectively.

In Practice depending on different investigations, this criterion may vary to some extent. For example, if DNS is only used for combustion problems, Δx, the grid size in the calculation domain, can be 3 or 4 times η. For some overall flowing properties, for instance, the mean kinetic energy and energy dissipation, 2 times η is a adequate choice for Δx to obtain the first and second order statistics, from the consideration of energy conservation [6]. Therefore it is claimed that requiring η to be resolved as the smallest scale is probably too stringent [5]. But these criteria above become loose when higher order derivatives or special quantities need to be investigated. More strictly, Sreenivasan [8] pointed out that because of intermittency, scales which are much smaller than η can be established locally in turbulence. To resolve also these local small scales or spots of violent events in a strict sense, the grid size must be determined in a more stringent way. Numerical tests [7, 4] have proved this statement. If we want to analyze the behaviors of dissipation elements in turbulence, a finer resolution becomes necessary.

To resolve the structure of dissipation elements, we have tested the following cases. For a DNS case with $Re_\lambda = 127$ and $\Delta x/\eta = 1.88$, there exist about 60000 extremal points and about 220000 dissipation elements. Once the grid number is doubled in each direction and the resolution becomes as high as $\Delta x/\eta = 0.94$, the number of extremal points and elements drop to about 20000 and 70000, respectively, because higher resolution can smooth the scalar field and many irrelevant kinks and burrs from numerical inaccuracy may be removed. Further more, comparing the results from two different fine resolution cases, $\Delta x/\eta = 0.44$ and $\Delta x/\eta = 0.85$ with the same Reynolds number $Re_\lambda = 75$, the number of extremal points remains almost the same ~ 2500, which suggests that a more stringent resolution level $\Delta x/\eta < 1.0$ may work as a resolution criterion to study dissipation elements in turbulence.

1.3 MPI Parallelization

MPI(Message Passing Interface) is a library of functions (in C) or subroutines (in Fortran) used together with the source code to perform data communication between processes on a distributed memory system. MPI is very effective and necessary for the huge jobs with special demand of memory. Because in this spectrum DNS code, most CPU time will be consumed in the frequently invoked FFT subroutines, to fit the code better with the architecture of the NEC vector machine, an effective MPI-parallelized FFT library is very important. Our great acknowledgment is given to Prof. D.Takahashi (University of Tsukuba, Japan) for the necessary instruction of a parallelized 3D FFT [9] library, which is specially developed for vector machines. Therefore the hardware resource can be effectively utilized. Under the control of MPI, the shared and private data on different processors communicate with each other for different calculations.

In the previous parallelized DNS code, the 3D arrays as large as (1024, 1024, 1024) have been decomposed in the z direction into 128 slabs and to dispatch

Table 1. MPI performance checking list

Indexes	Min [U,R]	Max [U,R]	Average
Real Time(sec)	85188.687[0,79]	85215.645[0,10]	85202.114
User Time(sec)	84722.290[0,0]	84984.510[0,54]	84909.579
System Time(sec)	42.854[0,113]	128.673 [0,0]	53.058
Vector Time(sec)	55638.568[0,15]	56724.445 [0,120]	55753.603
Instruction Count	$1.435 \cdot 10^{13}$[0,15]	$1.451 \cdot 10^{13}$[0,127]	$1.440 \cdot 10^{13}$
Vector Instruction Count	$2.493 \cdot 10^{12}$[0,15]	$2.525 \cdot 10^{12}$[0,120]	$2.502 \cdot 10^{12}$
Vector Element Count	$3.642 \cdot 10^{14}$[0,15]	$3.748 \cdot 10^{14}$[0,120]	$3.681 \cdot 10^{14}$
FLOP Count	$1.272 \cdot 10^{14}$[0,123]	$1.272 \cdot 10^{14}$[0,64]	$1.272 \cdot 10^{14}$
MOPS	4429.860 [0,15]	4554.280 [0,120]	4475.159
MFLOPS	1496.411 [0,54]	1501.026 [0,0]	1497.677
Average Vector Length	146.113 [0,15]	148.462 [0,120]	147.083
Vector Operation Ratio(%)	96.847 [0,15]	96.914 [0,120]	96.870
Memory size used(MB)	8691.033 [0,0]	8955.643[0,127]	8693.101
Global Memory size(MB)	16.000 [0,1]	32.000[0,0]	18.000
MIPS	169.015 [0,14]	171.004 [0,127]	169.562
Instruction Cache miss (sec)	78.181 [0,0]	82.531 [0,40]	80.749
Operand Cache miss (sec)	4492.328 [0,36]	4517.017 [0,24]	4500.858
Bank Conflict Time (sec)	4590.677 [0,56]	4618.210 [0,79]	4608.131

each of them to different processors. By using 16 nodes (128 processors) on NEC SX-8, the final performance is shown in Table 1.

Here $[U, R]$ specifies the Universe and the Process Rank in the Universe and the data above is measured from MPI_{Init} till $MPI_{Finalize}$.

An average performance of 1.5 GFLOPS per process on 128 processes and an Average Vector Length of 146 are reported. The total memory used is $1086GB$ or $8.6GB$ per process. This performance is reasonable for DNS codes based on FFT library, which usually is poorly vectorized to some extent, compared with finite difference schemes, because of the frequent invoking of small primary decomposition subroutines. Approximately 25 days are needed to reach a resolved turbulent field, from which the physical properties can be fully analyzed.

2 Conditional Statistics Along Gradient Trajectories

Once the DNS data are obtained, according to different theory and purposes, the data analysis will be the core of the work at next stages. One of the central topics in turbulence theory is the nonlocal behaviors, which can be investigated through two-point correlation functions or structure functions. The most prominent results derived from the Navier-Stokes equations are the Kolmogorov equation for the velocity structure function and the Yaglom equation for velocity-passive scalar coupled structure function. In the context of dissipation elements, to quantitatively parameterize these geometrical objects, two important parameters, the linear length between and the scalar difference at the extremal points, have been chosen. While it can be expected that the conditional scalar difference follows the 1/3 Kolmogorov scaling [2], numerical data show that the velocity difference between extremal points in direction of the line connecting the extremal points surprisingly follows a linear scaling [3]. In order to explain this finding, it is meaningful to study the two-point velocity difference affixed to gradient trajectories rather than doing that in the Cartesian coordinate.

Some selected gradient trajectories in 3D space are shown in Fig. 3. In 3D space gradient trajectories (yellow lines) have complicated geometrical shapes and their tangent directions **n** are time and space dependent. Each gradient trajectory connects one maximal (red) point and one minimal (blue) point and thus different trajectories can join at a same extremal point.

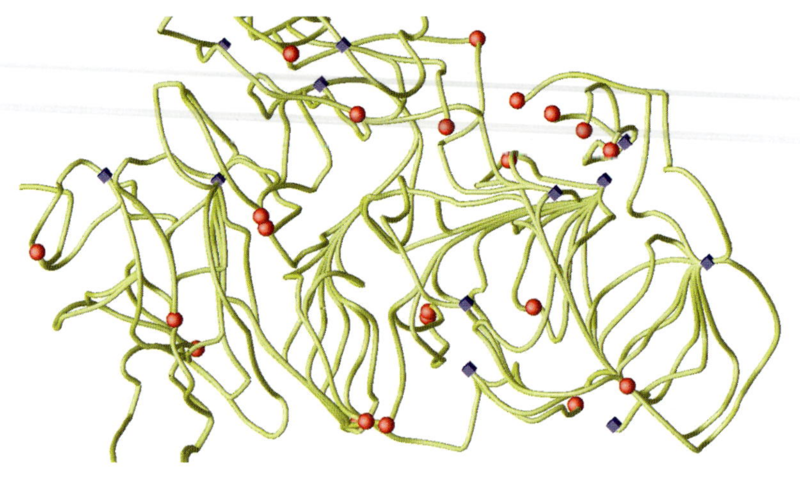

Fig. 3. Example of gradient trajectories (yellow lines) in a turbulent flow. Each trajectory connects one maximal (red) point and one minimal (blue) point. Different trajectories can join at a same extremal point

2.1 Theoretical Results and Numerical Verification

In a recent work [10], a new theory has been developed, which states that there is a linear scaling of the two-point velocity difference with respect to the separation arclength between along the same trajectories. As can be seen, this new scaling from the conditional statistics along gradient trajectories is essentially different from the classical Kolmogorov 1/3 scaling. To prove this important prediction, we need to analysis the statistics along trajectories. To achieve this, especially for large DNS data with higher Reynolds numbers, parallelization is necessary for the post-processing code. According to structure of the code, parallelization can be readily done with excellent performance.

We have compared the results for two different DNS cases, whose parameters are listed in Table 2.

Table 2. Characteristic parameters of two DNS cases

	case 1	case 2
mean shear $S = d\langle v_1\rangle/dx_2$	0.5	1.5
grid points	512^3	512^3
viscosity ν	0.003	0.003
turbulent kinetic energy k	0.827	2.62
dissipation ε	0.112	0.978
dissipation ε/k	0.135	0.373
scalar variance $\langle\phi'^2\rangle$	0.0622	0.0290
scalar dissipation $\langle\chi\rangle$	0.0188	0.0222
Kolmogorov scale η	0.022	0.0129
Taylor scale λ	0.471	0.283
$Re_\lambda = v_{rms}\lambda/\nu$	116.5	125.0
resolution $\Delta x/\eta$	0.558	0.950

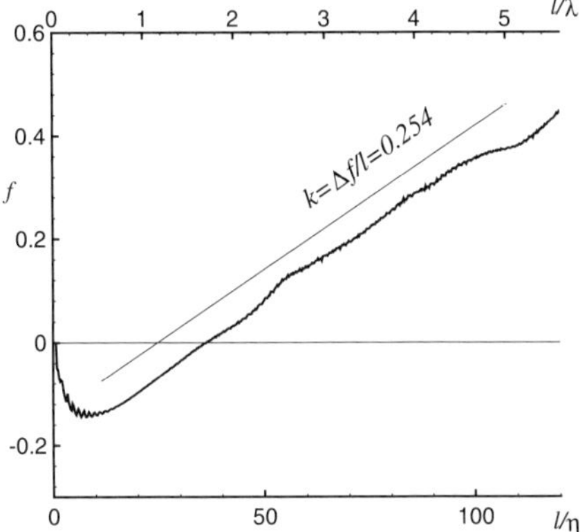

Fig. 4. Structure function along trajectories for case 1

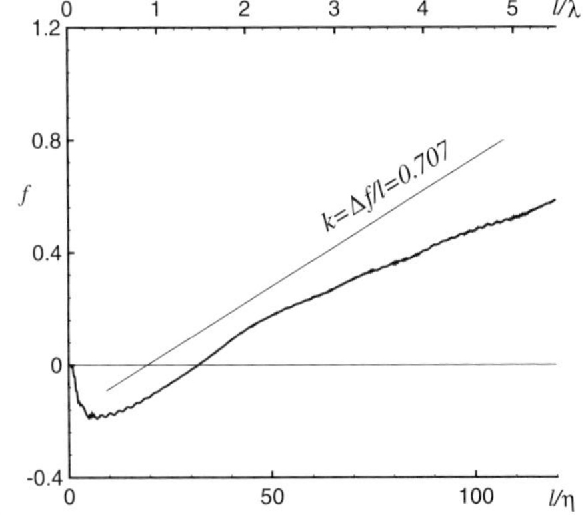

Fig. 5. Structure function along trajectories for case 2

The numerical Results are shown in Fig. 4 and Fig. 5 for case 1 and case 2, respectively. Differently from averaging in Cartesian coordinate where sample points are equally weighted, sample points along gradient trajectories are not equally volume-weighted and this weighting effect has been considered in the numerical analysis. It can clearly be seen the linear variation tendency with respect to the separation arclength between two points along the same trajectories in the inertial range, in satisfactory agreement with the theoretical analysis.

2.2 Computational Performance

Because of the relatively simply structure in the post-process code, after OpenMP parallelization, the scaling tested on asama tx-7 is almost linear up to 16 CPUs. For cases with 512^3 grid points, the memory needed is about $7G$ bytes, while for cases with $1024 \times 2048 \times 1024$ grid points, the memory will increase up to about $85G$ bytes.

Acknowledgement. The project of HRC-DNS has been continuously sponsored by HLRS. The author greatly appreciates the necessary support from HLRS. Special thanks go to Mr. Stefan Haberhauer (NEC) and Dr. D.Takahashi in Japan for the detailed technical help. Furthermore the motivation of the research by Prof. N.Peters(RWTH-Aachen, Germany) and the founding from Deutsche Forschungsgemeinschaft under Grant Pe 241/38-1 are acknowledged.

References

1. Wang, L., High-Resolution Direct Numerical Simulation of Homogeneous Shear Turbulence. High Performance Computing in Science and Engineering'08, Edited by W. Nagel, D. Kroener and M. Resch, Springer (2008)
2. Wang, L., Peters, N., The length-scale distribution function of the distance between extremal points in passive scalar turbulence. J.Fluid Mech., **554**, 457–475 (2006)
3. Wang, L., Peters, N., Length scale distribution functions and conditional means for various fields in turbulence. J. Fluid Mech., **608**, 113–138 (2008)
4. Donzis, D.A., Yeung, P.K.et al., Dissipation and enstrophy in isotropic turbulence: Resolution effects and scaling in direct numerical simulations. Physics of Fluids, **20**, 045108 (2008)
5. Moin, P., Mahesh, K., Direct Numerical Simulation: A Tool in Turbulence Research. Annual Review of Fluid Mechanics, **30**, 539–578 (1998)
6. Pope, S.B., Turbulent flows. Cambridge University Press (2000)
7. Schumacher, J., Sresnivasan, K.R., Yeung, P.K., Very fine srtuctures in scalar mixing. J. Fluid Mech., **531**, 113–122 (2005)
8. Sreenivasan, K.R., Schumacher, J, Yakhot, V., Intermittency and Direct Numerical Simulations. 58th Annual Meeting of the Division of Fluid Dynamics, APS, Chicago, USA (2005)
9. http://www.ffte.jp
10. Wang, L., Scaling of the two-point velocity difference along scalar gradient trajectories in fluid turbulence. Physical Review E, **79**, 046325 (2009)

Direct Numerical Simulation of Single Gaseous Bubbles in Viscous Liquids

Hendrik Weking, Christian Huber, and Bernhard Weigand

Institut für Thermodynamik der Luft und Raumfahrt, Universität Stuttgart, Pfaffenwaldring 31, 70569 Stuttgart, Germany
hendrik.weking@itlr.uni-stuttgart.de
christian.huber@itlr.uni-stuttgart.de

Summary. The rise behavior of single air bubbles of different size in viscous liquids (water/glycerol mixtures) is investigated using a 3D Direct Numerical Simulation method. The simulations where performed by the ITLR inhouse code Free Surface 3D (FS3D) which solves the incompressible Navier-Stokes equations using a Volume-of-Fluid (VOF) technique. The computational effort is reduced by using a moving frame of reference. This moving frame allows the simulation of a large scale rising trajectory. The numerical results, which comprise the terminal rise velocity, the aspect ratio of the deformed bubbles and the resulting bubble shape are compared to experimental data from literature. The simulations were performed on the NEC SX-8 and NEC SX-9 platforms of the HLRS.

1 Introduction

The motion of gaseous bubbles in a sourrounding liquid is part of many industrial applications, either as individual particles or as bubble clouds. An optimization of these applications requires a deeper understanding of the relevant hydrodynamics, e.g. the inner and outer flow field and the deformation of the bubble. Due to the non-linear character of the Navier-Stokes equations a general analytical solution can only be derived for bubbles of spherical or ellipsoidal shape or with slight deformation in creeping flows or high Reynolds number flows. Therefore experimental and numerical approaches have been made to investigate the rise behavior of single gaseous bubbles in a quiescent liquid. A wide range of contributions to the subject of the motion of bubbles, drops and particles is contained in the book of Clift, Grace and Weber [1]. Maxworthy et al. [2] presented an experimental investigation of single air bubbles in water/glycerol mixtures. A study of water/glycerol mixtures comprising experimental and numerical data was published by Raymond and Rosant [3].

In this paper we present a numerical approach for the rise behavior of single air bubbles in viscous liquids, i.e. water/glycerol mixtures investigated by

W.E. Nagel et al. (eds.), *High Performance Computing in Science and Engineering '09*, DOI 10.1007/978-3-642-04665-0_20,

Raymond and Rosant, using direct numerical simulation (DNS). The results are obtained using the ITLR inhouse code Free Surface 3D (FS3D), which allows the computation of large rising trajectories at acceptable computational effort using a moving frame of reference. The results which comprise the terminal rise velocity and the deformed shape are compared to experimental data from the literature. Additionally, a performance study regarding the computational speedup performance of FS3D on the NEC SX-8 platform has been made.

2 Numerical Method

The inhouse 3D CFD program FS3D solves the Navier-Stokes equations for incompressible flows with free surfaces using direct numerical simulation. The Volume-of-Fluid (VOF) method by Hirt and Nichols [4] is employed to describe the motion of the gaseous bubbles in liquids. Therefore an additional field variable f, representing the volume of fluid fraction, is used to distinguish between the liquid and the gaseous phase. The volume fraction is defined as:

$$f(\mathbf{x},t) = \begin{cases} 0 & \text{in the gaseous phase} \\ 0 < f < 1 & \text{in cells containing a part} \\ & \text{of the interface} \\ 1 & \text{in the liquid phase} \end{cases} \tag{1}$$

The movement of the volume fraction is traced by an additional transport equation

$$\frac{\partial f}{\partial t} + \nabla \cdot (f\mathbf{u}) = 0 \tag{2}$$

with the velocity vector $\mathbf{u}$ and the time t. In the calculations presented in this paper the flow is considered isothermal and incompressible and is governed by the conservation equations for mass

$$\nabla \cdot \mathbf{u} = 0 \tag{3}$$

and momentum

$$\frac{\partial(\rho\mathbf{u})}{\partial t} + \nabla \cdot (\rho\mathbf{u}) \otimes \mathbf{u} = -\nabla p + \rho\mathbf{k}_b + \nabla \cdot \mu \left[\left(\nabla\mathbf{u} + (\nabla\mathbf{u})^T\right)\right] + \mathbf{f}_\gamma \tag{4}$$

for two immiscible fluids, where ρ denotes the density, p the pressure and μ the dynamic viscosity. Due to the numerical setup presented in the next section the gravity $\mathbf{k}_g$ is replaced by the buoyancy $\mathbf{k}_b$, which differs from zero only in the gaseous phase. The relation between the buoyancy and the gravity reads

$$\mathbf{k}_b = \frac{(\rho_g - \rho_l)\,(1 - f(\mathbf{x},t))}{\rho(\mathbf{x},t)}\mathbf{k}_g, \tag{5}$$

where the subscript g denotes the gaseous phase and l the liquid phase. The density and the viscosity in the flow field are calculated by

$$\rho(\mathbf{x},t) = f(\mathbf{x},t)\rho_l + (1 - f(\mathbf{x},t))\rho_g \tag{6}$$

$$\mu(\mathbf{x},t) = f(\mathbf{x},t)\mu_l + (1 - f(\mathbf{x},t))\mu_g. \tag{7}$$

The surface tension at the interface is stated by the capillary stress tensor $\mathbf{f}_\gamma$. It is computed by the conservative continuous surface stress (CSS) model by Lafaurie et al. [5].

FS3D uses a structured finite volume scheme on a staggered (MAC) grid. The spatial discretization is of second order accurate, the accuracy in time is of first order in the considered cases. The reconstruction of the interface used for the calculation of the interface convection is based on the piecewise linear interface calculation (PLIC) method by Rieder and Kothe [6]. The fully conservative momentum convection and volume fraction transport, as well as the momentum diffusion, the buoyancy and the surface tension are treated explicitly.

3 Numerical Setup

3.1 Discretization

On the upper side of the 3D computational domain a uniform inflow condition is used as shown in Fig 1. On all other boundaries including the outflow a continuous (Neumann) condition is employed. A damping zone in front of the outflow boundary suppresses undesired back flow. The gravity pointing towards the outflow boundary leads to a strong acceleration of the liquid, which "falls" out of the domain, taking the bubble with it. This can be avoided

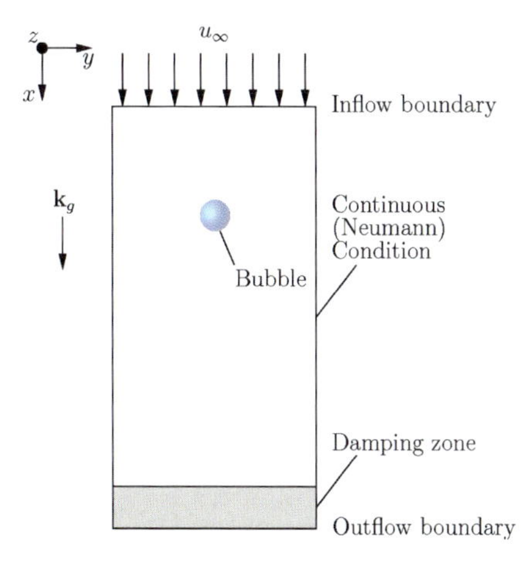

Fig. 1. Numerical Setup and coordinate system

by replacing the gravity by the buoyancy in equation (4). In the presented study a variety of bubble diameters were investigated. Therefore the extent of the computational domain varies depending on the initial bubble diameter. The computational domain has an overall length of 20 bubble diameters in the direction of the main flow and 10 bubble diameters between the lateral boundaries. The resolution is chosen to be 512 cells in main flow direction and 256 cells in the lateral direction, which makes about $33 \cdot 10^6$ computational cells in the whole domain. At this resolution one bubble diameter is resolved by 25.6 cells.

In all cases the bubble is initialized as a stationary sphere surrounded by quiescent liquid at a distance of 4.5 bubble diameters away from the inflow and 5 diameters away from the lateral boundaries.

3.2 Moving Frame of Reference

To investigate the bubble motion over a larger time scale and thus over a longer rising path without a large computational domain and the resulting computational effort, a moving frame of reference is employed based on the moving frame used by Graf [7] for falling droplets. This technique induces a counterflow in the domain. The counterflow is adapted dynamically over time. Therefore motion of the center of mass of the bubble is considered as a damped oscillation of a point mass around its initial position. Depending on the displacement and the velocity of the center of mass of the gaseous phase a correction body force $\rho \mathbf{k}_c$ is imposed to the flow field by adding it to the right hand side of (3) inducing the counter flow in the computational domain. This keeps the bubble close to its initial position, or from the view of a resting observer, moves the whole domain along with the bubble. This correction force is computed for all three spatial directions in every timestep by

$$\mathbf{k}_c = 2\omega \frac{\mathbf{P}_g}{m^*} + c_2 \omega^2 \Delta \mathbf{x}_g \tag{8}$$

with the momentum of the bubble $\mathbf{P}_g$ divided by the bubble virtual mass $m^* = m_g (\rho_f / \rho_g)$ and the displacement vector $\Delta \mathbf{x}_g$. The momentum and the displacement vector refer to the moving frame of reference. To avoid a build up in the displacement the value of c_2 is switched from 1 to 0 if the center of mass is approaching the initial position. The angular frequency governing the hardness of the dampening and thus the magnitude of the oscillation around the initial position is chosen to be

$$\omega = c_1 \sqrt{\frac{\rho_f}{\rho_g}} \pi \omega_{res} = c_1 \sqrt{\frac{\rho_f}{\rho_g}} \pi \sqrt{\frac{12\sigma}{d_e^3 \rho_f}} \tag{9}$$

where ω_{res} is the first resonance mode of the bubble surface. ω_{res} is multiplied by π to avoid that ω becomes a higher mode of that surface resonance frequency and thus prevent a build up of surface oscillations which could destroy the free surface. A value of $c_1 = 10$ was chosen in this study. Note, that this formulation does not fix the bubble at the start position but allows a displacement.

3.3 Investigated Cases

The motion of rising bubbles is investigated for air bubbles in different mixtures of water and glycerol. These systems were examined experimentally by Raymond et al. [3]. The material properties of air at $T = 20°\mathrm{C}$ are $\rho_g = 1.2\mathrm{kg/m}^3$ for the density and $\mu_g = 18 \cdot 10^{-6}\mathrm{Pas}$ for the viscosity. The properties of the water/glycerol mixtures are given in Table 1. The Morton number

$$Mo = \frac{\Delta\rho|\mathbf{k}_g|\mu_l^4}{\rho_l^2\sigma^3} \tag{10}$$

is defined as the ratio of viscous and surface tension forces. For each liquid the equivalent bubble diameter is varied between $d_e = 2mm$ and $d_e = 5mm$ in steps of $1mm$.

Table 1. Physical properties of the investigated liquids

Water/Glycerol $Vol\%$	Density $\rho_l[kg/m^3]$	Viscosity $\mu_l[Pas]$	Surface Tension $\sigma[N/m]$	Morton number Mo	System
40/60	1150	0.013	0.064	$10^{-6.032}$	$W40$
31/69	1172	0.024	0.064	$10^{-4.975}$	$W31$
24/76	1190	0.042	0.064	$10^{-4.01}$	$W24$

4 Results

4.1 Terminal Rise Velocity

As the bubbles are initialized at zero velocity, they experience a short phase of acceleration before they reach a terminal rise velocity. For the given diameter range, the lower Mo systems are settled in the region, where transition to a zigzagging or spiraling trajectory may occur. Although, in all investigated cases the rising path stayed rectilinear, i.e. no transition and therefore no strong oscillation of the rise velocity was observed. The numerical results for the terminal rise velocity are compared to experimental data from Raymond and Rosant [3] in Fig. 2. Raymond and Rosant noted that they had not made special efforts to ensure the purity of the investigated systems. Nevertheless, the experimental data show the behavior of non-contaminated fluids. For the higher Mo systems there is good agreement between the numerical data and

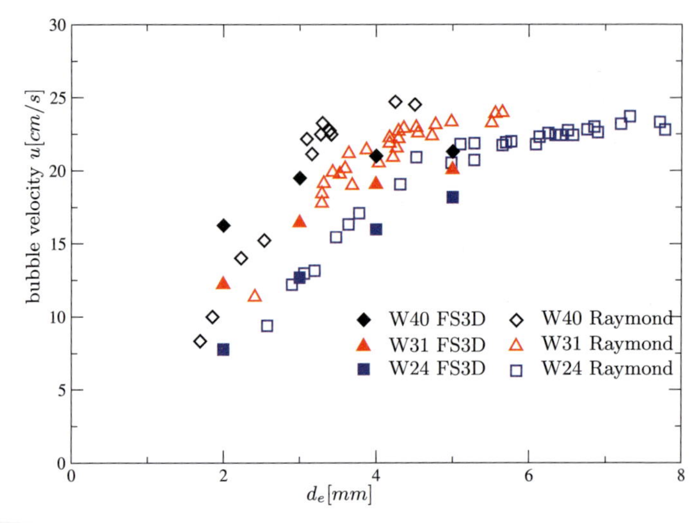

Fig. 2. The terminal rise velocity of bubbles vs. the equivalent diameter

the experiments regarding the qualitative trend and the absolute value. In the less viscous systems, the steep rise of the velocity with increasing bubble diameter, which is observed in the spherical regime, could not be reproduced by FS3D in this study. The simulation of bubbles with smaller diameters will be the aim of further investigation. They were not considered in this study due to the rising influence of spurious currents, which are non-physical accelerations mainly induced by the surface tension model.

For the larger bubble diameters, the $W40$ system computed using FS3D seems to approach an asymptotic value. This behavior is as well observed by Raymond and Rosant, but at a different value of the terminal rise velocity.

4.2 Bubble Shape

The deformation of the bubbles after initial acceleration phase, i.e. when the rectilinear rising path with the terminal velocity is reached, is presented in Fig. 3 as the aspect ration of the bubble height to the bubble width h/w. After reaching the terminal velocity, all of the presented bubbles developed a rotational symmetric oblate shape ($h/w \leq 1$). Especially the computed aspect ratios of the larger bubbles match well with the experimental results. For each Morton number the trend of the aspect ratio is nearly linear. The only exception is the $d_e = 5mm$ case of $W40$ which indicates the approach to an asymptotic value which is given as $h/w = 0.24$ by Clift et al. [1]. The computed bubble shapes of the $W40$ and $W24$ systems are compared in Fig. 5. Additionally, the internal flow field in the plane of symmetry of the bubble is shown. The velocity vectors plotted refer to the moving frame of reference.

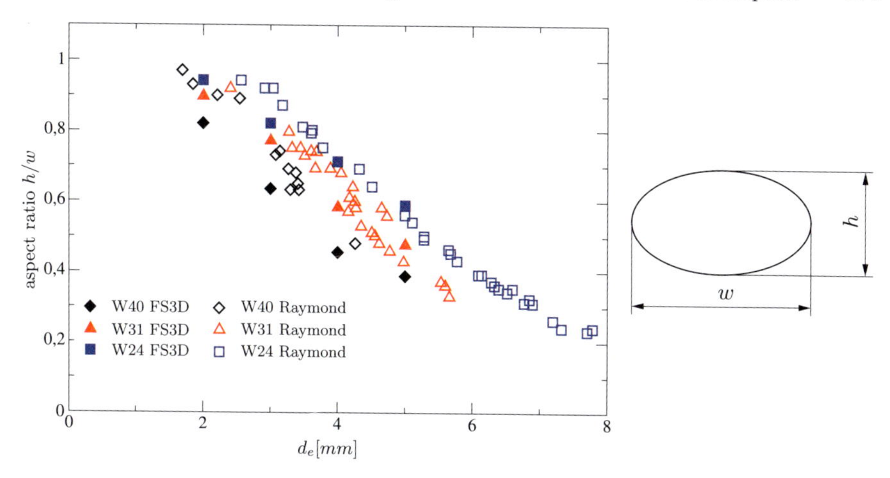

Fig. 3. Bubble aspect ration h/w over equivalent diameter d_e (left), h and w of an ellipsoidal shaped bubble (right)

4.3 Drag Coefficient

The drag force $|\mathbf{F}_D|$ affecting the rising bubble is characterized by the drag coefficient given by

$$C_D = \frac{|\mathbf{F}_D|}{\frac{\rho_l}{2} u_\infty^2 A_e}. \tag{11}$$

where the equivalent area is the projected area of the initial sphere $A_e = \pi d_e^2/4$. The computed terminal rise velocity shown in Fig. 2 is chosen as the reference velocity u_∞. The results are plotted over the Reynolds number

$$Re = \frac{\rho_f U_\infty d_e}{\mu} \tag{12}$$

in Figure 4. In the spherical regime, i.e. at lower Reynolds numbers, the computed drag coefficients show good agreement to the correlation for spherical bubbles at $Re > 2$ given by Clift et al. [1]. For larger diameters and thus a rising deviation from the spherical shape the numerical results defect from that correlation. The same trend is predicted by the correlation of Moore [8] and can also be seen in the experimental data of Raymond and Rosant, but the absolute values obtained by FS3D exceed the data of Raymond and Rosant.

5 Performance

All performance results were computed on the NEC SX-8 platform at HLRS. In order to analyse the performance a three dimensional test case with pure water as liquid and an ascending air bubble has been chosen, which results in a

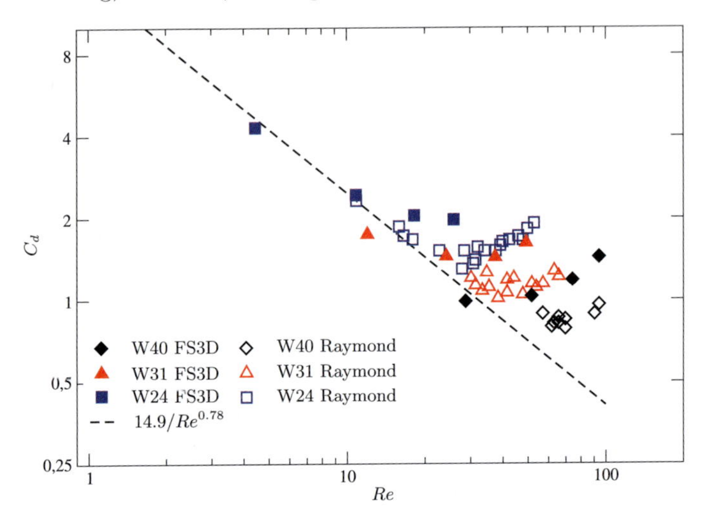

Fig. 4. Drag Coefficient over Reynolds number

Morton number of Mo$=10^{-10.6}$. The initial diameter of the bubble $d_e = 4mm$ and the total domain extension of $8cm \times 4cm \times 4cm$ yield a diameter ratio of $\frac{d_{Domain}}{d_e} = 10$. The simulation domain was discretized with a grid resolution of $512 \times 256 \times 256$ cells ($33 \cdot 10^6$ cells), resulting in 25.6 cells/diameter and a spatial resolution of $156\mu m$.

To get information about the actual speedup performance four computations were conducted, each of them using a different number of CPUs (1, 2, 4, and 8 CPUs). Therefore a parallel version of FS3D using the (-P auto) option was used and all computations were conducted on exclusive nodes. Additionally the program was compiled using the default optimization level (-C vopt) and maximum inlining. To get meaningful performance information the F_PROGINF=DETAIL option was utilized and all operations regarding I/O were disabled during the computation. In all four cases the computations were stopped after 2000 iteration cycles to get comparable results; furthermore the number is high enough to be sure that the initialization process plays a negligible role.

All computations achieved average vector lengths of about 208 and vector operation ratios of about 98, 1%. Also a computational performance in the range of 3, 5 GFLOPS per processor (on 8 CPUs) up to 6, 2 GFLOPS per processor (on 1 CPU) was attained.

In Fig. 6 the achieved relative speedup S, which is defined by

$$S = \frac{T_1}{T_N} \tag{13}$$

is depicted against the ideal speedup. Here T_N is the total execution time on N processors and T_1 is the total execution time of the parallel code on only one processor. In this case for two processors a speedup of 1.75 is achieved as with

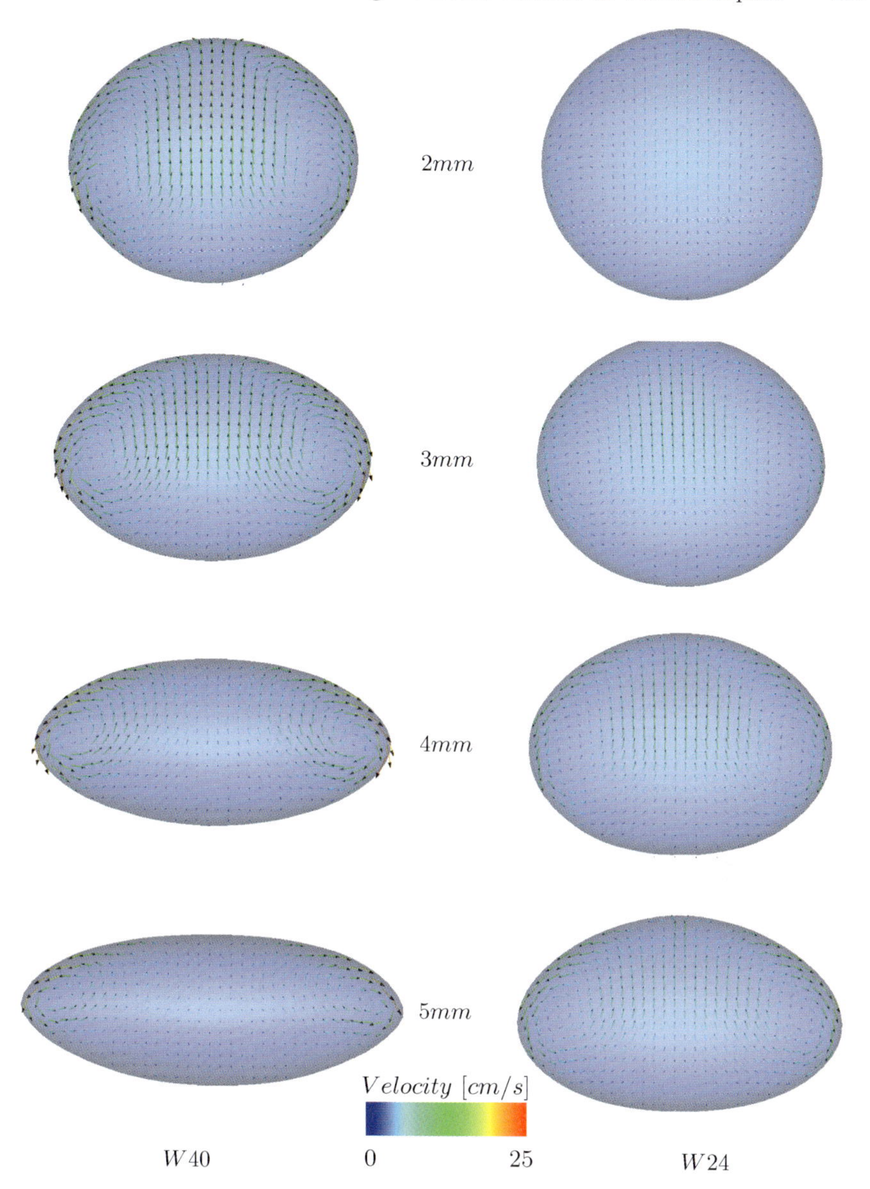

Fig. 5. Bubble shape and internal flow field

four processors the speedup increases up to 2.81 showing a great difference compared to the linear speedup. The speedup on eight processors shows with a value of 3.85 an even higher deviation from the linear speedup. The same effect can be seen when examining the achieved efficiency E for N processors

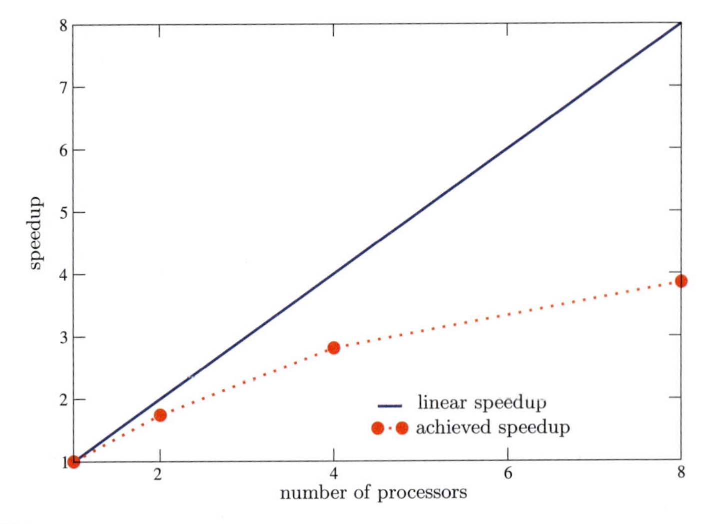

Fig. 6. Achieved relative speedup over the total number of processors

$$E = \frac{T_1}{T_p N} = \frac{S}{N} , \tag{14}$$

depicted in Fig. 7. Apart from the efficiency on eight processors which is reaching only 48,1%, the efficiencies for two and four processor systems all lie above 70%. Comparing the above findings with the absolute calculation times of the different cases, as shown in Fig. 8, it can be seen that the steepest drop in computational time occurs in the region of two and four processors. The time advantage above four processors is only marginal compared to the previous mentioned, but nevertheless from major importance for long-lasting simulations.

The speedup deficits, especially in case of the computation on eight processors can be better interpreted when the serial fraction, which denotes the percentage of time spent on the serial part of the code, is taken into account. Therefore the empirical serial fraction F_S was determined experimentally by using the Karp-Flatt Metric which is based on Amdahl's law [10]:

$$F_S = \frac{\frac{1}{S} - \frac{1}{N}}{1 - \frac{1}{N}} . \tag{15}$$

Herewith for all multiprocessor computations a constant value of about 15% was achieved for the serial fraction of the code. As the serial fraction stays constant using a different number of processors the loss in efficiency and so the lower speedup values are due to the inherent serial fraction instead of the parallel overhead. Under the assumption that the parallel part of the code performs with an ideal, linear speedup the maximum achievable speedup S_T on an infinite number of processors is limited to 6.67 for this case. Therefore equation (15) reads as follows:

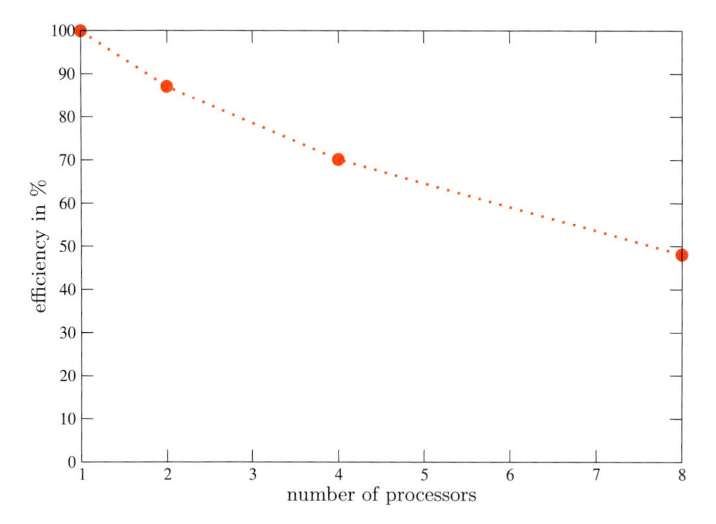

Fig. 7. Efficiency in % over the total number of used processors

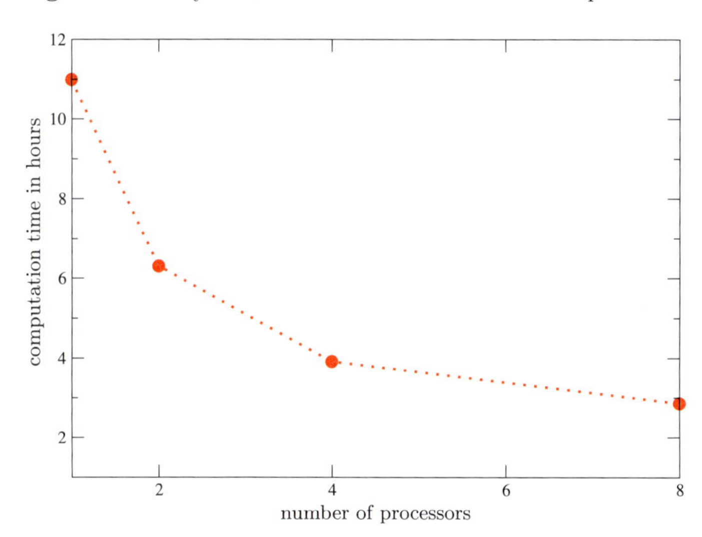

Fig. 8. Computation time in hours over the total number of used processors

$$S_T = \frac{1}{F_S} \ . \tag{16}$$

Altogether a respectable high computational performance was achieved. Summarized, the reduced speedup, like in the case of the computation on eight processors, can be interpreted as an effect of the time spent on the serial parts of the code. This can be further improved by reducing the inherent serial fraction of the code and further improvements of the already parallelized code parts. A major role hereby will be the further optimization of the implemented multigrid solver which plays a central role in FS3D. Beside the here treated case it is also expected that with a higher grid resolution the multigrid solver, which requires a significant high amount of the total computational time [9], gets more efficient.

6 Conclusion

A numerical study of the motion of single air bubbles in viscous liquids has been performed. To minimize the computational effort, a moving frame of reference technique was employed. The rise velocity and the deformation of the free surface obtained by using FS3D match the experimental results for larger Morton number systems published by Raymond and Rosant. For the spherical regime, the drag coefficients calculated correspond to the correlation given by Clift. Furthermore the computational speedup performance of FS3D on the NEC SX-8 platform has been investigated, showing the current parallel computing abilities of FS3D. For the examined case a maximum computational performance per processor of 6, 2 GFLOPS was achieved.

Acknowledgments. The authors greatly appreciate the *High Performance Computing Center Stuttgart* (HLRS) for support and computational time on the NEC SX-8 and NEC SX-9 platforms under the Grant No. FS3D/11142.

References

1. Clift, R., Grace, J.R., Weber, M.E.: Bubbles, Drops, and Particles. Dover Publications Inc., Mineola, New York, USA (2005)
2. Maxworthy, T., Gnann, C., Kürten, M., Durst, F.: Experiments on the rising of air bubbles in clean viscous liquids. Journal of Fluid Mechanics, **321**, 421–411 (1996)
3. Raymond, F., Rosant, J.: A numerical and experimental study of the terminal velocity and shape of bubbles in viscous liquids. Chemical Engineering Science, **55**, 943–955 (2000)
4. Hirt, C.W., Nichols, B.D.: Volume of Fluid (VOF) Method for the Dynamics of Free Boundaries. Journal of Computational Physics, **39**, 201–225 (1981)
5. Lafaurie, B., Nardon, C., Scardovelli, R., Zaleski, S., Zanetti, G.: Modelling merging and fragmentation in multiphase flows with SURFER. Journal of Computational Physics, **113**, 134–147 (1994)
6. Rieder, W.J., Kothe, D.B.: Reconstructing volume tracking. Journal of Computational Physics, **141**, 112–152 (1998)
7. Graf, F.: Numerische Untersuchungen angeströmter Flüssigkeitstropfen. Studienarbeit, Institut für Thermodynamikder Luft- und Raumfahrt, Universität Stuttgart, Germany (2000)
8. Moore, D.W.: The velocity of rise of distorted gas bubbles in a liquid of small viscosity. Journal of Fluid Mechanics, **23**, 749–766 (1965)

9. Sander, W., Weigand, B.: On the influence of the Nozzle Design for Water Sheets. In: High Performance Computing in Science and Engineering '07. Springer, Berlin Heidelberg New York (2008)
10. Bengel, G., Baun, C., Kunze, M., Stucky, K.U.: Masterkurs Parallele und Verteilte Systeme. Vieweg+Teubner, Wiesbaden (2008)

Wall Heat Load in Unsteady and Pulsating Combustor Flow

D. Panara, B. Noll, and M. Kindler

DLR-Institut für Verbrennungstechnik,
Pfaffenwaldring 38-40, 70569 Stuttgart, Germany

Summary. Despite the high interest in unsteady heat transfer phenomena, there is a fundamental lack of experimental and numerical work in this field. Moreover, the development of turbulence heat transfer models rely on the availability and accuracy of near wall data which are very difficult or even impossible to be measured. In recent years DNS has shown the capability to be used as a so called "numerical experiment" and has been applied to fill the gap between models and accurate data. In the present project DNS simulations were performed for a simple channel configuration with heat transfer. For the steady channel flow, the DNS results are validated against experimental data and very fine resolved DNS simulations available in the literature. Results are also shown and discussed for an unsteady pulsating flow. Neither numerical nor experimental results are available for the pulsating case involving heat transfer. In order to address the needs of turbulence modelling of unsteady near wall turbulent scalar transport the budget of the turbulent kinetic energy and its dissipation rate together with the budget of the turbulent temperature variance and its dissipation rate were determined at different time phases and for different values of the amplitude and frequency parameters. Thus, the present DNS database can be highly useful for future unsteady heat transfer closure development.

1 Introduction

Low emission lean premix combustion technology with high power density has led to the appearance of combustion instabilities which may cause severe damages to the combustor system. In order to prevent components failures, accurate wall heat load prediction in unsteady combustor flow conditions are required for a correct estimation of the component high-cycle thermal fatigue.

In the framework of the DFG-SFB-606 project (Deutsche Forschungsgemeinschaft, Sonderforschungsbereich 606: "Non-Stationary Combustion: Transportphenomena, Chemical Reactions, Technical Systems") researches were undertaken in order to address various aspects of unsteady combustion processes. Subproject C1 of the SFB 606 is devoted to research on high frequency oscillations in gas turbine combustion chambers. Within this subproject research

W.E. Nagel et al. (eds.), *High Performance Computing in Science and Engineering '09*, DOI 10.1007/978-3-642-04665-0_21,

on numerical simulation of high frequency combustion oscillations was performed. One important aspect of this work is the near wall behavior of flow and combustion in the presence of pressure oscillations. Here, the non understood phenomenon of near wall scalar transport as well as increased thermal wall load in presence of high frequency combustion oscillations are of high interest.

Computational Fluid Dynamics (CFD) simulations can help to investigate unsteady heat transfer processes. However, very often, CFD methods strongly rely on the accuracy of the near wall turbulence models which are often validated and developed for steady state flow conditions. The validity, limitations and the improvement of traditional turbulence models in the study of combustion chamber heat load and high-cycle thermal fatigue is an issue which requires a systematic analysis. Neither experimental nor numerical databases were available at the beginning of this project for the validation of URANS or hybrid LES/RANS models in non-isothermal unsteady flow conditions.

The main goal of the present project "WalLUP" thus was to apply accurate DNS simulations in order to address the effect of flow unsteadiness and pulsations on the turbulent wall heat transfer. The results obtained will serve as a database for the development of accurate URANS and hybrid LES/RANS methods for the numerical analysis of high-cycle thermal fatigue in presence of thermoacoustic combustion instabilities.

In the following, DNS results are presented for a simple channel configuration with heat transfer. The temperature has been treated as a passive scalar. For the steady channel flow, the DNS results are validated against experimental data and very fine resolved DNS simulations available in the literature [1, 2, 3]. Results are also shown and discussed for the unsteady pulsating flow. Unfortunately neither numerical nor experimental results are available for the pulsating case involving heat transfer. In this sense the present work can be considered original and could be used as a first numerical database for the development of better unsteady heat transfer models.

2 Test Cases and System of Equations

A simplified geometry and flow boundary conditions have been chosen to perform accurate DNS simulations of turbulent wall heat transfer with and without flow unsteadiness. The testcase geometry (see fig. 1) consists of an infinite flat channel in which the solid boundaries are kept at two different constant temperatures. For the investigations concerning unsteady heat transfer the turbulent channel flow pulsates due to a superimposed pulsating pressure source term with prescribed amplitude, frequency, and mean intensity. The DNS computational domain consists of a three dimensional channel with periodic boundaries in the span and stream-wise directions. No slip isothermal boundary conditions are prescribed at the solid walls. Temperature is treated as a passive scalar and thus no dependence of fluid property on temperature is

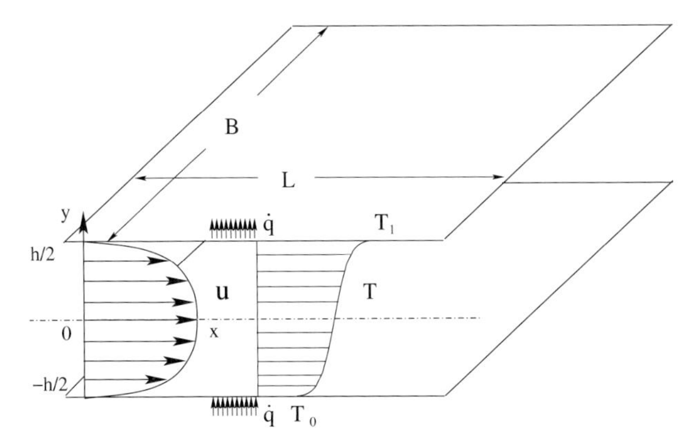

Fig. 1. Geometry, instantaneous velocity and temperature profile sketch

considered. The flow considered is air with a constant Prandtl number of 0.7. Considering the Temperature as a passive scalar in an incompressible flow, the governing equations can be simplified as follows:

$$\nabla \cdot (\vec{u}) = 0 \tag{1}$$

$$\frac{\partial \vec{u}}{\partial t} + \nabla \cdot (\vec{u}\,\vec{u}) - \nabla \cdot \frac{\tau}{\rho} + \frac{\nabla p}{\rho} = 0 \tag{2}$$

$$\frac{\partial T}{\partial t} + \nabla \cdot (T\vec{u}) = \frac{\nu}{\mathrm{Pr}} \nabla^2 T\,, \tag{3}$$

where

$$\frac{\tau}{\rho} = \nu \nabla \vec{u} + \nu \nabla \vec{u}^T - \nu \frac{2}{3} \mathrm{Tr}\left(\nu \nabla \vec{u}^T\right)\,. \tag{4}$$

No further assumptions have been made to model the turbulent flow. Each term of the above equations has been discretized by a finite volume approach. The CFD code used (OpenFoam) solves the coupled momentum-continuity system using a PISO strategy. The time discretization used is a typical backward Euler scheme (implicit, second order accurate in time). The gradient terms were discretized using a Gauss linear scheme (second order accurate). The divergence and Laplacian terms were discretized using a Gauss linear corrected scheme (unbounded second order, conservative) [4, 5].

3 Steady Channel Flow with Heat Transfer

The flow considered in this section is the Steady Channel Flow with Fixed Wall Temperature Difference (SChF-FWTD). A fully developed thermal and velocity field occurs between two infinite parallel plates (B and $L >> h$, see fig. 1) which are kept at different constant temperatures ($T1 > T0$). The flow Reynolds number (Re_{Dh}) based on the bulk flow velocity U_b and the hydraulic diameter $D_h = 2h$ was set to 13636. Using common turbulent channel flow correlations, it corresponds to a value of the friction velocity u_τ of 4.43. Consequently, the Reynolds number based on the friction velocity and half wall distance is $Re_{u_\tau} = 216$.

The computational grid consists of a cubic domain which extends about 430 wall units in the three dimensions. The finest grid used consisted of 100x160x100 points and was been stretched towards the center of the channel in the wall normal direction (Y - coordinate) with a geometrical progression of ratio 1.03. The resulting minimum and maximum cell dimensions in the Y-direction are respectively 0.6 and 6 wall units (cell ratio: $R = 10$). The grid is equally spaced in the spanwise (Z) and streamwise (X) directions with a resulting point interval of 4.3 wall units. A coarser grid consisting of 50x80x50 points has also been employed and comparisons are given with the finer grid showed very little discrepancies. Like the finer grid, the coarse grid is equally spaced in the span and streamwise direction. The stretching ratio is different and has been set to $R = 20$ giving at the wall values of y_+ comparable to the finer grid. No-slip and isothermal boundary conditions were used at the upper and lower solid walls. Periodic boundary conditions were imposed on the remaining boundaries. The flow motion was realized using a constant pressure source term in the momentum equation to balance the channel viscous losses.

3.1 Results

For the validation of the code, in fig. 2 the DNS temperature profile is compared with experimental values from Page et al. [6] at similar Reynolds Number. The data from [6] have been non-dimensionalized using the half channel

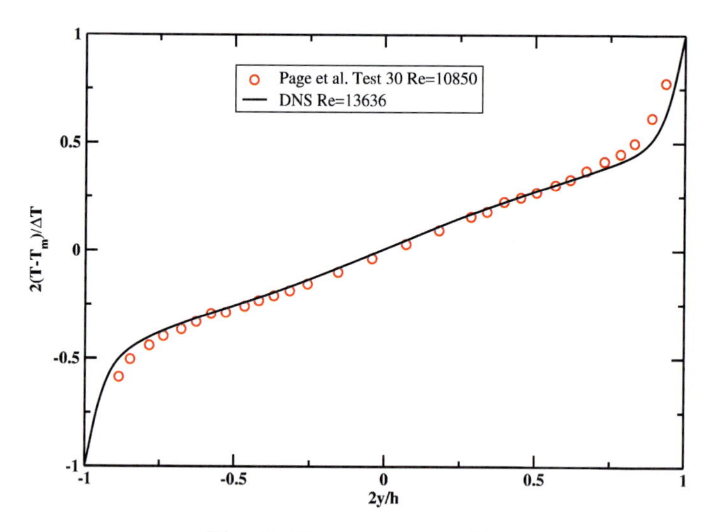

Fig. 2. Temperature profiles

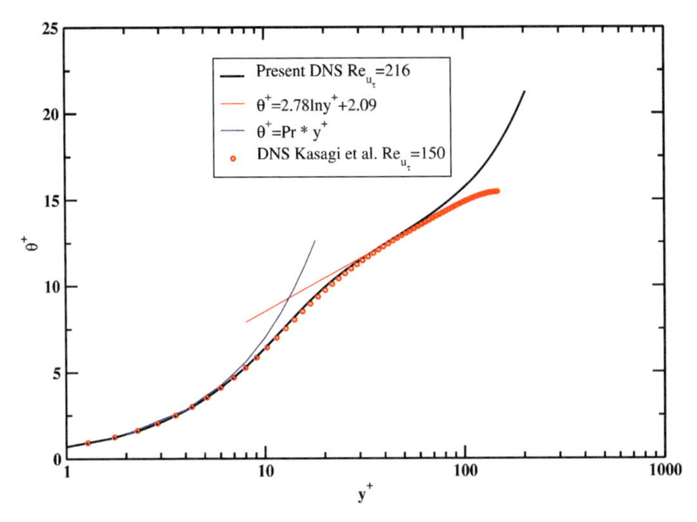

Fig. 3. Dimensionless temperature profiles. Comparison with DNS data from [1]

distance and wall difference. The DNS results and the experiments agree very well. In fig. 3 the temperature results are shown over wall units and are validated against the DNS results from Kasagi et al. [1]. In [1] the computational domain is analogous to the one used in the present work but the temperature boundary conditions are different. In [1] instead of prescribing a constant temperature difference between the channel walls, the flow is heated with a positive constant heat flux coming from the solid walls. This kind of heating configuration will be hereinafter referred as Steady Channel Flow with Constant Heat Flux (SChF-CHF) configuration and can easily be described using a classical Nusselt correlation for pipe flows.

Despite the boundary condition difference, both approaches are equivalent when the temperature profile is expressed in wall units:

$$\theta^+ = \frac{\overline{T} - T_w}{T_\tau}, \qquad \overline{T}_\tau = -\frac{\overline{q}_w}{\overline{\rho} c_p \overline{u}_\tau} \,. \tag{5}$$

In both configurations, in the near wall region, the temperature wall law is valid. Differences occur approaching the channel center due to the different heating also shown by Kawamura [7]. A more direct validation of the numerical results can be done against the numerical data from [3]. In [3] both geometry and thermal boundary conditions are similar to the present case. The only difference is in the value of the appropriate Reynolds number which, in the present case is 50% higher than in [3]. However, as shown by [7] for $Pr = 0.71$ (as in the present case) no appreciable Reynolds number effects are expected on the temperature wall profile which practically do not differ from the classical temperature wall law.

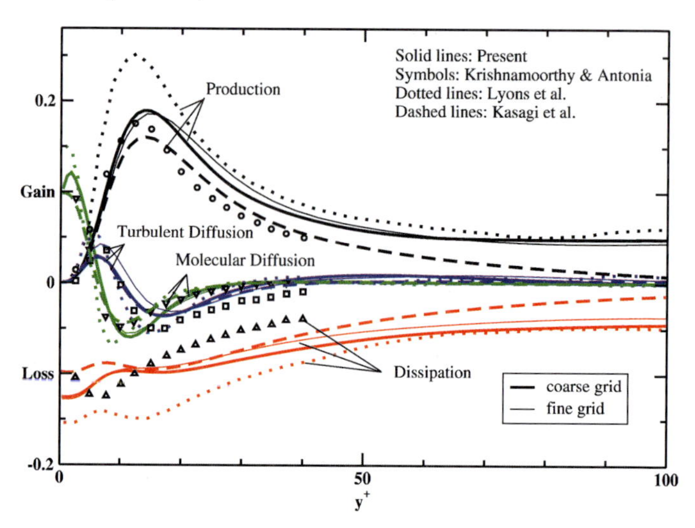

Fig. 4. Dimensionless k_Θ budget. Comparison with DNS data from Lyons et al. [3], Kasagi et al. [1] and experiments from Krishnamoorthy and Antonia [2]

In the present study a higher steady Reynolds number flow has been considered in order to consider later the effect of superimposed pulsations on the turbulence level. For this comparisons were made with LES and DNS results of [8] which were performed at a higher Reynolds number. In fig. 4, the budget of the temperature variance k_Θ is reported. Comparisons were made with the DNS data from [3] and the experiment from [2]. As pointed out by [3], the results obtained in the SChF-FWTD configuration compare well, in the near wall region ($y_+ < 40$), with the experiments or DNS in steady channel or at plate configurations with constant wall temperature or heat transfer. In the SChF-FWTD configuration the presence of a non-vanishing temperature gradient at the center of the channel determines a positive contribution of the production term in the k-equation which needs to be balanced by a non-vanishing dissipation term of the same intensity. In [1] and in the experiment of Krishnamoorthy and Antonia [2] instead, (see fig. 4) both the production and the dissipation term rapidly decrease towards zero value approaching the center of the channel. Similar comparisons were done for the ϵ_θ budget. Another interesting budget equation which can directly be compared with the DNS results in [1] is the turbulent kinetic energy budget k and has been reported in fig. 5. The good agreement between the present results with the data from [1] and [3] represents a good validation of the solver and the grid resolution used. The difference between the coarse and fine mesh results in the prediction of the budged quantities reported in figs. 4 and 5 is quite small. The budget behaviour is similar as well as the position of maxima and minima of each budget term. This allows using the coarser resolution to study qualitatively the dynamic behaviour of the near wall turbulence.

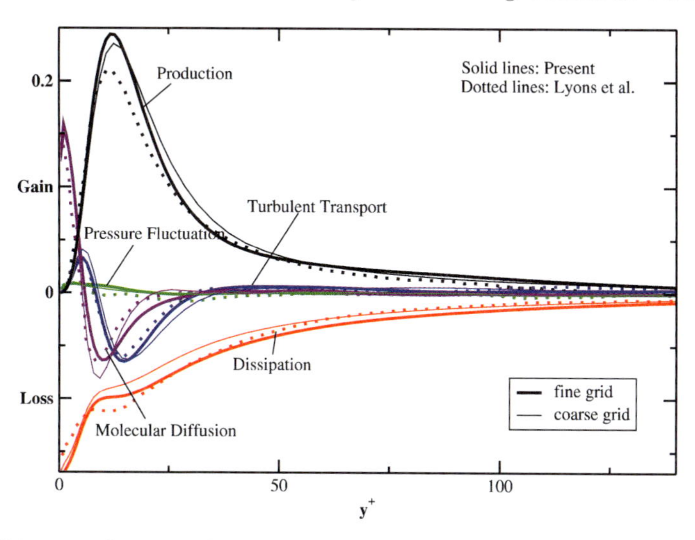

Fig. 5. Dimensionless turbulent kinetic energy budget. Comparison with DNS data from, Kasagi et al. [1]

4 Unsteady Channel Flow with Heat Transfer

In order to realize a pulsating flow within the channel, a pressure source term of the form

$$\frac{1}{\rho}\frac{\partial p}{\partial x} = K_m - K_{\mathrm{OSC}} \cdot sin(\omega t) \tag{6}$$

has been used. The flow geometry and boundary conditions considered are the same as in the SChF-FWTD configuration. In order to distinguish the pulsating channel from its steady counterpart the pulsating case will be referred hereinafter as the PulsChF-FWTD case.

As in [8], the oscillating pressure term K_{OSC} has been set in order to have at the channel centerline a pulsating amplitude ratio of 0.7 ($a_{uc} = 0.7$):

$$\langle u_c \rangle (t) = \overline{u}_c \left(1 + a_{uc} sin(\omega t + \phi_{uc})\right) \tag{7}$$

where $\langle u_c \rangle (t)$ represents the phase-locked ensemble average of the axial velocity component at the channel centerline.

Depending on the parameter auc, two flow regimes can be distinguished: the wave dominant ($a_{uc} >> 1$) and the current dominant ($a_{uc} << 1$). In the current dominant regime, the turbulent structures are only weakly affected by the variation of auc. The flow seems to be controlled mainly by the value of the non dimensional frequency ω_+:

$$\omega^+ = \frac{\omega \nu}{\overline{u}_\tau^2} \tag{8}$$

Instead of ω^+ often the Stokes length

$$l_s^+ = \frac{\delta_s \overline{u}_\tau}{nu}, \qquad \omega^+ = \frac{2}{l_s^+} \tag{9}$$

is used for flow characterization. For low values of ω_+ the flow can be considered quasi steady. The turbulence has time to relax to the local spectral equilibrium. The flow can be studied as a succession of steady states and the boundary layer is not affected by flow pulsations.

When the frequency is increased the turbulence production and dissipation start to show a phase lag. The stabilizing effect of acceleration can relaminarize the flow. The acceleration and deceleration phases are not symmetric and frequency effects on amplitude and phase with respect of the outer flow velocity are measured for the wall shear stress. This flow regime is often referred to as the intermediate frequency range.

In order to investigate the effect of pulsation frequency on the viscous-thermal boundary layer behaviour, three different values of the non dimensional frequency parameter ω_+ have been considered. Each value considered is representative of a different pulsation regime: quasi-laminar, intermediate frequency range and quasi steady. Moreover, in order to explain the non-linear effects of overall heat transfer enhancement in presence of very high amplitude oscillations reported by Ishino et al. [9], additional computations were performed at high values of pulsating Reynolds number ratio ($Re_b = Re_m = 6$). In table 1 a survey of the investigated parameter field together with the different pulsating flow regimes is given.

As it will be shown later on, the unsteady results in the Low-Reverse Flow case have a resolution comparable with other DNS and very accurate LES cases. In the case of Wave Dominated Flow, the high velocity amplitude involved determines a radical change in the scale of the turbulent structures. Keeping the grid fixed, at high velocity, i.e. high Reynolds Number, the accuracy of the calculation is considerably reduced. For this reason, the present high amplitude simulations should be regarded rather as a large eddy simulation LES than as a DNS.

Table 1. DNS computations, pulsating parameter and investigated flow regimes

	Quasi-Laminar Regime	Intermediate	Quasi-Steady Regime
Low Reverse Flow	$a_{uc} = 0.7, l_s^+ = 7$	$a_{uc} = 0.7, l_s^+ = 14$	$a_{uc} = 0.7, l_s^+ = 35$
Wave Dominated Flow	$a_{uc} = 6, l_s^+ = 7$	$a_{uc} = 6, l_s^+ = 14$	-

5 Low-Reverse Flow Results

The non-dimensional parameter considered by [8] (a_{uc} and l_s^+) have been limited to the case of Low-Reverse Flow. Despite the small difference in the mean flow Reynolds number, the near wall boundary layers are comparable and the turbulent Reynolds shear stresses seem to exhibit a similar behaviour. This was expected since the pulsation parameter were considered equal and the difference in Reynolds number affected just the ratio between the channel height and the Stokes length.

In fig. 6, the phase-locked averaged Nusselt number is reported for different frequency parameter values and in the case of small amplitude oscillations. The increased frequency seems to have a higher impact on the phase and amplitude of the Nusselt number than on its mean value. In fig. 6 the phase is reported again with respect to the pressure source term oscillation. According to the quasi steady theory, an analytical behaviour of the oscillating Nusselt number can be derived. Thus, at each time t the Nusselt number can be expressed as:

$$\mathrm{Nu}\left(t\right)=0.021\mathrm{Pr}^{0.5}\left(\frac{2hU_{b}\left(1+a_{uc}cos(\omega t)\right)}{\nu}\right)^{0.8} \tag{10}$$

The quasi-steady line is than in phase with the velocity pulsation and presents a 90° phase shift with respect to the pressure term. The Nusselt number behaviour at $l_s^+ = 35$ presents a similar behaviour. The maximum and minimum Nusselt number values are however delayed of $\pi/4$ and $\pi/2$ with respect of the quasi-steady line, respectively. Increasing the frequency, the Nusselt number amplitude decreases and the phase shift increases till the velocity and the heat transfer reach practically an opposition of phase. It is evident that the changing in amplitude and phase cannot be predicted within the quasi steady theory.

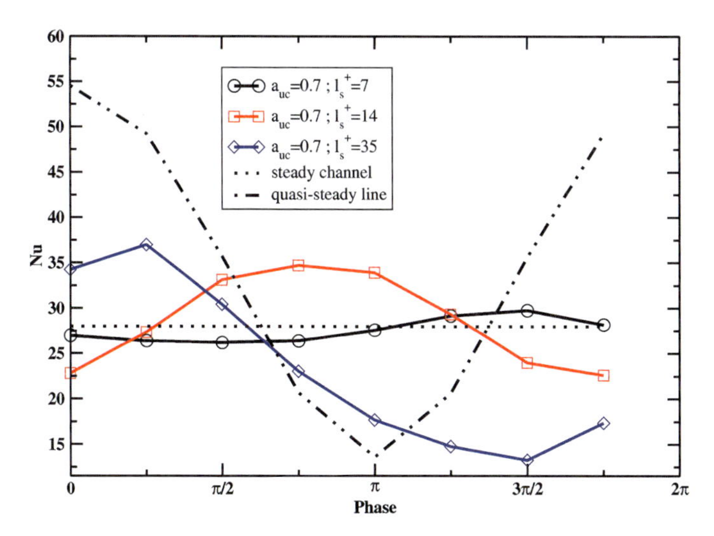

Fig. 6. Pulsating channel flow, Phase-locked averaged Nusselt number comparison

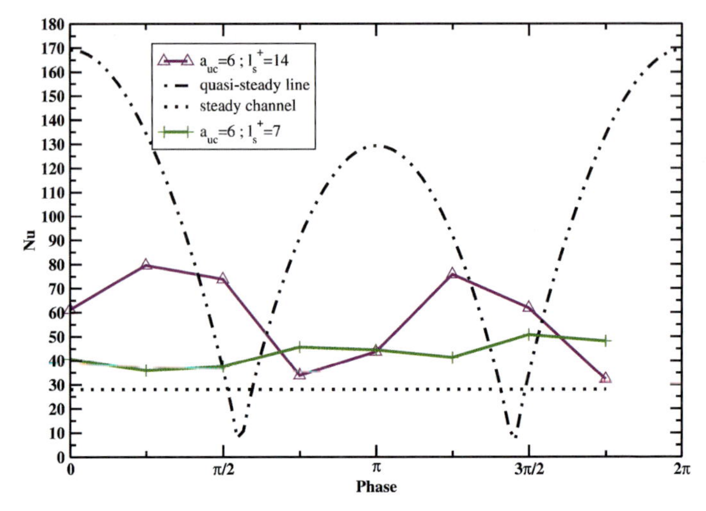

Fig. 7. Pulsating channel flow, Phase-locked averaged Nusselt number comparison, wave dominated flow

5.1 Wave Dominated Flow

In the present section the results in the Wave Dominated Flow conditions are reported. In fig. 7 the phase-locked averaged Nusselt number is shown. As expected, there is a big difference between the present results at $l_s^+ = 7$ and $l_s^+ = 14$ and the quasi-steady theory. The increased frequency seems to reduce not only the Nusselt number amplitude but also its overall mean value. This trend seems to be opposite to the experimental results in [9]. This has to be clarified in future investigations. Due to the mentioned reduced accuracy of the simulations no-budgets are evaluated here.

6 Computational Resources

The computational code used has been developed using the OpenFoam CFD Toolbox version 1.3 (http://www.opencfd.co.uk/openfoam/). Computations were conducted for CFD code validation and optimization on 2 and 4 nodes of the HP XC4000 system. Calculations on two different computational grids were done: 50x80x50 (200.000 points) and 100x160x100 (1.600.000 points). Since comparisons of the results on the two grids showed only little discrepancies most calculations were done on the coarse grid. For the coarse mesh the program executed on 4 nodes and thus the typical total CPU times required for each run was about 5 to 6 days.

In order to save disk space statistical averaging was performed in run time. Different averaged quantities were recorded (mean velocity, mean pressure, mean temperature, rms of pressure and velocity, 1 tensor field for $\overline{u_i' u_j'}$, 1 vector field for $\overline{T'u'_j}$, 5 scalar fields for the turbulent kinetic energy budget, 4 scalar fields for the fluctuating temperature budget, 7 scalar field for the budget of its dissipation rate). For steady state simulations generally the output is recorded every 10.000 iterations for a total of 4.2GB of temporary disk space (846MB of permanent disk space). For unsteady state simulations the phase-locked average is performed in runtime for 12 phases. The space required for each output is than about 8GB.

7 Conclusions

A DNS simulation campaign has been undertaken in order to establish a database for the validation of turbulence models for the calculation of unsteady wall heat transfer processes. The DNS results indicated in the intermediate frequency range an imbalance between turbulence production and dissipation terms. Moreover the DNS results showed a different time response of temperature fluctuation to flow unsteadiness with respect to turbulence production and dissipation.

From the simulation results the following conclusions can be drawn:

- Depending on the characteristics of the flow pulsations, the near wall velocity profiles can be strongly affected by the outer flow pulsations. Flow reversal can occur at high frequency also in the case of relatively small pulsation amplitudes.
- Flow pulsations affect not only the mean velocity profiles but also the turbulent structures and hence the level of near wall velocity fluctuations. As it is shown by the behaviour of the turbulent shear stress the stabilizing and destabilizing effects of flow acceleration and deceleration have a strong impact on the turbulent velocity fluctuations.
- It was found that the mean and phase-locked averaged turbulent Prandtl number is clearly affected by flow pulsations. In order to correctly predict the overall heat transfer in presence of flow pulsations thus the assumption of a constant turbulent Prandtl number should be released.
- The unsteady heat transfer is strongly affected by flow pulsations especially in the case of high pulsation amplitude. However, the pulsation frequency has a small effect on the overall value of heat transfer in the case of small pulsation amplitude. On the contrary, in the case of high amplitude pulsations the mean value of the Nusselt number can highly increase.

Acknowledgements. The always helpful support and the provision of resources on high performance computers of the High-Performance Computing-Centre Karlsruhe are gratefully acknowledged.

References

1. N. Kasagi, Y. Tomita, and A. Kuroda. Direct numerical simulation of passive scalar field in a turbulent channel flow. Journal of Heat Transfer, 114:598–606, August 1992.
2. L. V. Krishnamoorthy and R. A. Antonia. Temperature-dissipation measurements in a turbulent boundary layer. Journal of Fluid Mechanics, 176:265, 281, 1987
3. S. L. Lyons, T. J. Hanratty, and J. B. McLaughlin. Direct numerical simulation of passive heat transfer in a turbulent channel flow. International Journal of Heat and Mass Transfer, 34(4/5):1149, 1161, 1991.
4. The open source cfd toolbox, user guide. http://www.opencfd.co.uk/openfoam/index.html.
5. T. von Karman. Mechanical similitude and turbulence. NASA Technical Memo randums, (611), March 1931. Reprint from Nachrichten der Gesellschaft der Wissenschaften zu Göttingen, 1930.
6. F. Page, W. H. Corcoran, W. G. Schlinger, and B. H. Sage. Temperature and velocity distributions in uniform flow between parallel plates. Industrial and Engineering Chemistry, 44(2):419, 423, February 1952
7. H. Kawamura, H. Abe, and K. Shingai. DNS of turbulence and heat transport in a channel flow with different Reynolds and Prandtl numbers and boundary conditions. In 3rd International Symposium on Turbulence, Heat and Mass Transfer, 15:32, 2000
8. A. Scotti and U. Piomelli. Numerical simulation of pulsating turbulent channel flow. Physics of Fluids, 13(5):1367–1384, May 2001
9. Y. Ishino, M. Suzuki, T. Abe, N. Ohiwa, and S. Yamaguchi. Flow and heat transfer characteristics in pulsating pipe ows (effects of pulsation on internal heat transfer in a circular pipe flow). Heat Transfer-Japanese Research, 25(5):323

Implicit LES of Passive-Scalar Mixing in a Confined Rectangular-Jet Reactor

A. Devesa, S. Hickel, and N.A. Adams

Technische Universität München, Institute of Aerodynamics, 85747 Garching, Germany
Antoine.Devesa@aer.mw.tum.de

Summary. Recently, the implicit SGS modeling environment provided by the Adaptive Local Deconvolution Method (ALDM) has been extended to Large-Eddy Simulations (LES) of passive-scalar transport. The resulting adaptive advection algorithm has been described and discussed with respect to its numerical and turbulence-theoretical background by Hickel *et al.*, 2007. Results demonstrate that this method allows for reliable predictions of the turbulent transport of passive-scalars in isotropic turbulence and in turbulent channel flow for a wide range of Schmidt numbers. We now use this new method to perform LES of a confined rectangular-jet reactor and compare obtained results with experimental data available in the literature.

1 Introduction: Experimental Configuration

The mixing of passive-scalars at very high Schmidt numbers was recently studied under the conditions of laboratory experiments by the group of Rodney Fox at Iowa State University [1].

In this experiment, simultaneous velocity and scalar concentrations were carried out in confined rectangular jet and wake flows. For the purpose of our study, we will first focus on the confined planar jet. A bi-dimensional sketch of this test rig is displayed in Figure 1. The flow system was designed to provide a shear flow with a Reynolds number based on the channel hydraulic diameter between $5,000$ and $100,000$. Computations corresponding to the Reynolds number $50,000$ are envisaged in our work. The Schmidt number of the flow is $1,250$.

The measurements are carried out in a Plexiglas test section with a rectangular cross-section measuring 60 mm (height) by 100 mm (wide) and with an overall length of 1 m. The width of each of the inlet channels, separated by two splitters, is 20 mm and the aspect ratio of the rectangular jet is 5. Volumetric flow rates are respectively 1.0, 2.0 and 1.0 L/s, corresponding to free stream velocities of respectively 0.5, 1.0, and 0.5 m/s in the top, center and bottom inlet channels.

W.E. Nagel et al. (eds.), *High Performance Computing in Science and Engineering '09*, DOI 10.1007/978-3-642-04665-0_22,

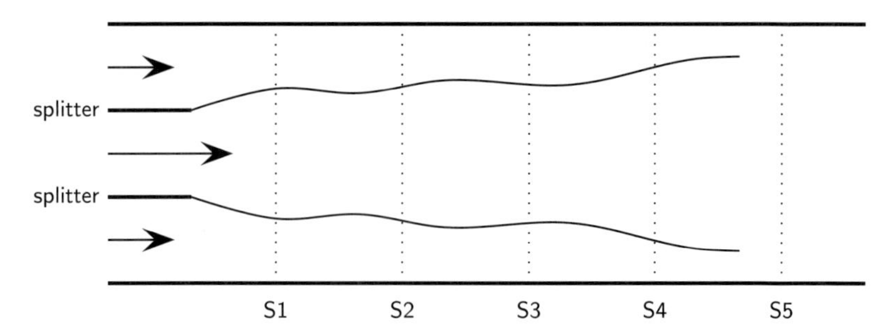

Fig. 1. Schematic of the confined rectangular jet experimental setup

Particle Image Velocimetry (PIV) was used to measure the instantaneous velocity field in 5 planar cross sections of the observed flow, corresponding to the 5 stations shown on Figure 1, S1 to S5. Scalar concentration measurements were carried out simultaneously at the same locations using Planar Laser-Induced Fluorescence (PLIF), so that velocity-scalar correlations were computed. Therefore, an experimental database was generated, including first- and second-order moments for velocity and scalar concentration, as well as cross-correlations.

For further details of the experimental configuration, please refer to the latest publications of the group on this topic [1, 2].

In close collaboration with the experimentalists and in the framework of the validation and assessment of implicit LES, which basic concept is presented in Section 2, we perform LES of the mixing and transport of a passive-scalar in the confined rectangular-jet reactor configuration. Our implicit LES code is based on the Adaptive Local Deconvolution Method (ALDM), presented in Section 3. In Section 4, some details of the extension of our subgrid-scale modelling approach to passive-scalar transport are given, before dealing with the performance of our code on the HLRS Supercomputers, Section 6, and presenting the results of the computations, Section 7.

2 Implicit LES

We consider LES of turbulent flows which are governed by the Navier-Stokes equations and by the incompressible continuity equation. A finite-volume discretization is obtained by convolution with the top-hat filter $\boldsymbol{G}$:

$$\frac{\partial \bar{\boldsymbol{u}}_N}{\partial t} + \boldsymbol{\nabla} \cdot \bar{\boldsymbol{N}}_N(\boldsymbol{u}_N) - \nu \boldsymbol{\nabla} \cdot \boldsymbol{\nabla} \bar{\boldsymbol{u}}_N = -\boldsymbol{\nabla} \cdot \bar{\boldsymbol{\tau}}_{SGS} \tag{1}$$

$$\boldsymbol{\nabla} \cdot \bar{\boldsymbol{u}}_N = 0 \tag{2}$$

where an overbar denotes the filtering $\bar{\boldsymbol{u}} = \boldsymbol{G} * \boldsymbol{u}$. The nonlinear term is abbreviated as $\boldsymbol{\nabla} \cdot \boldsymbol{N}(\boldsymbol{u}) = \boldsymbol{\nabla} \cdot \boldsymbol{u}\boldsymbol{u} + \boldsymbol{\nabla} p$, where $\boldsymbol{u}$ is the velocity field and p is the pressure. The employed filter approach [3] implies a subsequent discretization of the filtered equations. The subscript N indicates the resulting grid functions obtained by projecting continuous functions on the numerical grid. This projection corresponds to an additional filtering in Fourier space with a sharp cut-off at the Nyquist wavenumber $\xi_C = \pi/h$, where h is a constant grid spacing. The subgrid-stress tensor:

$$\boldsymbol{\tau}_{SGS} = \boldsymbol{N}(\boldsymbol{u}) - \boldsymbol{N}_N(\boldsymbol{u}_N) \tag{3}$$

originates from the discretization of the non-linear terms and has to be modeled in order to close Eq. (2). To certain extents, common explicit models are based on sound physical theories. Solved numerically, however, the discrete approximation of the explicit SGS model competes against the truncation error of the underlying numerical scheme. A theoretical analysis [4] comes to the conclusion that even a fourth-order central difference discretization has a numerical error which can have the same order of magnitude as the SGS model. This fact is exploited for implicit large-eddy simulation where no SGS model terms are computed explicitly. Rather the truncation error of the numerical scheme is used to model the effects of unresolved scales. A recent review on previous implicit LES approaches is provided by Grinstein and Fureby [5].

The Modified Differential Equation (MDE) for an implicit LES scheme is given by:

$$\frac{\partial \bar{\boldsymbol{u}}_N}{\partial t} + \widetilde{\boldsymbol{G}} * \widetilde{\boldsymbol{\nabla}} \cdot \widetilde{\boldsymbol{N}}_N(\widetilde{\boldsymbol{u}}_N) - \nu \boldsymbol{\nabla} \cdot \boldsymbol{\nabla} \bar{\boldsymbol{u}}_N = \boldsymbol{0} \tag{4}$$

$$\boldsymbol{\nabla} \cdot \bar{\boldsymbol{u}}_N = 0 \tag{5}$$

where $\widetilde{\boldsymbol{u}}_N$ denotes an approximant of the velocity $\boldsymbol{u}_N$. The local Riemann problem is solved by a consistent numerical flux function $\widetilde{\boldsymbol{N}}_N$. The symbols $\widetilde{\boldsymbol{G}}$ and $\widetilde{\boldsymbol{\nabla}}$ indicate that $\boldsymbol{G}$ and $\boldsymbol{\nabla}$ are replaced by their respective numerical approximations. In fact $\widetilde{\boldsymbol{G}} * \widetilde{\boldsymbol{\nabla}}$ can be a nonlinear operator. The truncation error is accordingly:

$$\boldsymbol{\mathcal{G}}_N = \boldsymbol{G} * \boldsymbol{\nabla} \cdot \boldsymbol{N}_N(\boldsymbol{u}_N) - \widetilde{\boldsymbol{G}} * \widetilde{\boldsymbol{\nabla}} \cdot \widetilde{\boldsymbol{N}}_N(\widetilde{\boldsymbol{u}}_N) \tag{6}$$

For implicit SGS modeling the discretization scheme is specifically designed so that the truncation error $\boldsymbol{\mathcal{G}}_N$ has physical significance, i.e.:

$$\boldsymbol{\mathcal{G}}_N \approx -\boldsymbol{G} * \boldsymbol{\nabla} \cdot \boldsymbol{\tau}_{SGS} \tag{7}$$

3 The ALDM Approach

With the adaptive local deconvolution method (ALDM) the local approximation $\widetilde{\boldsymbol{u}}_N$ is obtained from a solution-adaptive combination of deconvolution polynomials. Numerical discretization and SGS modeling are merged entirely.

This is possible by exploiting the formal equivalence between cell-averaging and reconstruction in finite-volume discretizations and top-hat filtering and deconvolution in SGS-modeling. Instead of maximizing the order of accuracy, deconvolution is regularized by limiting the degree of local interpolation polynomials and by permitting lower-order polynomials to contribute to the truncation error. Adaptivity of the deconvolution operator is achieved by weighting the respective contributions by an adaptation of WENO smoothness measures [6]. The approximately deconvolved field is inserted into a consistent numerical flux function. Flux function and nonlinear weights introduce free parameters. These allow the control of the truncation error which provides the implicit SGS model.

The performance of the optimized implicit model was evaluated by simulations of different flow configurations. Large-scale forced and decaying three-dimensional homogeneous isotropic turbulence were considered at a wide range of Reynolds numbers. For transitional flows the model performance was tested on the simulation of the instability and breakdown of the 3D Taylor-Green vortex. For all test cases the implicit model shows an excellent agreement with theory and experimental data. It is demonstrated that ALDM performs at least as well as established explicit models [7].

4 Implicit Subgrid-Scale Modeling for Passive-Scalar Transport

We consider the turbulent transport of passive-scalars, which do not measurably affect the velocity field. This case represents a one-way coupling of the scalar to the fluid. Hence, the closure problem is restricted to the scalar transport equation, where the flux function $\mathbf{F}$ is formally linear in c:

$$\frac{\partial c}{\partial t} + \nabla . \mathbf{F}(\mathbf{u}, c) = 0 \quad \text{with:} \quad \mathbf{F}(\mathbf{u}, c) = \mathbf{u}c - \frac{1}{ReSc}\nabla c \tag{8}$$

Turbulence modeling and discretization for the momentum equations remain unchanged [7]. However, the evolution of a non-uniform scalar field is subject to the velocity dynamics. Small-scale fluctuations of velocity and scalar are correlated in the presence of a scalar concentration gradient. The projection of Eq. (8) on a grid with finite resolution results in the modified equation:

$$\frac{\partial c_N}{\partial t} + \nabla . \mathbf{F}_N(\mathbf{u}_N, c_N) = \tau_{SGS} \tag{9}$$

The subgrid tensor: $\tau_{SGS} = \mathbf{F}(\mathbf{u}, c) - \mathbf{F}_N(\mathbf{u}_N, c_N)$ originates from the grid projection of advective terms and represents the effect of the action of subgrid scales and has to be approximated by a SGS model.

The various regimes (Fig. 2) that exist for the variance spectrum of passive-scalars at different Schmidt numbers [8] have to be recovered by the SGS

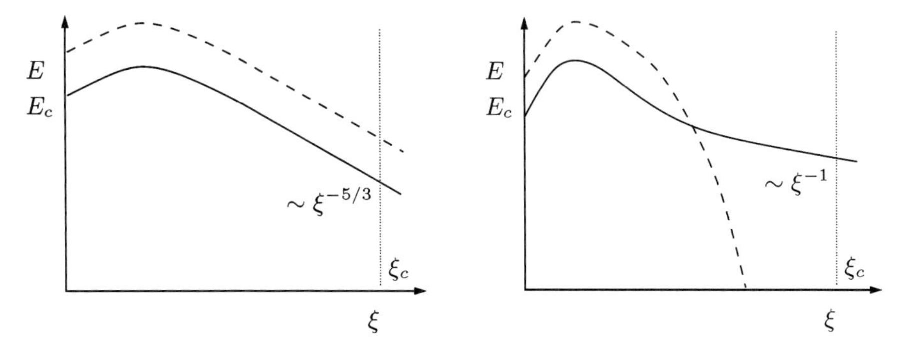

Fig. 2. Critical test cases for predicting the proper subgrid diffusion in Large-Eddy Simulations of scalar mixing. *Left:* Low Schmidt number regime. *Right:* High Schmidt number regime at moderate Reynolds number. —— scalar variance; - - - - kinetic energy; · · · · · numerical cutoff wavenumber

model. In the ALDM framework, implicit SGS modeling is accomplished by calibrating free discretization parameters. An analysis of the typical wave numbers, for low Schmidt number and high Schmidt number scalars, revealed that different parameters are required for each of these regimes. This approach for passive-scalar mixing has already been validated for several canonical flows by [9]. The computation of the experimental setup presented here will then assess the high Schmidt number model in a complex configuration.

5 Numerical Method and Computational Domain

The flow is described by the incompressible Navier-Stokes equations. The equations are discretized on a staggered Cartesian mesh. For time advancement the explicit third-order Runge-Kutta scheme of [10] is used. The time-step is dynamically adapted to satisfy a Courant-Friedrichs-Lewy condition with CFL = 1.0. The pressure-Poisson equation and diffusive terms are discretized by second-order centered differences. The Poisson solver is based on the stabilized bi-conjugate gradient (BiCGstab) method [11]. The Poisson equation is solved at every Runge-Kutta substep. All results presented in this document are obtained by the Simplified Adaptive Local Deconvolution (SALD) method [12] that represents an implementation of the Adaptive Local Deconvolution Method (ALDM) with improved computational efficiency. The validity of this numerical methodology has been established for plane channel flow [13], for turbulent boundary layer separation [14], and for passive scalar mixing [9]. The undergoing work represents namely the further step in the validation of the code in this latter field.

The computational domain used for reproducing the experimental configuration described in Section 1 is divided in two parts. The inlet part, composed by three inlet channels, are located upstream from the measurements domain. Each of the inlet channels is periodic in streamwise and spanwise directions, while bounded by solid walls in the transverse direction, and contains $16 \times 90 \times 160$ points. In the wall-normal direction a hyperbolic stretching is used to increase resolution near the wall, while the distribution is equidistant in the two other directions. The measurement section of the domain is confined by side walls in spanwise and transverse directions, while inlet is given by the flow conditions at the exit of the inlet channels, and oulet condition is a Neumann boundary condition for the pressure. This section contains $316 \times 270 \times 160$ finite volumes. In total, the adequate grid size for the computation of such flow requires approximately 14.4×10^6 points, based on a resolution of about $\Delta y^+ = 1$ close to the walls.

6 Performance

The flow solver used, INCA [12], is optimized for parallel high-performance vector computers. The present simulations are performed on a NEC SX-8 cluster at the Stuttgart high-performance computing center (HLRS). The CPU-time consumption is distributed as follows: 99 percents are used by the iterative solver for the pressure Poisson equation, 1 percent is dedicated to output and statistics purposes, as well as the rest of the computational routines, such as the calculation of the right-hand side of the Navier-Stokes equation.

The performance of the Poisson solver, that has then the most significant influence on the computation, is about 6 GFlops, which is approximately the same value as the overall performance of the computation. The computations use approximately 15 GBytes RAM and the calculation of one flow-through time requires 4600 CPU hours (cf. extent of the computational domain in Section 5).

7 Numerical Results

In this section, the main results obtained from the implicit LES are presented and compared to the measurements.

A qualitative analysis is first undertaken. Figs. 3 and 4 show instantaneous velocity and scalar concentration fields, respectively, from the implicit LES. The three inlet channels (bottom) and the long reactor part (long upper part) of the computational domain can be identified on these snapshots. The jet expansion close to the splitter plates and its further development downstream towards a flat channel flow are observed. The two mixing layers generated by the splitter plates reach a stable position after approximately $x/d = 10$, due to the confinement between the top and bottom walls.

A further analysis is concerned with the comparison between experimental data and numerical results. Since the flow is confined in the spanwise direction as well, there is no homogeneous direction at all in this configuration.

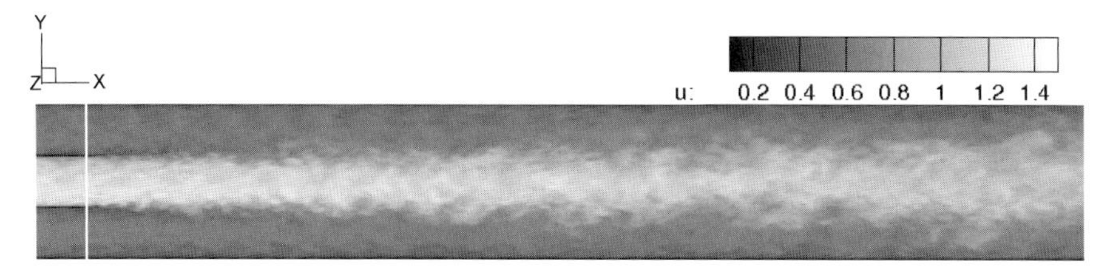

Fig. 3. Snapshot of the streamwise velocity field

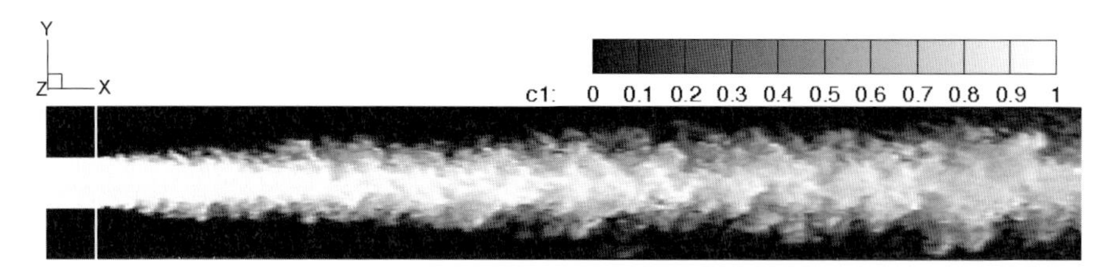

Fig. 4. Snapshot of the scalar concentration field

The numerical quantities presented here are hence only temporal statistics over about 1800 samples, and no spatial averaging is used. Despite this high number of samples, the statistics appear to be not totally smooth for some second-order moments. Among the five available observation positions of the experimental setup, only three are analyzed in the following: S1 ($x/d = 1$), S3 ($x/d = 7.5$) and S5 ($x/d = 15$).

Figs. 5 and 6 depict the comparison between experimental data and numerical results based on the averaged streamwise velocity and scalar concentration respectively. The flow topology and expansion of the jet are found to be in agreement with the experiment. First order moments for both dynamics and passive-scalar concentration from the implicit LES match globally very well with the experimental data. Indeed, comparisons at the two other measurement stations of the test rig exhibit the same concordance.

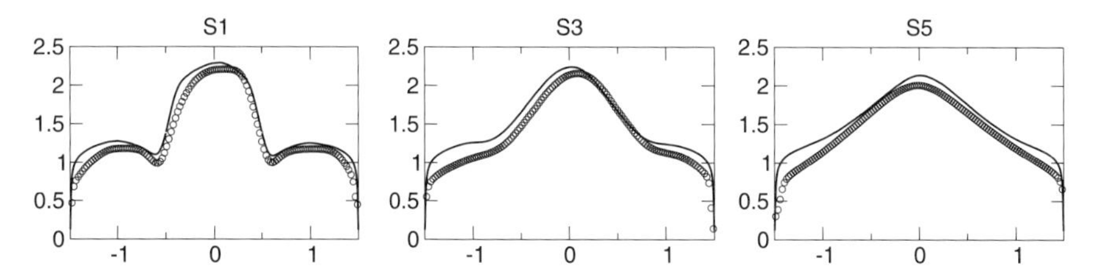

Fig. 5. Mean streamwise velocity profiles at the three measurement stations: S1, S3, S5. Symbols: experiment; solid line: ILES

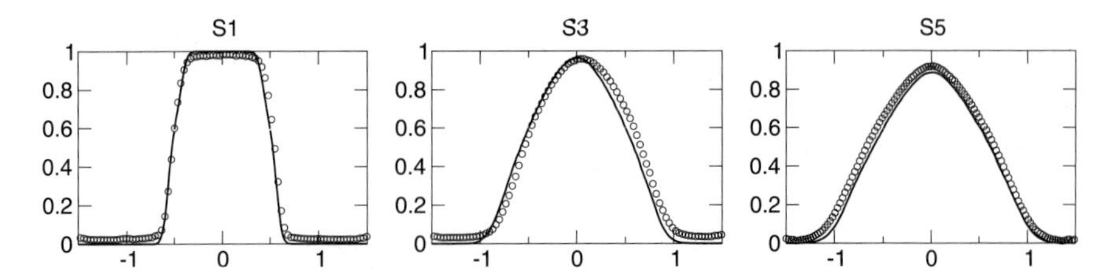

Fig. 6. Mean scalar concentration profiles at the three measurement stations: S1, S3, S5. Symbols: experiment; solid line: ILES

The main challenge of this work is, however, the comparison of second-order moments. Even though the flow-dynamic quantities, namely the stream-wise and cross-stream Reynolds stresses Figs. 7 and 8, as well as the shear-stress Fig. 9, are not presented here for the sake of concision, numerical results were found to match satisfactorily the measured data. The two peaks of scalar concentration variance (Fig. 10) at the first observation station S1 shows the high mixing regions corresponding to the position of the splitter plates. At this station, the resolution in the experiment is not sufficient to capture the real amplitude of those peaks. Therefore we think that the discrepancy with the implicit LES is not dramatic at this position. Further downstream, where the mixing is weaker, both numerical and experimental data agree quite well.

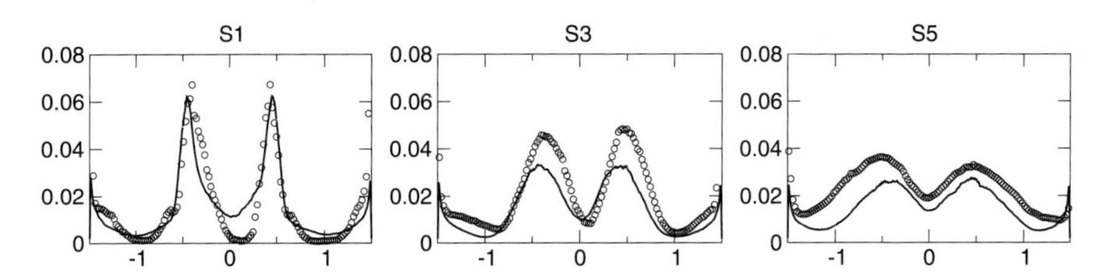

Fig. 7. Streamwise Reynolds stress profiles at the three measurement stations: S1, S3, S5. Symbols: experiment; solid line: ILES

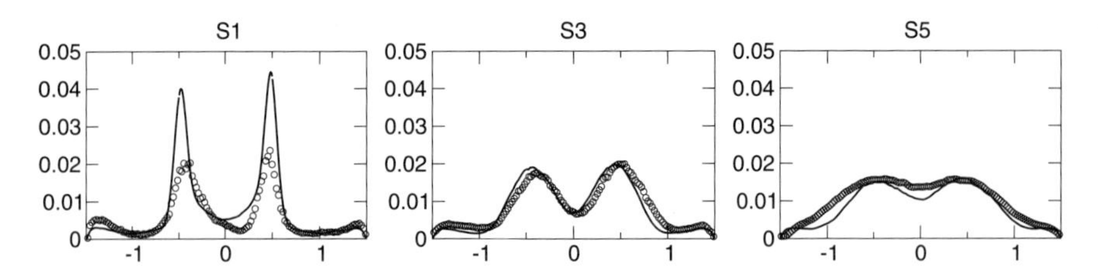

Fig. 8. Transversal Reynolds stress profiles at the three measurement stations: S1, S3, S5. Symbols: experiment; solid line: ILES

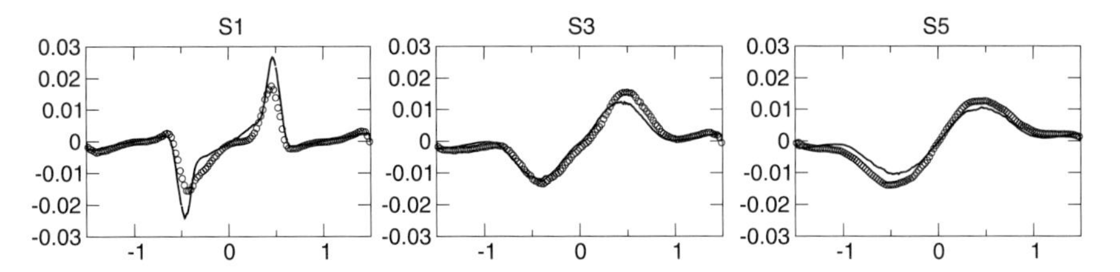

Fig. 9. Shear-stress profiles at the three measurement stations: S1, S3, S5. Symbols: experiment; solid line: ILES

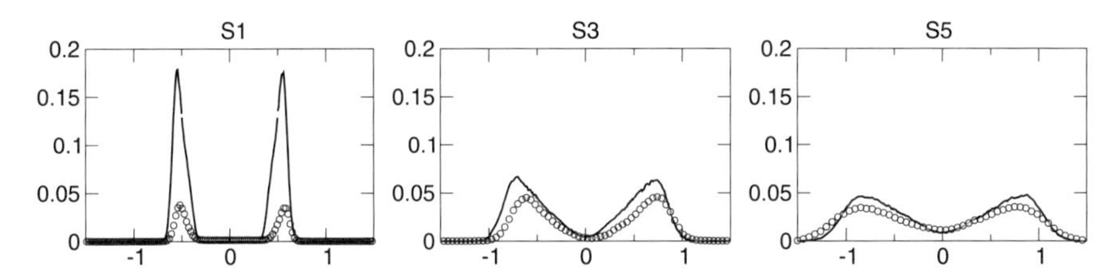

Fig. 10. Scalar variance at the three measurement stations: S1, S3, S5. Symbols: experiment; solid line: implicit LES

A further analysis is concerned with the velocity-scalar concentration correlations, that experimentalists could measure using simultaneously PIV / PLIF acquisition chains. Both correlations $< u'c' >$ and $< v'c' >$ have been normalized by the rms velocity and scalar concentration fluctuations, yielding to the velocity-concentration correlation coefficients:

$$< u'c' >_N = \frac{< u'c' >}{< u' >_{rms} < c' >_{rms}} \tag{10}$$

$$< v'c' >_N = \frac{< v'c' >}{< v' >_{rms} < c' >_{rms}} \tag{11}$$

The comparison of both quantities with the numerical results is shown in Figs. 11 and 12 and demonstrate an almost perfect agreement. The key-features of the flow present in the experiment are found in the simulation as well. At the closest position to the splitter plate S1, scalar concentration and velocity are totally uncorrelated in the core regions of the three inlet channels, giving levels close to 0. The longitudinal coefficient (Fig. 11) exhibits two separated high correlation regions, in the whole confined reactor, even if at S5, these regions almost take the form of two plateaus. The lateral coefficient (Fig. 12) is antisymmetric and the amplitudes in the zones of high correlation remain quite the same along the observation domain.

From the turbulent fluxes, the resolved turbulent viscosity and diffusivity can be computed, based on the assumption that the turbulent fluxes are pro-

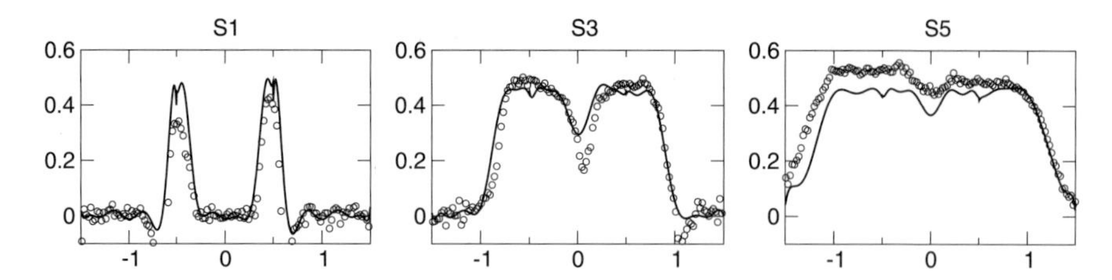

Fig. 11. Normalized $< u'c' >$ at the three measurement stations: S1, S3, S5. Symbols: experiment; solid line: implicit LES

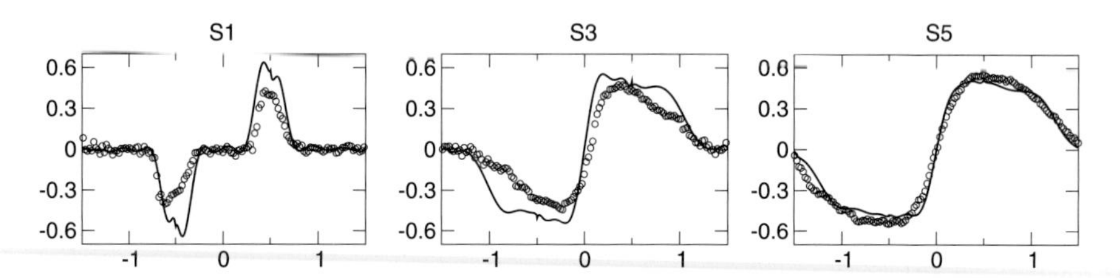

Fig. 12. Normalized $< v'c' >$ at the three measurement stations: S1, S3, S5. Symbols: experiment; solid line: implicit LES

portional to the velocity or scalar concentration gradients (Batchelor, 1949). These proportionality ratios are such that:

$$\nu_T = \frac{-< u'v' >}{\partial U/\partial y} \tag{12}$$

$$\Gamma_{22} = \Gamma_T = \frac{-< v'c' >}{\partial c/\partial y} \tag{13}$$

where the subscript 22 corresponds to the component (2,2) of the turbulent diffusivity tensor, which is not diagonal in this case, since the mean scalar concentration gradient is not aligned with the turbulent flux vector. The profiles are shown in Figs. 13 and 14. The agreement between experiment and simulation is once again very satisfactory, and the main features of the flow are recovered, even if the results at the station S1, the closest one to the splitter plates, are not appropriate for an analysis. Indeed, due to the presence of the division by the velocity or scalar concentration gradient respectively, these profiles show non-physical peaks. However, at S3 and S5, the maxima appear always at the location of the shear-layers, whereas these quantities tend to 0 close to the top and bottom walls. The amplitudes of the resolved part of the turbulent viscosity ν_T are in good agreement with the experiment for the stations upstream, but not perfectly recovered at the furthest one, S5. The trend for the turbulent diffusivity Γ_T is opposite: the prediction of the implicit LES is better far from the splitter plates as at the first stations S1 and S3.

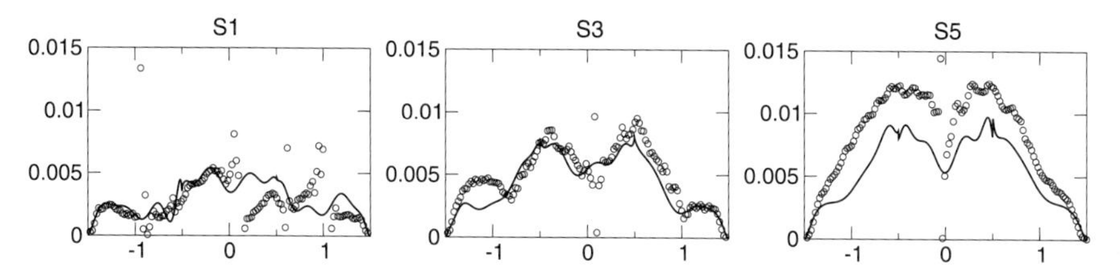

Fig. 13. Turbulent viscosity at the three measurement stations: S1, S3, S5. Symbols: experiment; solid line: implicit LES

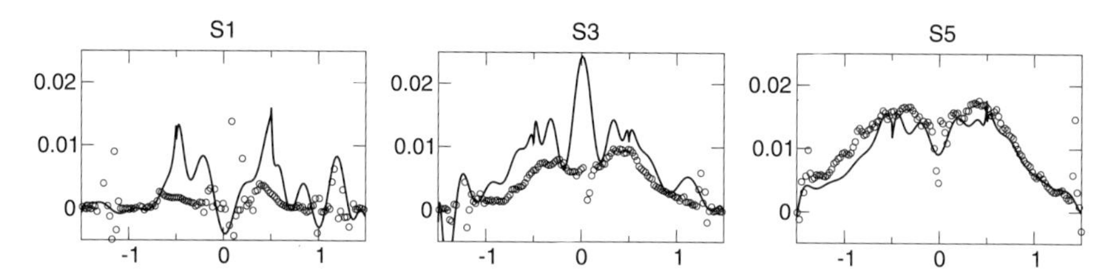

Fig. 14. Turbulent diffusivity at the three measurement stations: S1, S3, S5. Symbols: experiment; solid line: implicit LES

A measure of the turbulent Schmidt number can consequently be estimated:

$$Sc_T = \frac{\nu_T}{\Gamma_T} \tag{14}$$

The most classical subgrid-scale is based on the assumption of a constant turbulent Schmidt number of the order of 1. The results, which is not depicted here for the sake of concision, show that this assumption could hold, in this case, far away from the high mixing regions, but not in the intense shear and transport regions. In those places, the turbulent Schmidt number can clearly exhibit variations of more than several hundreds percent from a constant Schmidt number assumption ($Sc_T \approx 1$). Indeed, experimental values for the turbulent Schmidt number were for instance around 4 at the station S3, in the both shear-layers. Numerical values at these places were however remained in a smaller range of variation from the constant turbulent Schmidt number hypothesis.

8 Conclusions

In this paper, the numerical results from an implicit LES of the mixing of a passive-scalar ($Sc = 1,250$) in a confined coplanar reactor configuration are presented and analyzed. The agreement with experimental data for first- and second-moments is very satisfactory in the whole measurement region of the experiment. Some relevant quantities for the modeling of the passive-scalar transport equation, such as the resolved turbulent viscosity and diffusivity, are computed and described, as well as the turbulent Schmidt number. The very good overall agreement between experiment and simulation validates the presented implicit LES modeling environment for passive-scalars at high Schmidt numbers.

References

1. H. Feng. *Experimental study of turbulent mixing in a rectangular reactor.* PhD thesis, Iowa State University, Ames, Iowa, 2006.
2. H. Feng, M.G. Olsen, Y. Liu, R.O. Fox, and J.C. Hill. Investigation of turbulent mixing in a confined planar-jet reactor. *AIChE J.*, 51:2649–2664, 2005.
3. A. Leonard. Energy cascade in large eddy simulations of turbulent fluid flows. *Adv. Geophys.*, 18A:237–248, 1974.
4. S. Ghosal. An analysis of numerical errors in large-eddy simulations of turbulence. *J. Comput. Phys.*, 125:187–206, 1996.
5. F.F. Grinstein and C. Fureby. From canonical to complex flows: Recent progress on monotonically integrated LES. *Comp. Sci. Eng.*, 6:36–49, 2004.
6. C.-W. Shu. Essentially non-oscillatory and weighted essentially non-oscillatory schemes for hyperbolic conservation laws. *Tech. Rep. 97-65, ICASE, NASA Langley Research Center, Hampton, Virginia*, 1997.
7. S. Hickel, N.A. Adams, and J.A. Domaradzki. An adaptive local deconvolution method for implicit LES. *J. Comput. Phys.*, 213:413–436, 2006.
8. G. Batchelor. Small-scale variation of convected quantities like temperature in turbulent fluid. part 1. general discussion and the case of small conductivity. *J. Fluid Mech.*, 5:113–133, 1959.
9. S. Hickel, N.A. Adams, and N.N. Mansour. Implicit subgrid-scale modeling for large-eddy simulation of passive-scalar mixing. *Phys. Fluids*, 19:095102, 2007.
10. C.-W. Shu. Total-variation-diminishing time discretizations. *SIAM J. Sci. Stat. Comput.*, 9(6):1073–1084, 1988.
11. H.A. van der Vorst. Bi-CGSTAB: A fast and smoothly converging variant of Bi-CG for the solution of nonsymmetric linear systems. *SIAM J. Sci. Statist. Comput.*, 13:631–644, 1992.
12. S. Hickel and N.A. Adams. Efficient implementation of nonlinear deconvolution methods for implicit large-eddy simulation. In W.E. Nagel, W. Jäger, and M. Resch, editors, *High Performance Computing in Science and Engineering*, pages 293–306. Springer, 2006.
13. S. Hickel and N.A. Adams. On implicit subgrid-scale modeling in wall-bounded flows. *Phys. Fluids*, 19:105106, 2007.
14. S. Hickel and N.A. Adams. Large-eddy simulation of turbulent boundary-layer separation. In *5th International Symposium on Turbulence and Shear Flow Phenomena (TSFP5), Munich, Germany*, 2007.

Numerical Simulation of Riblet Controlled Spatial Transition

Stephan Klumpp, Matthias Meinke, and Wolfgang Schröder

Institute of Aerodynamics, RWTH Aachen University,
Wüllnerstr. 5a, 52062 Aachen, Germany
s.klumpp@aia.rwth-aachen.de

Summary. To analyze the fundamental physical mechanism which determines the damping effect of a riblet surface on three-dimensional transition several numerical simulations of spatial transition in a flat plate zero-pressure gradient boundary layer above a riblet wall are performed in this study. Two types of transition are investigated. The first type of transition, namely K-type transition, is induced by a dominant two-dimensional Tollmien-Schlichting (TS) wave and a weak spanwise disturbance. The second type of transition is purely excited by two oblique waves. The two-dimensional TS waves are found to be amplified by riblets, whereas three-dimensional structures, i.e., Λ-, hairpin, and streamwisely aligned vortices, are damped and their breakdown to turbulence is delayed compared to transition on a clean surface. The investigation of the near wall flow structure reveals secondary flows induced by the riblets and reduced wall normal ejections as well as a reduced downwash. Overall, especially the oblique transition is delayed by the riblets.

1 Introduction

Surface structures like riblets, which consist of tiny grooves aligned with the main flow direction, are well known for reducing friction drag in turbulent flows. Various experimental and numerical investigations [1, 2] show drag to be reduced if the riblet spacing in non-dimensional wall units $s^+ = su_\tau/\nu$ is below 30. At an optimum riblet geometry a drag reduction of about 10% has been achieved.

Although numerous studies exist on the drag reducing effect of riblet surfaces only a few investigations concerning the influence of riblets on laminar-turbulent transition have been performed. If a low degree of freestream turbulence exists natural transition sets in [3], associated with the development of two-dimensional Tollmien-Schlichting (TS) waves. In the beginning, these waves are linearly amplified followed by a three-dimensional stage, in which Λ- and hairpin vortices occur, and finally, the turbulent breakdown is observed. Neumann and Dinkelacker [4] have found the transition of the flow on a body of revolution to be delayed if the body was covered by riblets of a certain size

W.E. Nagel et al. (eds.), *High Performance Computing in Science and Engineering '09*, DOI 10.1007/978-3-642-04665-0_23,

compared to a smooth surface. Ladd *et al.* [5] have experimentally investigated the effect of riblets on transition of a zero-pressure gradient flow and have determined an accelerating influence. In these investigations [4, 5] neither the different stages of transition have been separated nor a more detailed analysis on how the riblets effect the transition has been performed. Grek *et al.* [6] have measured the influence of riblets on TS waves as well as on three-dimensional instabilities. They have found the TS waves to be amplified by riblets, whereas three-dimensional Λ- and hairpin vortices have been damped. Further experimental analyses of Litvinenko *et al.* [7] and Chernorai *et al.* [8] indicate riblets not only to damp Λ- and hairpin vortex structures but streamwisely aligned vortices to occur in laminar boundary layers in general. Luchini and Trombetta [9] have predicted a destabilizing effect of riblets on TS waves but a stabilizing on the Görtler instability based on a standard e^9-model. In [10] Ehrenstein shows by linear stability analysis the laminar channel flow over riblets to be more unstable than the parabolic Poiseuille profile at smooth walls.

In the experimental investigations, where a damping effect of riblets at the three-dimensional stage of transition has been found, riblets with a non-dimensional spacing $s^+ = su_\tau/\nu$ greater than 14 have been used, where u_τ denotes the local friction velocity. On the other hand, Ladd *et al.* [5] have investigated riblets at $s^+ < 10$, which might explain the acceleration of transition. Belov *et al.* [11] have found an acceleration of transition for riblets at $s^+ \approx 5$ also. The damping effect of riblets on streamwisely aligned vortex structures in laminar flow has been found in several investigations. However, to the best of the authors' knowledge in no paper a detailed near wall flow field at the transition stage is evidenced, since the flow structure has not been highly resolved in the extremely near-wall region.

To gain insight in the damping effect of the riblets, and to reveal the flow structure in the extremely near-wall region, which is a precondition to understand the physical mechanism of the damping, two types of transition in a zero-pressure gradient boundary layer are numerically simulated in this investigation. First, K-type transition, mainly following the experimental setup of Grek *et al.* [6], is investigated on a clean and two riblet surfaces. At this type of transition, the impact of riblets on two-dimensional TS waves and three-dimensional disturbances is analyzed. Since previous investigations of the impact of riblets on K-type transition reveal a damping effect on the three-dimensional disturbances and an amplification on two-dimensional TS waves, transition solely excited by two oblique waves on the clean and one riblet surface is investigated next. Under these conditions Λ-vortices will immediately occur and transform into streamwisely aligned vortices further downstream before breakdown to turbulence takes place. Since all perturbation structures of this type of transition are damped by riblets, the pure damping mechanism can be analyzed.

The paper is organized as follows. In section 2, the large-eddy simulation (LES) method is briefly outlined. Next, the flow configuration and the boundary conditions for the transition analyses are given. In section 5, the results of K-type transition are discussed first, followed by an analysis of the findings for oblique transition. Finally, the essential results are summarized.

2 Governing Equations

Since the continuum assumptions hold for the flows regarded here, the Navier-Stokes equations are an appropriate mathematical model for their simulation. Written in tensor notation and in terms of dimensionless conservative variables for a Cartesian coordinate system, they read:

$$\frac{\partial \boldsymbol{Q}}{\partial t} + (\boldsymbol{F}_\beta^C - \boldsymbol{F}_\beta^D)_{,\beta} = 0 \quad , \qquad \boldsymbol{Q} = [\varrho, \varrho u_\alpha, \varrho E]^T, \tag{1}$$

where $\boldsymbol{Q}$ is the vector of the conservative variables, and $\boldsymbol{F}_\beta^C$ denotes the vector of the convective and $\boldsymbol{F}_\beta^D$ of the diffusive fluxes:

$$\boldsymbol{F}_\beta^C - \boldsymbol{F}_\beta^D = \begin{pmatrix} \varrho u_\beta \\ \varrho u_\alpha u_\beta + p\delta_{\alpha\beta} \\ u_\beta(\varrho E + p) \end{pmatrix} - \frac{1}{Re} \begin{pmatrix} 0 \\ \sigma_{\alpha\beta} \\ u_\alpha \sigma_{\alpha\beta} + q_\beta \end{pmatrix} . \tag{2}$$

The stress tensor $\sigma_{\alpha\beta}$ is written as a function of the strain rate tensor $S_{\alpha\beta}$

$$\sigma_{\alpha\beta} = -2\,\mu(S_{\alpha\beta} - \tfrac{1}{3} S_{\gamma\gamma}\delta_{\alpha\beta}) \qquad \text{with} \qquad S_{\alpha\beta} = \tfrac{1}{2}(u_{\alpha,\beta} + u_{\beta,\alpha}). \tag{3}$$

The dynamic viscosity μ, is assumed here to be only a function of the temperature. Fourier's law of heat conduction is used to compute the heat flux q_β:

$$q_\beta = -\frac{k}{Pr(\gamma - 1)} T_{,\beta} \tag{4}$$

where Pr is the Prandtl number. The described system is closed with the equation of state for ideal gases

$$p = \frac{1}{\gamma}\varrho T \quad , \quad p = (\gamma - 1)\left(\varrho e - \frac{\varrho}{2} u_\beta u_\beta\right), \tag{5}$$

where γ is the ratio of the specific heat capacities and T the temperature.

For the LES, an implicit grid filter of width Δ is assumed to be applied to eqs. (1). A sub-grid scale model and the truncation error are coupled to the spatial steps of the computational grid in case of a finite volume scheme. Both terms lead to a damping of the smallest resolved scales and can be shown to be of the same order of magnitude for second-order accurate schemes, at least in the time averaged sense [12].

The discretization of the inviscid terms consists of a mixed centered-upwind AUSM (advective upstream splitting method) scheme [13] at second-order accuracy, whereas the viscous terms are discretized second-order accurate using a centered approximation. The temporal integration is done by a second-order explicit 5-stage Runge-Kutta method.

Table 1. Definition of the inflow perturbations; the quantity $\hat{u}_{2d}$ denotes the amplitude of the two-dimensional TS wave at frequency ω_{2d}; the three-dimensional waves are given by their amplitude $\hat{u}_{3d}$ and their frequency $\omega_{3d,alpha}$ in the streamwise and wavenumber β_{3d} in the spanwise direction

Type of transition	surface	$\hat{u}_{2d}/U_\infty$	$\omega_{2d}\delta_i/U_\infty$	$\hat{u}_{3d}/U_\infty$	$\omega_{3d,\alpha}\delta_i/U_\infty$	$\beta_{3d}\delta_i$
K-type	clean	3×10^{-2}	0.088	1×10^{-3}	0.088	0.242
K-type	riblet 1	3×10^{-2}	0.088	1×10^{-3}	0.088	0.242
K-type	riblet 2	3×10^{-2}	0.088	1×10^{-3}	0.088	0.242
oblique	clean	0	–	2×10^{-2}	0.088	±0.242
oblique	riblet 1	0	–	2×10^{-2}	0.088	±0.242

3 Boundary Conditions

The inflow boundary condition, which contains the defined disturbances, is given by

$$\mathbf{u}(0,y,z,t) = \mathbf{u}_{lam}(y) + \hat{u}_{2d} \times \mathbf{u}_{2d}(y,t) + \hat{u}_{3d} \times \mathbf{u}_{3d}(y,z,t), \quad (6)$$

where $\mathbf{u}_{lam}$ is the laminar Blasius boundary layer profile and $\mathbf{u}_{3d}$ the three-dimensional disturbances. The profiles of the amplitudes of the superposed waves are obtained by solving the Orr-Sommerfeld and Squire equations via a standard Chebyshev collocation method [14]. The quantities of the inflow boundary conditions for each investigation are summarized in Table 1 along with the prescribed surface. Besides the inflow boundary condition of the velocity all boundary conditions are equal at all investigations. The density is set to a constant value which corresponds to an isotropic acceleration to the used freestream Mach number of $Ma = 0.2$. A vanishing pressure gradient normal to the boundary is prescribed at the inlet.

In the spanwise direction periodic boundary conditions and on the wall no-slip and isothermal conditions are used. A zero pressure-gradient is assumed in the wall normal direction. On the outflow boundaries the velocity and density gradients are set zero and the freestream pressure is prescribed. A sponge layer is used to damp numerical reflections on the upper and outflow boundaries, as illustrated in Figure 1. In this sponge layer source terms are added to the right-hand side of the governing equations to drive the instantaneous solutions of density and pressure to the desired target solutions.

4 Flow Configuration and Computational Mesh

Three surface configurations have been investigated, one clean surface and two riblet surfaces. The riblet geometry was similar to that from Litvinenko *et al.* [7]. On the one hand, its transition delay capability has been proven and on the other hand, Hirt and Thome [15] have shown that the geometry can be manufactured for real technical applications.

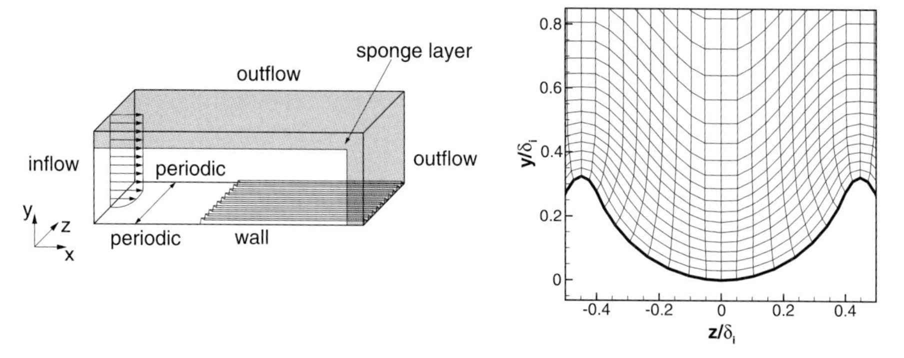

Fig. 1. Computational domain (left); detail of the mesh for riblet geometry (right)

The computational domain is shown in Figure 1. It has identical dimensions for all cases. The extension in the streamwise x, wall normal y, and spanwise direction z is $L_x/\delta_i = 440$, $L_y/\delta_i = 14$, and $L_z/\delta_i = 26$, where the reference length δ_i denotes the displacement thickness at the inflow boundary. The Reynolds number $Re_{\delta,i} = U_\infty \delta_i/\nu$ based on the freestream velocity and the displacement thickness is $Re_{\delta,i} = 618$.

A Cartesian grid with constant grid spacing in the streamwise and spanwise direction is used for the clean wall. The two riblet configurations differ by the distance of the beginning of the riblet structure from the inflow boundary. A wall adapted grid is required for the riblet surface, which is shown in Figure 1. Furthermore, the mesh in the streamwise direction has to be refined in the area where the riblet structure develops. Following the setup of Grek *et al.* [6] the riblet spacing is $s/\delta_i = 0.9$. The grid parameters of all three surfaces are summarized in Table 2. Here wall units base on the friction velocity is $u_\tau = 0.05U_\infty$ which is typical for the considered turbulent flow.

5 Results and Discussion

In this section the results of the performed numerical simulations are presented. First, an analysis of the data of K-type transition on the clean and

Table 2. Grid parameters; N_x, N_y, and N_z denote the grid points in the streamwise, wall-normal, and spanwise direction; the corresponding step sizes in each direction are given in wall units. The location where the riblet structure starts is denoted by x/δ_i.

Surface	$N_x \times N_y \times N_z$	Δx^+	Δx^+_{min}	Δy^+_{wall}	Δz^+	x/δ_i
clean	$831 \times 42 \times 581$	16.4	16.4	0.83	1.39	-
riblet 1	$859 \times 42 \times 581$	16.4	1	0.83	1.39	47
riblet 2	$859 \times 42 \times 581$	16.4	1	0.83	1.39	160

both riblet surfaces is given in section 5.1. Then, in section 5.2 the results discuss the oblique transition setup on the clean and one riblet surface.

5.1 K-type Transition

In the following, the results of spatially evolving K-type transition in a boundary layer are presented. First, the clean surface is considered. To check the quality of the simulation the streamwise development of the time and spanwise averaged displacement thickness δ_1/δ_i, the momentum thickness δ_2/δ_i, and the instantaneous skin-friction coefficient $c_f = \bar{\tau}_w/(1/2\rho_\infty U_\infty^2)$, where $\bar{\tau}_w$ denotes the spanwise averaged wall friction, are compared with empirical laws for laminar and turbulent boundary layers [16]. The leading edge of the laminar boundary layer is located at $x_{le,lam}/\delta_i = -208$, whereas the virtual leading edge of the turbulent boundary layer is at $x_{le,tur}/\delta_i = 50$ to best fit the computed displacement thickness with the empirical law. The shape factor $H_{12} = \delta_1/\delta_2$ is used to define the state of the boundary layer to be laminar ($H_{12} = 2.6$) or turbulent ($H_{12} = 1.4$). Next, the impact of riblets on two-dimensional TS waves is discussed based on the flow structure caused by riblet surface 1. Then, a detailed analysis of the impact on three-dimensional disturbances and the turbulent breakdown is presented by juxtaposing the data of the clean surface and the riblet configuration 2.

Clean Surface Configuration

The flow structure of the clean surface is visualized by λ_2-contours [17] in Figure 2. The different stages of K-type transition consisting of the two-dimensional TS waves, which deform into Λ-structures, then into hairpin vortices, and finally breakdown into turbulence, are clearly visible. The streamwise development of the time and spanwise averaged displacement thickness δ_1/δ_i and the momentum thickness δ_2/δ_i are shown in Figure 3. Overall, a very good agreement in the laminar and turbulent state is evident, i. e., the displacement thickness and the momentum thickness closely match the empirical laws of laminar and turbulent boundary layers. The development of the shape factor H_{12} is shown in Figure 4, which clearly indicates laminar-turbulent transition. Finally, the development of the skin-friction coefficient is compared with the empirical law in Figure 5. In the range of $0 \leq x/\delta_i \leq 180$ the c_f-distribution oscillates around the laminar value induced by the two-dimensional TS wave. At $x/\delta_i \approx 180$ the coefficient deviates from the laminar state and reaches the turbulent state at $x/\delta_i \approx 280$. The illustrations of the results in Figures 2 to 5 evidence a fully turbulent state near the outflow region of the computational domain.

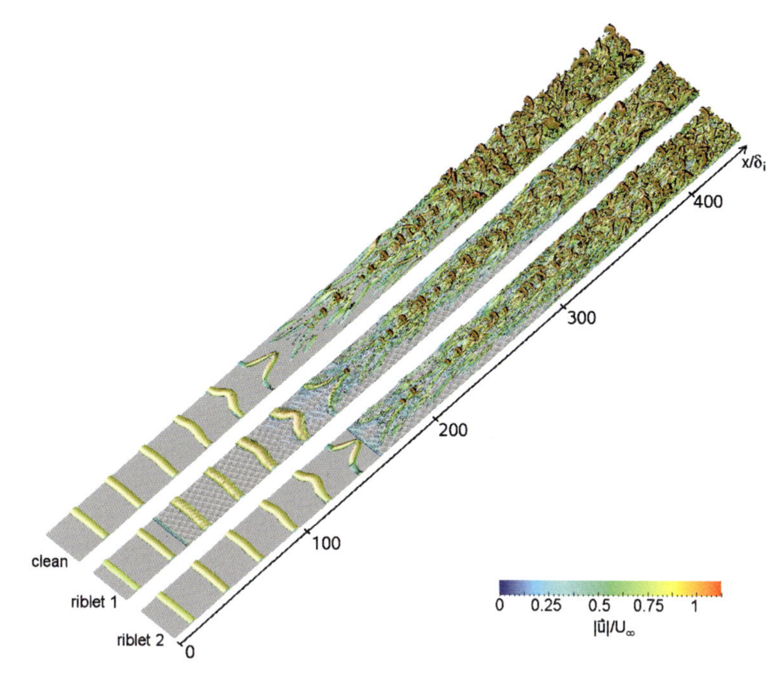

Fig. 2. λ_2-contours for the clean and the riblet configurations at K-type transition. The color represents the velocity magnitude

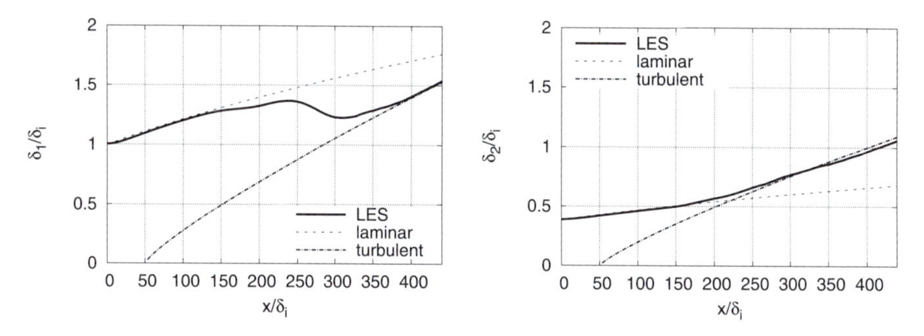

Fig. 3. Displacement thickness (left) and momentum thickness (right) vs x/δ_i on the clean surface at K-type transition

Riblet Surface Configuration

The following two sections present the impact of a riblet surface on the two-dimensional TS waves first. Next the impact on the three-dimensional disturbances is considered.

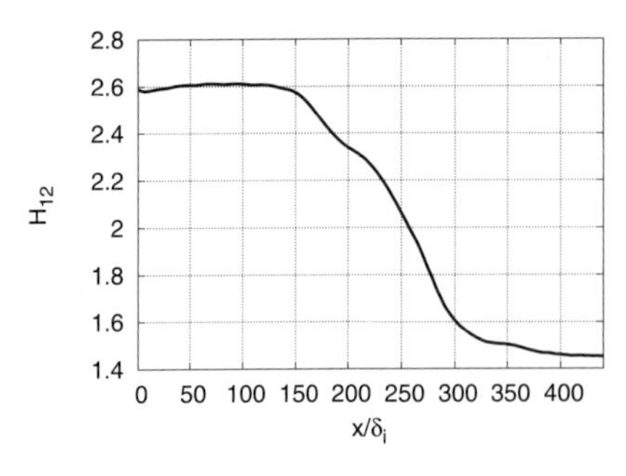

Fig. 4. Shape factor for the clean surface problem at K-type transition

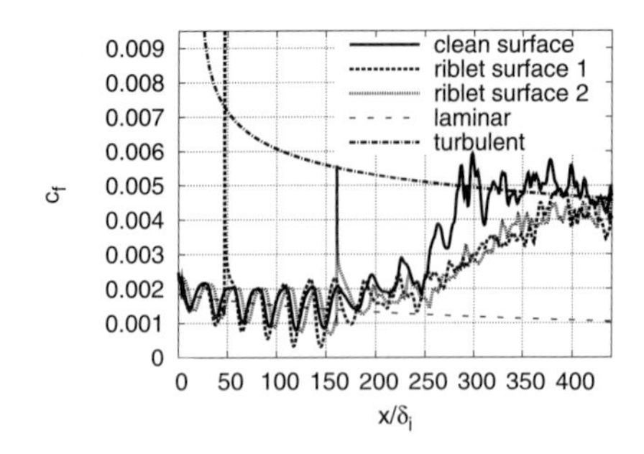

Fig. 5. Skin-friction coefficient vs x/δ_i at K-type transition.

Impact on Tollmien-Schlichting Waves

Figure 2 shows the riblet structure of the riblet configuration 1 to develop at $x/\delta_i = 47$ where the disturbances are mainly two-dimensional. The maximum instantaneous wall normal velocity $|v|_{max}$ in each wall-normal cross section is compared with the clean wall distribution from $x/\delta_i = 0$ to $x/\delta_i = 200$ in Figure 6. The peaks at $x/\delta_i = 47$ are caused by the rising riblets. The maximum wall-normal velocity illustrates the growth of the TS-waves. In the range of $50 \leq x/\delta_i \leq 150$ the wall-normal velocity's amplitude on riblet surface 1 is about 15% higher than on the clean surface. This is illustrated by comparing the peak values in the referred range in Figure 6. These results of an amplified two-dimensional TS wave agree qualitatively with the findings of Grek *et al.* [6], Ehrenstein [10], and Luchini and Trombetta [9].

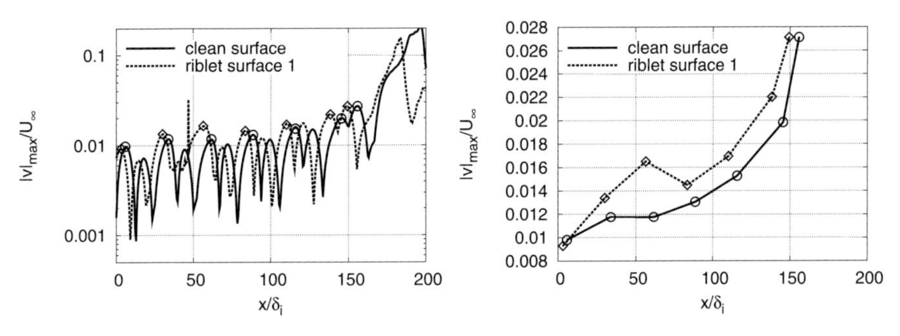

Fig. 6. Maximum wall-normal velocity $|v|_{max}$ (left) and peak values of $|v|_{max}$ (right) vs x/δ_i at K-type transition; symbols are at the same x/δ_i location in both figures

Impact on Three-dimensional Disturbances

To isolate the effects of riblets on the different transition stages a second riblet surface having riblets rising further downstream, i.e., at a streamwise position $x/\delta_i = 160$ where Λ-structures are already established, is used. Figure 2 evidences the near-wall flow structure. The flow upstream of the riblets is identical to that above the clean surface. This is indicated by equal fluctuations of c_f in Figure 5. The peak in the distribution of the skin-friction coefficient is caused by the rising riblets. Downstream of this position the c_f-distribution of the clean surface differs remarkably from the two riblet configurations, while the riblet configurations possess a likewise c_f behavior in this region. That is, the amplification of the TS waves above riblet surface 1 is not strong enough to dominate the three-dimensional state of transition. For the clean wall configuration c_f reaches the turbulent state at $x/\delta_i \approx 300$. Since the gradients of the riblet configurations are much smaller, the riblet configurations reach a turbulent level not before $x/\delta_i \approx 390$. The delay of the turbulent breakdown causes a remarkable skin-friction reduction in this area of up to 40%. Therefore, the total drag could be clearly reduced despite the relatively small area affected by this effect.

The analysis of the maximum spanwise and wall-normal velocity of the clean surface and riblet configuration 2 in Figure 7 evidences the main differences in the range of $270 \leqslant x/\delta_i \leqslant 330$, where according to Figure 5 also c_f differs the most. In this area both velocity components are approximately 40% smaller when the riblet configuration is considered. The direct impact of the riblets on the spanwise velocity is illustrated in Figure 8, where the flow structure is visualized by streamlines above the riblet surface. The large vortex above the riblets represents one leg of a hairpin vortex. Besides the expected wave-like overflow of the riblet tips some recirculation zones are observed resulting in high energy dissipation.

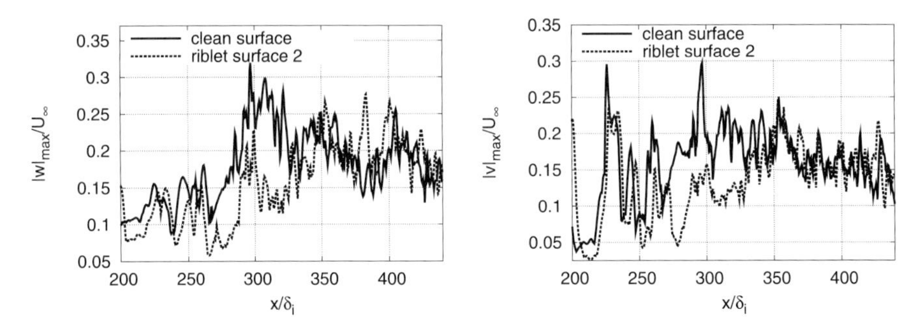

Fig. 7. Maximum spanwise velocity (left) and maximum wall-normal velocity (right) vs x/δ_i at K-type transition

The damping of the spanwise velocity component in the three-dimensional state causes an immediate decrease of the maximum wall-normal velocity since the maximum wall-normal velocity is located between the Λ- and hairpin-vortex legs and is driven by these counter-rotating vortices. The reduction of the wall-normal velocity between these legs lowers the transport of low-speed

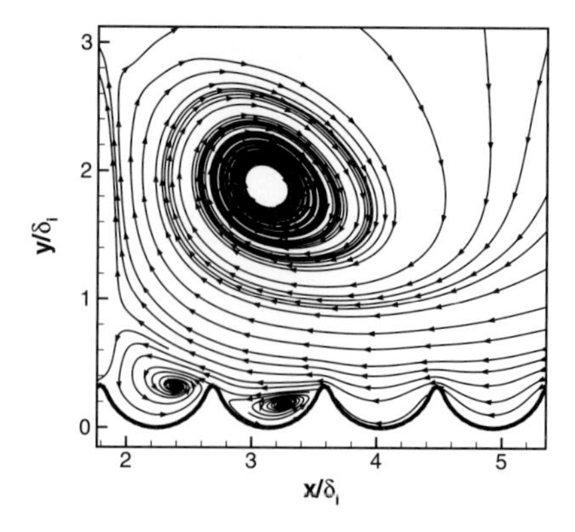

Fig. 8. Streamlines on riblet surface 2 at $x/\delta_i = 190$ at K-type transition

fluid from the wall to the outer region of the boundary layer such that the breakdown into a turbulent spot is delayed [6].

5.2 Oblique Transition

Next, the results of oblique transition are discussed. At oblique transition all occurring disturbances contain a spanwise velocity component and are therefore damped by the riblets. First, the data of the clean wall problem is presented and compared with the empirical turbulent boundary layer laws. Subsequently, the results on the riblet and the clean surfaces are juxtaposed.

Clean Surface Configuration

The flow structure of oblique transition on the clean surface is shown in Figure 9. From the inflow boundary to $x/\delta_i \approx 80$ Λ-vortices are generated by the two incoming oblique waves the parameters of which are given in Table 1. Further downstream, the Λ-vortices deform into streamwisely aligned vortices followed by turbulent breakdown. The distribution of the shape factor and spanwise averaged skin-friction coefficient c_f over the streamwise position are given in Figure 10. Downstream of $x/\delta_i \approx 80$ the skin-friction coefficient deviates from the laminar state caused by the downwash of faster fluid towards the wall by the streaks, which is typical for oblique transition [18]. The breakdown to turbulence at $x/\delta_i \approx 300$ is evidenced by a dramatic increase of c_f which is emphasized by an overshoot compared to the empirical c_f-distribution of a turbulent flat plate flow.

Riblet Surface Configuration

At oblique transition the riblet surface damps the development of transitional structures and thus delays transition. In Figure 9, the flow structures on the

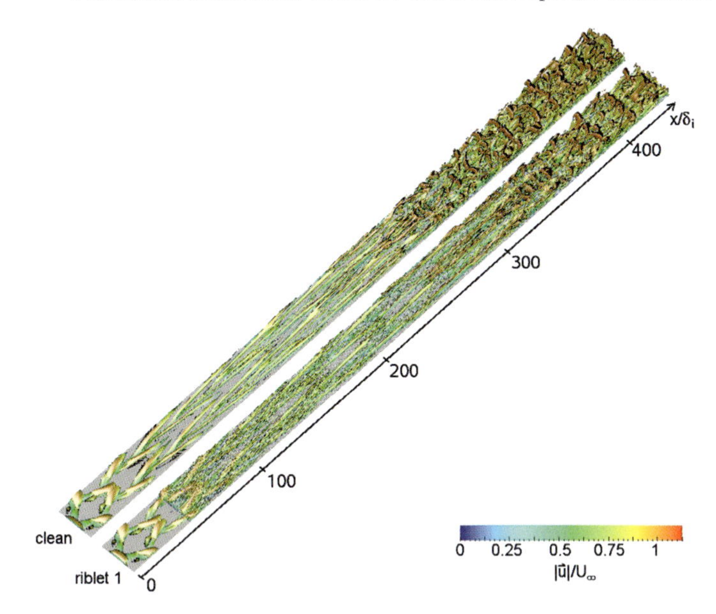

Fig. 9. λ_2-contours for the clean and the riblet configuration at oblique transition. The color represents the velocity magnitude

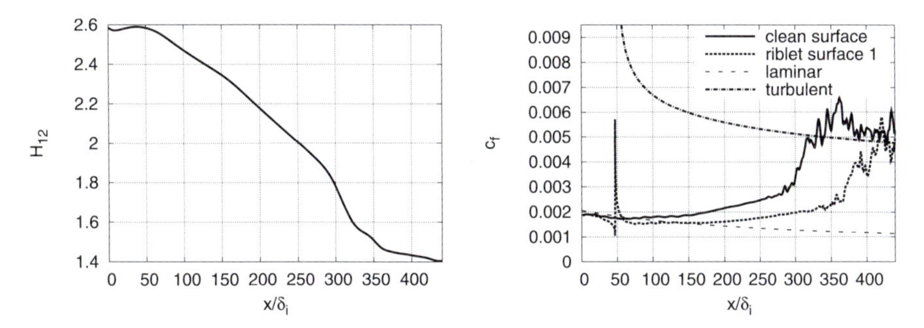

Fig. 10. Shape factor (left) and skin-friction coefficient on the clean and riblet configuration 1 (right) vs x/δ_i at oblique transition

clean and the riblet surface are juxtaposed. The states of transition are observed for both configurations, however, the breakdown into turbulence occurs further downstream on the riblet surface. Like at K-type transition a secondary flow structure, i. e., areas of recirculation between the tips of the riblet, develops at oblique transition.

The damping effect caused by this secondary flow is evidenced by comparing the development of the maximum spanwise $|w|_{max}$ and wall-normal velocity $|v|_{max}$ on the clean and the riblet surface in Figure 11. In the region $50 \leq x/\delta_i \leq 250$, which is characterized by streamwisely aligned vortex structures, the spanwise velocity on the riblet surface is on the average

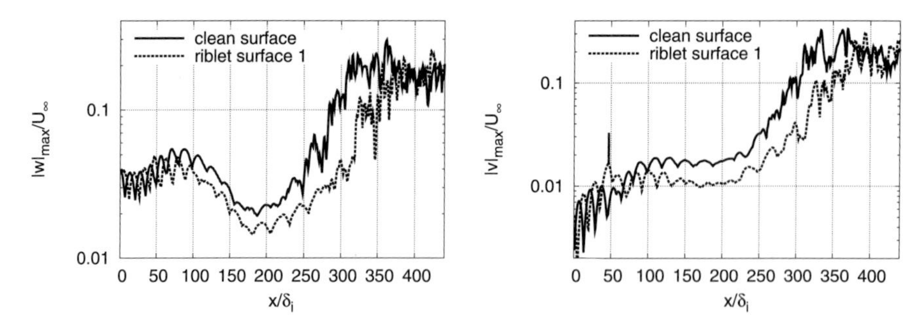

Fig. 11. Maximum spanwise velocity (left) and maximum wall-normal velocity (right) vs x/δ_i at oblique transition

about 25% lower than that on the clean surface and the wall-normal velocity even 35%. Since the maximum wall-normal velocity represents the strength of the near-wall vortex structures, a damped wall-normal velocity means a damped downwash of faster fluid and as such a reduced drag in the interval $50 \leq x/\delta_i \leq 250$.

The development of the skin-friction coefficient c_f is shown in Figure 10. The peak at $x/\delta_i = 47$ is caused by the first appearance of the riblets. At $x/\delta_i \approx 350$ the skin-friction coefficient increases dramatically evidencing the breakdown to turbulence. Compared to the clean surface the development of the turbulent state is delayed by $\Delta x/\delta_i \approx 200$, which yields a maximum drag reduction of 77% at $x/\delta_i \approx 350$. The total drag of the flat plate is lowered by approximately 30%.

6 Computational Resources

The simulations were carried out on the NEC SX-8 installed at the HLR Stuttgart. The results presented in chapter 5 were computed on a domain of integration that is divided into 40 blocks, while each block resides on a single CPU. The meshes for case "clean" and "riblet" contain about 25 million mesh points, such that each node computes the solution for a subdomain with about 600.000 points. Data between the blocks is exchanged via MPI (message passing interface). The computational details are given in Tab. 3. Furthermore, the same code was used for a large-eddy simulation around a DRA2303 profile at a Reynolds number of 2.6 million, a transonic Machnumber of 0.7 and an angle of attack of 2.3^o, which is a case for which shock buffeting occurs. This simulation is currently run on the NEC SX-9 on two nodes with 16 CPUs each. The domain of integration is divided into 32 blocks where the block size varies by about 15%. The overall number of meshpoints is 101 million. Although the simulation results are not presented in this paper Tab. 3 presents the details of the computational performance on NEC SX-9.

Table 3. Sample performance on NEC SX-8 and SX-9

Case	"clean"	"riblet"	DRA2303
Hardware	SX-8	SX-8	SX-9
Number of CPUs	40	40	32
Number of Nodes	5	5	2
grid points/CPU	$5.07 \cdot 10^5$	$5.24 \cdot 10^5$	$31.6 \cdot 10^5$
total grid points	$20.1 \cdot 10^6$	$20.9 \cdot 10^6$	$101.1 \cdot 10^6$
Avg. User Time [s]	84121	82758	41095
Avg. Vector Time [s]	78241	76252	39604
Vector Operations Ratio [%]	99.6	99.5	99.765
Avg. Vector Length	238.1	238.9	254.4
Memory/CPU [MB]	505	515	2125
total Memory [GB]	19.2	20.125	99.688
Avg. MFLOPS/CPU	4001	3956	15100
Max. MFLOPS/CPU	4022	3975	16460
total GFLOPS	148.82	158.278	483.185

7 Conclusions

A set of numerical simulations of a spatially evolving zero-pressure gradient boundary layer undergoing laminar-turbulent transition has been performed. K-type transition and oblique transition have been initiated by superposed typical disturbances. For the streamwisely aligned riblets an amplification of the two-dimensional TS waves has been found at K-type transition. The development of the three-dimensional Λ- and hairpin vortex structures is weakened resulting in a delayed breakdown of these structures into turbulence. The damped strength of the streamwise vortices leads to a reduced wall-normal velocity such that the transport of low-speed fluid into outer regions of the boundary layer is lowered and as such breakdown to turbulence is delayed. At oblique transition, breakdown to turbulence is delayed even further downstream, since at all transition stages essential disturbances such as Λ-vortices and streamwisely aligned streaks are damped by the riblets.

In brief, when riblets are applied on technical surfaces to delay transition the type of transition and the stage of transition should be known to optimally distribute the riblets. Since riblets do amplify two-dimensional TS waves, an acceleration of transition by inappropriately placed riblets is possible. However, the investigation does show an enormous potential of drag reduction if riblets are properly distributed.

Acknowledgement. The authors gratefully acknowledge the financial support of the joint cooperative project "RibletSkin" by the Volkswagen Foundation, Hannover, Germany.

References

1. Bechert, D.W., Bruse, M., Hage, W., van der Hoeven, J.G.T., Hoppe, G.: Experiments on drag-reducing surfaces and their optimization with an adjustable geometry. J. Fluid Mech. **338** (1997) 59–87
2. Choi, H., Moin, P., Kim, J.: Direct numerical simulation of turbulent flow over riblets. J. Fluid Mech. **255** (1993) 503–539
3. Schlichting, H., Gersten, K.: Grenzschicht-Theorie. Springer-Verlag, Berlin Heidelberg (2006)
4. Neuman, D., Dinkelacker, A.: Drag measurements on v-grooved surfaces on a body of revolution in axial flow. Applied Scientific Research **48** (1991) 105–114
5. Ladd, D.M., Rohr, J.J., Reidy, L.W., Hendricks, E.W.: The effect of riblets on laminar to turbulent transition. Exp. in Fluids **14** (1993) 1–2
6. Grek, G.R., Kozlov, V., Titarenko, S.: An experimental study of the influence of riblets on transition. J. Fluid Mech. **315** (1996) 31–149
7. Litvinenko, Y.A., Chernoray, V.G., Kozlov, V.V., Loefdahl, L., Grek, G.R., Chun, H.H.: The influence of riblets on the development of a Λ structure and its transformation into a turbulent spot. Physics - Doklady **51** (2006) 144–147
8. Chernorai, V., Kozlov, V., Loefdahl, L., Grek, G., Chun, H.: Effect of riblets on nonlinear disturbances in the boundary layer. Thermophys. and Aeromech. **13** (2006) 67–74
9. Luchini, P., Trombetta, G.: Effects of Riblets upon Flow Stability. Applied Scientific Research **54** (1995) 313–321
10. Ehrenstein, U.: On the linear stability of channel flow over riblets. Phys. Fluids **8** (1996) 3194–3196
11. Belov, I.A., Enutin, G.V., Litvinov, V.N.: Influence of a flat plate streamwise and spanwise ribbed surface on the the laminar-turbulent transition. Uch. Zap. TsAGI **17** (1990) 107–111
12. Ghosal, S., Moin, P.: The Basic Equations for the Large Eddy Simulation of Turbulent Flows in Complex Geometry. J. Comput. Phys. **118** (1995) 24–37
13. Liou, M.S., Steffen Jr., C.J.: A New Flux Splitting Scheme. J. Comput. Phys. **107** (1993) 23–39
14. Schmid, P.J., Henningson, D.S.: Stability and transition in shear flows. Springer Verlag, New York (2001)
15. Hirt, G., Thome, M.: Rolling of functional metallic surface structures. CIRP Annals - Manufacturing Technology **57** (2008) 317–320
16. Ducros, F., Comte, P., Lesieur, M.: Large-eddy simulation of transition to turbulence in a boundary layer developing spatially over a flat plate. J. Fluid Mech. **326** (1996) 1–36
17. Jeong, J., Hussain, F.: On the identification of a vortex. J. Fluid Mech. **285** (1995) 69 – 94
18. Berlin, S., Lundbladh, A., Henningson, D.: Spatial simulation of oblique transition in a boundary layer. Phys. Fluids **6** (1994) 1949–1951

Understanding the Dynamics and Control of a Turbulent Impinging Jet via Pulsation and Swirl Using Large Eddy Simulation

Naseem Uddin[1,2], Sven Olaf Neumann[1], and Bernhard Weigand[1]

[1] Institut für Thermodynamik der Luft- und Raumfahrt, Universität Stuttgart, Pfaffenwaldring 31, Stuttgart 70569, Germany
Olaf.Neumann@itlr.uni-stuttgart.de
[2] Mechanical Engineering Department, NED University of Engineering and Technology, Karachi 75270, Pakistan

Summary. Impinging jets are used in a variety of engineering applications like in chemical reactors, mixing devices, drying and cooling applications. Good quality simulations of this highly complex flow field is a challenging task. In this work, the flow field and heat transfer of turbulent pulsating and swirling jet impingements are investigated by Large Eddy Simulation (LES). The benchmark case of turbulent impinging jet with out swirl and excitation, recommended by ERCOFTAC, is simulated first. In all investigations the jet's Reynolds number (Re) is 23000 and jet outlet-to-target wall distance (H/D) is 2. The agreement between experimental data and simulation gives encouragement for further investigation of complex flow of jets impingement with excited inlet velocity profiles and swirl. The pulsating jets are investigated at four different excitations modes. Where as the swirl is introduced at two different swirl numbers (S). The correlation between the heat transfer mechanism, flow kinematics and turbulence quantities is investigated.

1 Introduction

The control of flow and heat transfer mechanism of turbulent impinging jet is a crucial research area, where the jet's inlet velocity field is modulated through swirl and pre-defined excitations. Impinging jets are used in a variety of engineering applications like, drying, spraying, drilling and gas turbine internal blade cooling. The accurate simulation of this complex flow field is a challenging task. The complexity of the turbulent impinging jet can be enhanced further by the addition of pulsation and swirl.

It is known that through jet excitation or pulsation the jet behavior can be altered significantly. Pulsations affect not only the flow structures but also

W.E. Nagel et al. (eds.), *High Performance Computing in Science and Engineering '09*, DOI 10.1007/978-3-642-04665-0_24,

the entrainment effects. In the case of limited mass flow rate, the possibility of jet heat transfer enhancement is very important. For case of excitation of the jet, not only the frequency of excitation is important but the amplitude of the excitation is important as well. Further, if the jet is impinging, the target to wall distance (H/D) is also playing a significant role. The pioneering work of experimental investigation of an excited jet impingement had been done by Nevins and Ball [26]. They found no significant heat transfer enhancement due to excitation. Liu and Sullivan [23] examined the effect of excitation on heat transfer at target to wall distances smaller than one. It is found that when the jet is forced close to natural frequency, intermittent vortex pairing is obtained, which gives rise to less organized eddy-like structures which contain high turbulence and subsequently improve heat transfer. Hwang and Cho [16] investigated the effect of excitation on heat transfer. They found that for jet excitations at the fundamental frequency of the initial vortex generation, the potential core is small and the jet's turbulence level is high. The subharmonic of the fundamental frequency excitation promotes vortex pairing. It is found that for $H/D < 4$, increase in heat transfer is possible in the wall jet zone, but overall this increase is small. On the other hand, for larger H/D ratios (H/D 12 or higher) heat transfer decreases. Mladin and Zumbrunnen [25] found that the deciding factors for heat transfer enhancement is the product of the frequency and amplitude. At one particular Reynolds number and H/D ratio the product of Strouhal number and amplitude is important. For the higher values of the product of frequency and amplitude, enhancement in heat transfer is expected and for lower values of the product, the reduction in heat transfer is anticipated. Hwang and Cho [16] have also noticed that both frequency and amplitude effect the resulting heat transfer. Camci and Herr [4] have investigated the oscillating planar turbulent jet with heat transfer. They found that the oscillation in the jet, compared with the stationary jet case, can improve heat transfer up to 70 %. Azevedo et al. [1] have investigated the heat transfer due to impinging jet. It is observed that the pulsations reduce the heat transfer, which according to them is due to relatively low-magnitude, small-scale turbulent fluctuations superimposed on the instantaneous periodic flow.

There are few numerical studies devoted to the effect of velocity field excitation of impinging jet on heat transfer. Hofmann [14] investigated the effect of excitation on impinging jets with heat transfer using Reynolds averaged Navier Stokes equations (RANS) models. In the prediction of the flow and thermal field of an impinging jet some RANS based turbulence models unveil their difficulties in representing the occuring complex flow phenomena. Jiang et al. [17] used LES to investigate the rectangular jet behaviour. They found that strong vortical development is exhibited when the forced frequency is almost the same as the characteristic frequency. Chung et al. [5] performed DNS simulations of a laminar impinging jet at Reynolds numbers of 5300, 500, and 1000. Jiang et al. [18] investigated the effect of perturbations on Nusselt number of turbulent impinging jets using Direct Numerical Simulations (DNS). They found no significant increase in heat transfer, as the near wall flow structures were unaffected by the excitation of the jet inflow profile. The amplitude of excitation was 4%. It is also important to understand the influence different jet modes on the heat transfer mechanism. In most of the experimental studies the jet issues from the converging nozzles, in which the issuing jet has a flat velocity profile with a thin boundary layer near the nozzle lip. The most important jet mode in this case is the shear layer mode. The most amplified frequency in the jet scaled with the initial momentum layer thickness (δ^{**}) and jet exit velocity, gives the range of Strouhal number for this mode as [15].

$$St_{\delta^{**}} = \frac{f^{'}\delta^{**}}{U_o} \tag{1}$$

where $f^{'}$ is the most amplified frequency and U_o is mean the velocity. However, if the jet emerges in a fully developed flow state from the pipe, the momentum thickness is of the order of the jet radius and the jet does not exhibit the shear-layer mode, instead it will attain the jet column mode. The mode represented by Strouhal number (scaled with diameter) is called jet column mode or preferred mode. The preferred mode of the jet is defined as:

$$St_D = \frac{f_o D}{U_o} \tag{2}$$

Gutmark and Ho [11] have found that the preferred mode depends upon the initial conditions and can occur in the range of Strouhal number from 0.3 to 0.85.

In this paper, the effect of excitation of inlet velocity profile of turbulent impinging jet on heat transfer is investigated using LES. The jet issues from a pipe. The target-to-wall distance is two. The jet is then excited at four different frequencies. These frequencies correspond to preferred mode (of unexcited jet) and subharmonic & first and higher harmonic of this mode of the jet. The amplitude of excitation is kept at 50% of the inlet bulk velocity. Furthermore the computed results of the heat transfer for the exited jet configuration are compared to results for the swirling jet, which is an alternative way of modulating the jet inlet as the azimuthal velocity component introduces additional instabilities in the jet. Hereby the characterizing swirl number S in the incompressible jet is defined by the following equation and will be referenced in the following text:

$$S = \frac{G_\theta}{RG_{ax}} = \frac{1}{R}\frac{\int_0^R r^2 u_{ax} u_\theta dr}{\int_0^R r u_{ax}^2 dr} \tag{3}$$

where, G_θ and G_{ax} are the tangential and axial momentum fluxes, u_θ and u_{ax} are the tangential and axial velocities of the swirling jet and R is the radius of the jet at the inlet.

The simulation conditions are tabulated below, the Reynolds number corresponds to the axial bulk velocity at the inlet:

Table 1. LES Simulations

Cases	modulation	Re
Case-0-I	No-modulation	23000
Case-EV-I	Subharmonic excitation ($f_{n/2}$)	23000
Case-EV-II	Preferred mode excitation (f_n)	23000
Case-EV-III	Harmonic excitation (f_{2n})	23000
Case-EV-IV	Higher harmonic excitation (f_{5n})	23000
Case-S-I	Weak swirl (S=0.2)	23000
Case-S-II	Strong swirl (S=0.47)	23000

2 Computational Details

2.1 Numerical Grid

The geometry of the computational domain used for the non-pulsating and pulsating jet impingement is shown in Figure 1(a). The domain consists of a circular impingement zone and a circular pipe of length 6D. The target to wall distance (H) is 2D. The domain is confined by an adiabatic, no-slip wall at the top. Also the boundary condition at the pipe wall is adiabatic. The domain is 16D in length. The computational domain is modeled via hexahedral structured O-grids, developed via ICEM-CFD of ANSYS, Inc. The most important domain region is the stagnation and wall jet region, in which the Δr^+ is $\approx$ 36.3 and $(r\Delta\theta)^+ \approx 20$, where $()^+ =(.)u_\tau/\nu$. The mean grid spacing inside the jet is 6.6 times the length of the Kolmogorov scale. The grid used for the pulsating jet has 5×10^6 control volumes.

Figure 1(b) shows the schematic representation of the domain used for the investigation of the heat transfer due to impingement of the swirling jet at bulk Reynolds number of 23000. The grid spacing in dimensionless units is $\Delta r^+ \approx 27$, $(r\Delta\theta)^+ \approx 20$ and $\Delta y^+/\eta^+$ is 6, where, η is the length of the Kolmogorov scale. In the investigations, a hexahedral structured grid was used. The grid used for the swirling jet has 6.4×10^6 control volumes.

The heat flux of 1000 W/m^2 is applied at the impingement wall for both swirling and pulsating jet cases.

2.2 Inflow Conditions

Pulsating Jet Inflow Conditions

The inflow conditions are very important for the LES. For inflow conditions, the digital-filter based turbulence generation procedure proposed by Klein et al. [20] was used.

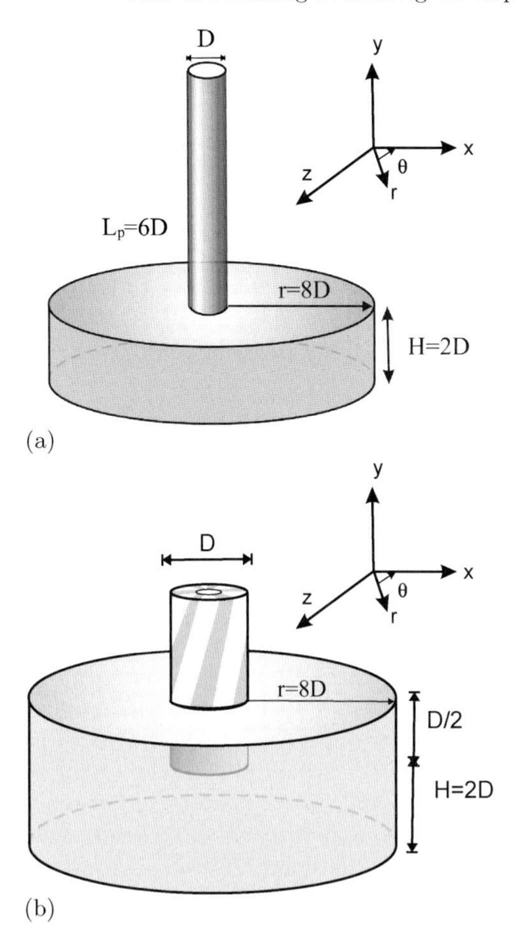

Fig. 1. Schematic diagram of the computational domain used for the simulation of (a) pulsating and (b) swirling jet impingement cases. The origin is fixed at the geometric stagnation point. The jet enters at y=2D. A swirling device is shown for clarification but is not geometrically modeled in the simulation

In the jet's axial direction, a fully developed turbulent pipe flow mean velocity profile, together with velocity fluctuations varying in time are prescribed at the inlet. The turbulence inflow conditions are generated, based on the procedure by Klein et al. [20]. The velocity fluctuations generated are based on digital filtering of random data. In this procedure, the inflow turbulent velocity generated is based on the relation:

$$u_i = < u_i > + a_{ij} u_j^{'}, \tag{4}$$

where $< u_i >$ is the mean axial velocity, $u_j^{'}$ are velocity fluctuations and a_{ij} is a tensor related to the Reynolds stress tensor (see [20]). According to this procedure, the prescribed auto-correlation function R_{uu} for homogeneous turbulence in later stages is used to describe the turbulence:

$$R_{uu}(\hat{r}) = exp\left(\frac{-\pi\hat{r}^2}{4L(t)^2}\right), \tag{5}$$

where $\hat{r}$ is the position vector and L(t) is the integral length scale at the inflow plane.

The mean velocity profile of fully developed turbulent flow is altered based on the following relation:

$$U_{in} = U_b + A_N sin(2\pi f t) \tag{6}$$

where U_{in}, U_b and A_N are the jet's inlet velocity, jet's bulk velocity and the amplitude of excitation respectively, A_N is taken as 50% of the U_b in this section.

2.3 Outflow Conditions

At the outlet, the Convective boundary conditions are used. As this allow a negligible influence on the evolution of the flow structures in the finite-size computational domain. It is experienced that the use of von Neumann type boundary conditions with a full three-dimensional domain simulation, for the impinging jet case, generates a non-physical feedback, which strongly affects the global stability of the simulation.

The convective boundary condition is spatially and temporally local, and defined as ([6])

$$\frac{\partial U}{\partial t} + \Lambda\frac{\partial U}{\partial r} = 0. \tag{7}$$

The quantity Λ is taken to be a constant convection velocity of the large-scale structures, which is set to a constant as per global mass conservation requirements.

3 Numerical Code & its Performance

3.1 FASTEST

The computations were performed with the CFD code FASTEST (Flow Analysis Solving Transport Equations Simulating Turbulence). The code is based on a finite volume discretization with pressure velocity coupling done through a SIMPLE algorithm. For LES computations, a top hat filter was used. The three dimensional filtered Navier-Stokes equations along with the subgrid model proposed by Germano [9] were solved. The second order implicit Crank-Nicolson method is used for time discretization and for space discretization of the convective terms a second order central differencing scheme is used. The resulting set of equations is solved using a SIP solver [27].

3.2 Code Performance & Solution Control

Computations are done on two different computing clusters available at Höchstleistungsrechenzentrum (HLRS), Stuttgart, Germany. For computations related to jet's excitation, the CRAY Opteron Cluster is used. The noteworthy improvements in the code are the vectorization of the subroutines related to LES subgrid scale model. The FASTEST has two different loop structures available in the code, which could be used for computations, depending on the architecture of the computing system. The routines related to LES had the three times nested loop version only. The use of a collapsed single loop is a natural choice for vector machines as it gives better vectorization. The single loop version is now also available for the LES-model. The code has attained 5.6 Gflops and high vectorization rates (>99.8%). For Case-S-II, the vectorized version of the code is used and computations are done on the NEC SX-8.

The number of processors and computing platforms used are listed below:

Table 2. Computational Platforms

Cases	Re	Modulation	CPUs	Computing Platform
Case-0-I	23000	No-modulation	20	Cray Opteron
Case-EV	23000	Exciation	20	Cray Opteron
Case-S-I	23000	Weak swirl (S=0.2)	20	Cray Opteron
Case-S-II	23000	Strong swirl (S=0.47)	4	NEC SX-8

The computational domain for pulsating and swirling jet is divided into 42 and 90 blocks, respectively. This domain decomposition helps in computational load balancing on the processors. The inter-processor communications were performed with standard Message Passing Interface (MPI). The simulation of the pulsating jet requires 10000 CPU hours and on average the simulation of swirling jet requires 15000 CPU hours on Cray Opteron clusters. The computations on NEC SX-8 are much faster and require three to four times less computational time than the one on the Cray Opteron Platform.

On average total simulation time for each case is equivalent to 20 cycles, where one cycle corresponds to the natural frequency of a pulsating jet. The flow becomes statistical stationary after about 9.3 cycles. The simulation is controlled in such a way that the CFL number remains less than one. The dimensionless time step used is $\Delta t D/U_{ax}$ and equals to 7E-07.

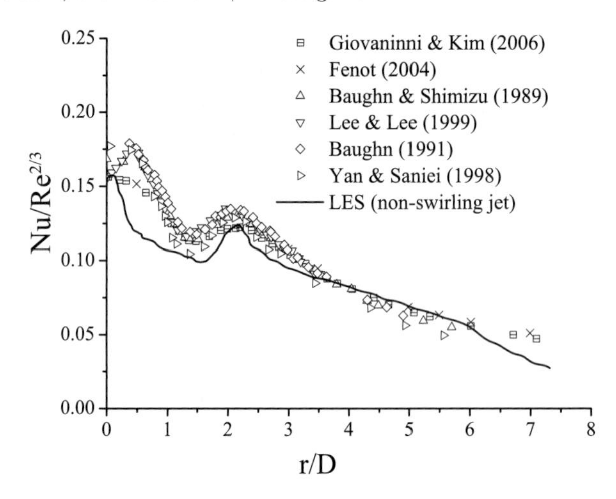

Fig. 2. The radial distribution of Nusselt number in case of an impinging jet with out swirl and pulsation

4 Results

The Nusselt number distribution is defined as

$$Nu = \left(\frac{q_w^{''}}{T_w - T_j} \right) \frac{D}{\lambda}, \tag{8}$$

where $q_w^{''}$ is the heat flux at the target wall. T_w is the temperature attained by target wall after jet impingement and T_j is the jet's inlet temperature. In Figure 2, the heat transfer for a pulsating jet is compared with experimental data. As in some experiments, the Reynolds number is different from 23000, the Nusselt number distribution is therefore normalised by $Re^{2/3}$, which is recommended by Martin [24]. Nusselt number from the LES simulation is compared with the experimental data of Giovannini and Kim ($Re = 23000$) [10], Fenot ($Re = 23000$) [8], Baughn ($Re = 23750$) [2], Lee et al. ($Re = 20000$) [22], Baughn and Shimizu ($Re = 23300$) [3] and Yan and Saniei ($Re = 23000$) [28].

It is found that LES can correctly predict the stagnation point Nusselt number and also the secondary peak is well captured. However, for $r/D<1$ the LES under predicts the experimental Nusselt number distribution. One probable reason for this difference can be the artificial inflow turbulence conditions. Figure 2 shows the radial distribution of the Nusselt number. The case is used for a benchmark testing of the LES set-up.

Figure 3 shows the energy spectrum of the velocity signal in case of excited and unexcited jet at the same position in the jet. Only the large scales are affected by the excitation and become more energetic. However, the small scales are unaffected by the excitation. It is interesting to note that the slope of the spectrum is independent of the excitation. Experimentally, Azevedo et al. [1] observed that the excitation only affect the large scale structures. As the present simulations endorse these findings, the excited jet simulation on the same grid as used for non-excited jet is justified.

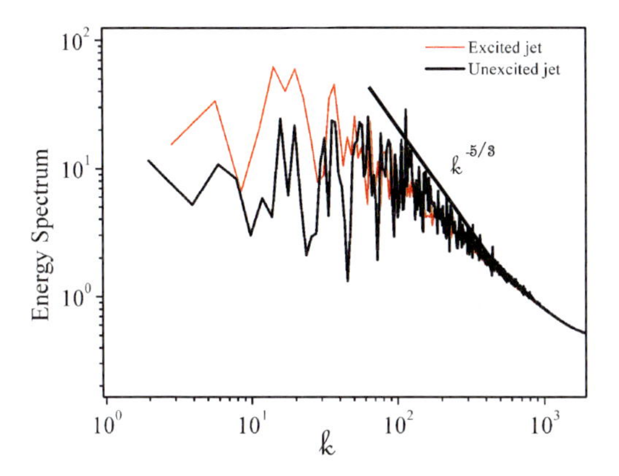

Fig. 3. The turbulence spectrum in case of excited and unexcited jet

Instead of exciting the inlet velocity field at random frequencies, it is decided to excite the jet at frequencies corresponding to the jet's preferred mode (f_n), subharmonic of preferred mode ($f_{n/2}$), twice of preferred mode (f_{2n}) and five times of Preferred mode (f_{5n}). The Strouhal numbers of the jet are:

- St=0.16 (Naturally occurring subharmonic)
- St=0.32 (natural/preferred)
- St=0.6 (harmonic)
- St=1.45 (higher harmonic)

The instantaneous entrainment behavior at different excitational frequencies is shown in Figure 4. Figure 5 shows the entrainment behavior in case of a natural jet. It is found that the fast change in velocity increases the structures renewal effect but it also makes the flow structures smaller, making them less effective for heat transfer enhancement.

Over the years, researchers with different objectives, investigated many interesting phenomena in excited jet cases, for example, large scale structural dynamics, vortex pairing, vortices collective interaction and unsteady separation of the wall boundary layer. In case of the excited jet impingement, the understanding of these phenomena is important, as they may affect the heat transfer distribution on the target wall. It is therefore interesting to discuss the jet's flow field under excitation effects, which would help in understanding the differences between the heat transfer distributions discussed in this paper.

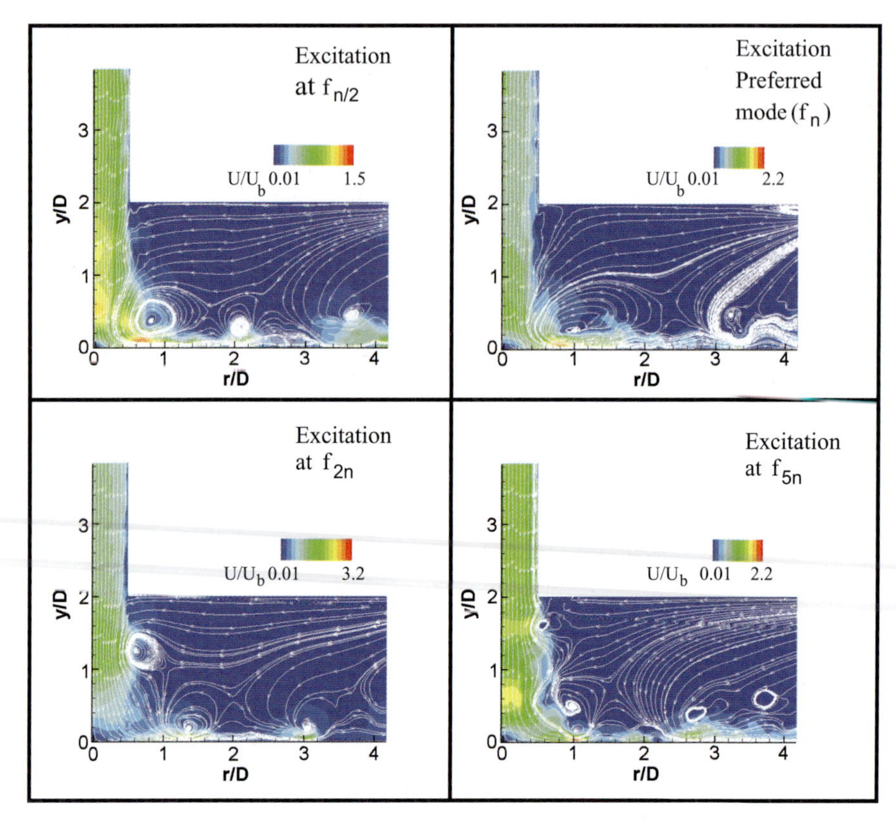

Fig. 4. Instantaneous pathlines in case of jet inlet velocity excitation, at different time steps

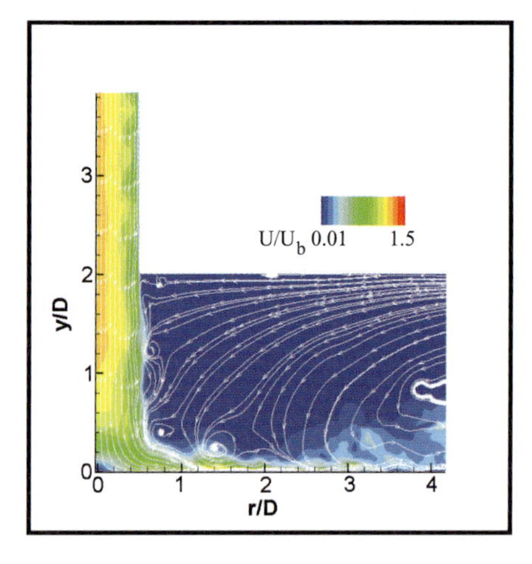

Fig. 5. Instantaneous pathlines in case of unexcited (natural) jet

Kataoka et al. [19] have discovered the interesting phenomenon of surface renewal effects in case of impinging jets. According to them the flow over the target surface is renewed by the large scales structures depending on the H/D ratios, Reynolds number and velocity fluctuations. The large scale structures arise from the flow emerging from the nozzle and flow impact on the surface. These structures effect the whole flow field and control the impinging jet dynamics. The phenomenon in the stagnation zone is quantified through the surface renewal parameter, defined as:

$$Sr = \frac{\sqrt{(\overline{u'u'})_{stg}}}{U_{CL}} \frac{f_e D^2}{\nu} \tag{9}$$

where f_e is the frequency of the large scale structure passage frequency. It can be interpreted that the surface renewal parameter is the product of the turbulent Reynolds number and Strouhal number. Kataoka et al. [19] found experimentally that as the surface renewal parameter increases the heat transfer on the target wall in the stagnation region increases. However, the investigations and reasoning proposed by Kataoka et al. [19] were valid only for the case of naturally occurring unexcited jet cases. However, as in case of jet excitation the turbulent coherent structures can be controlled it can be imagined that the surface renewal effect is controlled as well.

Another, important phenomenon in axisymmetric jets is vortex pairing in which two discrete vortices coalesce together. It can occur in free as well as impinging jets. Under controlled excitation multiple coherent structures merge together and it is called collective interaction [13]. Two of the characteristic features associated with the collective interaction are the sharp drop in passage frequency and relatively high shear-layer growth [12].

Several researchers observed these phenomena in their experiments and its role in the heat transfer enhancement was investigated (see [23, 16]). In these investigations, in case of turbulent jet (Re=23000) at the amplitude of excitation of 50% with outlet-to-target wall distance (H/D) of two, no pairing and collective interaction pheomena could be observed. However, the phenomenon of unsteady separation is observed. Didden and Ho [7] investigated the axisymmetric jet excitation at a Reynolds number of 19000 and the frequency of excitation corresponds to the preferred mode of the unforced jet. Landreth and Adrian [21] investigated the impingement of a low Reynolds number turbulent circular jet using Particale Image Velocimetry (PIV). They found that the primary vortex or ring vortex causes the formation of secondary vortices at the wall, which are lifted away from the wall jet. Later both primary and secondary vortices merge together. A similar phenomenon was observed through simulation. The sharp rise in the displacement thickness is an indication of unsteady separation. The displacement thickness is defined as:

$$\delta^* = \int_0^{\delta} \left[1 - \left(\frac{u}{U_\infty}\right)\right] dy \tag{10}$$

where δ is the boundary layer thickness, U_∞ is the velocity at the edge of the wall jet boundary layer.

At one particular instant, Figure 6 shows the displacement thickness at different radial positions in case of excitation at the preferred mode. The sharp rise in the thickness is a sign of the unsteady separation. It is observed that in case of excitation at subharmonic frequency the unsteady separation is delayed.

The contours of turbulent kinetic energy for the excited and the unexcited jets, are shown in Figures 7 and 8. In case of non-excited jet the turbulent kinetic energy is highest in the wall jet. The same trend can be seen in case of excitation at preferred mode and its subharmonic frequency. The change in trend is observed in case of excitation at frequencies higher than the preferred mode, the turbulent kinetic energy becomes higher more closer to the jet and later in case of higher harmonic excitation it is high at the jet outlet, with significant reduction in the amount of turbulent kinetic energy near the wall. Further, it is found that in case of a natural jet the turbulent kinetic energy is concentrated at the location of the secondary peak. However, in case of excitation the turbulent kinetic energy is high in the wall jet zone but the distribution is uniform, causing the secondary peak to be flattened or removed.

The radial distribution of the Nusselt number in case of velocity field excitation is plotted in Figure 9.

It is found that the excitation of the inlet velocity field at high amplitude can improve the heat transfer. Also the heat transfer at the stagnation point varies with the increase in excitational frequencies. The investigations at four different frequencies show that the increase in surface averaged heat transfer is very significant at the frequency corresponding to the half of the

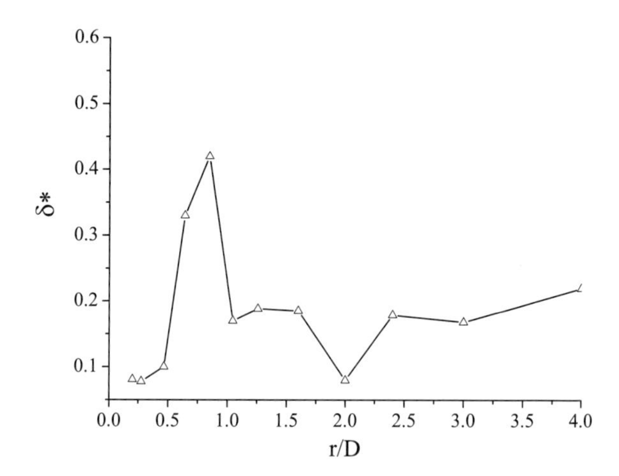

Fig. 6. Displacement thickness (in mm) computed for wall jet boundary layer (excitation at preferred mode)

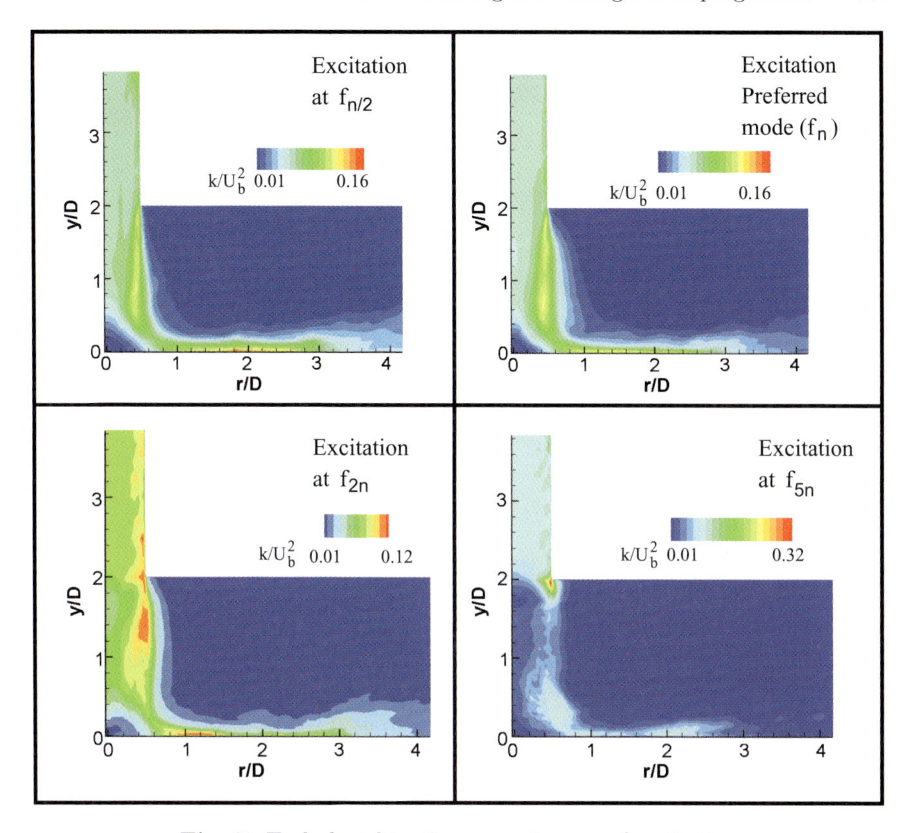

Fig. 7. Turbulent kinetic energy in case of excitation

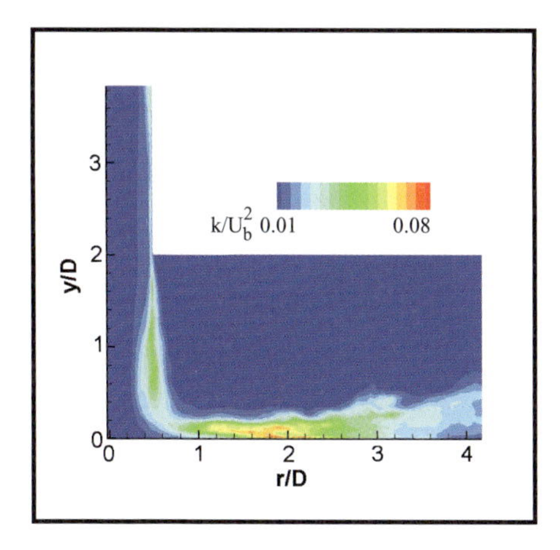

Fig. 8. Turbulent kinetic energy in case of unexcited jet

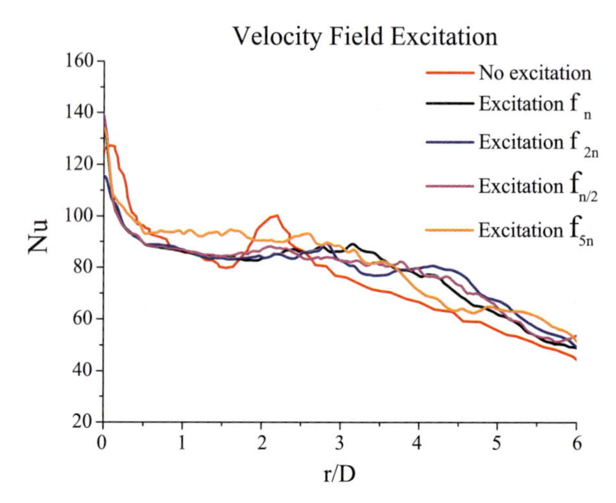

Fig. 9. Radial distribution of Nusselt number in case of velocity field excitation

preferred mode. It is interesting to note that a heat transfer enhancement can be achieved with excitation of the inlet velocity field at the large amplitude (in our case 50%). However, care must be taken in selecting the excitational frequencies. The frequencies depend on the geometric configuration of the jet outlet, which in fact affects the jet flow instabilities growth and jet structure.

Figure 10 shows a comparison of the two investigated control mechanisms applied for heat transfer enhancement in the impinging jet: A superposition of swirl causes a drop of the Nusselt number right at the stagnation point, whereas in the inner core region ($r/D \leq 2$) the heat transfer enhancement can be larger than the one achived by the investigated velocity exitation. Further

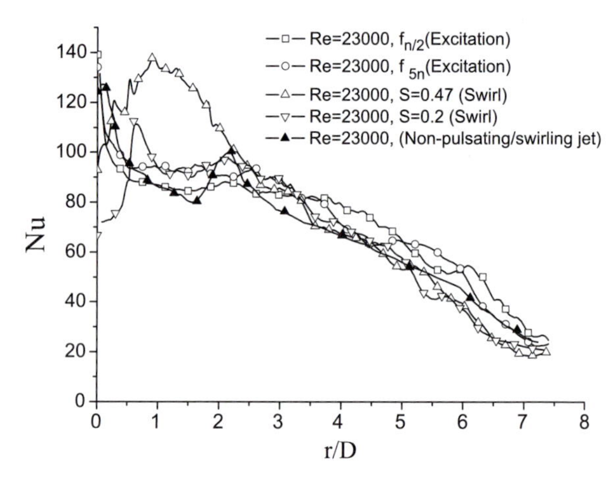

Fig. 10. Radial distribution of Nusselt number in case of velocity field excitation ($f_{n/2}$ & f_{5n}) and swirl (S=0.2 & 0.47)

outward ($r/D > 3$) the effect is vice versa and the velocity pulsations remains superior with regard to the heat transfer intensification. The location of the peak in the radial distribution of Nusselt number shifts with the increase in swirl number.

5 Conclusions

The series of LES investigations of turbulent pulsating and swirling jet impingement are performed on the HLRS computing platforms. It is found that:

1. The exited impinging jet reveals a pronounced energy receptivity at small wave numbers, which correspond to large scale excitations. The slope of the spectrum is independent of the excitation.
2. The excitation at higher frequencies give rise to non-linear interactions between the scales and the global jet receptivity becomes lower. This causes a reduction in the most amplified frequency. In case of the excitation at a preferred mode and its subharmonic the resulting structures are big. The excitation causes the merging of the vortices in the jet and big chunks of vortical structures with high turbulence strike the surface.
3. Appart from the stagnation point a superposition of swirl can lead to a Nusselt number enhancement in the inner core region ($r/D \leq 2$) of the jet which is larger than the one achived by the investigated velocity exitation. Further outward ($r/D > 3$) the findings are vice versa and the effect of the velocity pulsations remains superior but at smaller heat transfer enhancement rates.

Acknowledgement. The authors would like to thank Prof. Dr. M. Schäfer and Dr.-Ing. D. Sternel, Fachgebiet für Numerische Berechnungsverfahren im Maschinenbau (FNB), Technische Universität, Darmstadt, Germany, for providing the FASTEST code and helpful discussions. The first author wants to thank the Higher Education Commission (HEC), Pakistan and DAAD, Germany for providing a PhD fellowship. The support provided by Dr.-Ing. P. Lammers, Höchstleistungs-Rechenzentrum (HLRS), Stuttgart is gratefully acknowledged.

References

1. L. F. A. Azevedo, B. W. Webb, M. Queiroz, Pulsed air jet impingement heat transfer, Experimental Thermal and Fluid Science **8**, 206–213 (1994).
2. J. W. Baughn, S. Shimizu, Heat transfer measurement from a surface with uniform heat flux and an impinging jet, Int. J. of Heat Transfer, **111**, 1096–1098 (1989).
3. J. W. Baughn, A. E. Hechanova, X. Yan, An experimental study of entrainment effects on the heat transfer from a flat surface to a heated circular impinging jet, Journal of Heat Transfer, **113**, 1023–1025, (1991).

4. C. Camci, F. Herr, Forced Convection heat transfer enhancement using a self-oscillating impinging planar jet, J. Heat Transfer, **124**, 770–782 (2002).
5. Y. M. Chung, K. H. Luo, N. D. Sandham, Numerical study of momentum and heat transfer in unsteady impinging jets, Int. J. of Heat and Fluid Flow, **23**, 592–600, (2002).
6. T. Colonius, Numerically non-reflecting boundary and interface conditions for compressible flow and aeroacoustic computations, AIAA, **35**, 7, 1126–1133 (1997).
7. N. Didden, C. Ho, Unsteady separation in a boundary layer produced by an impinging jet, J. Fluid Mech, **160**, 235–256 (1985).
8. M. Fenot, Etude du refroidissement par impact de jets. Application aux aubes de turbines, Universite de Poitiers, France (2004).
9. M. Germano, U. Piomelli, P. Moin, W. H. Cabot, A dynamic subgrid-scale eddy viscosity model, Phy. Fluid, **3**, 1760–1765 (1991).
10. A. Giovannini, N. S. Kim, Impinging jet: Experimental analysis of flow field and heat transfer for assesment of turbulence models, Annals of the Assembly for International Heat Transfer Conference 13, TRB-15 (2006).
11. E. Gutmark, C. Ho, Preferred modes and the spreading rates of jets, Phy. Fluids, **26**, 10, 2932–2938 (1983).
12. C. M. Ho, N. S. Nossier, Dynamics of impinging jet. Part 1, the feedback Phenomenon, J. Fluid Mechanics, **105**, 119–142 (1981).
13. C. M. Ho, L. S. Huang, Subharmonics and vortex merging in mixing layers, J. Fluid Mechanics, **119**, 443–473 (1982).
14. H. M. Hofmann, Wärmeübergang beim pulsierenden prallstrahl, Dissertation, Universität Karlsruhe (2005).
15. A. K. M. F. Hussain, Coherent structures - reality and myth, Phy. Fluids, **26**, 10, 2816–2850 (1983).
16. S. D. Hwang, H. H. Cho, Effects of acoustic excitation positions on heat transfer and flow in axisymmetric impinging jet: main jet excitation and shear layer excitation, Int. J. of Heat and Fluid Flow, **24**, 199–209 (2003).
17. P. Jiang, Y. C. Guo, C. K. Chan, W. Y. Lin, Frequency characteristics of coherent structures and their excitations in small aspect-ratio rectangular jets using large eddy simulation, Computers & Fluids, **36**, 611–621 (2007).
18. X. Jiang, H. Zhao, K. H. Luo, Direct computation of perturbed impinging hot jets, Computers & Fluids, **36**, 259–272 (2007).
19. K. Kataoka, M. Suguro, H. Degawa, K. Maruo, I. Mihata, The effect of surface renewal due to large scale eddies on jet impingement heat transfer, Int. J. Heat and Mass Transfer, **30**, 3, 559–567 (1987).
20. M. Klein, A. Sadiki, J. Janicka, A digital filter based generation of inflow data for spatially direct numerical or large eddy simulations, Journal of Computational Physics, **18**, 652–665 (2003).
21. C. C. Landreth, R. J. Adrian, Impingement of a low Reynolds number Turbulent circular jet onto a flat plate at normal incidence, Experiments in Fluids **9**, 74–84 (1990).
22. J. Lee, S. Lee, Stagnation region heat transfer of a turbulent axisymmetric jet impingement, Experimental Heat Transfer, **12**, 137–156 (1999).
23. T. Liu, J. P. Sullivan, Heat transfer and flow structures in an excited circular impinging jet, Int. J. Heat and Mass Transfer, **17**, 3695–3706 (1996).
24. H. Martin, Wärmeübergang bei Prallströmung, VDI-Waermeatlas, VDI (1998).

25. E. C. Mladin, D. A. Zumbrunnen, Alterations to coherent flow structures and heat transfer due to pulsations in an impinging air-jet, Int. J. Therm. Sci., **39**, 236–248 (2000).
26. R. G. Nevins, H. D. Ball, Heat transfer between a flat plate and a pulsating impinging jet, National Heat Transfer Conference Boulder, CO, ASME (1961).
27. H. L. Stone, Iterative solution of implicit approximations of multidimensional partial differential equations, SIAM J. Numer. Anal., **5**, 3, 530–558 (1988).
28. X. Yan, N. Saniei, Heat Transfer Measurements From a Flat Plate to a Swirling Impinging Jets, Proceedings of 11th International Heat Transfer Conference, Kyonju, Korea (1998).

Stability Analysis of a Coupled Helmholtz Resonator with Large Eddy Simulation

Balázs Pritz[1], Franco Magagnato[2], and Martin Gabi[3]

[1] Department of Fluid Machinery, University of Karlsruhe, 76128 Karlsruhe, Germany pritz@ism.uka.de
[2] Department of Fluid Machinery, University of Karlsruhe, 76128 Karlsruhe, Germany magagnato@ism.uka.de
[3] Department of Fluid Machinery, University of Karlsruhe, 76128 Karlsruhe, Germany gabi@ism.uka.de

Summary. Lean Premixed combustion, which allows for reducing the production of thermal NOx, is prone to combustion instabilities. There is an extensive research to develop a reduced physical model, which allows - without time-consuming measurements - to calculate the resonance characteristics of a combustion system consisting of Helmholtz-resonator-type components (burner plenum, combustion chamber). For the formulation of this model numerical investigations by means of compressible Large Eddy Simulation (LES) are carried out. In these investigations the flow in the combustion chamber is isotherm, non-reacting and excited with a sinusoidal mass flow rate. The foregoing investigations concentrated on the single combustion chamber as a single resonator.

In this paper the results of the numerical investigations of a coupled system will be presented. The reduced physical model was extended for the coupled system of burner and combustion chamber. By means of numerical simulation and the physical model the resonance characteristics of a combustion system can be predicted already during the design phase. In order to predict the resonant characteristics of the coupled system and to provide an insight into the flow mechanics 10 compressible LES were carried out. The results are in very good agreement with the experimental investigations.

1 Introduction

It is well known that in order to fulfil the stringent demands for low emissions of NOx, the lean premixed combustion concept is commonly used. However, lean premixed combustors are susceptible to thermo acoustic instabilities driven by the combustion process and possibly sustained by a resonant feedback mechanism coupling pressure and heat release [1, 2]. This resonant feed back mechanism creates pulsations typically in the frequency range of several hundred Hz and which reach high amplitudes so that the system has to be

W.E. Nagel et al. (eds.), *High Performance Computing in Science and Engineering '09*, DOI 10.1007/978-3-642-04665-0_25,

shut down or is even damaged. Although the research activities of the recent years have contributed to a better understanding of this phenomenon the underlying mechanisms are still not well enough understood.

For the prediction of the stability of technical combustion systems regarding the development and maintaining of self-sustained combustion instabilities the knowledge of the periodic-non-stationary mixing and reacting behaviour of the applied flame type (flame model) [3, 4, 5] and a quantitative description of the resonance characteristic of the gas volumes in the combustion chamber is conclusively needed. Combustion instabilities are characterised by time-dependent heat release rate of the flame as well as by the time-dependent static pressure within the combustion chamber or the connected volumes like mixing unit, burner plenum or exhaust gas system. A possibility to describe periodic combustion instabilities is to calculate the observed frequency by using the geometric parameters of the resonating volume, i.e. the eigenfrequency. In this calculation the damping in the system is neglected [6, 7]. In real systems there is, however, always a certain amount of damping. It has an important role to limit the amplification of pulsation in the range of the resonant frequency [8, 9].

At the University of Karlsruhe a physical model to predict the resonance characteristics of real, damped combustion systems was developed [10]. As the result of the foregoing research the model is able to describe the resonant characteristics of a single Helmholtz resonator type combustor for different operation conditions and geometries [11]. It is important to mention that this model describes the system in the low frequency range and do not cover the high frequency instabilities.

In order to understand the mechanisms related to pulsations in a complex combustion system it is necessary but not sufficient to predict the resonant characteristics only of the single components. The model was extended therefore for the coupled system of burner and combustion chamber. This is an indispensable step to enable the prediction of the resonance characteristic of combustion systems existing of numerous different volumes like mixing unit, air/fuel supply, burner plenum, combustion chamber and the components of the exhaust gas system. In order to prove the prediction of the model for the coupled system different geometric parameters such as burner volume, resonator geometry and, furthermore, operating parameters such as mean volume flow rate were varied under isothermal conditions [12].

By means of numerical simulation and the physical model the resonance characteristics of a combustion system can be predicted already during the design phase. The main goal of the numerical investigation is to predict the damping coefficient of the system which is an important input for the physical model. In order to provide an insight into the flow mechanics inside the system LES were carried out. LES is an approach to simulate turbulent flows based on resolving the unsteady large-scale motion of the fluid while the impact of the small-scale turbulence on the large scales is accounted for by a sub-grid scale model. By the prediction of flows in complex geometries, where large, anisotropic vortex structures dominate, the statistical turbulence models often fail. The LES approach is for such flows more reliable and more attractive as it allows more insight into the vortex dynamics. In recent years the rapid increase of computer power has made LES accessible to a broader scientific community. This is reflected in an abundance of papers on the method and its applications.

The solution of the fully compressible Navier-Stokes equations was essential to capture the physical response of the pulsation amplification, which is mainly the compressibility of the gas volume in the chamber. Viscous effects play a crucial role in the oscillating boundary layer in the neck of the Helmholtz resonator and, hereby, in the damping of the pulsation. The pulsation and the high shear in the resonator neck produce highly anisotropic swirled flow. Therefore it is improbable that an Unsteady Reynolds Averaged Navier-Stokes Simulation (URANS) can render such flow reliably.

The results of the investigation of the single resonator [13] showed that LES can predict accurately the resonant characteristics of the single resonator and the damping, respectively. In the present work the ability of LES predicting the resonant characteristics of the coupled system was investigated. The mesh for the LES was optimized on the findings of the investigations of the single resonator in the previous phase. The resonator characteristics predicted by the numerical simulation were in very good agreement with the data from the experiments.

2 Simulated Configuration

For the prediction of the self-sustained combustion instabilities and the realization of a stability analysis it is not sufficient to implement the resonance characteristic of a single, separated combustion chamber. Therefore, the described mathematical model was extended to two coupled Helmholtz-resonators in order to derive knowledge about complex interaction of real combustion chambers with more than one resonating subvolume.

The investigations were separated into an experimental and a numerical part. By the use of a number of suppositions and simplifications equations describing a coupled system of two Helmholtz-resonators were generated in analogy to a mechanical oscillator model (coupled stiff-mass-damper-system, see Figure 1). The predictions from the equations (resonance characteristics of the system of the coupled Helmholtz-resonators) were proved by the experimental and numerical results.

In order to modelling a coupled system the burner plenum was added upstream to the combustion chamber. On the right hand side of Fig. 1 the investigated system of two coupled Helmholtz-resonators is indicated. The coupling of the resonating elements is located at the exit of the burner plenum, which is the inlet of the combustion chamber (see Fig. 1: resonator neck). The mean mass flow rate is pulsated partially by the pulsator. The oscillating mass flow rate passes through the burner plenum, reaches the combustion chamber through the resonator neck and leaves the system at the end of the exhaust gas pipe. The oscillating mass flow rate is measured at the entrance of the system and at the end of the exhaust gas pipe to determine the resonance behaviour of the coupled resonator system.

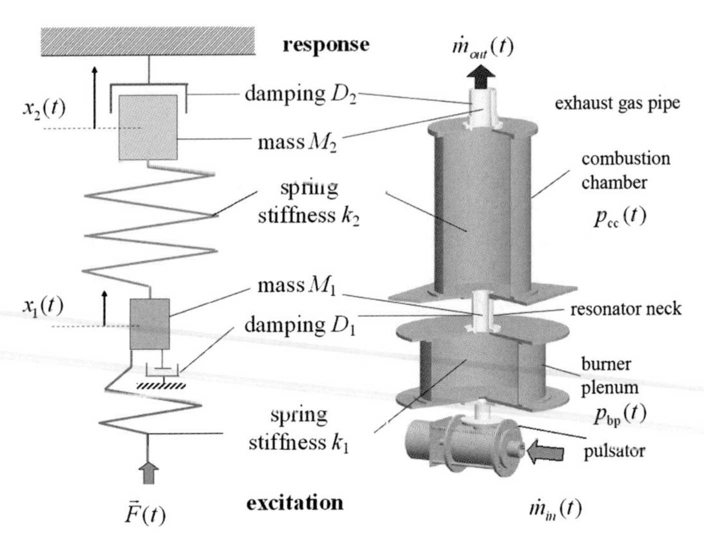

Fig. 1. Coupled Helmholtz-resonators and oscillating masses connected with springs and damping elements

2.1 Numerical Method

The Large Eddy Simulations (LES) of this system were carried out with the in-house developed parallel flow solver called SPARC (Structured PArallel Research Code) [14]. The code is based on 3-dimensional block structured finite volume method and parallelized with the message passing interface (MPI). In the case of the combustor the compressible Navier-Stokes equations are solved. The spatial discretization is a second-order accurate central difference formulation. The temporal integration is carried out with a second-order accurate implicit dual-time stepping scheme. For the inner iterations the 5-stage Runge-Kutta scheme was used. The time step was $\Delta t = 2 \times 10^{-6}$ s. The Smagorinsky-Lilly, the dynamic Smagorinsky and the Monotonically Integrated Large Eddy Simulation (MILES) approach were tested to modelling the subgrid-scale (SGS) structures [15, 16, 17]. The results show no significant differences in the predicted damping parameters. By means of the full multigrid method it was possible to study the effect of grid refinement.

2.2 Boundary Conditions

The geometry of the configuration chosen for the numerical investigation is illustrated in Figure 2. The observation windows and the inserted baffle plates increased the complexity of the geometry and hence the generation of the mesh significantly. The baffle plates were placed into the burner plenum and into the combustion chamber to avoid the jet of the nozzle and of the resonator neck to flow directly through the system, furthermore to achieve a homogeneous distribution of the velocity in the cross-section of the measuring point at the end of the exhaust gas pipe. The definition of the computational domain is akin to the case of the single Helmholtz-resonator [13]. The inlet surface is placed upstream of the nozzle. For the inlet condition the pulsating mass flow rate is used. The use of non-reflecting boundary condition could be omitted similar to the investigation of the single resonator. The effect of the turbulence at the inlet was investigated by means of the Synthetic Eddy Method (SEM) [18]. This method was recently extended for compressible flows by the authors and implemented in SPARC [19]. The SEM is more universal then e.g. the slicing method and it is less complicated to combine with the non-reflecting layer. The investigation confirmed the expectation that the nozzle decreases strongly the turbulence level downstream and therefore in this case the inflow turbulence has no influence on the resonant characteristic of the system.

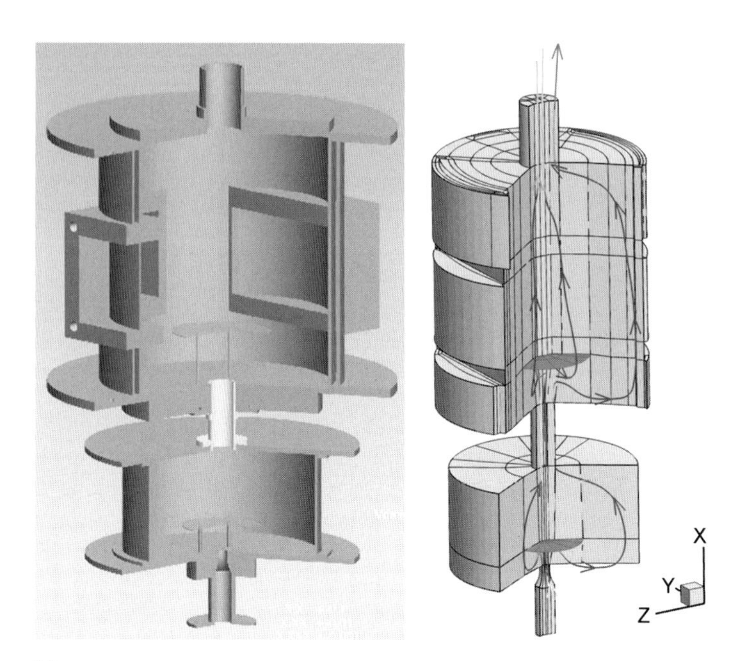

Fig. 2. CAD model and block structure of the burner plenum and the combustion chamber

The outlet boundary was placed in the far field. The size of this outflow region is $50 \times D_{\mathrm{egp}}$ in axial direction and $40 \times D_{\mathrm{egp}}$ in radial direction (see Figure 3). In this region the cells are stretched near to the outlet surface in order to alleviate reflections. The outlet surface at $x = 5m$ is inclined based on the observation explained next. In order to obtain a statistically steady solution before applying the excitation at the inlet a long calculation on the multigrid level 4 (coarsest mesh) and 3 was carried out. The entropy waves generated by the transient of the initialization must be convected through the burner plenum and the combustion chamber and finally out of the system. As the velocity behind the baffle plates is quite small this needed a relative long time. After the acoustic waves generated also by the transient of the initialization were decayed, it was detected, that acoustic waves of a discrete frequency were amplified to extreme high amplitudes. The wave length coincided width the length of the computational domain. After the outlet surface was slanted these standing waves decayed.

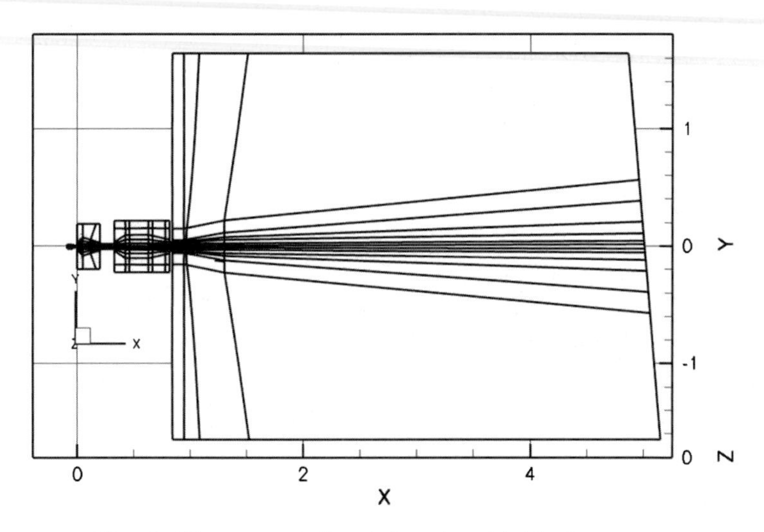

Fig. 3. The computational domain with block structure

At the surfaces the no-slip boundary condition and an adiabatic wall are imposed. For the first grid point normal to the wall $y^+ < 1$ is obtained. For the distribution of the computational cells the findings of the previous phase was applied. Thus much more nodes were arranged in the regions of the resonator neck and of the exhaust gas pipe, respectively, and in this case around the baffle plates too.

The final version of the mesh consists of app. 27×10^6 control volumes distributed in 612 blocks.

3 Results and Discussion

3.1 Flow Features

The visualization of the flow by means of the vorticity magnitude is depicted in Figure 4 and by means of the Q-criterion in Figure 5. The plots demonstrate the high shear in the resonator neck and in the exhaust gas pipe, which agree well with the results of the single resonator. The high shear is due to the high acceleration during the pulsation. The deceleration phase of the pulsation period induces a strong separation of vortices from the shear layer into the middle part of the pipes and makes the flow highly turbulent there. This increases the momentum transport crosswise to the mean flow and implies an increasing of the hydraulic resistance. The impingement of the jets on

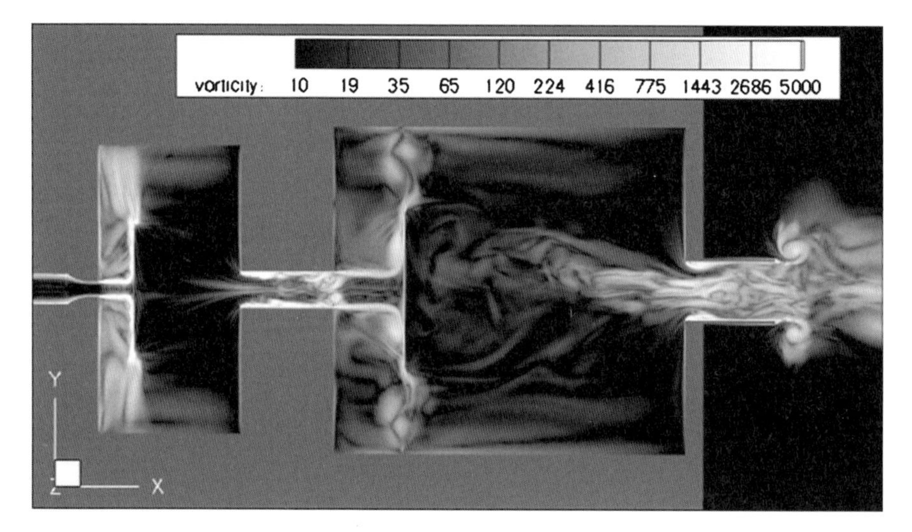

Fig. 4. Instantanieous vorticity distribution in the symmetry plane with exponential distribution of the colour levels

the baffle plates generates additional shear layers and vortex shedding which increase the damping effect. Behind the plates the flow motion is very slow and the flow is strongly accelerated again at the inflow into the resonator neck and into the exhaust gas pipe respectively. The 3-dimensional plot demonstrates in addition the influence of the observation windows. A part of the flow diverted by the baffle plate impinges on the lower part of the window and is diverted in negative x direction.

3.2 Comparison of the Calculations with Experiment

Initially, the experimental and numerical investigations were started simultaneously. Therefore no experimental data for the computation were available.

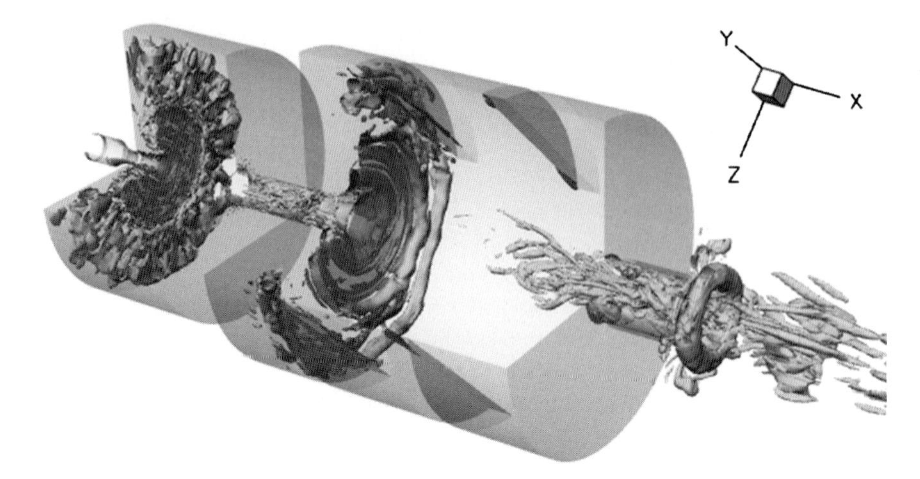

Fig. 5. Isosurfaces of the Q-criterion at $10^4 s^{-2}$

As the resonant characteristic of the given configuration was unknown the eigenfrequencies of the components had been calculated first as rough estimation. On the basis of this data 10 LESs were carried out on the coarser mesh levels with different excitation frequencies. These calculations were relative fast because of the moderate number of grid points on these levels. Fortunately, the estimations were acceptable and the computations could be continued on the finer grid levels. The computations were carried out on the HP XC4000 at the Scientific Supercomputing Centre on the University of Karlsruhe.

The results of the computation on different grid levels are plotted in Figure 6 and Figure 7. The difference in the resonance characteristic on the finest and second finest grid is negligible. It was tested only at the lower resonant frequency, at the highest amplitude ratio, because the calculation on the finest mesh is very time consuming. The higher is the amplitude ratio the higher is the mesh dependency. The result shows that the flow phenomena, which are influencing the damping, are adequate resolved on the second finest mesh. It is important to take into consideration that the mesh was optimized on the results of the investigation of the single resonator.

The results plotted in Fig. 6 and 7 show generally a very good prediction of the resonance frequencies and of the phase shift, respectively. In Figure 6 there is, however, a discrepancy of app. 20% in the prediction of the amplitude ratio at the highest peak. As discussed above, there is a significant shearing on the baffle plates. Unfortunately, in the experiments the plates were perforated. This was necessary to achieve the best velocity distribution at the outlet for the measurement with hot wire. In the simulations the wall condition was used for the plates. The resolution of the holes would yield a tremendous large number of the grid points. A boundary condition which can model this effect was not available. By the time the geometry data of the configuration

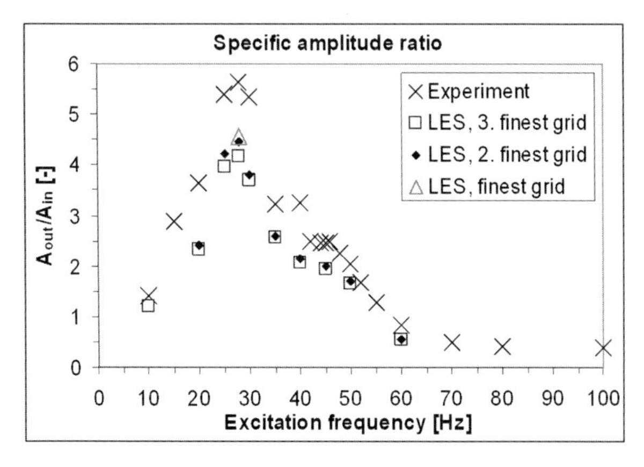

Fig. 6. Amplitude response of the coupled resonators

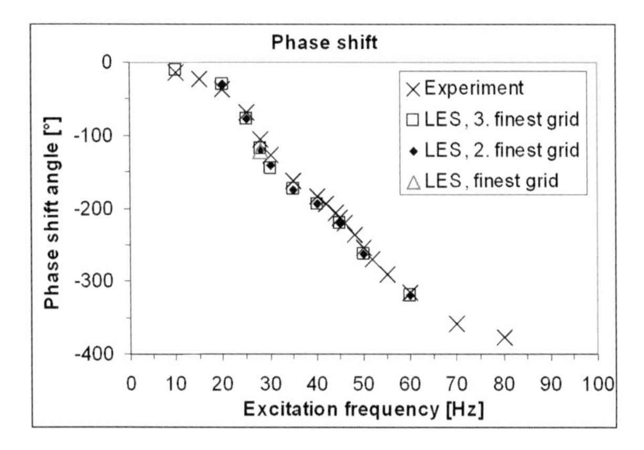

Fig. 7. Phase transfer function of the coupled resonators

were received, it was not possible to replace the plates any more. Probably this difference plays the major part in the underprediction of the amplitude ratio. The implementation of a boundary condition to simulate the perforated plates is in progress.

4 Computational Efficiency

In the computation of these results we have been using up to 108 Opteron processors of the HP XC4000 in Karlsruhe. The in-house developed code Sparc is parallelized with the MPI 1.2.7 software. The computational time for one point of the excitation frequency was about $400h$. Since we were using 612 blocks of the finite volume scheme we could efficiently distribute the blocks on the 108 processors with the domain decomposition technique. The load balancing was at about 97%. The parallel efficiency was also very good. Since in

Karlsruhe the communication is done with the InfiniBand 4X DDR Interconnect the parallel efficiency was close to 98%. From our recent investigations we know that a higher resolution of the computational mesh is required. We think that using about 24 million points in the next phase we be adequate for a well resolved Large Eddy Simulation.

5 Summary

In this paper the focus was on the numerical investigation of a coupled resonating combustion systems. The foregoing investigations started with the test, that the single components (burner plenum, combustion chamber) have resonance characteristics of single Helmholtz-resonators. The coupled system consisting of two Helmholtz-resonators was investigated under isothermal conditions. The dependence of the resonance characteristics of the coupled system from the resonance characteristics of the single units was investigated experimentally with the variation of parameters like the volume, which can realise different eigenfrequencies of the components. The results show a very good agreement between experimental data and theoretical considerations. The ability of the derived physical model to describe coupled combustion systems with Helmholtz-characteristics was proven by the measurements.

The numerical investigations showed the capability of predicting the damping factor accurately by means of Large Eddy Simulation for a single resonator and for coupled systems as well. Using the full multigrid method it was possible to study the effect of the grid refinement. The computational domain of the coupled system is significantly larger than the domain of the single resonator. The results show, however, that the resonant characteristic is predicted sufficiently good already on the second finest grid level of the optimized mesh. The number of the computational cells on the second finest grid level is about the same as on the finest mesh of the single combustion chamber. Therefore it is expected that the prediction of the damping parameter of a more complex combustion system by means of Large Eddy Simulations is realizable within the limits of the computational resources. The results can, however, strongly depend on special boundary conditions as roughness or perforated plates. The effect of roughness is simulated with the Discrete Element Method successfully [20].

Acknowledgements. The present work is a part of the subproject A7 of the Collaborative Research Centre (CRC) 606 – "Unsteady Combustion: Transportphenomena, Chemical Reactions, Technical Systems" at the University of Karlsruhe. The project is supported by the German Research Foundation.

Nomenclature

Roman Symbols	
F	force (N)
$\dot{m}$	mass flow rate (kg/s)
p	pressure (Pa)
t	time (s)
x, y, z	Cartesian coordinates (m)
T	temperature (K)
y^+	dimensionless distance from the wall
Δt	time step (s)
Subscript	
1,2	first and second resonator
bp	burner plenum
cc	combustion chamber
egp	exhaust gas pipe
in	at the inlet of the test section
out	at the outlet of the test section

References

1. Lefebvre, A.H., 1999, *Gas Turbine Combustion*, Taylor & Francis, Philadelphia.
2. Lieuwen, T., Yang, V., (Eds.), 2006, *Combustion Instabilities in Gas Turbine Engines*, American Institute of Aeronautics & Astronautics (AIAA), U.S.
3. Külsheimer, C., Büchner, H., 2002, "Combustion Dynamics of Turbulent Swirling Flames", *Combustion and Flame*, Vol. 131, Nr. 1-2, pp. 70–84.
4. Büchner, H., Lohrmann, M., 2003, "Coherent Flow Structures in Turbulent Swirl Flames as Drivers for Combustion Instabilities", *Proceedings of the International Colloquium on Combustion and Noise Control*, ISBN 1-871315-82-4.
5. Lohrmann, M., Büchner, H., 2003, "Influence of the Air Preheating Temperature on the Flame Dynamics of Kerosene-LPP Swirl Flames", *Proceedings of the European Combustion Meeting 2003.*
6. Richards, G.A., Gemmen, R.S., Yip, M.J., 1997, "A Test Device for Premixed Gas Turbine Combustion Oscillations", *ASME J Eng for Gas Turbines and Power*, Vol. 119, pp. 776–782.
7. Hantschk, C.C., 2004, "Self-exited Pressure Pulsations in Combustion Chambers of Steam Generators", *VGB PowerTech*, Vol. 4, pp. 104–107.
8. Dupere, I.D.J., Dowling, A.P., 2005 "The Use of Helmholtz Resonators in a Practical Combustor", *ASME J Eng for Gas Turbines and Power*, Vol. 127, pp. 268–275.
9. Gysling, D.L., Copeland, G.S., McCormick, D.C., Proscia, W.M., 2000, "Combustion System Damping Augmentation with Helmholtz Resonators", *ASME J Eng Gas Turbines and Power*, Vol. 122, pp. 269–274.
10. Büchner, H., 2001, "Strömungs- und Verbrennungsinstabilitäten in technischen Verbrennungssystemen", Habilitation, Universität Karlsruhe (TH).

11. Arnold, G., Büchner, H., 2003, "Modelling of the Transfer Function of a Helmholtz-Resonator-Type combustion chamber", *Proceedings of the European Combustion Meeting 2003.*
12. Russ, M., Büchner, H., 2007, "Berechnung des Schwingungsverhaltens gekoppelter Helmholtz-Resonatoren in technischen Verbrennungssystemen", in Verbrennung und Feuerung, *VDI-Berichte zum 23. Deutscher Flammentag.*
13. Magagnato, F., Pritz, B., Büchner, H., Gabi, H., 2005, "Prediction of the Resonance Characteristics of Combustion Chambers on the Basis of Large-Eddy Simulation", *Journal of Thermal Science*, Vol. 14, pp. 156–161.
14. Magagnato, F., 1998, "KAPPA-Karlsruhe Parallel Program for Aerodynamics", *TASK Quarterly*, Vol. 2, pp. 215–270.
15. Boris, J.P., Grinstein, F.E., Oran, E.S., Kolbe, R.L., 1992, "New Insights into Large-Eddy Simulation", *Fluid Dyn. Res.*, Vol. 10, pp. 199–228.
16. Smagorinsky, J., 1963, "General Circulation Experiments with the Primitive Equations", *Monthly Weather Review*, Vol. 91, pp. 99–164.
17. Lilly, D.K., 1967, "The Representation of Small-Scale Turbulence in Numerical Simulation Experiments", *Proc. IBM Scientific Computing Symposium on Environmental Sciences*, Yorktown Heights, N.Y., IBM form no. 320-1951, White Plains, New Yok, pp. 195–210.
18. Jarrin, N., Benhamadouche, S., Laurence, D., Prosser, R., 2005, "A Synthetic-Eddy-Method for Generating Inflow Conditions for Large-Eddy Simulations", *Symposium on Hybrid RANS-LES Methods*, Stockholm.
19. Magagnato, F., Pritz, B., Gabi, M., 2006, "Inflow Conditions for Large-Eddy Simulation of Compressible Flow in a Combustion Chamber", *Proceedings of the 5th International Symposium on Turbulence, Heat and Mass Transfer*, Dubrovnik, Croatia.
20. Pritz, B., Magagnato, F., Gabi, M., 2008, "Investigation of the Effect of Surface Roughness on the Pulsating Flow in Combustion Chambers with LES", *Proceedings of the EU-Korea Conference on Science and Technology*, Heidelberg, Germany.

Diffusers with Three-Dimensional Separation as Test Bed for Hybrid LES/RANS Methods

Dominic von Terzi[1], Hayder Schneider[1], and Jochen Fröhlich[2]

[1] Institut für Thermische Strömungsmaschinen, Univeristät Karlsruhe, Kaiserstr. 12, D-76131 Karlsruhe
vonterzi@kit.edu, hayder.schneider@its.uka.de
[2] Institut für Strömungsmechanik, Technische Univeristät Dresden, George-Bähr-Str. 3c, D-01069 Dresden
jochen.froehlich@tu-dresden.de

Summary. The turbulent flow in two asymmetric diffusers with complex three-dimensional separation was computed employing Large-Eddy Simulations (LES) and Reynolds-Averaged Navier–Stokes (RANS) calculations. The computational setup matches existing experiments in the literature. The objective of the present study is to obtain reference data to be used for assessing the performance of newly developed hybrid LES/RANS techniques.

1 Introduction

For the simulation of turbulent flows of engineering interest, Reynolds-Averaged Navier–Stokes (RANS) calculations constitute the workhorse. In this approach all turbulent fluctuations are represented by a statistical model and integration of the governing equations yields the mean velocities of the flow field of interest. For a variety of turbulent flows these models have proven to be adequate and, as a consequence, RANS calculations are fairly successful in predicting the mean flow behavior of "simple" turbulent flows at reasonable computational costs. An alternative method is Large-Eddy Simulation (LES). This semi-deterministic method does not rely on modeling the entirety of the turbulent fluctuations. Instead, the largest flow structures are computed directly and only the impact of the unresolved small scales on the resolved flow field is represented by a model. Since these effects are arguably easier to model, LES often delivers more accurate and more reliable results than a RANS prediction for complex turbulent flows [1]. The downside of the method is a substantially higher computational cost making LES still too expensive for routine application in engineering. Hence, for a given complex turbulent flow, a practitioner of Computational Fluid Dynamics (CFD) may face the predicament that economical RANS calculations produce unsatisfactory results whereas LES remains unaffordable. A possible remedy for this

W.E. Nagel et al. (eds.), *High Performance Computing in Science and Engineering '09*, DOI 10.1007/978-3-642-04665-0_26,

dilemma is to employ a so-called hybrid LES/RANS method that combines both methodologies in a single simulation. The motivation to construct a hybrid is then "to perform LES only where it is needed while using RANS in regions where it is reliable and efficient" [2].

Several distinct approaches to devise LES/RANS hybrids exist and a host of individual methods has been proposed having met with varying degrees of success and popularity. In [2], the fundamental principles of various hybrid methods are formulated and categorized, supplemented with characteristic examples and with assessments. The present study is part of a project with the goal to develop hybrids that follow the paradigm of "segregated LES-RANS modeling" [3]. Segregated modeling is based on decomposing the entire domain into clearly identifiable regions for RANS and LES before starting the simulation. The connection between the distinct zones during the simulation is then established via explicit coupling of the solution at the interfaces. The aim of the approach is to compute all models in their regime of validity: steady RANS for flows with stationary statistics and unsteady LES with high resolution where it is needed. Therefore one can choose the best-suited turbulence modeling for each sub-domain without considering compatibility issues and without fear of inconsistencies in their use. In particular, the occurrence of a so-called gray zone where the resolved motion has to accomplish some quasi-physical transition between LES and RANS character is an often encountered disadvantage with non-segregated approaches. The price to pay is the need for comparatively complex coupling conditions.

During the first part of the project reported on here, different coupling conditions have been derived (see [3, 4] and references therein). These were calibrated with the help of turbulent channel flow simulations and some of them were then applied to the more challenging flow over periodic hills. The latter test case was selected, since it was designed to exhibit the complex physics, like massive separation and formation of large coherent structures, that are known to challenge traditional RANS models. And, indeed, RANS calculations failed to deliver accurate predictions of this flow [5]. The calibrated coupling condition of the new hybrid LES/RANS method, on the other hand, performed well. However, so did other hybrids in the literature [6]. Moreover, for this by design cost-effective setup with artificial periodic streamwise boundary conditions, there can be hardly any cost advantage of a hybrid method over traditional LES.

The next level up in complexity to canvass the performance of newly developed hybrid LES/RANS techniques is then a flow with intricate three-dimensional regions of separated flow. A suitable test case should cater for three needs: (1) reveal, again, the limitations of the RANS approach, (2) allow for straight-forward insertion of the segregated coupling or any other hybrid technique whose capabilities should be analyzed, and (3) still be accessible to LES to allow for direct comparison. In the following, we will report on reference LES and RANS simulations of a potential candidate.

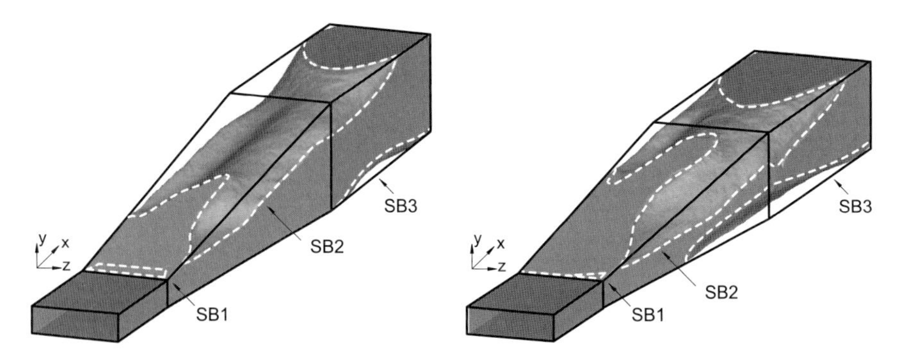

Fig. 1. Mean streamwise velocity iso-contour $U/U_b = 0$ for diffusers D1 (simulation IIb, left) and D2 (simulation IIIb, right); white dashed lines indicate separation bubbles (SB1–3); no presence of separation in regions not shown; figure reproduced from [7]

2 Test Case Selected and Setup

Recently, experiments of incompressible flow in two 3D diffusers were performed at Stanford University [8, 9] with the objective to provide a new test case for complex 3D separated flows. The two diffusers have the same rectangular inlet section with fully-developed turbulent channel flow and the same expansion ratio, but different aspect ratios at the outlet. The observed resulting flow fields are vastly different as is illustrated using LES data in Fig. 1. The challenge for turbulence models is then twofold: (1) Ideally, a simulation should be able to predict within a desired accuracy at reasonable cost the extent and the location of the areas of separated flow and hence the performance of such a pressure recovery device. (2) To be useful for any design process, the computation should capture, at least qualitatively, the effect of changes in the geometry.

2.1 Experiments Simulated

The measurements [8, 9] were performed in a recirculating water channel with Magnetic Resonance Velocimetry (MRV). Figure 2 shows a schematic view of the simulated flow region that consists of an inlet channel (IC), the diffuser (D1 or D2) and an outlet duct (OD). The length of the inlet channel L_{IC} was chosen such that a fully developed turbulent profile was attained prior to the expansion. The Reynolds number based on the inlet channel height and the bulk velocity was 10,000 for both diffusers. The origin of the Cartesian coordinate system was placed at the intersection of the two non-expanding walls and the beginning of the expansion. All values reported were made dimensionless using as reference values the bulk velocity $U_b = 1$ m/s and inlet channel height $H = H_{IC} = 1$ cm, inlet channel width $B = W_{IC} = 3.33$ cm or

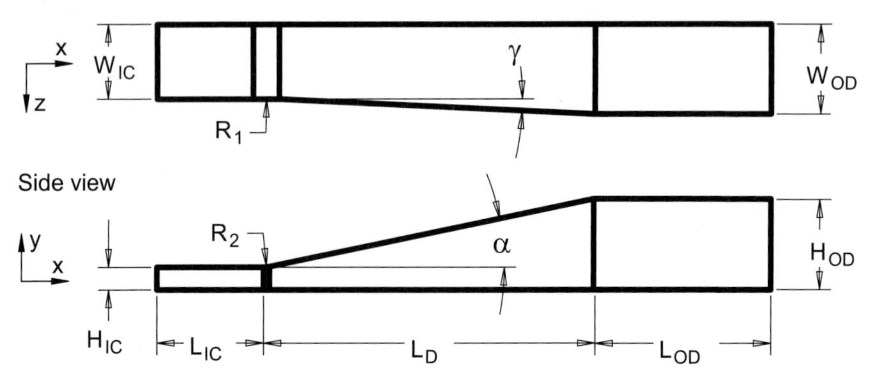

Fig. 2. Diffuser design; flow in positive x-direction (from left to right)

Table 1. Dimensions of diffusers D1 and D2 (lengths are made dimensionless with $\mathrm{H_{IC}}$)

	Dimension	D1	D2
Inlet channel	height ($\mathrm{H_{IC}}$)	1	
	width ($\mathrm{W_{IC}}$)	3.33	
Diffuser	length ($\mathrm{L_D}$)	15	
	top angle of expansion (α)	11.3°	9°
	side angle of expansion (γ)	2.56°	4°
	radius (R1 = R2)	6	2.8
Outlet duct	height ($\mathrm{H_{OD}}$)	4	3.37
	width ($\mathrm{W_{OD}}$)	4	4.51

diffuser length $L = L_D = 15$ cm. Table 1 summarizes the geometric features of the two diffusers studied.

2.2 Computational Setup

Various parameters of the computational setup are compiled in Tab. 2. Simulations were carried out on five different grids (G1–G5) having between 1.6 and 7.4 million grid cells and covering different solution domain lengths. All grids feature equidistant grid spacing with 16 cells per channel height H in the streamwise direction. Four grids (G1–G4) consist of 60, and one grid (G5) of 120 cells in both the vertical and lateral directions. While one grid (G1) is stretched in the two wall-normal directions in order to allow for wall-resolving simulations, the other grids (G2–G5) are equidistant for use with wall-functions.

LES was performed for both diffusers (D1 and D2) using grids G2–G5 and the standard Smagorinsky model with $C_s = 0.065$ and van Driest damping. Unsteady turbulent inflow data were generated in a synchronized computa-

Table 2. Domain sizes and grid parameters

Grid	G1	G2	G3	G4	G5
Domain length $x/H \in$ (with inflow generator)	$[-5; 23]$		$[-5; 28]$	$[-8; 28]$	$[-5; 28]$
$N_x \times N_y \times N_z$	$448 \times 60 \times 60$		$528 \times 60 \times 60$	$576 \times 60 \times 60$	$528 \times 120 \times 120$
Total number of cells	$\approx 1.6 \times 10^6$		$\approx 1.9 \times 10^6$	$\approx 2.1 \times 10^6$	$\approx 7.4 \times 10^6$
Inflow location $x/H =$	-2			-5	-2
Grid spacing	stretched	equidistant			

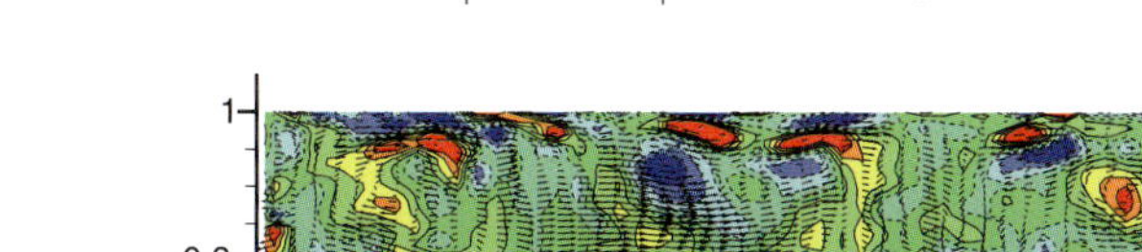

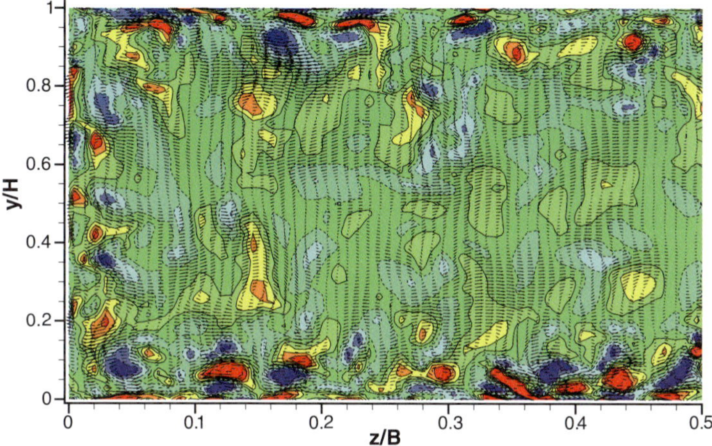

Fig. 3. Fully developed turbulent flow in the inlet duct: Velocity vectors and color contours of streamwise vorticity ω_x in a cross-section of the inflow data generator (LES, case IId); only half of the plane shown for clarity; $-5 \leq \omega_x \leq 5$; $x = -3.5$

tion running in parallel. To this end, in front of the domain periodicity was enforced within a section of length $l/H = 3$ and a controller was used to enforce the nominal experimental mass flux. Within this inflow data generator, a fully-developed turbulent flow is obtained that serves to provide the main simulation with time-dependent realistic flow structures (see Fig. 3). The effective inlet planes for the various diffuser simulations are then at $x/H = -2$ for grids G1–G3 and G5, and at $x/H = -5$ for grid G4. At the outlet a convective boundary condition (BC) is enforced. In grid G4 a buffer zone in which the viscosity is increased by a factor of 100 has additionally been included within $x/H \in [27; 28]$. At walls, an adaptive wall-function developed by Hinterberger [10] is used. The wall-function is based on the velocity profile obtained with Direct Numerical Simulation (DNS) of fully-developed turbulent channel flow. It hence covers the entire boundary layer and switches itself off if the wall is well resolved. Adaptive time-stepping ensured a CFL limit of less than 0.65 (with $\Delta t \approx 0.006 - 0.01$). In total 600,000 time steps were computed. Averaging started after roughly 150 H/U, resulting in an averaging time of more than 5000 H/U and 2500 H/U, or 200 and 100 flow-through times for the simulations on grids G2–G4 and G5, respectively.

RANS simulations were only carried out for diffuser D1 and the results shown were obtained with the standard two-equation k-ω model of Wilcox as turbulence closure. The simulations were wall-resolving on the stretched grid (G1). Two inlet BC were tested: a uniform inlet velocity profile as well as a fully developed channel-flow profile. The no-slip condition was enforced at the wall and a homogeneous Neumann BC at the outlet. Using the implicit time–scheme, the time step was set to 0.01 and the simulations were converged to steady-state. 20,000 time-steps were computed for each simulation.

Table 3. Overview of simulations discussed and computational costs

Simulation	Diffuser	Model	Grid	Inflow BC	Outflow BC	CPUhrs
Ia	D1	RANS	G1	uniform	Neumann	1.5×10^3
Ib	D1	RANS	G1	turbulent	Neumann	2.2×10^3
IIa	D1	LES	G2	turbulent	convective	15×10^3
IIb	D1	LES	G3	turbulent	convective	18×10^3
IIc	D1	LES	G4	turbulent	buffer zone	20×10^3
IId	D1	LES	G5	turbulent	convective	120×10^3
IIIa	D2	LES	G2	turbulent	convective	15×10^3
IIIb	D2	LES	G3	turbulent	convective	18×10^3

The different geometries, turbulence models and grids used for the various simulations are given in Tab. 3. Simulation Ia used a uniform velocity profile as the inlet condition while all other simulations used a fully developed channel profile. Simulations with other RANS models using LESOCC2 and commercial flow solvers were conducted as well, but are not reported, since they did not yield any better results than those presented here.

2.3 Flow Solver Employed

The simulations presented here were performed with the Finite Volume CFD–code LESOCC2 developed at the University of Karlsruhe and described in [10]. This FORTRAN 95 program solves the incompressible, three-dimensional, time-dependent, filtered and/or Reynolds-Averaged Navier-Stokes equations on body-fitted, collocated, curvilinear, block-structured grids. The viscous fluxes are always discretized with second-order accurate central differences whereas the convective fluxes are approximated either with the same method for LES or a monotonic second-order upwind scheme for RANS calculations. Time advancement is accomplished by an explicit, three-step low-storage Runge–Kutta method for LES. For RANS calculations, a second-order implicit multi-step method was used. Conservation of mass is achieved by the SIMPLE algorithm with the pressure–correction equation being solved by the strongly implicit procedure (SIP) of Stone. The momentum interpolation

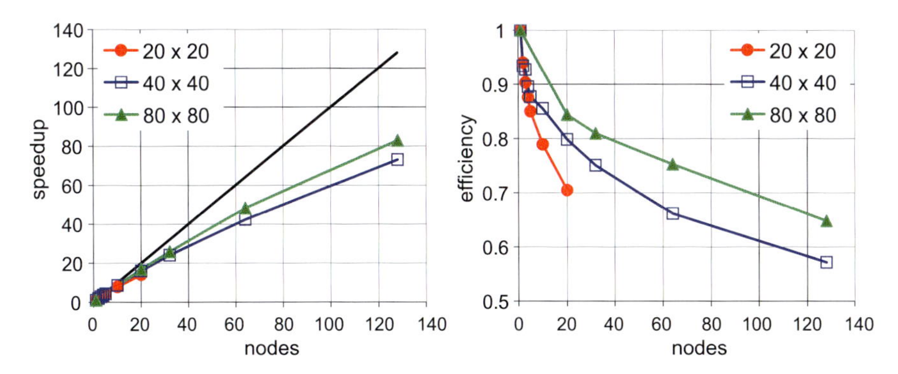

Fig. 4. Speedup (left) and efficiency (right) over nodes for the parallelization of the flow solver used on the HP–XC-4000 for a model problem (one node has four processor cores)

method of Rhie and Chow is employed to prevent pressure–velocity decoupling and associated oscillations.

The code has been used in numerous studies and substantial experience has been gained with respect to its numerical properties and its performance on various hardware platforms including the NEC SX-8, SGI-Altix, and several Linux clusters. Parallelization is accomplished via domain decomposition and MPI for the data transfer. To achieve better performance on a given hardware, different implementations of the computationally most intensive algorithms exist. The user can decide between highly vectorized versions and unvectorized versions of subroutines or can choose to trade off between memory requirements and number of operations in order to fit a run on a specific machine or possibly take advantage of cache effects. Scaling of LESOCC2 on the HP–XC-4000 (the Linux cluster used for the present investigation) was tested for up to 512 processor cores and satisfactory results were obtained (see Fig. 4). For the test shown in Fig. 4, the problem size per processor was kept constant. As expected, results improved for larger numbers of grid points per processor, i.e. smaller relative communication overhead. For the study presented here, the problem sizes per processor was several times larger than the largest case shown in Fig. 4. The data presented were obtained using between 112 and 144 processor cores, i.e. 28 and 36 nodes, respectively. The computational costs of the simulations are compiled in Tab. 3.

3 Selected Results

3.1 Simulating Diffuser D1

Diffusers are pressure recovery devices and as such the wall-pressure coefficient C_p is a quantity that directly reflects the performance of a diffuser. For incompressible flow, C_p is defined as $C_p = 2(p - p_{ref})/(\rho U_{ref}^2)$,

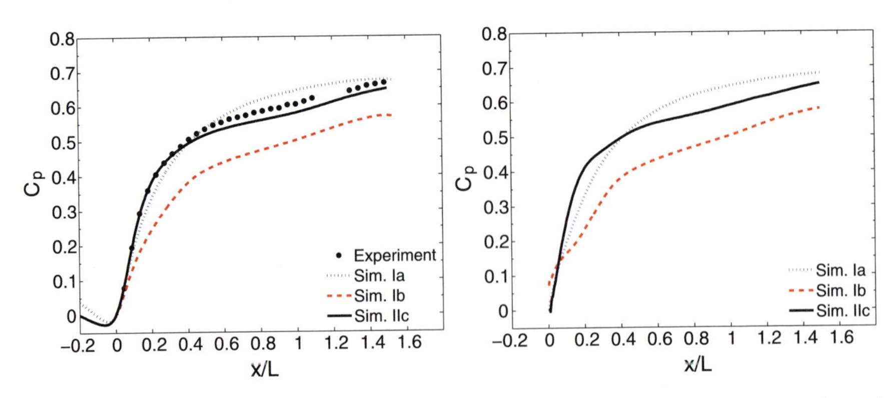

Fig. 5. Mean wall-pressure coefficient C_p at $z/B = 0.5$ (left) and $z/B = 1$ (right) along the lower wall of diffuser D1

where p is the mean pressure, p_{ref} the pressure at a reference point, here $(x/H, y/H, z/B) = (0, 0, 0.5)$, and $U_{ref} = U_b$. Figure 5 shows C_p for diffuser D1 as measured in the experiments (data available only at $z/B = 0.5$ [9]) and predicted by RANS calculations and LES. RANS simulation Ia seems to yield qualitatively acceptable results, whereas simulation Ib delivers a too low pressure recovery. LES, on the other hand, yields excellent agreement with the experimental data.

The good qualitative performance of simulation Ia regarding C_p, however, is misleading, since it is the result of a fortuitous prediction of the velocity magnitude and gradient close to the lower wall along the center-plane that is likely caused by error balancing. This is corroborated by simulation Ib with a fully turbulent inflow profile for the mean values, k and ω. This calculation yields a more realistic prediction of the turbulent kinetic energy distribution and the mean flow profiles prior to separation (see [7]), but fares worse for C_p. Farther downstream both RANS calculations fail in a similar manner as is demonstrated in Fig. 6. In these plots, the mean streamwise velocity contours are shown at a critical location in the diffuser. The experimental data show that the recirculation bubble is horizontally aligned at the upper wall and all LES (simulation IIa shown) yield reasonable agreement for both the extent of the recirculation bubble and the location of maximum forward velocity. On the other hand, the RANS simulations (Ia and Ib) shift the region of separated flow to the right wall leading to a completely different flow field. This demonstrates that RANS models have great difficulty in capturing the shape and extent of the separated flow region and, hence, are not likely to reliably predict the performance of such diffusers. Results obtained with other RANS models and other flow solvers are similar (or even worse) than those presented here [11].

The instantaneous and mean flow results of LES IId in a x-y plane are shown in Fig. 7 to give an impression of the flow field. In [7] additional stud-

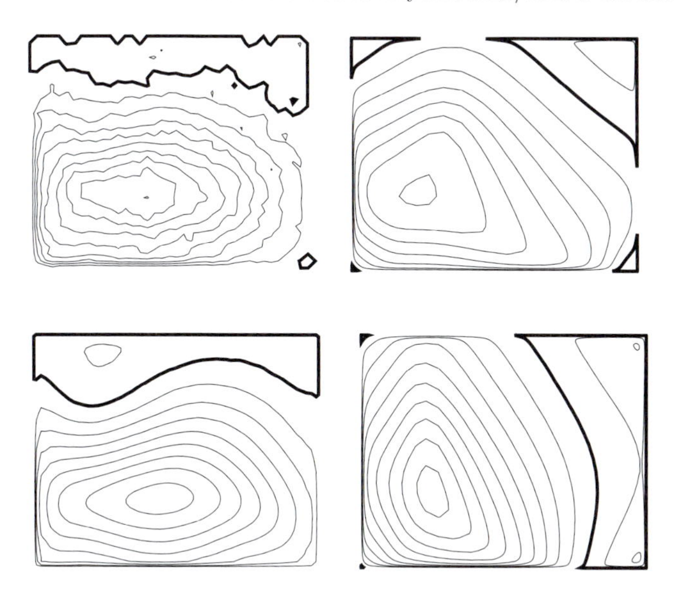

Fig. 6. Mean streamwise velocity U/U_b contours of experiments (top left), LES (bottom left, case IIa) and RANS simulations Ia and Ib (top and bottom right) in the cross-section at $x/H = 10$ for diffuser D1; same velocity contours shown for all plots with thicker line indicating the zero-velocity contour

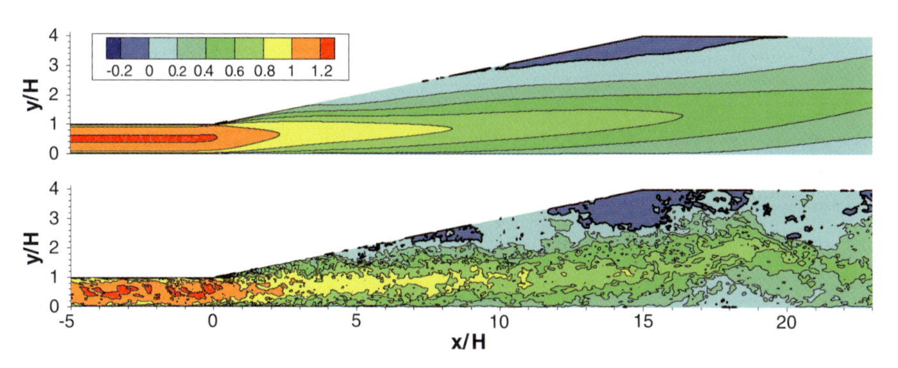

Fig. 7. Mean (top) and instantaneous (bottom) streamwise velocity contours obtained by LES (case IId) in the cross-section at $z/B = 0.5$ for diffuser D1; same velocity contours shown for all plots with thicker line indicating the zero-velocity contour

ies are reported on the impact of turbulence models, boundary conditions, domain length and grid size on the results of diffuser D1. Also an assessment of computational costs for different modeling techniques is given. A compilation of data from simulations of several research groups (including results from the present authors) with different flow solvers and turbulence modeling approaches can be found in [11] with a comparison of LES and hybrid RANS/LES results being presented in [12].

3.2 Simulating Diffuser D2

To our knowledge, the first and so far only computations of diffuser D2 published have been computed by the first two authors [11, 7]. In this section, a selection of additional, so far unpublished results of these LES (cases IIIa and IIIb) will be shown. However, diffuser D2 will be subject of a forthcoming workshop [13] and by then RANS, LES and hybrid methods will have been applied to this setup by the present authors and several other international research groups.

In Fig. 8, profiles of selected mean and r.m.s. velocities are shown for diffuser D2 at various streamwise locations for the centerplane $z/B = 0.5$. Simulations IIIa and IIIb are in good agreement with the experimental data (only mean values are available) close to the upper wall and overpredict the velocity magnitude at the lower diffuser wall for increasing streamwise location. Both simulations IIIa and IIIb yield the same results and confirm the negligible influence of the outlet BC for diffuser D2.

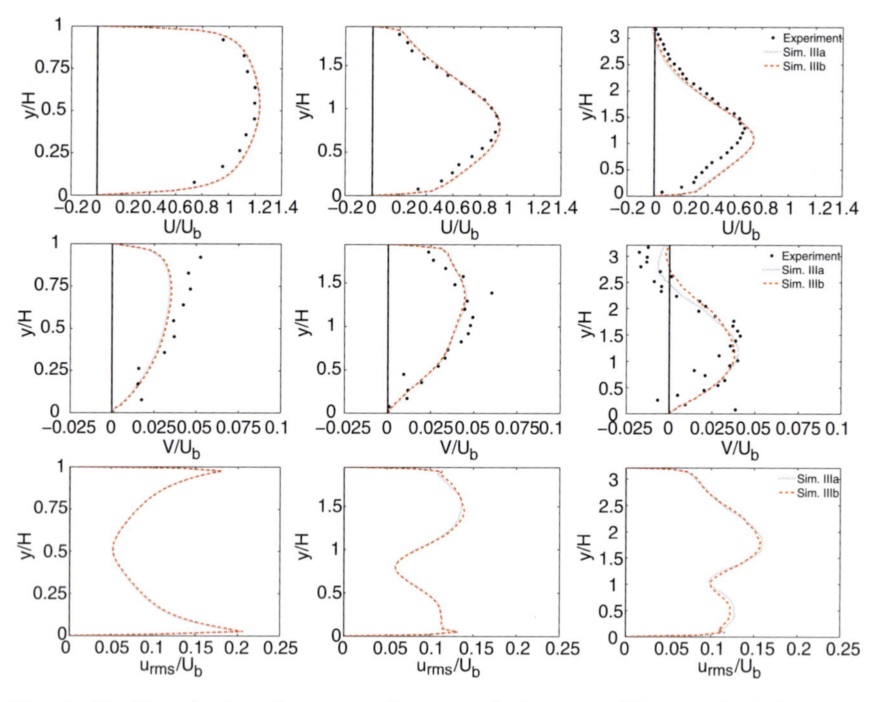

Fig. 8. Profiles of selected mean and r.m.s. velocities at $z/B = 0.5$ for diffuser D2; from top to bottom: U/U_b, V/U_b and u_{rms}/U_b; from left to right: $x/H = 0$, 6 and 14

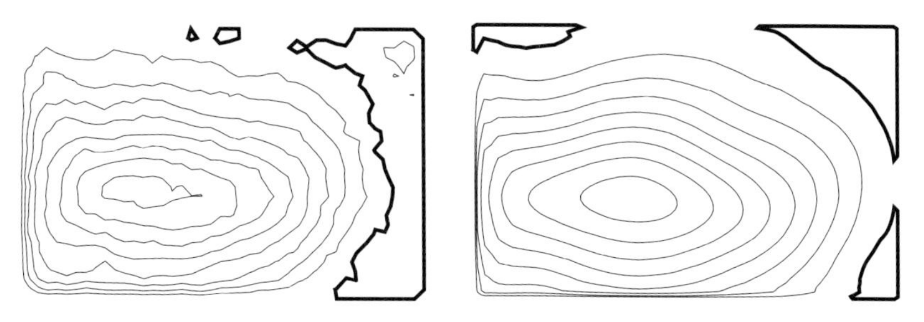

Fig. 9. Mean streamwise velocity U/U_b contours of experiments (left) and LES (case IIIa, right) in the cross-section at $x/H = 10$ for diffuser D2; same velocity contours shown for all plots with thicker line indicating the zero-velocity contour

The contour plots of U/U_b shown in Fig. 9 reveal the difference in the characteristics of the flow field of diffuser D2 compared to diffuser D1 for the same location. The experimental data show that, due to the change in the outlet geometry, the separation bubble moves from a horizontal alignment at the top wall (see Fig. 6) to a predominantly vertical orientation on the right side. The LES data exhibit the same trend and show good agreement with the region of maximum forward flow. The good agreement of the LES with the experimental data becomes even clearer when scrutinizing the velocity profiles at various spanwise locations (see [7]).

Note that the RANS simulations for diffuser D1 shown above and in [7] as well as other RANS calculations of this diffuser [11] predict results resembling those of diffuser D2 while attempting to compute diffuser D1. In contradistinction, the LES succeeds in predicting the flow features in both diffusers and, consequently, it can be used to study the impact of geometric changes.

3.3 Capturing the Geometric Sensitivity

In the above, it has been established that RANS calculations have severe problems in predicting accurately and reliably the flow field of asymmetric diffusers with three-dimensional separation. In contrast, LES can deliver such results and, consequently, can be employed to study the impact of variations in the geometry on the dominant features of the flow field. To this end, we return to Figure 1, where iso-contours of the mean streamwise velocity $U/U_b = 0$ are shown for the two diffusers D1 (left) and D2 (right). The LES data is particularly well suited for 3D visualization of the separated zones highlighted by the white dashed lines. Note the complicated three-dimensionality of these lines. The variation in the top-to-side angle of expansion ratio between the two diffusers determines the differences in the flow field in which three different separation bubbles can be identified. The first one (SB1) is small in size and extent and is located at the beginning of the diffusers. This is an artifact of the present numerical set-up, where rounded corners in the experiments were replaced by sharp corners in order to facilitate easier grid generation.

Apparently, this simplification had no effect on the downstream results. The second separation bubble (SB2) close to the upper wall is rather large and ranges from the middle of the diffuser until the beginning of the outlet duct. The third one (SB3) is found at the edge of the lower and expanding side wall in the region of the outlet duct. For both diffusers, SB2 is geometrically triggered at the corner of the two expanding walls and develops as a function of the top-to-side angle of expansion ratio. In diffuser D1 a higher ratio is realized resulting in a larger separation bubble close to the upper wall which is nearly horizontally aligned. In contrast, for diffuser D2, SB2 is less dominant close to the upper wall but is shifted more to the expanding side wall. SB3 shows to have a much larger upstream effect in diffuser D2 and extends until the middle of the diffuser where it nearly merges with SB2.

A quantitative assessment of the ability of LES to capture accurately and reliably the geometric sensitivity of the flow in asymmetric diffusers can be provided by calculating the fraction of area with reverse flow of the total area in cross-sections along the diffuser. The outcome is plotted in Fig. 10 for both experiments and LES. The experimental uncertainty is estimated by the dot-dashed lines and was obtained by calculating the same fraction of cross-sectional area using $U/U_b = \pm 0.05$ instead of $U/U_b = 0$ as a threshold. The estimate is based on information provided in [8] and by inspection of the experimental data itself and, in reality, is likely to be even higher. The characteristics of diffuser D1 are matched by all LES (simulations IIb and IId are shown) and deviations reside within estimated experimental uncertainties beyond the location of the maximum in this plot. Until $x/L = 0.4$ the simulations are in agreement with the experimental data. From that point on simulations IIb predicts marginally higher separation until $x/L = 1.6$, where the effect of the outflow BC becomes noticeable. On the other hand, simulation IId gives a lower fraction of area with separated flow until $x/L = 1.1$. From this location on the predictions are higher and yield fully attached flow

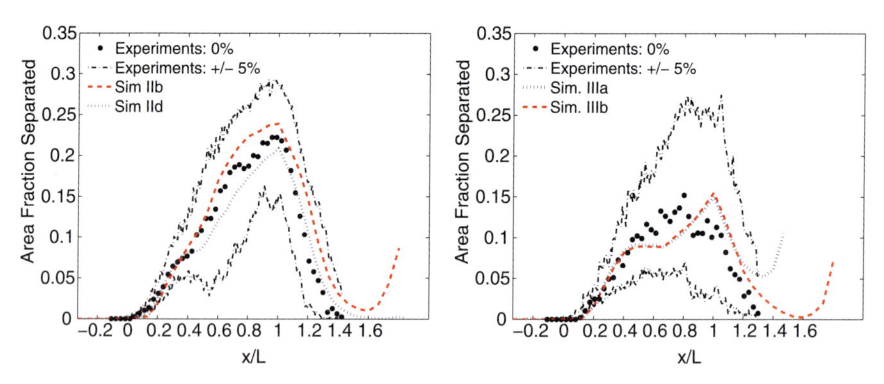

Fig. 10. Fraction of cross-sectional area separated in experiments and simulations IIa, IIb and IIc for diffuser D1 (left) and D2 (right); RANS data of diffuser D1 deviates substantially (not shown); figure reproduced from [7]

at $x/L = 1.6$. The results of simulations IIa and IIc (not shown in Fig. 10) are in close agreement to those obtained by simulation IIb. The amount of area fraction separated along the diffuser computed by simulation IIa is almost identical to the one obtained by simulation IIb until $x/L = 1.3$, where the effect of the outflow BC becomes evident due to the shortened domain length. The same identity is true for simulation IIc, except that from $x/L = 1.3$ the amount of separated flow is further diminished until fully attached flow is attained at $x/L = 1.6$. This improved behavior may be due to employing a buffer zone for this simulation. Both the amount and streamwise distribution of reverse flow per cross-sectional area in diffuser D2 (Fig. 10 right) are well matched by the LES IIIa and IIIb and deviations reside again within measurement uncertainty. Again until $x/L = 0.4$ the computational results are in agreement with the experimental data. Between $x/L = 0.4$ and 0.8, the LES results underpredict and then overpredict the backward flow until $x/L = 1.3$. Thereafter, again the effect of the outflow BC is visible for simulation IIIa.

4 Conclusions

LES and RANS calculations were performed for incompressible, fully developed turbulent duct flows expanding into diffusers with two deflected walls. The asymmetric expansions yields complicated three-dimensional flow separation within the diffusers with strong sensitivity to the geometric variations. Findings in the literature were confirmed that these flows pose a severe challenge to the RANS methodology and accurate and reliable results can not be expected from such simulations. On the other hand, it could be demonstrated that LES is appropriate for this situation. The simulations reported here provides an extensive database for validation of enhanced RANS models and alternative turbulence modeling techniques like the hybrid LES/RANS methods envisioned for the current project.

Acknowledgements. The work reported here was carried out within the "Research Group Turbo-DNS" at the Institut für Thermische Strömungsmaschinen. Its financial support by means of the German Excellence Initiative and Rolls–Royce Deutschland is gratefully acknowledged. The hybrid methods to be tested have been developed within a project funded by the German Science Foundation under contract number Fr 1593 / 1-1,2. The authors are grateful for the provision of computer time by the Steinbuch Centre for Computing.

References

1. J. Fröhlich. *Large Eddy Simulation turbulenter Strömungen.* Teubner Verlag, 2006.
2. J. Fröhlich and D. A. von Terzi. Hybrid LES/RANS methods for the simulation of turbulent flows. *Prog. Aerospace Sci.*, 44:349–377, 2008.

3. D. A. von Terzi, I. Mary, and J. Fröhlich. Segregated LES/RANS coupling conditions for the simulation of complex turbulent flows. In C. Brun, D. Juve, M. Manhart, and C.-D. Munz, editors, *Numerical Simulation of Turbulent Flows and Noise Generation*, volume 104 of *Notes on Numerical Fluid Mechanics and Multidisciplinary Design*, pages 231–252. Springer, 2009. ISBN 978-3-540-89955-6.
4. D. A. von Terzi, J. Fröhlich, and W. Rodi. Hybrid techniques for Large-Eddy Simulations of complex turbulent flows. In W. E. Nagel, D. B. Kröner, and M. M. Resch, editors, *High Performance Computing in Science and Engineering '08*, pages 317–332. Springer, 2009. ISBN 978-3-540-88301-2.
5. S. Jakirlić, R. Jester-Zürker, and C. Tropea, editors. *9th ERCOFTAC / IAHR / COST Workshop on Refined Turbulence Modelling*. Darmstadt University of Technology, 2001.
6. S. Šarić, S. Jakirlić, M. Breuer, B. Jaffrézic, G. Deng, O. Chikhaoui, J. Fröhlich, D. A. von Terzi, M. Manhart, and N. Peller. Collaborative assessment of eddy-resolving strategies for the flow over periodic hills. In preparation.
7. H. Schneider, D. A. von Terzi, H.-J. Bauer, and W. Rodi. Reliable and accurate prediction of three-dimensional separation in asymmetric diffusers using Large-Eddy Simulation. *ASME paper GT2009–60110*, 2009.
8. E. M. Cherry, C. J. Elkins, and J. K. Eaton. Geometric sensitivity of three-dimensional separated flows. *Int. J. Heat and Fluid Flow*, 29(3):803–811, 2008.
9. E. M. Cherry, C. J. Elkins, and J. K. Eaton. Pressure measurements in a three-dimensional separated diffuser. *Int. J. Heat and Fluid Flow*, 30(1):1–2, 2009.
10. C. Hinterberger. *Dreidimensionale und tiefengemittelte Large–Eddy–Simulation von Flachwasserströmungen.* PhD thesis, Institute for Hydromechanics, University of Karlsruhe, 2004.
11. G. Brenn, S. Jakirlić, and H. Steiner, editors. *Proceedings of the 13th SIG15 ERCOFTAC / IAHR Workshop on Refined Turbulence Modelling, 25–26 September 2008, Graz University of Technology, Graz, Austria*, 2008.
12. S. Jakirlić, G. Kadavelil, S. Šarić, M. Breuer, M. Kornhaas, D. C. Sternel, M. Schäfer, H. Schneider, D. von Terzi, and W. Rodi. Comparative LES and hybrid RANS/LES study for predicting turbulent flow separation in a 3d diffuser. In *EUROMECH Colloquium 504 Large Eddy Simulation for Aerodynamics and Aeroacoustics, March 23–25, Munich*, 2009.
13. *Proceedings of the 14th SIG15 ERCOFTAC / IAHR Workshop on Refined Turbulence Modelling, 18 September 2009, Sapienza Universita di Roma, Rome, Italy*, 2009.

Final Report on Project GGT0607: Lattice Boltzmann Direct Numerical Simulations of Grid-Generated Turbulence

K. Beronov[1] and N. Özyilmaz[2]

[1] Leibniz–Rechenzentrum, Boltzmannstr. 1, 85748 Garching
beronov@lrz.de
[2] LSTM, Universität Erlangen–Nürnberg, Cauerstr. 4, 91058 Erlangen

Summary. The largest known direct numerical simulations of grid-generated turbulence were performed, on one hand resolving the grid placed in an incompressible fluid flow while on the other hand following reliably the evolution of flow structure both in the direction of the mean flow and across it well into a region of self-similar behaviour. The excellent scaling of the employed lattice Boltzmann algorithm allowed to perform a series of DNS runs and quantify several parametric effects. Several major qualitaitve results were obtained, as well, having immediate impact on turbulence modeling: A new algebraic law (of Kolmogorv type) was discovered for the decay of turbulent kinetic energy in low-intensity grid-generated turbulence. Furthermore, the assumption of the existence of a single decay rate for the anisotropy of weak "homogeneous" turbulence was shown to be qualitatively inappropriate, while the assumption of an "inertial range" for the spatial development of such turbulence – on which the former assumption should be based – appears to be valid.

1 Problem Formulation

Grid-generated turbulence is a classical experimental flow [1] producing data for validation and calibration of homogeneous turbulence theories. Direct numerical simulations (DNS) since the 1980s deliver data for same purpose, but have not been used to investigate GGT directly. The project presented here and the simultaneous but independent work from [2] are the first DNS of this kind. They provide qualitatively new data for practically relevant modeling of inhomogeneous turbulence. They are orders of magnitude more demanding than homogeneous turbulence DNS. The present work treats a larger domain and several Reynolds numbers, thus providing more reliable and in-depth insights than [2], while both studies use essentially the same lattice Boltzmann (LBM) scheme for the DNS. This numerical method and the performance of the NEC SX-8 at HLRS were prerequisits for the feasibility of the simulations.

W.E. Nagel et al. (eds.), *High Performance Computing in Science and Engineering '09*, DOI 10.1007/978-3-642-04665-0_27,

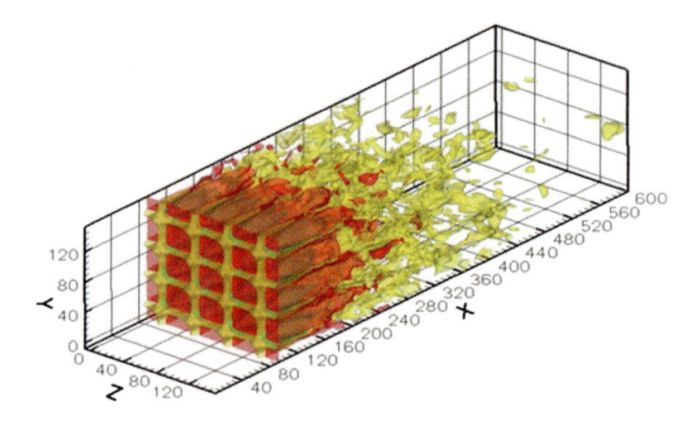

Fig. 1. Snapshot of simulated turbulence visualized by isosurfaces of the velocity component aligned with the channel axis

The flow was driven by a constant pressure drop in the main flow direction. The shape of the turbulence generating grid and the typical flow field in its vicinity can be inferred from figure 1, where the instantaneous streamwise velocity component is visualized.

The main objective was to capture in detail the generation as well as the subsequent, supposedly self-similar decay of turbulence. This required a length of at least $40M$ [1] for the resolved part of the flow domain in streamwise direction, where M is the grid stride. Based on this length and the bulk flow velocity U_m, a "mean Reynolds" number Re_m can be defined. Most simulations were carried out at $Re_m \approx 1400$, a few more either at $Re_m \approx 2300$ or at rather lower values.

A technically interesting aspect to be clarified was the influence of grid "porosity" on turbulence characteristics, incl. Re_m for given driving force, decay rate and inhomogeneity. The porosity ϵ is undestood as the ratio between the open and the full cross-sectional area in the direction of mean flow, at the (narrowest section of the) grid. To that end, a number of cases, namely $\epsilon = 0.53$, 0.64, 0.72, and 0.82, were run at the same Re_m. Keeping the grid stride fixed, at $M = 40$ in the units of the regular numerical mesh used (see below), the thickness of the rods was changed to obtain the desired porosity.

The effect of (weak) asymmetry in the grid geometry was investigated for the first time in the literature on grid turbulence, not only the numerically based one. To that end, a square formed of four adjacent grid openings was made to have individually different sizes from the size given by the regular grid stride. The best result hoped for in this context would be an independence of all calculated statistics that involve averaging over the full cross-seciton of the flow, on the size of individual grid openings at fixed overall porosity.

In order to find out the effect of the size of the computational domain, an extra simulation was conducted with the $\epsilon = 72$ grid and the same driving force, but with a numerical domain approximately 6 times longer in the streamwise direction. Similarly, the effect of lateral domain size was evaluated by comparing results for cross-sections of $4M$ and $10M$ at the same forcing and streamwise domain length.

2 Numerical Method

An explicit solver with constant time step was chosen for the simulation, taking full advantage of its speed per time step. This made sense in terms of performance, since the Kolmogorov scale was comparable to the numerical step size over essentiall the whole computational domain. The cross-sectional size of that domain was 160 to 400 grid points, its streamwise length 2000 to 12000 points, corresponding to a total of 50 to 400 million "fluid points" in the LBM mesh.

2.1 Flow Solver

For all DNS in this project, one and the same LBM-based solver was used. The origin of the lattice Boltzmann equation (LBE) is based on the kinetic theory, more specifically Boltzmann equation and it describes basically the evolution of the velocity distribution function on a regular lattice in such a way that the macroscopic fluid dynamics behavior, i.e. the (weakly compressible)Navier-Stokes equations (with Newtonian viscosity) are recovered via the Chapman-Enskog expansion. In the absence of external forces, the LBM equation reads

$$f_\alpha(x + \xi_\alpha \delta_t, t + \delta_t) = f_\alpha(x, t) + \Omega_\alpha(f)\,, \tag{1}$$

where f_α denotes the single-particle distributions, ξ the microscopic velocity, α is the direction of these discrete velocities, and Ω is the collision integral. The macroscopic quantities of interest, fluid density ρ and velocity $\mathbf{u}$, are $\rho = \sum_\alpha f_\alpha$ and $\rho\mathbf{u} = \sum_\alpha \xi_\alpha f_\alpha$. The collision operator $\Omega(f)$ can be linearized, using either one relaxation time sclae (BGK or single-relaxation-time, SRT model) or more such scales (multiple-relaxation-time, MRT models). Here, the simplest collision model, the BGK method was employed. This has been found in earlier work [3] to provide very good approximation to incompressible turbulence as far as velocity and pressure statistics are concerned and at least a moderate resolution (including a numerical Mach number below 0.1) is uniformly guaranteed. Experiments with some MRT schemes have shown no decisive advantage with respect to effective performance and no noticeable difference in the turbulence statistics obtained, when compared to the BGK simulations.

The standard LBM schemes require a regular grid with equal spatial steps in all three coordinate directions and equal time steps. Here, the D3Q19 LBM scheme was used - a three-dimensional stencil with 18 out of 26 lattice points lying on the sides of a cube centerd at a given grid point and extending one grid step in each positive and negative coordinate direction. In this scheme, the pressure is calculated as $p = p_{ref} + c_s^2 \rho$ where, in lattice units of time and length, $c_s^2 = 1/3$ and the equilibrium has $\rho = 1$.

Schemes with 15 or even 13 lattice points exist, but our experience has shown that the transformations to and from momentum space that they require cause an overhead that does not really pay off via the reduced number of variables per point or via a significant gain in stability. The explicit, straightforward version of D3Q15 is less stable than that of D3Q19 used here and in [2]. We employed the LBM-solver BEST, developed at LSTM-Erlangen. It is MPI parallelized and has versions optimized for various cache and vector-based platforms, including the NEC SX-8 at HLRS and the SGI ALTIX at HLRB-II. This solver has been verified for many different applications, including among others, porous media and fixed-bed reactors as well as developed plane channel turbulence [3]. It was shown to be more efficient and at the same time more accurate, for transient and DNS computations compared to standard (finite-volume-based) CFD tools. It is also much easier to extend, maintain, parallelize, and apply, especially for relatively simple geometries like the ones considered here.

An extensive performance test for that particular code [4] has directly preceded the production runs in this project. Its results showed, in comparison with IBM Power5 and SGI Altix architectures, a markedly better and problem-size-independent performance on the NEC SX-8 at HLRS. In particular, more than 60% and typically almost 75% of the theoretical peak performance are achieved at all relevant problem sizes (per CPU and in total). This corresponds to some 10 GFlop/s on each processor.

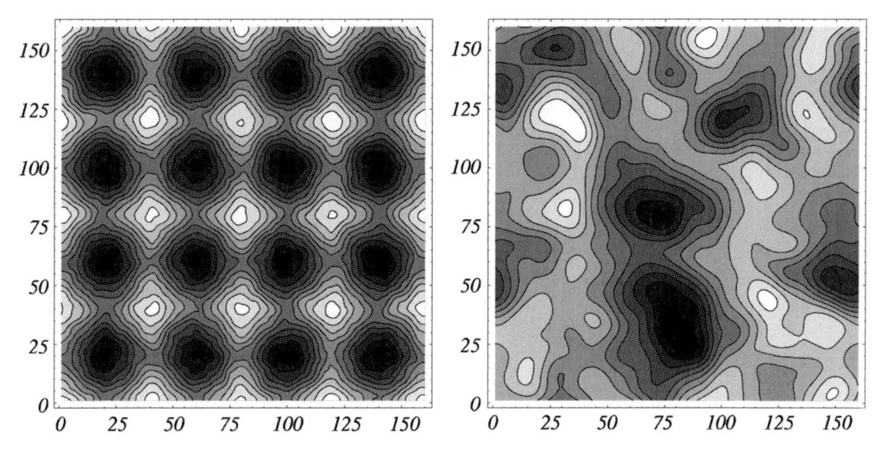

Fig. 2. Snapshots of the streamwise velocity component at two distances downstream of the grid: $x/M = 2.5$ in the close region, and $x/M = 42$ (which is some 1700 mesh points away) in the self-similar region. The latter location is beyond the first measuring station in [1] and thus definitely representative of the "nearly isotropic turbulence" region of the experimental literature

2.2 Boundary Conditions

The LBM-specific implentation of non-slip boundary conditions at the rods of the turbulence generating grid was by means of the standard bounce-back scheme. This assured robust simulation at relatively high local Reynolds numbers and - because the rods are alligned with the coordinate directions of the numerical mesh - a sufficient accuracy.

In the case of plane channel flows, as the statistics obtained from experiments are independent of the streamwise and spanwise directions in the fully-developed region of the channel and the flow is bounded by two parallel walls. Therefore, applying periodic boundary conditions along these flow directions and a non-slip velocity condition at the walls correspond exactly to the underlying physical problem. In unbounded flows, including free turbulent flows, however, the treatment of the unavoidable numerical boundary conditions is less straightforward. In order to simulate the whole flow field, the computational domain must be oversized in comparison with the flow structures and domeins of interest. Some kind of artificial boundary conditions is needed, which permit the use of a finite computational domain as an acceptable approximation. Grid-generated turbulence belongs to this kind of turbulent flows. Spanwise structures are comparable to the mesh stride close to the (turbulence generating) grid but grow algebraically in size with increasing distance downstream. Streamwise structures are even longer. We are interested here not only and not primarily with the region close to the grid, but rather with the range of self-similar decay of turbulent energy commencing shortly further downstream.

There are only a few numerical works on free turbulent flows in the literature. In their majority, an inflow velocity profile was imposed at the inlet of the computational domain, while at the outlet, either a zero-gradient or a convective boundary condition was implemented. One may apply a zero-gradient condition at the outflow boundary instead of a convective one, expecting more realistic flow there. This was done by Djenidi [2] who also simulated grid-generated turbulence. He compared convective and zero-gradient conditions. However, he did not observe any significant change in the results.

Another way to handle far-field boundary conditions when simulating unbounded turbulent flows, not as common as the above mentioned types of inflow/outflow conditions yet, is becoming more popular now, especially in the spectral method community, where boundary conditions are traditionally an uneasy issue. The idea of this method is the following. Since it is practically never possible to avoid non-physical results close to artificial boundaries, some neighborhood of the computational domain boundaries is consciously sacrificed with regard to physical results. So far, there is no qualitative difference from what should be properly done when the other types of far-field conditions are used. But this alternative approach then calls to use periodic boundary conditions for relevant couples of domain boundaries. Spectral codes treat such conditions in a natural way. But they are physically correct only along so called homogeneous spatial directions. Now the numerical advantages are used even along the inhomogeneous directions, at the price of neglecting a part of the computed results. This approach is known as the "fringe region" method. By introducing a region adjacent to the outflow domain, in which the turbulent quantities are damped, the governing equations are modified in such a way over that region, that the full numerical solution is forced to become spatially periodic. This idea was first suggested by Spalart in 1988, in the context of turbulent boundary layer DNS using spectral methods. The approach worked well enough in a number of subsequent studies.

According to that method, the Navier-Stokes equations are modified by adding an artificial forcing term to the physical flow driving. It acts to suppress flow disturbances with respect to a prescribed in/outflow condition $\mathbf{U}$. It is designed to vanish outside the fringe region, allowing a natural flow almost everywhere outside. We have verified, that the mainly compressible disturbances propagate only negligibly upstream and are damped out shortly before reaching the (next periodic copy of the) turbulence generating grid downstream. The "infinitesimal" disturbances left over in the vicinity of that grid are in fact welcome for generating physically adequate turbulent flow in the simulation. Below, The additional force term has the form of a linear restoring force, $G = -\gamma\,(\mathbf{u} - \mathbf{U})$, where the flow is driven (perhaps only) by prescribing $\mathbf{U}(x_1, t)$, an effective mass flow rate, and velocity fluctuations are suppressed using a prescribed linear damping function $\gamma(x_1) \geq 0$. Denoting the right hand side of the Navier Stokes equation $N[\mathbf{u}]$,

$$\partial \mathbf{u}/\partial t\,(x_1, x_2, x_3, t) = N[\mathbf{u}] + \gamma(x_1)\,(\mathbf{U}(x_1, t) - \mathbf{u}) \qquad (2)$$

The "fringe function" is effective, i.e. $\gamma(x_1) \neq 0$, only for $x_b < x_1 < x_b + L$. The region is thus a slab orthogonal to the mean flow direction x_1. In principle, no other forcing needs to be applied in the above equation:

If the mean flow is known e.g. to be spatially evolving in the downstream direction x_1, like in turbulent boundary layer, then $\mathbf{U}(x_1, t)$ would have a non-trivial x_1-dependence with appropriate inflow/outflow behaviour. In the present work, this was not the case, however, and the following simple form of a damping function was applied: $\gamma(x_1) = (1 - \cos(2\pi(x_1 - x_b)/L))/\,2$. In all computations, the streamwise extend of the computational mesh had the same number of points. Thus, the fringe region had always the same length. It started from $x_b = 2000$ and had a length of $L = 400$ in lattice units. This range corresponds to $50 < x/M < 60$. Therefore, only the results up to $x_b/M = 50$ will be discussed.

3 Physical Results

The available grid resolution allowed the simulation of flow evolution over substantially longer downstream distances than any previous work. This was sufficient to check the hypothesis that decaying grid-generated turbulence is self-similar sufficiently far from the grid, which is the justification e.g. for using universal "return to isotropy" (and "eddy viscosity") values in turbulence models. The numerical mesh size allowed, however, a reliable resoltion of the flow only at the modest Reynolds numbers cited in the Introduction. These correspond to maximum Kolmogorov scales of $1 < \eta < 5$ immediately at the grid (corresponding to Taylor microscale based Reynolds numbers of 40 or less) and approximately 0.2 to 1.0 away from it. The downstream spatial profile of $\eta(x_1)$ is sharply peaked at the grid, falling rapidly off to below 20% of its maximum within about $5M$ and then undergoing a very weak further decay downstream.

In order to quantify the spatial structure of the simulated turbulence, various turbulence statistics were measured as functions of longitudinal and, in some cases, also of spanwise location. Both "local" and "usual" version of Reynolds stress components and turbulent kinetic energy,

$$\overline{u_i u_j}(x_1, x_2, x_3) = \frac{1}{t_1 - t_0} \sum_{t=t_0}^{t_1} u_i' u_j'(x_1, x_2, x_3, t) \tag{3}$$

$$k(x_1, x_2, x_3) = (\overline{u_1 u_1}(x_1) + \overline{u_2 u_2}(x_1) + \overline{u_3 u_3}(x_1))/\,2 \tag{4}$$

$$\langle u_i u_j \rangle (x_1) = \frac{1}{(t_1 - t_0) N_2 N_3} \sum_{t=t_0}^{t_1} \sum_{n_2=1}^{N_2} \sum_{n_3=1}^{N_3} \tilde{u}_i \tilde{u}_j (x_1, n_2 \Delta x, n_3 \Delta x, t) \tag{5}$$

$$K(x_1) = (\langle u_1 u_1 \rangle (x_1) + \langle u_2 u_2 \rangle (x_1) + \langle u_3 u_3 \rangle (x_1))/\,2 \tag{6}$$

were measured, where two different definitions of fluctuating velocity components are used, using the (global) bulk velocity U_m and the local time-averaged (as in (3) denoted by an overbar) axial velocity, respectively:

$$u_j'(x_1, x_2, x_3, t) = u_j(x_1, x_2, x_3, t) - \delta_{j1}\, U_m(x_1) \tag{7}$$

$$\tilde{u}_j(x_1, x_2, x_3, t) = u_j(x_1, x_2, x_3, t) - \delta_{j1}\, \overline{u}_1(x_1, x_2, x_3) \tag{8}$$

The first of these two definitions involves much larger fluctuation magnitudes, especially in the downstream reagion close to the grid, where the cross-sectional variation of $\overline{u}_1$ is largest (between 0 and about U_m/ϵ).

Along with the standard definition of Reynolds stress anisotropy (here denoted A_{ij}), which is numerically confined between -1/3 and 2/3, we also considered a measure of anisotropy a_{ij} which reflects in a markedly quantitative way the transversal inhomogeneity of the flow:

$$A_{ij}(x_1) = \frac{1}{N_2 N_3} \sum_{n_2=1}^{N_2} \sum_{n_3=1}^{N_3} \frac{\overline{u_i u_j}}{2k}(x_1, n_2 \Delta x, n_3 \Delta x) - \frac{1}{3}\,\delta_{ij} \tag{9}$$

$$a_{ij}(x_1) = \frac{\langle u_i u_j \rangle}{2K}(x_1) - \frac{1}{3}\,\delta_{ij} \tag{10}$$

To directly measure the transversal inhomogeneity of a field $F(x_1, x_2, x_3)$, we considered

$$I[F](x_1, x_2) = \frac{N_2 \sum_{n_3=1}^{N_3} F(x_1, x_2, n_3 \Delta x)}{\sum_{n_2=1}^{N_2} \sum_{n_3=1}^{N_3} F(x_1, n_2 \Delta x, n_3 \Delta x)}, \tag{11}$$

$$\text{in particular,} \quad I_{ij}(x_1, x_2) = I\left[\overline{u_i u_j}\right], \quad I_m(x_1, x_2) = I\left[\overline{u}_1\right], \tag{12}$$

$$\text{and then,} \quad I_{ij}(x_1) = \frac{1}{N_2} \sum_{n_2=1}^{N_2} I_{ij}(x_1, n_2 \Delta x), \tag{13}$$

and verified that the choice of averaging direction (x_3 in the above definitions, or alternatively x_2) is not essential. Δx stays for the same uniform computational grid step used in LBM along any cooridinate.

3.1 Is Turbulence Self-Similar?

The spatial evolution of diagonal and off-diagonal components of the anisotropy tensor a_{ij} are shown in Figure 3 for a couple of runs having different Reynolds numbers. The streamwise velocity turbulent energy $\langle u_1^2/2 \rangle (x_1)$ exhibits an anisotropy distribution which allows for the clear distinction of several regions downstram of the grid: near-grid, self-similar anisotropy decay, and "developed" region. In the latter, the anisotropy appears to have reached an asymptotic but still nonzero value (see the Subsection on anisotropy below) while the turbulent energy continues to decay (see Figure 4).

A nearly levelling-off degree of anisotropy (and inhomogeneity) at larger downstream distances may appear unexpected. It would represent, when supported by other studies, a significant new aspect relevant to turbulence modeling. Considering the present result alone, the influence of outflow numerical boundary conditions needs to be excluded by independent simulations. On the other hand, the decay of anisotropy (and inhomogeneity) of both $\langle u_1^2 \rangle (x_1)$ and $\langle u_2^2 \rangle (x_1)$ are quite similar and qualitatively almost independent of the Reynolds number in the second and third ("far downstream") spatial ranges. In the latter region, their decay appears to follow an approximately constant algebraic rate. This is a qualitativ indication, that the numerical outflow does not destroy the physical nature of the simulated flow, at least up to $50M$.

The trends observed in Figure 3 and the width of the newly identified regions are not dependent of the Reynolds number, while amplitudes and (to a less significant extent) decay rates are. While the difference in decay rates can possibly be argued to be a low-Reynolds-number effect, the qualitative difference between the spatial regions is a qualitative observation that would probably not vanish in the high-Re limit. This results supports the intuitive expectation that the use of a single "anisotropy decay rate" everywhere, including the regions near the grid, is inappropriate. Simple decay rates should be applied "far from the grid" only. There, a self-similar behaviour of the spatial evolution of turbulence can be expected. The data shown are consistent with such behaviour for $x > 10M$.

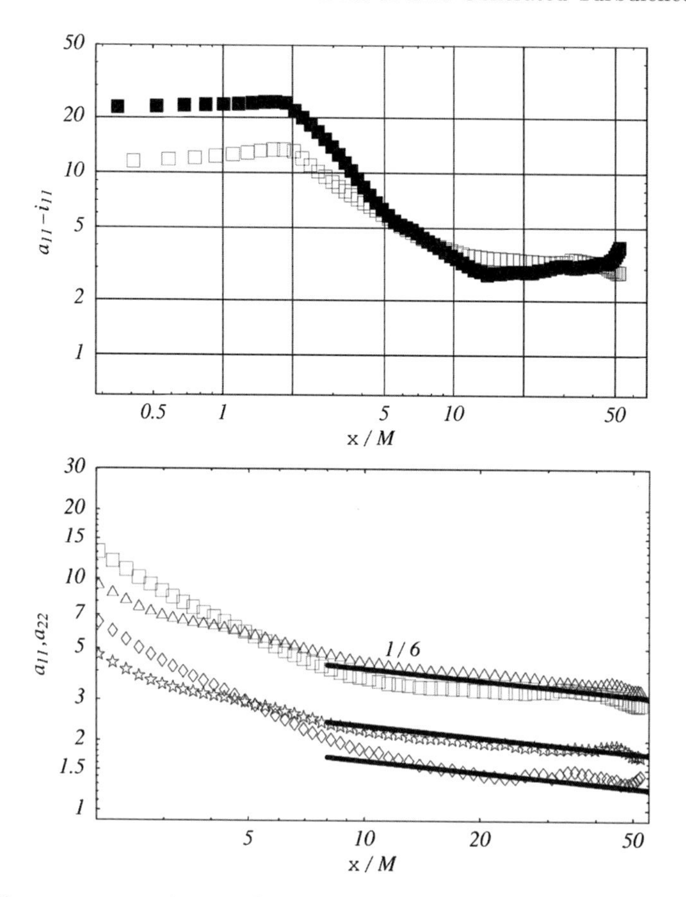

Fig. 3. Downstream evolution of anisotropy $a_{ij}(x_1)$ and inhomogeneity $I_{ij}(x_1)$ of velocity components — top: both measures applied to the streamwise velocity component in one representative run, bottom (note the different left end of the axial coordinate range): comparison of anisotropy between this and another run and between streamwise and transversal components

3.2 Decay Rates

The decay rate of anisotropy is low, for all Reynolds stress components, when looking at the "self-similar region" far from the grid. An algebraic decay exponent near $-1/6$ is suggested by Figure 3.

Precisely the opposite situation is observed for the decay of turbulent kinetic energy: Over the same self-similar region, a clear algebraic decay rate of $-5/3$ is observed, essentially independently of the Reynolds number. A theoretical derivation of this decay law, based on similar arguments as in the Kolmogorov law for the energy spectrum of isotropic turbulence in the inertial range (but with energy flux in streamwise direction rather than across wavenumber shells) and on the premise of weak turbulence will be provided

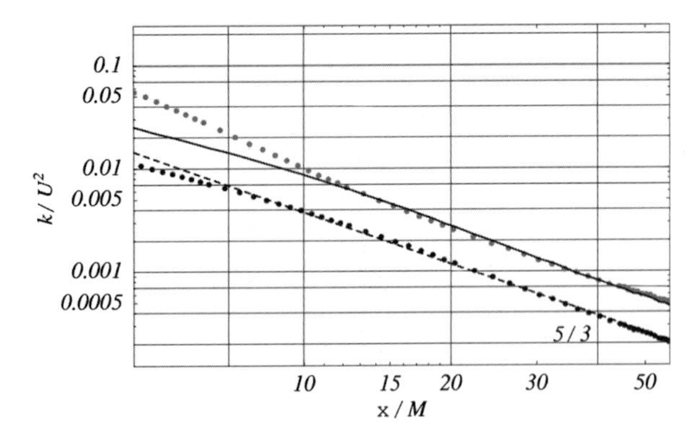

Fig. 4. Normalized turbulent kinetic energy $k(x_1)/U_m^2$ for two Reynolds numbers corresponding to different levels of turbulence after the grid (dots), and $\left\langle u_1^2 \right\rangle (x_1)/U_m^2$ (solid line), compared to algebraic decay laws $\sim x_1^{-5/3}$ (dashes)

elsewehre [9]. In the light of that argument, the main result from the presented simulations is the motivation and first "experimental" validation of a new "turbulence law", relevant to any weak turbulence which is influenced (as is usually the case) by anisotropy-generating factors.

An investigation into the energy decay was the main motivation for the present study, driven by the controversy about the algebraic decay rates, ongoing for some 40 years now. Algebraic decay-rate exponents obtained from wind tunnel and water channel measurements tend to be low, mostly between 1.1 and 1.5, while those obtained in more confined environments are significantly higher, up to 2. The present estimate of 5/3 lies in between; its validity can be claimed only for a very clearly defined setting and for low Reynolds numbers. It is thus not necessarily in contradiction with earlier, experimental measurements taken at typically much higher Reynolds numbers and under very specific test conditions. But the new rate is only the second one, after an analysis for liquid Helium experiments, which can be derived theoretically and has been so clearly demonstrated by measurements.

3.3 Spatial Inhomogeneity

The definition of "inhomogeneity" proposed in (12) excludes automatically the effect of downstream spatial decay of turbulent energy. It allows to simply ignore regions of numerical instability or with unphysical effects, such as the "fringe region" and its vicinity. A closer look at available experimental data, including very recent ones [8], reveals that most of the usual velocity statistics exhibit a significant spatial inhomogeneity long distance downstream of the grid. This observation appears to some extent to be in contradiction to the general expectation that grid-generated turbulence is a good approximation to the ideal of homogeneous isotropic turbulence. It was therefore of interest

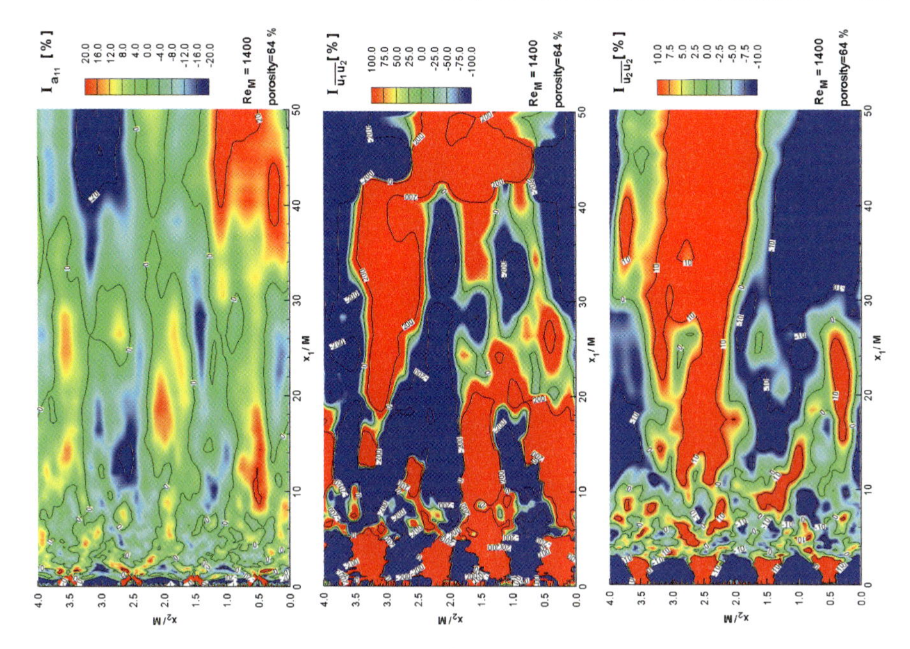

Fig. 5. Inhomogeneity measures $I_{11}(x_1, x_2)$ (left), $I_{22}(x_1, x_2)$ (middle), and $I_{12}(x_2, x_2)$ (right), as defined in (12) and obtained from the run with porosity 64%

to quantify the degree and spatial characteristics of inhomogeneity and to lookinto possible explanations for its appearance.

The spatial anisotropy along and across the mean flow direction (measured in one of the simulations but typical also for the rest) is shown in Figure 5. All components exhibit a spatially ordered hierarchy (in mean flow direction) of increasing inhomogeneity, reminiscent of the backward energy cascade in some kinds of anisotropic turbulence. But the $\langle u_1^2 \rangle$ Reynolds stress component undergoes that process at a slower rate, while the inhomogeneity of the other Reynolds stresses appears to be "driven" directly by that of the mean velocity U_m, shown in Figure 7. In all cases, this "cascade" is spontaneous, supporting the results of recent experiments [8].

3.4 Porosity

The effect of grid opening size, expressed through the nondimensional characteristic of grid porosity ϵ defined in the Introduction, is obvious with regard to flow resistence: for a given driving pressure force, a lower porosity leads to a lower Reynolds number. For the range of parameters investigated here, this is a nearly linear relationship. No parametric expression will be offered here, however, since no study of the influence of force magnitude has been made, including the way of transitioning to a high-Reynolds-number limit.

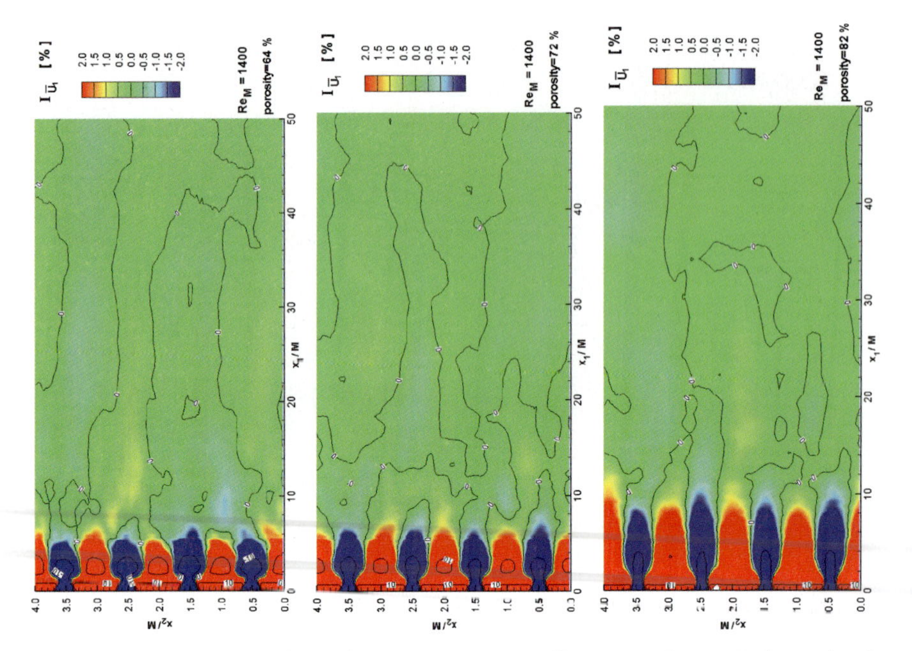

Fig. 6. Inhomogeneity of the (local, time-averaged) mean velocity $I_m(x_1, x_2)$, obtained at porosities (fr0m left to right) of 64%, 72%, and 82%

The effect on inhomogeneity, on the other hand, was investigated in sufficient detail. As can be seen from Figure 6, it is not significant. Qualitatively, lower porosity leads to higher fluctuations near the grid but also to more intensive mixing and thus shorter distance downstream of the grid, over which the mean flow becomes essentially homogeneous. Thus, a set of perfect grid and "perfect" boundary conditions appears to lead to a "perfectly homogeneous" mean flow. The results were supported by a simulation, in which the computational domain was widened 2.5 times, allowing for large-scale structures to develop, if a mechanism for their spontaneous generation would be present. As seen from Figure 7, there appears to be no such mechanism.

3.5 Asymmetry

To explain the significant inhomogeneity in the (time averaged) mean flow, observed in experiments, it was hypothesized that slight asymmetries in the set-up can lead to significant cross-flow variations of the mean fields. This hypothesis was tested by using a geometrically perturbed grid pattern, as explained in the Introduction. As seen from Figure 7, this can explain the inhomogeneity found in laboratory measurement data.

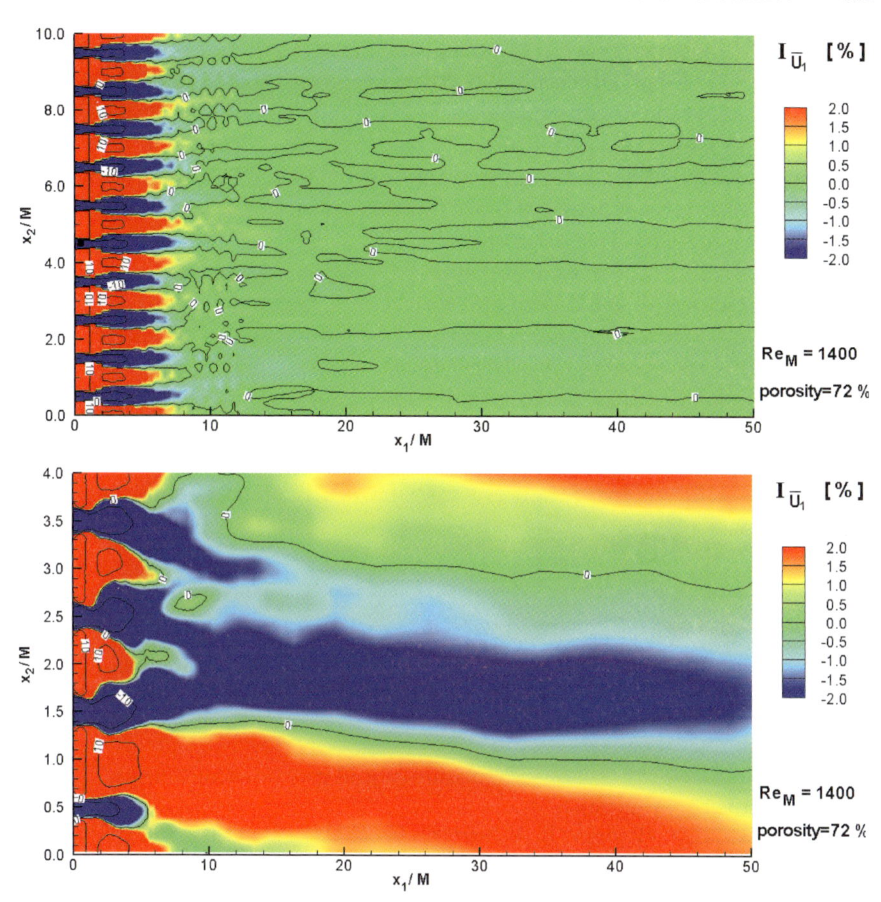

Fig. 7. The same statistics as in the previous figure, measured at 72% porosity in a run employing (left) a 2.5 times wider domain, or (right) a non-perfect grid with 10% enlargement (along each of the x_2 and x_3 coordinates) of one of the grid openings

4 Conclusion and Outlook

The largest direct numerical simulations of grid-generated turbulence known up to date were performed on the SX-8 of HLRS. The size of the computations was sufficient to resolve the grid and to allow for sufficient evolution of the flow structure both in the direction of the mean flow and across it. This was achieved using only moderately sized HLRS resources, in particular, typically up to 8 nodes of the SX-8. The excellent scaling of the employed lattice Boltzmann algorithm allowed to perform a series of DNS runs and thus a parametric study of several effects and good convergence of all velocity statistics of interest.

Several major results were obtained, having immediate impact on turbulence modeling. A new algebraic law (of Kolmogorv type) was discovered for the decay of turbulent kinetic energy in low-intensity grid-generated turbulence. Furthermore, the assumption of the existence of a single decay rate for the anisotropy of weak "homogeneous" turbulence was shown to be qualitatively inappropriate, while the assumption of an "inertial range" for the spatial development of such turbulence – on which the former assumption should be based – appears to be valid. The latter result could only be obtained using computational domains with a very long streamwise dimension; it is very important and useful for turbulence modeling and in particular for the analysis of the observed algebraic decay rates.

Future computational work would meaningfully try to extend both the streamwise dimension of the simulated flow domain and, more importantly, the range of Reynolds numbers. It can be expected, that after two or three such extensions, the major trends in decay rates for nearly homogeneous turbulence will be known.

References

1. G. Comte-Bellot, S. Corrsin, Simple Eulerian time-correlation of full- and narrow-band velocity signals in grid-generated, 'isotropic' turbulence. J. Fluid Mech., **48**, 273–337 (1971)
2. L. Djenidi, Lattice-Boltzmann simulation of grid-generated turbulence. J. Fluid Mech., **552**, 13–35 (2006)
3. P. Lammers, K.N. Beronov, R. Volkert, G. Brenner, F. Durst, Lattice BGK direct numerical simulation of fully developed turbulence in incompressible channel flow. Computers and Fluids, **35**(10), 1137–1153 (2006)
4. P. Lammenrs, U. Küster, Recent prerformance results of the lattice Boltzmann method. inSiDE, Spring 2006, **4**(1), 4–7 (2006)
5. N. Özyilmaz, K.N. Beronov, A.Delgado, Characterization of the dissipation tensor from DNS of grid-generated turbulence. In: High Performance Computing in Science and Engineering, Garching/Munich, Springer (2007)
6. N. Özyilmaz, K.N. Beronov, On the Reynolds stress anisotropy and inhomogeneity of grid-generated turbulence. Proc. TSFP5, München (2007)
7. N. Özyilmaz, K.N. Beronov, A. Delgado, Characterizing Reynolds-stress and dissipation-rate decay and anisotropy in DNS of grid-generated turbulence. Proc. Appl. Math. Mech. (2008)
8. Ö. Ertunc, N. Özyilmaz, H. Lienhart, F. Durst, K. Beronov, Homogeneity of turbulence generated by static-grid structures. Journal of Fluid Mechanics, to appear (2009)
9. K. Beronov, Inertial scaling in low-Re grid-generated turbulence. Phys. Rev. Lett., submitted (2009)

Application of FDEM on the Numerical Simulation of Journal Bearings with Turbulence and Inertia Effects

Torsten Adolph[1], Willi Schönauer[1], Roman Koch[2], and Gunter Knoll[2]

[1] Karlsruhe Institute of Technology, Forschungszentrum Karlsruhe GmbH, Steinbuch Centre for Computing, Hermann-von-Helmholtz-Platz 1, 76344 Eggenstein-Leopoldshafen, Germany
{torsten.adolph,willi.schoenauer}@iwr.fzk.de

[2] Institut für Maschinenelemente und Konstruktionstechnik, University of Kassel, Mönchebergstraße 3, 34125 Kassel, Germany
{roman.koch,gunter.knoll}@imk.uni-kassel.de

Summary. For the numerical simulation of journal bearings, current software solutions use the Reynolds differential equation where inertia terms are not included. The Finite Difference Element Method (FDEM) is a black-box solver for nonlinear systems of elliptic and parabolic partial differential equations (PDEs). Based on the general black-box we implement the Reynolds equation with inertia terms for the simulation of a journal bearing. We can easily implement different models for the turbulence factors and the dynamic viscosity, and we also consider cavitation. We give results for different Reynolds numbers, and we also give a global error estimate for each of the cases. This shows the quality of the numerical solution and is a unique feature of FDEM.

1 Introduction

Thermal turbo-engines, for example turbo-supercharger, have to be driven by high rotational speeds to realize good combustion efficiencies. These rotors are often pivoted by journal bearings, particularly when they are running with high rotational speeds. Because of stiffness and damping characteristics, journal bearings have advantages over ball bearings. In order to get similar properties, additional complex constructive measures have to be added on ball bearings, such as squeeze dashpots. Above a specific rotational speed, the rotors get instable, i.e. self-controlled vibrations are developed which can cause a total collapse of the system. This critical speed depends on different bearing and system parameters and can only be determined for simplified models.

Current software solutions to simulate journal bearings are based on the so-called Reynolds differential equation (1), which follows from the equation

W.E. Nagel et al. (eds.), *High Performance Computing in Science and Engineering '09*, DOI 10.1007/978-3-642-04665-0_28,

of momentum (3) and the continuity equation (4) under conditions of Stokes flows [2].

$$\frac{\partial}{\partial x}\left(\frac{\rho h^3}{12\eta}\frac{\partial p}{\partial x}\right)+\frac{\partial}{\partial y}\left(\frac{\rho h^3}{12\eta}\frac{\partial p}{\partial y}\right)-\frac{\partial}{\partial x}(\rho h u_m)-\frac{\partial}{\partial y}(\rho h v_m)-\frac{\partial(\rho h)}{\partial t}=0 \quad (1)$$

where ρ is the density, η is the dynamic viscosity, h is the height of the crack, p is the pressure, and u_m and v_m are computed from

$$u_m=\frac{u_1+u_2}{2}, \qquad v_m=\frac{v_1+v_2}{2} \quad (2)$$

with the velocities u_1, v_1 and u_2, v_2 in x- and y-direction on the upper and the lower boundary, respectively.

In comparison to the complex direct numerical simulation (DNS) based on the equation of momentum (3) and the continuity equation (4) [1]

$$\begin{aligned}
\rho\left(\frac{\partial u}{\partial x}u+\frac{\partial u}{\partial y}v+\frac{\partial u}{\partial z}w+\frac{\partial u}{\partial t}\right)&=-\frac{\partial p}{\partial x}+\frac{\partial \tau_{xx}}{\partial x}+\frac{\partial \tau_{xy}}{\partial y}+\frac{\partial \tau_{xz}}{\partial z}+F_x\,,\\
\rho\left(\frac{\partial v}{\partial x}u+\frac{\partial v}{\partial y}v+\frac{\partial v}{\partial z}w+\frac{\partial v}{\partial t}\right)&=-\frac{\partial p}{\partial y}+\frac{\partial \tau_{yx}}{\partial x}+\frac{\partial \tau_{yy}}{\partial y}+\frac{\partial \tau_{yz}}{\partial z}+F_y\,, \quad (3)\\
\rho\left(\frac{\partial w}{\partial x}u+\frac{\partial w}{\partial y}v+\frac{\partial w}{\partial z}w+\frac{\partial w}{\partial t}\right)&=-\frac{\partial p}{\partial z}+\frac{\partial \tau_{zx}}{\partial x}+\frac{\partial \tau_{zy}}{\partial y}+\frac{\partial \tau_{zz}}{\partial z}+F_z\,,
\end{aligned}$$

$$\frac{\partial(\rho u)}{\partial x}+\frac{\partial(\rho v)}{\partial y}+\frac{\partial(\rho w)}{\partial z}+\frac{\partial \rho}{\partial t}=0\,, \quad (4)$$

the solution of the Reynolds equation is more efficient in terms of computing time. Hence, to implement lubricant influences in multi-body systems, the Reynolds equation should be taken. In (3) and (4), u, v and w are the velocities in x-, y- and z-direction, p is the pressure, and the τ_{ij} are the shear stresses.

2 Topic

Depending on design and operating conditions of high speed journal bearings, turbulence and inertia effects can influence the system's motion.

At present, commercial software solutions include many mathematical models to simulate turbulent flows. In general, the so-called decomposed Navier-Stokes equations are used, in which the velocities are divided in average and fluctuating parts.

$$u=\bar{u}+u', \quad v=\bar{v}+v', \quad w=\bar{w}+w', \quad p=\bar{p}+p'\,. \quad (5)$$

This procedure leads to additional terms in the momentum equation, which stand for the turbulent fluctuations [1]

$$\rho\left(\frac{\partial\bar{u}}{\partial x}\bar{u}+\frac{\partial\bar{u}}{\partial y}\bar{v}+\frac{\partial\bar{u}}{\partial z}\bar{w}\right)=-\frac{\partial\bar{p}}{\partial x}+\eta\Delta\bar{u}+\rho\left(\frac{\partial\overline{u'^2}}{\partial x}+\frac{\partial\overline{u'v'}}{\partial y}+\frac{\partial\overline{u'w'}}{\partial z}\right),$$

$$\rho\left(\frac{\partial\bar{v}}{\partial x}\bar{u}+\frac{\partial\bar{v}}{\partial y}\bar{v}+\frac{\partial\bar{v}}{\partial z}\bar{w}\right)=-\frac{\partial\bar{p}}{\partial y}+\eta\Delta\bar{v}+\rho\left(\frac{\partial\overline{u'v'}}{\partial x}+\frac{\partial\overline{v'^2}}{\partial y}+\frac{\partial\overline{v'w'}}{\partial z}\right),\quad(6)$$

$$\rho\left(\frac{\partial\bar{w}}{\partial x}\bar{u}+\frac{\partial\bar{w}}{\partial y}\bar{v}+\frac{\partial\bar{w}}{\partial z}\bar{w}\right)=-\frac{\partial\bar{p}}{\partial z}+\eta\Delta\bar{w}+\rho\left(\frac{\partial\overline{u'w'}}{\partial x}+\frac{\partial\overline{v'w'}}{\partial y}+\frac{\partial\overline{w'^2}}{\partial z}\right).$$

These additional terms are called Reynolds stresses and include new unknown quantities. Straight forward, extra equations are needed to solve the system of equations, which is referred as the closure problem. To couple the fluid mechanics with multi-body dynamics, the described equation (6) leads to a very complex and computing-intensive model.

Under specific restrictions a similar equation, according to the classical theory of journal bearings, can be developed, which is based on the turbulence estimate (6) and results to an extended form of the Reynolds equation [3], [4]

$$\begin{aligned}
&\frac{\partial}{\partial x}\left(G_x\frac{\rho h^3}{\eta}\frac{\partial p}{\partial x}\right)+\frac{\partial}{\partial y}\left(G_y\frac{\rho h^3}{\eta}\frac{\partial p}{\partial y}\right)\\
&\quad-\frac{\partial}{\partial x}\left(G_x\rho h u_m\right)-\frac{\partial}{\partial y}\left(G_y\rho h v_m\right)-\frac{\partial(\rho h)}{\partial t}= \qquad (7)\\
&\quad-\frac{\partial}{\partial x}\left(G_x\frac{\rho h^2}{\eta}\left(\frac{\partial(V_x h)}{\partial t}+\frac{\partial I_{xx}}{\partial x}+\frac{\partial I_{xy}}{\partial y}\right)\right)\\
&\quad-\frac{\partial}{\partial y}\left(G_y\frac{\rho h^2}{\eta}\left(\frac{\partial(V_y h)}{\partial t}+\frac{\partial I_{yx}}{\partial x}+\frac{\partial I_{yy}}{\partial y}\right)\right)
\end{aligned}$$

with the bulk velocities V_x and V_y and the momentum fluxes I_{xx}, I_{yy}, I_{xy} and I_{yx}. This is a 2-D PDE for the pressure. In comparison to (6) this yields a considerable decrease of numerical complexity, however stability of the numerical process is complicated to control.

The implementation of a stable software solution for the partial differential equation (7) is the main issue of the research activities, in which time efficient algorithms have to be used.

Different research activities in this field developed different turbulence models, which can be applied to special behaviours of lubrication. To take turbulence models into account (equation (7)), the parameters

$$G_x,\quad G_y,\quad G_{Jx},\quad G_{Jy} \qquad (8)$$

have to be used. The general form follows experimentally measured constants and depends on the Reynolds number Re:

$$G = \frac{c_0}{c_1 + \frac{c_2}{c_3} Re_{eff}^{c_4}} \; . \tag{9}$$

The possibility to implement alternative turbulence models must exist. In addition, pressure dependent characteristics of lubricants must be considered. The dynamic viscosity η is a function of the pressure p. Initially, we implement the two different viscosity models of Barus [5] and Roelands [6]:

$$\eta_{Barus}(p) = \eta_0 e^{\alpha p} \; , \tag{10}$$

$$\eta_{Roelands}(p) = \eta_0 e^{(ln(\eta_0)+9.67)\left(-1+\left(1+\frac{p}{p_r}\right)^{\kappa}\right)} \tag{11}$$

where η_0 is the dynamic viscosity at ambient pressure and α is the viscosity-pressure coefficient in the Barus viscosity equation. κ is the pressure-viscosity coefficient and p_r is a reference pressure in the viscosity model of Roelands.

The density ρ also depends on the pressure p [7]

$$\rho(p) = \rho_0 \frac{5.9 \cdot 10^8 + 1.34p}{5.9 \cdot 10^8 + p} \tag{12}$$

where ρ_0 is the density at ambient pressure.

3 Finite Difference Element Method FDEM

FDEM is an unprecedented generalization of the FDM on an unstructured FEM mesh. It is a black-box solver for arbitrary nonlinear systems of 2-D and 3-D elliptic or parabolic PDEs. With certain restrictions it can be used also for hyperbolic PDEs. If the unknown solution is $u(t, x, y, z)$ the operator for PDEs and BCs (boundary conditions) is (2.4.1) and (2.4.2) in [8]:

$$Pu \equiv P(t, x, y, z, u, u_t, u_x, u_y, u_z, u_{xx}, u_{yy}, u_{zz}, u_{xy}, u_{xz}, u_{yz}) = 0 \; . \tag{13}$$

For a system of m PDEs, u and Pu have m components:

$$u = \begin{pmatrix} u_1 \\ \vdots \\ u_m \end{pmatrix} , \quad Pu = \begin{pmatrix} P_1 u \\ \vdots \\ P_m u \end{pmatrix} . \tag{14}$$

As we have a black-box solver, the PDEs and BCs and their Jacobian matrices of type (2.4.6) in [8] must be entered as Fortran code in prescribed frames.

The geometry of the domain of solution is entered as a FEM mesh with triangles in 2-D and tetrahedra in 3-D. The domain may be composed of subdomains with different PDEs and non-matching grid. From the element list and its inverted list, we select for each node more than the necessary number of nodes for difference formulas of a given consistency order q. By a sophisticated algorithm, from this set the necessary number of nodes is determined, see Sect. 2.2 in [8]. From the difference of formulas of different consistency order, we get an estimate of the discretization error. If we want e.g. the discretization error for u_x, and $u_{x,d,q}$ denotes the difference formula of consistency order q, the error estimate d_x is defined by

$$d_x := u_{x,d,q+2} - u_{x,d,q} \; , \tag{15}$$

i.e. by the difference to the order $q+2$. This has a built-in self-control: if this is not a "better" formula the error estimate shows large error.

With such an error estimate, we can explicitly compute the error of the solution by the error equation (2.4.8) in [8]. The knowledge of the error estimate allows a mesh refinement and order control in space and time (for parabolic PDEs), see Sect. 2.5 in [8].

A special problem for a black-box solver is the efficient parallelization because the user enters his domain by the FEM mesh. We use a 1-D domain decomposition with overlap to distribute the data to the processors, see Sect. 2.8 in [8]. We use MPI. A detailed report on the parallelization is [9]. The resulting large and sparse linear system is solved by the LINSOL program package [10] that is also efficiently parallelized for iterative methods of CG type and (I)LU preconditioning.

4 Implementation of the Reynolds Equation

To simplify the model we want to eliminate the time derivative from (7). Therefore, we assume that the density ρ and the bulk velocities V_x and V_y are not time dependent. The time derivative of the crack height h equals the velocity w in z-direction:

$$w(x,y) = \frac{\partial h(x,y)}{\partial t} \tag{16}$$

This velocity w is an input parameter, i.e. it is given in each grid node.

By this means, (7) becomes

$$\begin{aligned} &\frac{\partial}{\partial x}\left(G_x \frac{\rho h^3}{\eta}\frac{\partial p}{\partial x}\right) + \frac{\partial}{\partial y}\left(G_y \frac{\rho h^3}{\eta}\frac{\partial p}{\partial y}\right) \\ &- \frac{\partial}{\partial x}\left(G_x \rho h u_m\right) - \frac{\partial}{\partial y}\left(G_y \rho h v_m\right) - \rho w = \\ &- \frac{\partial}{\partial x}\left(G_x \frac{\rho h^2}{\eta}\left(V_x w + \frac{\partial I_{xx}}{\partial x} + \frac{\partial I_{xy}}{\partial y}\right)\right) \\ &- \frac{\partial}{\partial y}\left(G_y \frac{\rho h^2}{\eta}\left(V_y w + \frac{\partial I_{yx}}{\partial x} + \frac{\partial I_{yy}}{\partial y}\right)\right) . \end{aligned} \tag{17}$$

From this partial differential equation, we want to compute the pressure p.

For the computation of the bulk velocities and the momentum fluxes, we need the velocity profiles u and v which are defined by

$$u = \frac{z(z-h)}{2\eta}\frac{\partial p}{\partial x} + \left(1 - \frac{z}{h}\right) u_1 + \frac{z}{h} u_2 \,, \tag{18}$$

$$v = \frac{z(z-h)}{2\eta}\frac{\partial p}{\partial y} + \left(1 - \frac{z}{h}\right) v_1 + \frac{z}{h} v_2 \,, \tag{19}$$

Then the bulk velocities are defined by

$$V_x = \frac{1}{h}\int_0^h u\,dz = \frac{1}{h}\left(-\frac{h^3}{12\eta}\frac{\partial p}{\partial x} + u_m h\right) \,, \tag{20}$$

$$V_y = \frac{1}{h}\int_0^h v\,dz = \frac{1}{h}\left(-\frac{h^3}{12\eta}\frac{\partial p}{\partial y} + v_m h\right) \,, \tag{21}$$

the momentum fluxes I_{xx}, I_{yy}, I_{xy} and I_{yx} are defined by

$$I_{xx} = \int_0^h u^2\,dz, \quad I_{yy} = \int_0^h v^2\,dz, \quad I_{xy} = I_{yx} = \int_0^h uv\,dz \,. \tag{22}$$

Therefore, it holds

$$I_{xx} = \frac{h^5}{120\eta^2}\left(\frac{\partial p}{\partial x}\right)^2 - \frac{h^3}{6\eta} u_m \frac{\partial p}{\partial x} + \frac{h}{3}\left(u_1^2 + u_1 u_2 + u_2^2\right) \tag{23}$$

$$I_{yy} = \frac{h^5}{120\eta^2}\left(\frac{\partial p}{\partial y}\right)^2 - \frac{h^3}{6\eta} v_m \frac{\partial p}{\partial y} + \frac{h}{3}\left(v_1^2 + v_1 v_2 + v_2^2\right) \tag{24}$$

$$I_{xy} = \frac{h^5}{120\eta^2}\frac{\partial p}{\partial x}\frac{\partial p}{\partial y} - \frac{h^3}{12\eta}\left(u_m \frac{\partial p}{\partial y} + v_m \frac{\partial p}{\partial x}\right) + \frac{h}{6}(2u_1 v_1 + u_1 v_2 + u_2 v_1 + 2u_2 v_2) \,. \tag{25}$$

For the computation of the turbulence factors G, the Reynolds number Re is needed. It is defined by

$$Re = \frac{\rho V_x h}{\eta} \,. \tag{26}$$

For G_x and G_y we use

$$G_x = \frac{1}{1 + \frac{0.0136}{12} Re^{0.9}}, \quad G_y = \frac{1}{1 + \frac{0.0043}{12} Re^{0.96}} \,, \tag{27}$$

and it holds

$$G_{Jx} = G_{Jy} = 1 \,. \tag{28}$$

The "system" of PDEs consists of only a single PDE here. But from the inertia terms, the PDE includes third partial derivatives of the pressure. As FDEM can only cope with second partial derivatives, we have to introduce two auxiliary variables q and r with

$$q := \frac{\partial p}{\partial x}, \quad r := \frac{\partial p}{\partial y} . \tag{29}$$

Although the original problem consists of one PDE, we have to solve a system of three PDEs. The problem becomes very complicated because the terms in the PDE (17) contain products of terms that have to be differentiated with respect to x and y and that depend on the pressure p and its derivatives in x- and y-direction. In FDEM we have to compute the Jacobian matrices for the linearization, i.e. the matrix of all first-order partial derivatives of the PDE. The entries of these Jacobian matrices are lengthy terms because of the mentioned dependencies and therefore the implementation of the Reynolds equation and the corresponding Jacobian matrices is very sophisticated, above all if there is prescribed a CPU time limit for the execution of the code.

The velocities u_1, u_2, v_1, v_2 and w as well as the crack height h are functions of x and y and are given for each node.

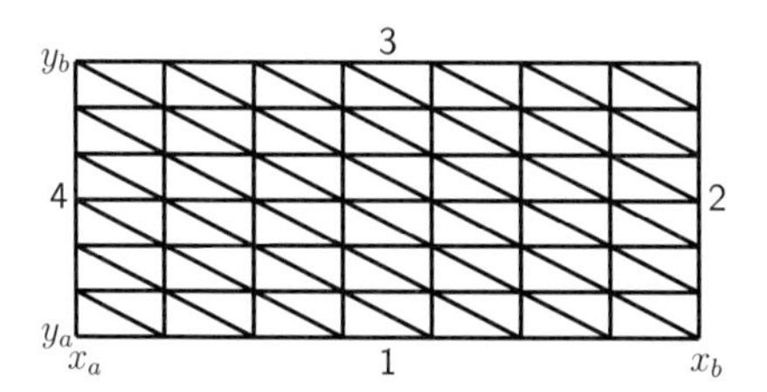

Fig. 1. Solution domain Ω and boundary numbering

The solution domain Ω is rectangular with $x \in [x_a, x_b]$, $y \in [y_a, y_b]$, see Fig. 1. On the 4 boundaries, we prescribe the boundary conditions for the pressure p given in Table 1.

Table 1. Boundary conditions for the boundaries given in Fig. 1

bd. no.	boundary condition
1	$p(x, y_a) = p_{y_a}(x)$
2	$p(x_b, y) = p_{x_b}(y)$
3	$p(x, y_b) = p_{y_b}(x)$
4	$p(x_a, y) = p_{x_a}(y)$

The pressure must not become negative. So we prescribe $p = 0$ in each node where we get $p \leq 0$ from (17). This means cavitation. As we do not know the values for the pressure in each node in advance, we must split the computation into two parts. First, we compute the pressure in each node with (17). This is denoted by test step. Then we see where we get negative pressure. We repeat the computation, but now we set $p = 0$ for these nodes in the so-called computation step, for the nodes with $p > 0$ in the test step we keep on computing the pressure from PDE (17).

5 Numerical Results

The implementation of the code and the computations have been carried out on the distributed memory parallel computer HP XC4000 at the Steinbuch Centre for Computing of the Karlsruhe Institute of Technology, Germany. We compute on AMD Opteron processors with 2.6 GHz frequency (5.2 GFLOPS peak performance) and InfiniBand 4X DDR interconnect.

The solution domain is $\Omega = [0, 2\pi] \times [0, 0.001]$, i.e. it holds

$$x_a = 0, \quad x_b = 2\pi, \quad y_a = 0, \quad y_b = 0.001 . \tag{30}$$

We use 11 nodes in x-direction and 21 nodes in y-direction, so the elements of the grid have very small angles which causes linear dependencies of the nodes that we collect for the difference stars. Here it is very important to have a robust algorithm for the selection of the nodes for the difference and error formulas.

On the four boundaries we set $p = 0$, i.e. it holds

$$p_{x_a}(y) = p_{x_b}(y) = p_{y_a}(x) = p_{y_b}(x) = 0 . \tag{31}$$

The crack height h is chosen so that we get cavitation, i.e. the crack must become thinner in the left half of the domain, and in the right half it gets wider again being axially symmetric to $x = \pi$. h is a function of x and does not depend on y. We use the following function depicted in Fig. 2:

$$h = 10^{-5} \cdot \left(1 + \frac{1}{10} \cos x\right) . \tag{32}$$

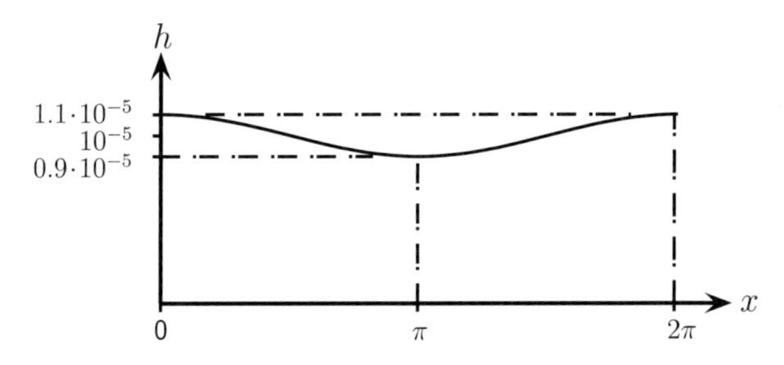

Fig. 2. Plot of the crack height $h(x)$

The given functions u_1, u_2, v_1, v_2 and w are constant, more precisely it holds

$$u_2 = v_1 = v_2 = w = 0 \,. \tag{33}$$

By the choice of u_1 we control the Reynolds number as Re depends on the bulk velocity V_x (26) that is in turn dependent on u_m (20). We carry out 8 computations with different Reynolds numbers. We compute with values for u_1 between 1 m/s and 10^5 m/s which results in a maximum Reynolds number between 0.2 and 24500.

We use the Barus viscosity equation (10) with $\eta_0 = 20\,\text{mPas}$ and the pressure coefficient $\alpha = 1.4 \cdot 10^{-8}$. Furthermore, it holds $\rho_0 = 890\,\text{kg/m}^3$ for the density at ambient pressure (12).

In Fig. 3 we depict the solution resulting from the test step for $Re = 24500$. We see that the pressure in the right half of the solution domain—where the crack height h is increasing again—is negative. So the pressure is set to zero for the computation step in each node in the right half.

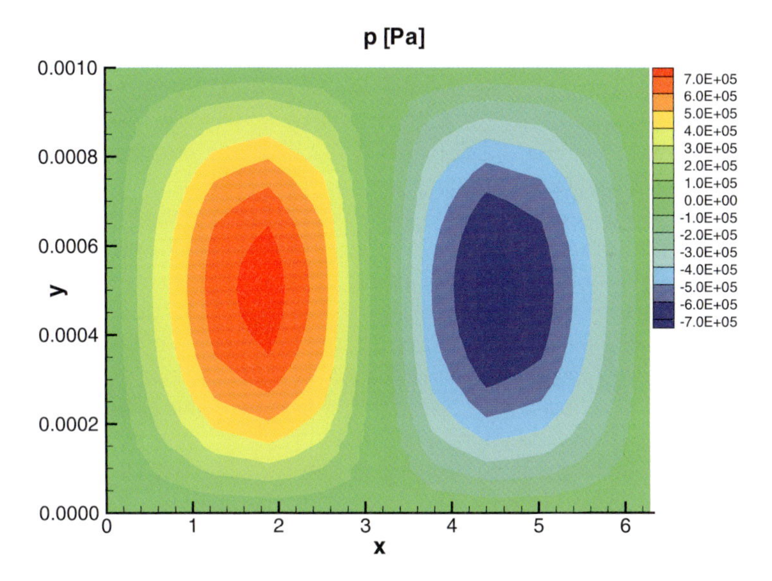

Fig. 3. Contour plot of the pressure p for the test step for $Re = 24500$

In Table 2 we present the results for the computation step of the 8 computations. In the second and third column we give the values for u_1 and the Reynolds number Re, respectively. In column 4 you see the maximum pressure in the solution domain. Columns 5 and 6 give the maximum and the mean global relative estimated error, respectively. "Mean" means the arithmetic mean of all nodes. To get aware of the location of the maximum pressure and the maximum error you may have a look on Figs. 4 (for the pressure) and 5 (for the maximum error). The last column gives the computation time

Table 2. Results for different Reynolds numbers on 11×21 grid (computation step)

No.	u_1 [m/s]	Re	p_{max} [Pa]	rel. estimated error max.	rel. estimated error mean	CPU time [s]
1	1	0.2	$0.122 \cdot 10^1$	$0.347 \cdot 10^{-7}$	$0.544 \cdot 10^{-8}$	0.086
2	5	1.2	$0.611 \cdot 10^1$	$0.365 \cdot 10^{-7}$	$0.560 \cdot 10^{-8}$	0.086
3	10	2.4	$0.122 \cdot 10^2$	$0.389 \cdot 10^{-7}$	$0.580 \cdot 10^{-8}$	0.085
4	50	12	$0.613 \cdot 10^2$	$0.578 \cdot 10^{-7}$	$0.749 \cdot 10^{-8}$	0.086
5	100	24	$0.123 \cdot 10^3$	$0.819 \cdot 10^{-7}$	$0.967 \cdot 10^{-8}$	0.085
6	1000	245	$0.130 \cdot 10^4$	$0.586 \cdot 10^{-6}$	$0.683 \cdot 10^{-7}$	0.086
7	10000	2450	$0.193 \cdot 10^5$	$0.132 \cdot 10^{-4}$	$0.172 \cdot 10^{-5}$	0.085
8	100000	24500	$0.758 \cdot 10^6$	$0.852 \cdot 10^{-3}$	$0.126 \cdot 10^{-3}$	0.090

on one processor for the whole computation with test and computation step. We see that the prescribed time limit objective (CPU time $\leq 1\,\mathrm{s}$) is easily achieved.

The pressure increases proportional with the Reynolds number up to $Re = 245$, for larger Reynolds numbers it increases disproportionately. The maximum errors are very small, even for the largest Reynolds number $Re = 24500$. The maximum errors occur for high pressures and at the boundary of the cavitation domain, see Fig. 5. This is because we use nodes from the left and the right of the cavitation boundary for the error formulas.

From Fig. 4 we see that the pressure equals zero in the right half of the domain after the computation step. Figure 5 shows that the error equals zero in this part of the domain. For the larger Reynolds numbers the grid is too coarse to get smaller errors because the derivatives of the pressure in x- and y-direction become very large. The grid we use has been provided by the IMK Kassel. If we completely refine the grid resulting in 21×41 nodes, the maximum error is reduced to $0.230 \cdot 10^{-3}$ which is about 1/4 of the original maximum error which is what we expected.

The main objective was to implement the code so that the computation time on one processor is below one second for the given problem. Although the code is fully parallelized, this is of no importance here because the serial code will be used on Linux workstations and Windows PCs. This is the reason why we forego scalability tests.

6 Concluding Remarks

We implemented an innovative code for the numerical simulation of a journal bearing which includes the Reynolds differential equation under consideration of inertia terms which are neglected in conventional software solutions. By the use of a finite difference method, the FDEM code is flexible to such an extent that we even can give a global error estimate. Furthermore, the code clearly

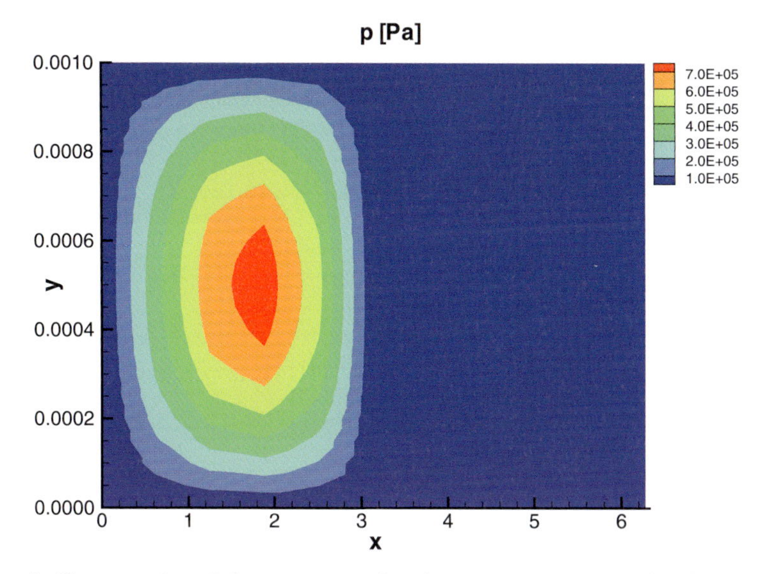

Fig. 4. Contour plot of the pressure p for the computation step for $Re = 24500$

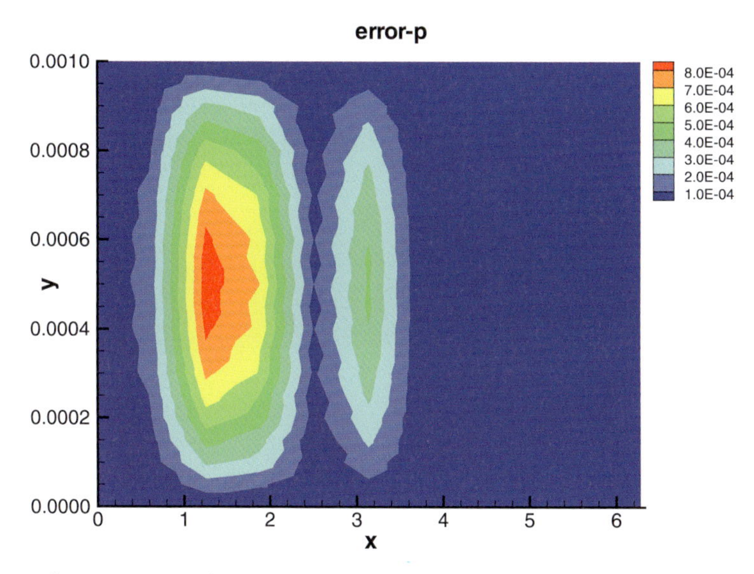

Fig. 5. Contour plot of the error of the pressure for the computation step for $Re = 24500$

meets the demand to run on Linux workstations and Windows PCs on one processor with a CPU time below one second. We demonstrated the usability of the FDEM code with the modifications for the numerical simulation of a journal bearing for different Reynolds numbers. Even for the largest Reynolds numbers the errors stay very small.

References

1. H. Schlichting, K. Gersten, Grenzschicht-Theorie. Springer-Verlag, 9. Auflage
2. O. Reynolds, On the Theory of Lubrication and its Application to Mr. Beauchamps Tower's Experiments, including an Experimental Determination of the Viscosity of Olive Oils. Phil. Trans., 177, pp. 157-234
3. G.G. Hirs, A Bulk-Flow Theory for Turbulence in Lubricant Films. ASME Journal of Lubrication Technology, pp. 137-146, 1973
4. L. San Andrés, Turbulent Hybrid Bearings With Fluid Inertia Effects. ASME Journal of Tribology, Vol. 112, pp. 699-707, 1990
5. C. Barus, Isothermals, Isopiestics and Isometrics relative to Viscosity. Am. J. of Sience, 45, pp. 87-96, 1893
6. C.J.A. Roelands, Correlational Aspects of the Viscosity-Temperature-Pressure Relationship of Lubricating Oils. Ph.D. Thesis, Technical University Delft, The Netherlands, 1966
7. D. Dowson, G.R. Higginson, Elastohydrodynamic Lubrication, The Fundamentals of Roller and Gear Lubrication. Pergamon Press, Oxford, Great Britain, 1997
8. W. Schönauer, T. Adolph, FDEM: The Evolution and Application of the Finite Difference Element Method (FDEM) Program Package for the Solution of Partial Differential Equations, 2005, available at http://www.rz.uni-karlsruhe.de/rz/docs/FDEM/Literatur/fdem.pdf
9. T. Adolph, The Parallelization of the Mesh Refinement Algorithm in the Finite Difference Element Method, Doctoral Thesis, 2005, available at http://www.rz.uni-karlsruhe.de/rz/docs/FDEM/Literatur/par_mra_fdem.pdf
10. LINSOL, see http://www.rz.uni-karlsruhe.de/rd/linsol.php

Numerical Analysis of Transition Effects in 3D Hypersonic Intake Flows

Martin Krause[1], Birgit Reinartz[2], and Marek Behr[2]

[1] Chair for Computational Analysis of Technical Systems,
RWTH Aachen University, Theaterplatz 14, 52062 Aachen, Germany
krause@cats.rwth-aachen.de

[2] reinartz@cats.rwth-aachen.de and behr@cats.rwth-aachen.de

Summary. A combined experimental as well as computational analysis of a complete scramjet demonstrator model has been initiated. The experimental tests will take place under real flight conditions at a hypersonic wind tunnel. Prior to those tests, a numerical analysis of the performance of two possible demonstrator geometries is conducted. In the current paper, the results of the performance analysis for two different two- and three-dimensional intakes employing single and double outer compression ramps as well as side wall compression are discussed. It is shown that the intakes, in combination with a subsequent isolator, are able to generate flow conditions required for stable supersonic combustion using a central strut injector. Furthermore, the effect of boundary layer transition is also discussed.

1 Introduction

For future, reusable space transportation systems as well as for hypersonic flight vehicles the use of an airbreathing propulsion system with supersonic combustion is a key technology concerning design and overall vehicle concept. In this context only the use of a scramjet propulsion system meets all the aerodynamic and gasdynamic requirements and offers a real alternative towards the classical rocket driven systems. Accordingly, the main scientific objective of the projects networked within the German Research Training Group "Aerothermodynamic Design of a Scramjet Engine for Future Space Transportation Systems" is the design and the development of a scramjet demonstrator engine using the necessarily different experimental and numerical procedures and tools provided by the involved scientists. Three different German universities (Stuttgart University, RWTH Aachen University and Technical University Munich) as well as the DLR are involved in this program. The developed demonstrator engine includes all the highly integrated components of a scramjet like inlet, isolator, combustion chamber and nozzle. Experimental testing of the scramjet demonstrator is scheduled for the second

W.E. Nagel et al. (eds.), *High Performance Computing in Science and Engineering '09*, DOI 10.1007/978-3-642-04665-0_29,

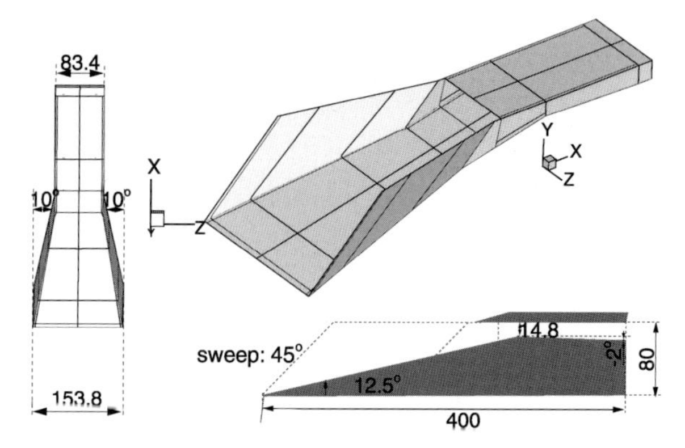

Fig. 1. Design studies of the new 3D inlet and isolator configuration

part of 2009 at the hypersonic wind tunnel AT-303 of the Institute of Theoretical and Applied Mechanics (ITAM) in Novosibirsk, Russia. The AT-303 is a first rate test facility specifically developed for scramjet testing and using a unique principle of multi-stage adiabatic compression to obtain the required high pressures and high temperatures for real flight conditions. Depending on the test conditions, the measurement times lie between 40 to 200 ms, allowing for steady supersonic combustion tests in spite of ignition delay times etc. One possible intake configuration used for the scramjet demonstrator (see Fig. 1) is a variation of a 3D mixed compression inlet tested before at ITAM [1]. Two phenomena characterize the technological problems of the intake: on the one hand, the interaction of strong shock waves with thick hypersonic boundary layers causes large separation zones that reduce the captured mass flow and thus the engine performance. On the other hand, the compression process has to be just right to obtain a temperature high enough for self ignition of the injected hydrogen without increasing the exit pressure too much. In this paper the performance analysis of two different intake concepts as well as the subsequent isolators at a flight Mach number of 8 are performed.

2 Physical Model

2.1 Conservation Equations

The governing equations for high-speed turbulent flow are the unsteady, compressible, Favre-averaged Navier-Stokes equations in integral form

$$\frac{\partial}{\partial t}\int_V \mathbf{U}\ \mathrm{d}V + \oint_{\partial V} \left(\mathbf{F}^{\mathrm{c}} - \mathbf{F}^{\mathrm{d}}\right) \mathbf{n}\ \mathrm{d}S = \mathbf{0} \tag{1}$$

where

$$\mathbf{U} = [\ \bar{\rho}\ ,\ \ \bar{\rho}\tilde{\mathbf{v}}\ ,\ \ \bar{\rho}\tilde{E}\ ]^{\mathrm{T}} \tag{2}$$

is the array of the mean conserved quantities: density, momentum density, and total energy density. The tilde and the bar over the variables denote the mean value of Favre-averaged and Reynolds-averaged variables, respectively. The quantity V denotes an arbitrary control volume with the boundary ∂V and the outer normal $\mathbf{n}$. The flux array is divided into its inviscid part and its diffusive part

$$\mathbf{F}^{\mathrm{c}} = \begin{pmatrix} \bar{\rho}\tilde{\mathbf{v}} \\ \bar{\rho}\tilde{\mathbf{v}} \circ \tilde{\mathbf{v}} + \bar{p}\,\mathbf{1} \\ \tilde{\mathbf{v}}(\bar{\rho}\tilde{E} + \bar{p}) \end{pmatrix} \quad \text{and} \quad \mathbf{F}^{\mathrm{d}} = \begin{pmatrix} \mathbf{0} \\ \bar{\boldsymbol{\sigma}} - \overline{\rho \mathbf{v}'' \circ \mathbf{v}''} \\ \tilde{\mathbf{v}}\bar{\boldsymbol{\sigma}} + \overline{\mathbf{v}''\boldsymbol{\sigma}} - \bar{\mathbf{q}} - c_p\,\overline{\rho \mathbf{v}'' T''} \\ -\frac{1}{2}\,\overline{\rho \mathbf{v}''\,\mathbf{v}'' \circ \mathbf{v}''} - \tilde{\mathbf{v}}\,\overline{\rho \mathbf{v}'' \circ \mathbf{v}''} \end{pmatrix}$$

where $\mathbf{1}$ is the unit tensor and $\circ$ denotes the dyadic product[3]. Currently, the airflow is still considered to be a calorically perfect gas with constant ratio of specific heats, $\gamma = 1.4$, and a specific gas constant of $R = 287\,\mathrm{J/(kgK)}$, which yields the following expression for the specific total energy:

$$\tilde{E} = c_v \bar{T} + \frac{1}{2}\tilde{\mathbf{v}}\tilde{\mathbf{v}} + k \ . \tag{3}$$

The last term represents the turbulent kinctic energy

$$k := \frac{1}{2}\frac{\overline{\rho \mathbf{v}''\mathbf{v}''}}{\bar{\rho}} \ . \tag{4}$$

For isotropic Newtonian fluids, the mean molecular shear stress tensor

$$\bar{\boldsymbol{\sigma}} = 2\bar{\mu}\,\bar{\mathbf{S}} - \frac{2}{3}\bar{\mu}\,\mathrm{tr}\big(\bar{\mathbf{S}}\big)\,\mathbf{1} \tag{5}$$

is a function of the mean strain rate tensor

$$\bar{\mathbf{S}} := \frac{1}{2}\left[\mathrm{grad}(\bar{\mathbf{v}}) + (\mathrm{grad}(\bar{\mathbf{v}}))^{\mathrm{T}}\right] \tag{6}$$

where the molecular viscosity $\bar{\mu} = \bar{\mu}(\bar{T})$ is determined by Sutherland's law. Similarly, the molecular heat flux is considered a linear isotropic function of the temperature gradient

$$\bar{\mathbf{q}} = -\frac{c_{\mathrm{p}}\bar{\mu}}{Pr}\,\mathrm{grad}(\bar{T}) \tag{7}$$

with the Prandtl number $Pr = 0.72$.

2.2 Turbulence Closure

To close the above system of partial differential equations, the Boussinesq hypothesis is used where the remaining correlations are modeled as functions

[3] Scalar Products of dyadics formed by two vectors $\mathbf{a}$ and $\mathbf{b}$ with a vector $\mathbf{c}$ are defined as usual, i.e., $\mathbf{a} \circ \mathbf{b}\,\mathbf{c} = \mathbf{a}(\mathbf{b}\mathbf{c})$, $\mathbf{c}\,\mathbf{a} \circ \mathbf{b} = (\mathbf{c}\mathbf{a})\mathbf{b}$.

of the gradients of the mean conservative quantities and turbulent transport coefficients. The Reynolds stress tensor thus becomes

$$-\overline{\rho\mathbf{v}''\circ\mathbf{v}''} = 2\mu_t\,(\bar{\mathbf{S}} - \frac{1}{3}\,\mathrm{tr}(\bar{\mathbf{S}})) - \frac{2}{3}\bar{\rho}k\,\mathbf{1}, \tag{8}$$

with the eddy viscosity μ_t, and the turbulent heat flux is

$$c_p\,\overline{\rho\mathbf{v}''T''} = -\frac{c_p\mu_t}{Pr_t}\,\mathrm{grad}(\bar{T}), \tag{9}$$

with the turbulent Prandtl number $Pr_t = 0.89$. Finally, for hypersonic flows the molecular diffusion and the turbulent transport are modeled as functions of the gradient of the turbulent kinetic energy

$$\overline{\mathbf{v}''\boldsymbol{\sigma}} - \frac{1}{2}\,\overline{\rho\mathbf{v}''\,\mathbf{v}''\circ\mathbf{v}''} = (\mu + \frac{\mu_t}{Pr_k})\,\mathrm{grad}(k), \tag{10}$$

with the model constant $Pr_k = 2$.

The turbulent kinetic energy and the eddy viscosity are then obtained from the turbulence model. In case of laminar flow, both variables are set to zero to regain the original transport equations. For this paper two turbulence models were used for the numerical simulations: the transition model from Menter/Langtry augmented with in-house correlations [2] and the SSG/LRR-ω Reynolds-stress model by Speziale-Sarkar-Gatski (SSG) in the version of Eisfeld [3].

3 Numerical Methods

The computations have been performed using the DLR FLOWer code, Versions 116.+ and the QUADFLOW code developed in SFB 401. Both solvers use finite-volume formulations. The well-established FLOWer code is an explicit structured solver operating on hexahedral multi-block grids and contains different flux formulations including central discretization and all important techniques for convergence accelerations [4]. QUADFLOW is a new unstructured adaptive, implicit solver with surface based data structure. Different upwind flux formulations are implemented as well as acceleration techniques. Multi-scale analysis and B-Spline techniques form the basis for automatic grid adaptation. Both solvers are capable of doing hypersonic, turbulent flow computations because upwind discretizations and advanced compressible turbulence models are implemented [5].

3.1 Spatial Discretization and Time Integration

The finite-volume discretization that is applied to (1) in both solvers, ensures a consistent approximation to the conservation laws. Starting from a formulation in general curvilinear coordinates ξ, η, ζ, the computational domain is divided into non-overlapping hexahedra which are in FLOWer bounded by plane quadrilateral surface elements. The integral formulation (1) is then applied to each cell (i, j, k) separately. In QUADFLOW these elements are curved instead of plane in order to ensure that new surface grid points introduced during adaptation are positioned on the true wetted surface. Standard central discretization schemes are used for diffusive terms in both codes. For the convective terms upwind discretization was applied in both codes to achieve the present hypersonic flow results. The AUSM (Advection Upstream Splitting Method) scheme was used. Higher-order accuracy and consistency with the central differences used for the diffusive terms is achieved by MUSCL (Monotonic Upstream Scheme for Conservation Laws) Extrapolation, and TVD (Total Variation Diminishing) property of the scheme is ensured by a modified van Albada limiter function in FLOWer. In QUADFLOW the Venkatakrishnan limiter was used. FLOWer solves the system of ordinary differential equations by an explicit five-stage Runge-Kutta time-stepping scheme of fourth order in combination with different convergence acceleration techniques, like multigrid and local time stepping for asymptotically steady-state solutions [6]. Additionally, because for inviscid non heat-conducting stationary flow the total enthalpy is a constant throughout the flow field, its numerical deviation is applied as forcing function to accelerate convergence. For turbulent flow, the time integration of the turbulence equations is decoupled from the mean equations and the turbulence equations are solved using a Diagonal Dominant Alternating Direction Implicit (DDADI) scheme.

QUADFLOW uses a fully implicit Euler method for time integration and a Newton-Krylov-Method for the linear equation system. Local time stepping can also be applied in QUADFLOW.

Implicity increases the numerical stability of turbulent flow simulations which is especially important since the low Reynolds number damping terms as well as the high grid cell aspect ratios near the wall make the system of turbulent conservation equations stiff. Due to the CFL condition for explicit schemes in FLOWer, the CFL number of the multi-stage Runge-Kutta scheme has an upper limit of 4. Implicit residual smoothing allows to increase the explicit stability limit by a factor of 2 to 3 [6]. Due to the complete implicit formulation with use of the Newton-Krylov-Method QUADFLOW does not have such restricted limits.

3.2 Boundary Conditions

At the inflow, outflow and other farfield boundaries, a locally one-dimensional inviscid flow normal to the boundary is assumed. The governing equations are linearized based on characteristic theory and the incoming and outgoing number of characteristics are determined for both solvers. For incoming characteristics, the state variables are corrected by freestream values using the linearized compatibility equations. Else the variables are extrapolated from the interior [6]. However, a certain difference between QUADFLOW and FLOWer exists in the formulation of the characteristic boundary conditions.

In QUADFLOW the characteristics are formulated in space and time, in contrast FLOWer only uses space dependent formulations. For turbulent flow, the turbulent freestream values are determined by specifying the freestream turbulence intensity Tu_∞: $k_\infty = 0.667 \cdot Tu_\infty v_\infty$ and $\omega_\infty = k_\infty/(0.001 \cdot \mu)$. This is done in both flow solvers.

For steady inviscid flow, it is sufficient to set $[\![\tilde{v}\tilde{n}]\!] = 0$ at slip surfaces. In the viscous case, the no-slip condition is enforced at solid walls by setting all velocity components to zero. Additionally, the turbulent kinetic energy and the normal pressure gradient are set to zero. The specific dissipation rate is set proportional to the wall shear stress and the surface roughness. The energy boundary condition is directly applied through the diffusive wall flux: either by driving to zero the contribution of the diffusive flux for adiabatic walls or by prescribing the wall temperature or wall heat flux when calculating the energy flux through wall faces. At the symmetry plane of the half configuration, the conservation variables are mirrored onto the ghost cells to ensure symmetry.

4 Results

This chapter will present results for two scramjet configurations that were examined numerically with both flow solvers. The computations were performed in 3D fully turbulent with the SSG-ω Reynolds-stress model in FLOWer and with the transition model augmented by in-house correlations in QUADFLOW. This was done to investigate the differences in the solution for a fully turbulent assumed flow field and a more realistic transitional one.

The first analysed intake concept is a 2D double ramp configuration as seen in fig. 2 with non-converging sidewalls. The second one (fig. 3) is a single ramp configuration with converging sidewalls to realise additional compression. These concepts have a defined leading edge radius on the first ramp and on the cowl of 0.5 mm. The sidewalls always have sharp leading edges.

Fig. 2. Hypersonic 2D inlet model with non converging sidewalls

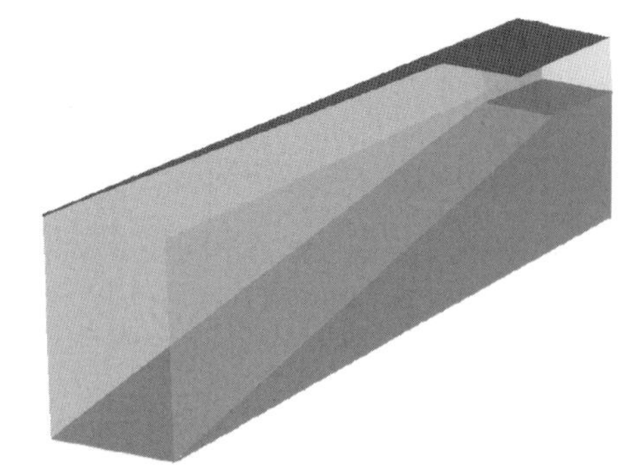

Fig. 3. Hypersonic 3D inlet model with converging sidewalls

Remarkable differences occur by using the 3D geometry with converging sidewalls. Due to the additional sidewall compression for the 3D intake, it can be shorter in comparison to a 2D one to get the same compression ratio. Thus, at first glance the boundary layer is expected to be thinner before entering the isolator. A thinner BL leads to more mass flow entering the combustion chamber because of the smaller displacement thickness. The sidewalls create vortices, which cause extra shocks as well as additional shock/boundary layer and shock/shock interactions, that enhance the separation regions and the thickness of the boundary layers.

4.1 Flow Conditions, Geometrical Description and Numerical Grids

In the following section the flow conditions, geometrical concepts and numerical grid aspects for the different test cases are introduced. The same inflow conditions will be used for the both cases. The assumed data correspond to real flight conditions for an altitude of 30 km. The flow conditions are listed in table 1. For describing the intake geometries a coordinate system is introduced. The x-axis points in free stream flow direction, z is defined along the leading edge of the ramp and y completes the right hand Cartesian reference system.

2D Intake with Non-Converging Sidewalls

The first intake concept has a similar configuration to the ones investigated at the Shock Wave Laboratory and at DLR. It consists of a double ramp followed by a convex curved ramp shoulder and a diverging isolator (angle of $-1°$). The first ramp is 712.9 mm long and has an angle of 7.5° to the x-axis. The second ramp is 96.5 mm long and has an angle of 19.0°. The cowl is at y=228 mm. The intake has a width of 76 mm. The convex ramp shoulder is defined by a third order polynomial given as follows:

$$y = 1.4134241 \cdot 10^{3} - 5.5189621 \cdot 10^{0} x + 7.4191574 \cdot 10^{-3} x^{2} - 3.1211094 \cdot 10^{-6} x^{3} \ . \tag{11}$$

The sidewall starts at x=y=0, has a height of 5 mm at the beginning and an angle of 16° to the x-axis. Its upper boundary is parallel to the x-axis and hits the cowl. In the following, this intake will be called 2D-I-NCS.

3D Intake with Converging Sidewalls

The second intake is similar to the configuration sketched in fig. 3. It shows a single ramp for compression and a convex edge to redirect the flow into horizontal direction followed by a short isolator. The ramp is 483.4 mm long and has an angle of 13.5° to the x-axis. The cowl is at y=150 mm. The intake is 106 mm wide at the inflow side and 76 mm wide at the isolator. The sidewall starts at x=y=0, has a height of 150 mm at the beginning with no angle to the x-axis and converges with 2.8° on each sidewall side. In the following, this intake will be called 3D-I-CS.

Flow Conditions

Table 1 shows the flow conditions. The free stream turbulence intensity was assumed to be 0.2 %.

Table 1. Flow conditions for the different investigated test cases

test-case	Mach	Re_l [1/m]	T_∞ [K]	p_∞ [Pa]	T_{wall} [K]	Tu_∞ [%]
2D-I-NOS	8	$2.945 \cdot 10^6$	226.7	1171.9	300	0.2
3D-I-CS	8	$2.945 \cdot 10^6$	226.7	1171.9	300	0.2

Numerical Grids

The grids used had an approximate total of 6 million points, a minimum resolution of $1 \cdot 10^{-6}$ in x, y and z direction on solid walls and a maximum resolution of $2 \cdot 10^{-3}$ at the symmetry plane in the middle of the intake or at outflow boundaries. This led to y^+ values below 1.

4.2 Numerical Results

2D Intake with Non-converging Sidewalls (2D-I-NCS)

In the following a comparison between the fully turbulent solution performed with the SSG-Reynolds-stress model in FLOWer and the one done with the SST transition model in QUADFLOW will be presented. The Mach contours for the fully turbulent solution are shown in fig. 4. Despite a very small sidewall, the ramp shocks are bent in y-direction yielding an increased loss in mass flow. Vortices are generated in the corners between sidewalls and ramps. The vortices lead to shock bending, very high heat loads on the surface and an inhomogeneous flow distribution in the isolator. All these effects can also be seen for the transitional computation. Nevertheless, differences occur due to the fact that the BL is laminar on the first ramp and a separation bubble is formed in the kink between the two ramps by using the transition model. The change from laminar to turbulent state occurs in the shear layer on top of the bubble.

Fig. 5 compares the temperature distribution at isolator exit for the FLOWer and QUADFLOW computation. The QUADFLOW solution shows a more homogeneous temperature distribution and bigger region of high temperature on the isolator bottom than the FLOWer solution. In the contrary FLOWer gives higher temperatures up to 1600 K, which is almost 250 K higher then the maximum predicted temperature by QUADFLOW. Such a hot

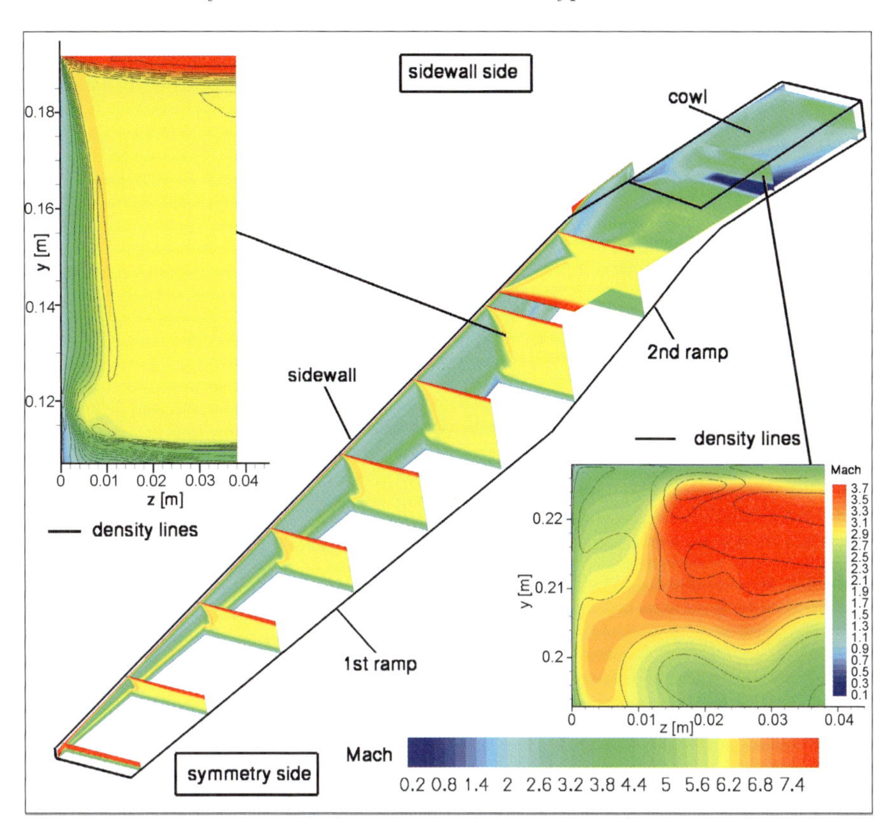

Fig. 4. Mach for SCRamjet intake 2D-I-NCS

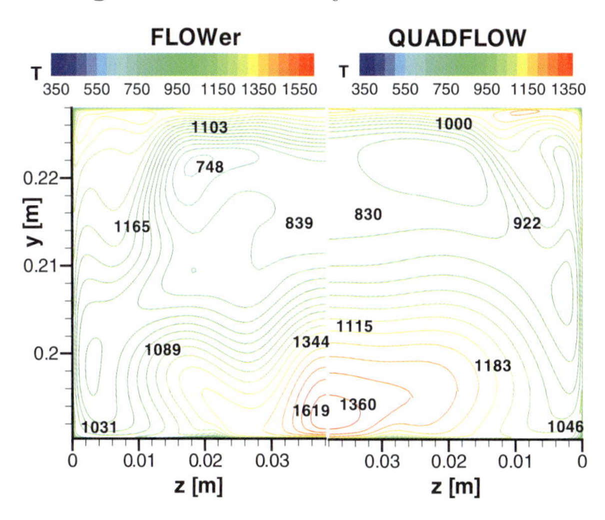

Fig. 5. Comparison of temperature distribution at isolator exit for SCRamjet intake 2D-I-NCS

spot might help to ignite the hydrogen in the combustion chamber, whereas the more homogeneous temperature distribution might support a more stable combustion.

It was found that the continuously curved ramp shoulder is favourable with respect to mass flow loss and to heat loads on the walls in comparison to a convex corner. It has been proven that moderate ramp angles to compress and redirect the flow into the inlet lead to a higher efficiency and a more homogeneous isolator flow. As can be seen in fig. 4 the sidewall was minimised to catch only the first ramp shock. This results in smaller sidewall BLs and thus leads to less shock bending and sidewall compression. This finally results in less mass flow loss and higher intake efficiency.

3D Intake with Converging Sidewalls (3D-I-CS)

To avoid the deficiencies of simply ramp compression (e.g. small compression ratio, long intake, high enthalpy and mass flow loss), studies of a 3D intake (as sketched in fig. 3) were conducted. The sidewalls along the ramp converge with an angle of 2.8° and thus contribute to the compression of the flow, resulting in a much shorter and a wider intake (intake length/width: 2D:L/W $\approx$ 14; 3D: L/W $\approx$ 4). The averaged Mach number and temperature at the isolator exit were approximately the same as computed for the 2D-I-NCS. The sidewalls begin with the first ramp and their leading edges are normal to the x-z plane and point in y direction (a coordinate system is shown in fig. 6). This leads to thick boundary layers at the sidewalls and to large vortices generated in the corners. These vortices reduce the BL thickness in some regions because slow wall material is transported away from the walls. At other locations these vortices lead to separation zones at the sidewalls and the ramp. These additional separations generate additional shock-shock interactions.

Fig. 6 shows the Mach number distribution for half of the intake. The sidewall compressions generates an oblique shock that interferes with the one created on the ramp. This creates vortex structures. In comparison to the intake with non-converging sidewalls these vortex structures are stronger and have even higher influence on the flow field. A thick hypersonic BL is created on the ramp surface. Its thickness is reduced due to the expansion over the convex corner. In contrary to the intake configurations examined before, this one shows no separation bubble in the isolator. It can be seen in fig. 6 that the ramp shock passes the cowl with quite a big distance. The reason for this is, that the design was done with oblique shock theory and a 2D method of characteristics from DLR Cologne [7]. Therefore, the effects of sidewall compression could not be taken into account. To improve the efficiency and to reduce the mass flow loss of the intake, the ramp can be shortened or the cowl lip can be moved further upstream.

Fig. 7 presents the effective intermittency distribution for several cut planes. The top right picture shows a view from the top (plane at 80% of intake height). The picture on the bottom right presents a view from the

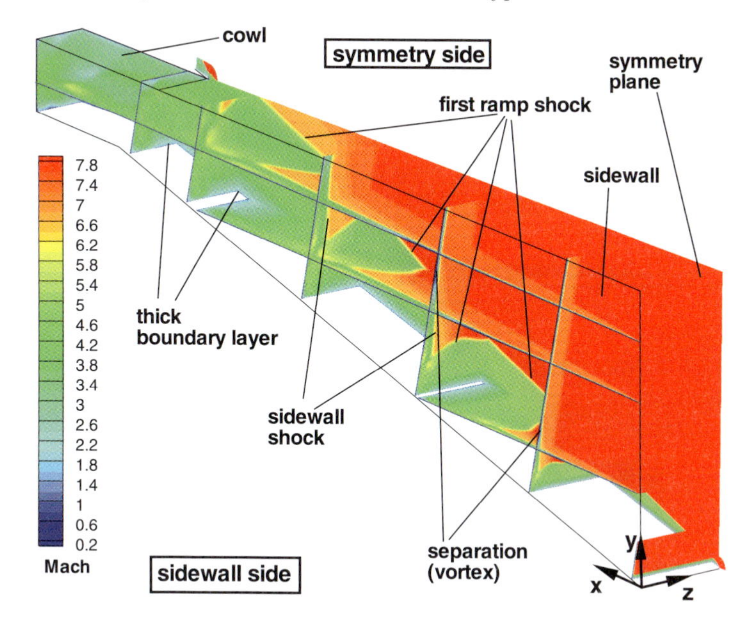

Fig. 6. Mach contours for 3D SCRamjet intake with sidewall compression (3D-I-CR)

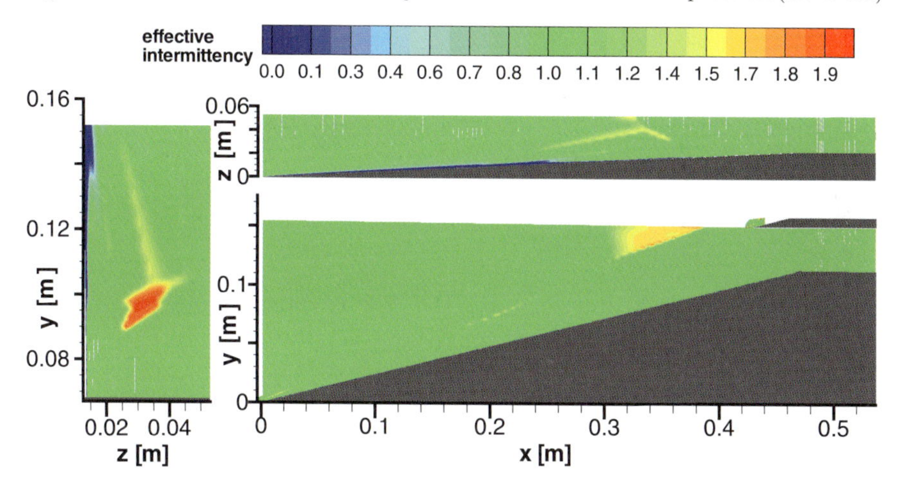

Fig. 7. Effective intermittency distribution for several cut planes of intake 3D-I-CR

side (symmetry plane) and the left one a view from the front (plane at 50% of intake length). The intermittency distribution displays a fully turbulent boundary layer on the ramp. The reason for this might be the higher deflection angle in comparison to the 2D intake configuration. The view from the top shows a partly laminar BL on the sidewall. At approximately x=0.24 m a small separation is created, which generates an additional shock in front of it. Downstream of the separation the flow reattaches and turns into turbulent state. The picture on the left of fig. 7 shows a similar behaviour. The bound-

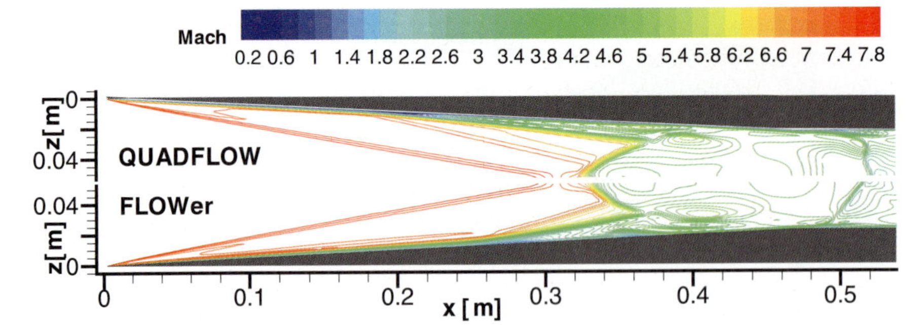

Fig. 8. Mach number distribution for QUADFLOW and FLOWer of intake 3D-I-CR (x-y plane)

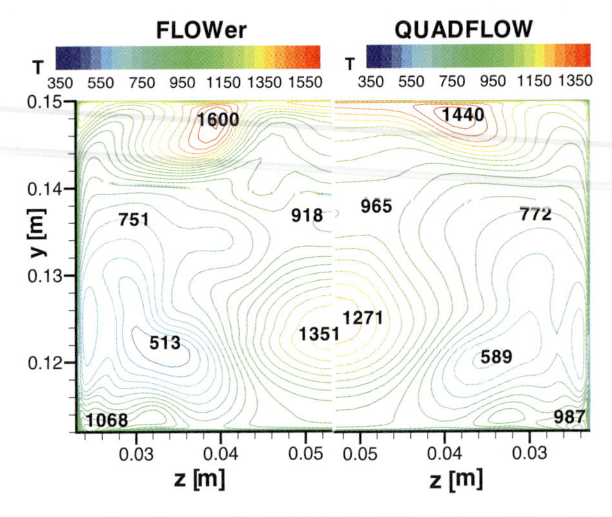

Fig. 9. Temperature distribution for QUADFLOW and FLOWer of intake 3D-I-CR (x-z plane)

ary layer is laminar before it passes the ramp shock. For this particular cut it means that the sidewall boundary layer becomes turbulent at y=0.085 m. Every part of the BL above this value is laminar. This changes the flow field in certain ways. This can be seen in figure 9.

Fig. 8 presents the Mach number distribution for a cut plane (x - z plane, view from the top) at 80% of the intake height, i.e. in the middle of the isolator. In the upper part of the picture the QUADFLOW solution is displayed, on the bottom the FLOWer one. It is recognisable that the oblique shocks generated at the sidewall leading edges have the same angles and strength. On the contrary, the BL thickness is different for both solutions. Due to the fact that the boundary layer is completely turbulent in the FLOWer solution, it is much thicker than in the QUADFLOW solution. Due to the laminar boundary layer a small separation is created in the QUADFLOW solution at $x \approx 0.2m$. This causes an additional shock in comparison to the FLOWer solution. As mentioned before, the vortex stuctures at the sidewalls are also different. That changes the flow field downstream of the interacting shock planes. This leads to different exit conditions.

The temperature and pressure distribution at the isolator exit are very important, because they define the inlet conditions for the combustion chamber. Hence, fig. 9 shows the temperature distribution for the isolator exit. In the left part of the picture the FLOWer solution can be seen, the QUADFLOW solution is on the right side. The pressure distribution on the ramp showes only minor differences in the numerical solutions, but the temperature distribution for the isolator exit changes strongly if the transition model is used. The amplitude of the vortex strength is smaller for the QUADFLOW solution, i.e. the maximum temperature is lower and the minimum temperature is higher. This means that the distribution is more homogeneous. Whether this is more appropriate than the more inhomogeneous distribution predicted by FLOWer is still to be tested by combustion simulations. The hot spots might ensure an ignition, but the more hogomeneous flow might start combustion in more spots.

Nevertheless, these computations show that it is absolutely necessary to consider transition and relaminarisation effects for hypersonic intake flows. On the one hand, it is essential to predict the right heat flux into the structure and to predict the right flow structure. On the other hand, it is necessary to define the right inflow conditions for the combustion chamber.

4.3 Comparison of the Different Intake Configurations

Table 2 shows a comparison between the results of the intakes: 2D-I-NCS and 3D-I-CS (for same free stream flow conditions and isolator inlet cross-section). Considering these results, the 3D intake seems to be superior at first glance, but there are several disadvantages. Due to the strong shock interactions and vortices the flow field is very inhomogeneous. There are large gradients within the flow quantities. It is also not clear how this complex flow system will react on disturbances in the free-stream conditions. Because of the inhomogeneous flow great differences in the heat loads on the surface of the structure exist. At the exit of the isolator zones with very high temperatures (about 1600 K in FLOWer and 1450 K in QUADFLOW) and zones with very low temperatures (about 550 K) are in close vicinity. It has to be decided by the combustion chamber specialists whether this is favourable or modifications of the compression flow are required. Further investigations have to be done for the 3D intake to get a broader basis for decisions by combustion chamber arguments. To avoid the difficulties that were mentioned above for a 3D intake, the 2D-I-NCS seems to be a good choice. The flow is more homogeneous at isolator exit. The heat loads on the surface are lower and there are less disturbances in the flow. Problems might occur with the overall dimensions of such an intake, especially for wind-tunnel or shock tube experiments. The

Table 2. Comparison of different intake configurations, values at isolator exit

	2D-I-NCS (F/Q)	3D-I-CS (F/Q)
mass flow possible [kg/s]	0.753	0.69146
loss [%]	20.88 / 14.6	15.23 / 17.29
averaged temperature [K]	1035 / 969.4	987 / 970.4
averaged contraction ratio	28.5 / 28.4	42.7 / 42.3
averaged Mach number	3.02 / 3.2	3.24 / 3.27

differences in the homogeneity of the flow at isolator exit can be seen by comparing fig. 9 and 5.

Table 2 shows a comparison of the flow quantities for the different intake configurations for both flow solvers. The first value in the column stands for the FLOWer solution (F) and the second for the QUADFLOW (Q) one. It can be recognised that the differences between the fully turbulent FLOWer solution and the transitional one with QUADFLOW are quite small and the reasons for them are manifold. For the intake 2D-I-NCS the transitional computation predicts 6% less mass flow loss, but the averaged temperature at the isolator exit is smaller than for the FLOWer simulation, whereas the Mach number is a little higher. The laminar boundary layers at the first ramp and on the sidewalls might be the reason for less spillage. A laminar BL is thinner and leads to a smaller shock distance to the leading edge and to a flatter shock angle, which results in less spillage. A partly laminar BL at the sidewalls was observed for the intake 3D-I-CS. The BL at the ramp is completely turbulent from the start. Due to the laminar BL parts at the sidewalls, additional separations occur together with a different vortex structure in comparison to the fully turbulent result. This leads to an increase in spillage and less contraction, followed by a bigger Mach number.

How big the influence of the flow solver, the turbulence model or the fact of considering transition is on the presented results, has to be pointed out doing more computations. It might be necessary to do either a computation with the SST turbulence model in QUADFLOW and FLOWer or to do a fully turbulent simulation in QUADFLOW using the SSG Reynolds-stress model. Additionally, it might be interesting to do a computation with the transition model coupled with the Reynolds-stress model. This will be part of future work. Furthermore, comparisons with experimental results will be done as soon as they are available.

5 Computational Considerations

The FLOWer computations are performed on a NEC SX-8 cluster using 16 processors. The used grids generally have a number of cells between three and four millions. The number will increase by approximately half a million cells once the isolator section has to be extended. So far, the memory requirements in the 3D cases are around 80 gigabyte. The vectorization level of the FLOWer program package is 99.7%. For a typical problem, answer time lies between five to seven days on the NEC SX-8, depending on how busy the machine is. A single batch job performs approximately 30000 time steps and requires 10 hours of CPU-time per node, after which it is resubmitted into the batch queue. FLOWer uses block based MPI parallelization.

6 Conclusions

In this paper the performance analysis of two intakes at a flight Mach number of 8 is performed. The role of the present numerical computations is to complement the experimental investigations and to enhance the understanding of the obtained results. Additionally, the numerical simulation completes the knowledge of the flow field in areas which are not accessible to measurements and allows for an overall performance analysis of the inlet geometry. The current analysis denotes the developing corner vortices as a key feature to ensure self ignition for the low temperature (T_0 = 1550 K) experimental test run.

References

1. Goonko, Y. P., "Investigation of a Scramjet Model at Hypersonic Velocities and High Reynolds Numbers," *AIAA*, Vol. 5273, 2002.
2. Krause, M. and Ballmann, J., "Application of a Correlation-Based Intermittency Transition Model for Hypersonic Flows," Deutscher Luft–und Raumfahrtkongress, paper no. DLRK2008–081235, September, 2008.
3. Eisfeld, B., "Die Eigenwerte des Systems aus RANS- und Reynolds-Spannungsgleichungen," Department for Aerodynamics and Fluid Mechanics, TU Braunschweig, Internal Report, IB 124-2003/1, 2003.
4. Kroll, N., Rossow, C.-C., Becker, K., and Thiele, F., "The MEGAFLOW Project," *Aerospace Science and Technology*, Vol. 4, No. 4, 2000, pp. 223–237.
5. Reinartz, B. U., van Keuk, J., Coratekin, T., and Ballmann, J., "Computation of Wall Heat Fluxes in Hypersonic Inlet Flows," AIAA Paper no. 02-0506, Proceedings of 40th AIAA Aerospace Sciences Meeting and Exhibit, Reno (NV), USA, 14-17 January, 2002.
6. Radespiel, R., Rossow, C., and Swanson, R., "Efficient Cell-Vertex Multigrid Scheme for the Three-Dimensional Navier-Stokes Equations," *AIAA Journal*, Vol. 28, No. 8, 1990, pp. 1464–1472.
7. Häberle, J. and Gülhan, A., "Experimental Investigation of a Two-Dimensional and a Three-Dimensional SCRamjet Inlet at Mach 7," *Journal of Propulsion and Power*, Vol. 24, No. 5, September–Oktober 2008, pp. 1023–1034.

Preinvestigations of a Redesigned HIRENASD Wing Model in Preparation for New Aero-Structural Dynamic Experiments in ETW

B.-H. Chen[1], L. Reimer[1], M. Behr[1], and J. Ballmann[2]

[1] Lehrstuhl für Computergestützte Analyse Technischer Systeme (CATS), RWTH Aachen, Schinkelstr. 2, D-52062 Aachen, Germany
http://www.cats.rwth-aachen.de
chen@cats.rwth-aachen.de
[2] Lehr- und Forschungsgebiet für Mechanik (LFM), RWTH Aachen, Schinkelstr. 2, D-52062 Aachen, Germany
http://www.lufmech.rwth-aachen.de
ballmann@lufmech.rwth-aachen.de

Summary. In this paper preliminary numerical investigations are presented which have been conducted at the Chair of Mechanics (LFM[3]) of RWTH Aachen University during the course of the new experimental/numerical project Aero-Structural Dynamics Methods for Airplane Design (ASDMAD). The goal was to study the dynamic vibrational decay behaviour of the HIRENASD wing modified by different wingtip devices. After a short introduction to the project specifications the numerical method and model is described. The calculated results are compared with those of the original HIRENASD model, and conclusions are drawn with respect to the differences between the two models.

1 Introduction

Today the development and design of future aircraft are more than ever influenced by reducing environmental effects, besides demanding increase in safety, comfort and above all efficiency. In this context methods of Computational Aero-Structural Dynamics (CASD), which have proven correct prediction ability for non-stationary aeroelastic experiments in realistic ranges of Mach and Reynolds numbers play a major role in achieving this goal.

The ASDMAD (Aero-Structural Dynamics Methods for Airplane Design) project aims are the further development of new computational and experimental methods of Aero-Structural Dynamics (ASD) as tools for the design of large passenger aircraft. The CASD package SOFIA developed at RWTH

[3] has meanwhile merged into Chair for Computational Analysis of Technical Systems (CATS)

W.E. Nagel et al. (eds.), *High Performance Computing in Science and Engineering '09*, DOI 10.1007/978-3-642-04665-0_30,

Aachen University within the frame of the Collaborative Research Centre "Flow Modulation and Fluid-Structure Interaction at Airplane Wings" (SFB 401) forms the basis for the computational tool and the preparation of new experiments in the European Transonic Wind Tunnel (ETW) under cryogenic conditions to achieve high Reynolds numbers. In the more recent past SOFIA has been used for the design and analysis of a wind tunnel wing model and validated successfully against *static* and *dynamic* aeroelastic experiments in *subsonic* flow, conducted within the SFB 401's central HIRENASD project *High Reynolds Number Aero-Structural Dynamics* [1–6]. Fig. 1 shows the numerical and experimental HIRENASD model on the left and right, respectively.

For the new experiments in ASDMAD project the SFB 401 will provide its elastic wing model, which has a 34 degrees leading edge sweep and was developed, built and equipped with measuring techniques within the frame of the HIRENASD project, and tested in ETW. The wing model has been redesigned in the tip region to be shortend first and thereafter equipped with different tip devices of which one will be a winglet with aerodynamic control surface (WACS) which can be actuated (ACS). The redesigned wing model will have the same span as the original one.

The model preparations for the experiments in ETW are currently in progress. In the run-up to the experiments numerous preliminary investigations have been performed, including static and dynamic aeroelastic simulations. During this reporting period four winglet shapes differing only in the winglet additional sweep angle (0°, 5°, 10°, 15°) have been designed and investigated regarding their aerodynamic performance considering static aeroe-

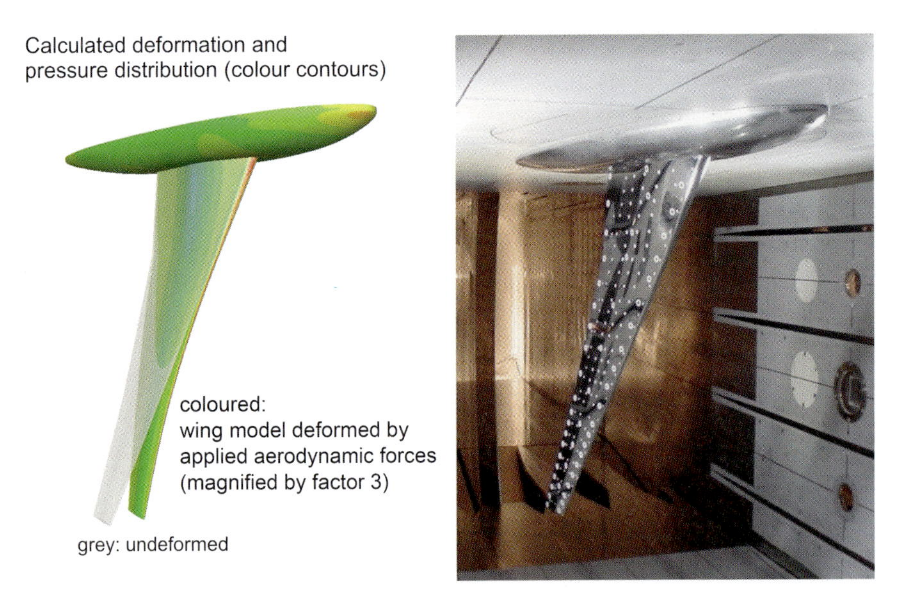

Fig. 1. HIRENASD wind tunnel model

lastic simulations. The configuration with about 10 degrees additional sweep was selected in a compromise decision between aerodynamic properties and required internal space for the ACS driving mechanism. For the thus modified wing model extensive static aeroelastic simulations were carried out with and without deflected WACS at angles between -5° and $+5^o$. Aeroelastic equilibrium configurations were investigated to determine the effects resulting from changes of the parameters Mach number Ma, dynamic loads factor which is dynamic pressure q over Young's modulus E, q/E, and angle of attack α on aerodynamic design loads. Furthermore, the influence of the WACS on structural loads and flow behaviour was analysed. The pre-examinations further comprised the determination of the vibrational decay behaviour of the wing as a function of the incident flow. In this process time histories of deformation, in particular vertical displacement u_{tip} and structural torsion $\phi_{T,tip}$, at wing tip and harmonic spectra are recorded. The latter enable the identification of the natural frequencies that are involved in bending and torsional vibration. The dynamic response simulations are also carried out with fixed WACS deflection. The results are analysed w.r.t. the aerodynamic influence of the winglet shape and the efficiency of the WACS with regard to its influence on the dynamic wing in comparison to the original HIRENASD model. In this report only results of the dynamic investigations are presented, with emphasis on the dynamic response behaviour of the wing-winglet model.

2 Numerical Method and Model Representation

2.1 CAE Package SOFIA

The aeroelastic software package SOFIA follows a partitioned approach, in which separate programs are operated iteratively for the solution of structural deformation and the flow solution on a deforming grid. The Aeroelastic Coupling Module (ACM) [1] which allows for distinct Euler- or Navier-Stokes flow solvers with distinct FE-based structural solvers forms the core of SOFIA. On the one hand ACM coordinates the sequence of solver calls, and on the other hand it transfers the aerodynamic loads and structural deformations in mutual directions between flow field and structural solver. For unsteady aeroelastic problems different loose coupling schemes with prediction-/correction steps are available, enhanced by extrapolation techniques for flow and structural deformation states as well as a tight coupling scheme. For steady aeroelastic problems an under-relaxed Gauss-Seidel scheme is used.

The deformation state of the structure is typically determined in SOFIA using the in-house FEAFA code (*Finite Element Analysis for Aeroelasticity*). It offers a full range of FE (*Finite Element*) types comparable to commercial CSM (*Computational Structural Mechanics*) packages. However, the slenderness of aircraft structures motivates, at least in parts, a discretisation with generalised Timoshenko beam elements provided also in FEAFA where this is admissible in terms of beam idealisation. In doing so in SOFIA, very accurate aeroelastic flow simulations for complete aircraft are possible with only slightly increased computation cost compared to pure CFD computations which ignore structural deformation. To integrate the system of equations of motion in time either a Bossak-α scheme is used or, after modal decomposition, the resulting decoupled equations of model motion are solved by applying the Newmark scheme. Resulting linear systems of structure equations are solved by applying the freely accessible SPARSKIT library [7] which is optimised for sequential sparse matrix operations. Due to the comparably low cost of the structural dynamics computation part, at least in comparison with the flow solution, FEAFA is still programmed as sequential code in FORTRAN90.

Although two other flow solvers are currently available in SOFIA, i.e. QUADFLOW [8] and TAU [9], the FLOWer code is preferred to be used at HLRS because of its optimisation for vector machines. The development of FLOWer was led by the German Aerospace Center (DLR) during projects MEGAFLOW I/II [9] and MEGADESIGN [10]. FLOWer solves the three-dimensional and time-dependent Favre- and Reynolds-averaged Navier-Stokes (RANS) equations for perfect gas on structured deformable multi-block grids. Either central differencing or upwinding can be used to discretise fluxes. Reynolds stress tensor can be approximated by numerous one- and two-equation turbulence models or optionally computed directly by different Reynolds stress models. Stationary flow problems are solved by a multi-step Runge-Kutta scheme. Convergence to steady state is accelerated applying multigrid, local time stepping and implicit residual smoothing. Since non-stationary flows can be computed using FLOWer by employing implicit Dual-Time Stepping, the time step size can be chosen based on the time scales according to the physical properties of the aeroelastic problem. Besides its vector optimisation, FLOWer offers high-performance computing techniques for shared and distributed memory machines, i.e. both coarse- and fine-grain parallelisation by using MPI and OpenMP libraries. FLOWer is written almost completely in FORTRAN77.

For the deformation of multiblock-structured grids (e.g. FLOWer grids) the in-house MUGRIDO code (*Multiblock Grid Deformation Tool*) [11, 12] is applied. It generates a fictitious framework of beams by modelling the CFD block boundaries and a given percentage of grid lines as massless linear elastic Timoshenko beams. These are considered rigidly fixed together in points of intersection and to the aerodynamic surface as well, such that cell angles are preserved where beams, i.e. grid lines, intersect or emerge from a solid surface. The right hand side for the resulting FE problem in MUGRIDO is supplied by the deflections of the wetted surface relative to the undeformed grid. A well shaped volume CFD grid is finally reconstructed from the deformed beam framework by two- and three-dimensional Transfinite Interpolation.

2.2 Wind Tunnel Model Specifiactions

Fig. 2 presents the computational model and the aerodynamic shape of all winglets investigated. In this figure the one with the 10° sweep angle additional to the 34° leading edge sweep at the very right was selected for the redesign of the wind tunnel model. The main wing part is the shortened HIRENASD model [6], whereas the half span of almost $1.30m$ of the new model is preserved. The winglet has a dihedral of 40° and a length of approximately $0.12m$. The geometric properties of the winglet are shown in Fig. 3. The proposed cut-off positions of the HIRENASD wind tunnel model for attaching the RWTH winglet are depicted in Fig. 4. The trailing edge of the winglet part belongs completely to the lower side of the HIRENASD model. The cut-off positions of the upper and lower shells of the HIRENASD wing are displaced in order to stiffen the mounting of the winglets. The additional winglet will also be manufactured from *C200 Maraging Steel* which is the same material as for the HIRENASD model. The concept of the control device actuation is aimed at a mechanical version, but has not been designed completely yet. The torque to be produced by the drive unit of the actuator mechanism to deflect the control device to a particular deflection angle was estimated using maximum control device loads computed in aforementioned RANS-based aeroelastic simulations using the SOFIA package.

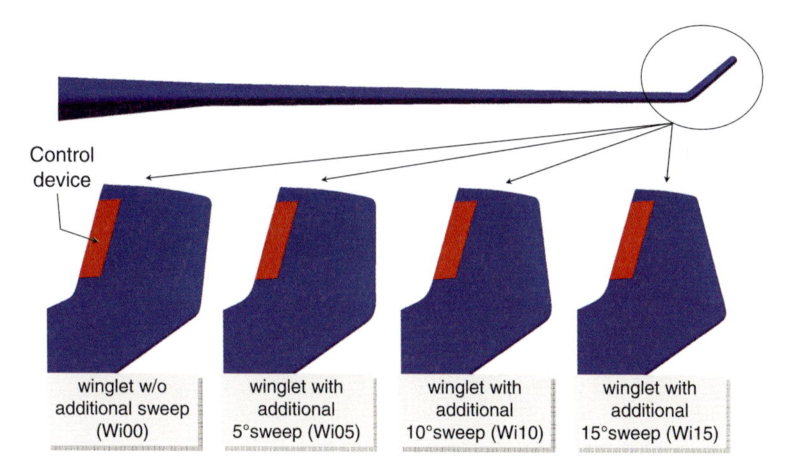

Fig. 2. Investigated winglet shapes with different sweep angles

2.3 Structural Model

The properties of the wing-winglet structure was idealised using an efficient beam model based on multi-axial Timoshenko beam elements. The structural

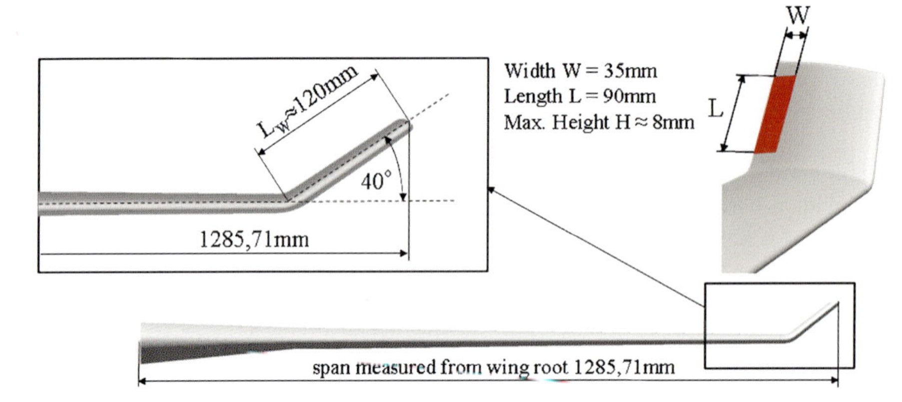

Fig. 3. Geomtric properties of the winglet

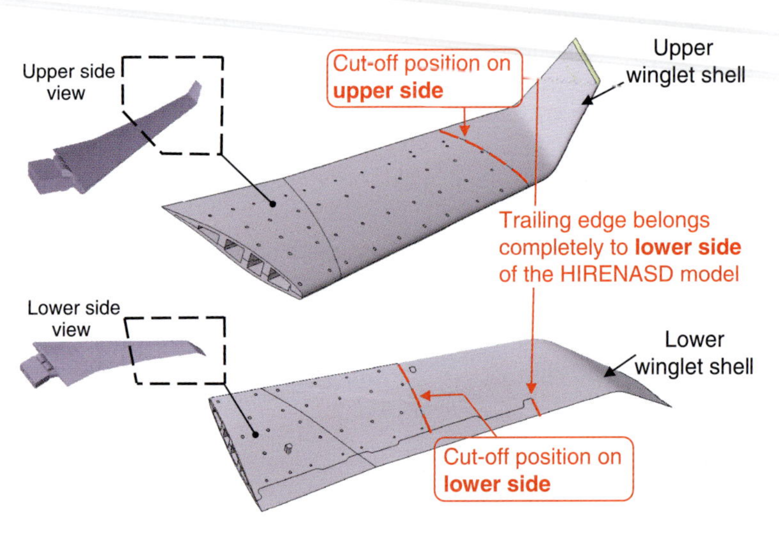

Fig. 4. Proposed cu-ff positions for the upper and lower parts of the wind tunnel model

representation has been generated by extending the already existing beam identification of the original HIRENASD wing model [1] by a very stiff winglet part. Fig. 5 shows the aerodynamic surface of the wing-winglet model and the beam model, which is divided into two parts. The left one represents the balance and the clamping region including the excitation mechanism and the right one the wing, respectively.

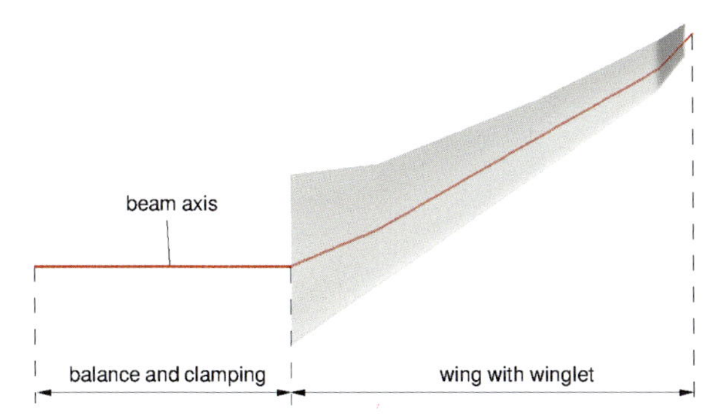

Fig. 5. Aerodynamic surface and beam model with additional consideration of balance inertia and elasticity

2.4 CFD Grid for Flow Simulation

For the numerical investigations a new CFD grid was designed. It constitutes a half-model, whereby only cases with flow conditions symmetrical to the fuselage mid-plane were considered.

The grid depicted in Fig. 6 was generated using the program ICEM CFD from ANSYS and has approximately 1.6 million grid points and 1.4 million cell volumes. The wetted surface has a no-slip condition applied on it. A symmetry plane normal to the wing mid-plane is assumed at the wing root corresponding to an inviscid plane wall. Farfield conditions are specified at the domain boundaries far away from the model. Therefore no influences from wind tunnel walls are considered in this study.

3 Prediction of Aeroelastic Model Behaviour

The numerical results presented here are obtained using the CAE Package SOFIA (see sec. 2.1) and comprise the predicted dynamic behaviour of the elastic wing-winglet model defined in sec. 2.2. The structural numerical model and flow grid used are described in sec. 2.3 and sec. 2.4, respectively. The Linearised Explicit Algebraic (LEA) model, based on the solution of two additional equations, one for the turbulent kinetic energy k and one for the specific dissipation rate ω was used in all CAE simulations to model the turbulence.

The dynamic preliminary investigations were conducted for six different transonic Mach numbers (Ma=0.75, 0.78, 0.80, 0.83, 0.85 and 0.88) and three angles of attack (α=0°, 2° and 4°). For each Mach number four different values of loading factors (q/E=$0.22 \cdot 10^{-6}$, $0.34 \cdot 10^{-6}$, $0.48 \cdot 10^{-6}$ and $0.60 \cdot 10^{-6}$) were examined, while keeping the Reynolds number constant at Re=$23.5 \cdot 10^6$. At this value of the Reynolds number transition is concentrated on the leading edge region such that the flow is fully turbulent over the whole wing surface.

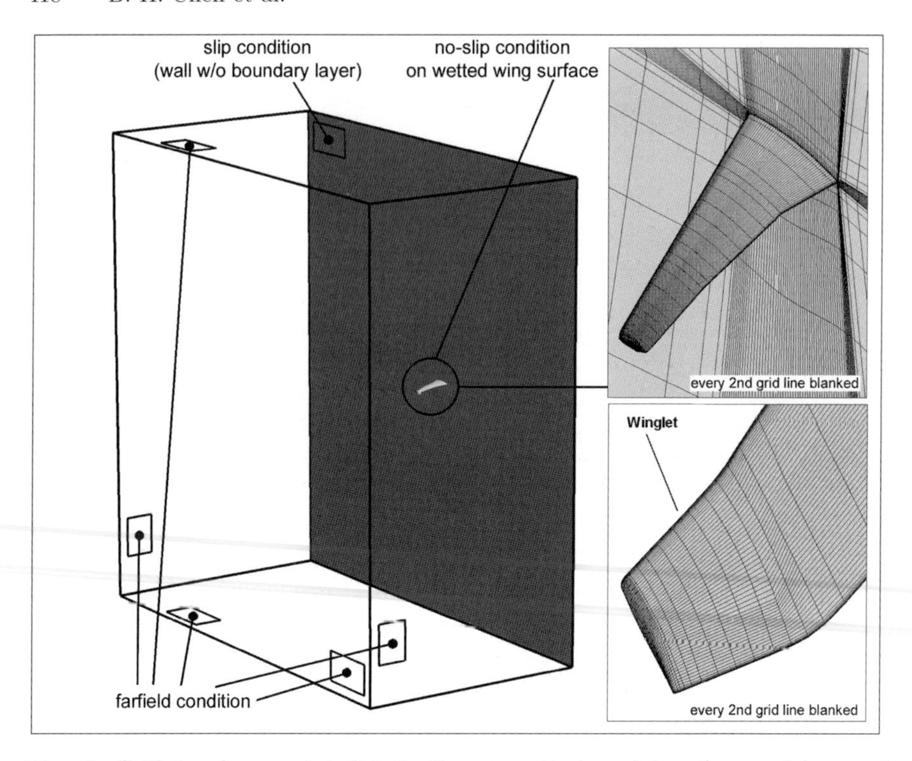

Fig. 6. Grid topology, point distribution on wetted model surface and imposed boundary conditions for the flow grid with wing-winglet configuration

The envelope for the experiments in ETW is depicted in Fig. 7 for Mach number Ma=0.80.

3.1 Dynamic Aeroelastic Response Behaviour

Dynamic aeroelastic problems are characterised by the interaction of the wing's elastic and inertial properties with the ambient flow. Compared to the values under vacuum condition, oscillation frequencies and damping of the structure change due to the interaction with the flow. It is therefore essential to verify that the wing model holds positive aerodynamic damping for all planned flow conditions. Hence, vibrational decay simulations of the wing-winglet model about the aeroelastic equilibrium configuration were conducted for the mentioned flow conditions.

In a preliminary step additional external loads were defined in order to obtain an AEC with additional deflection. The loads were defined such that the initiated shape deviation corresponds to combinations of 1st, 2nd flap-bending and 1st torsion dominated mode shape. These loads were then applied during

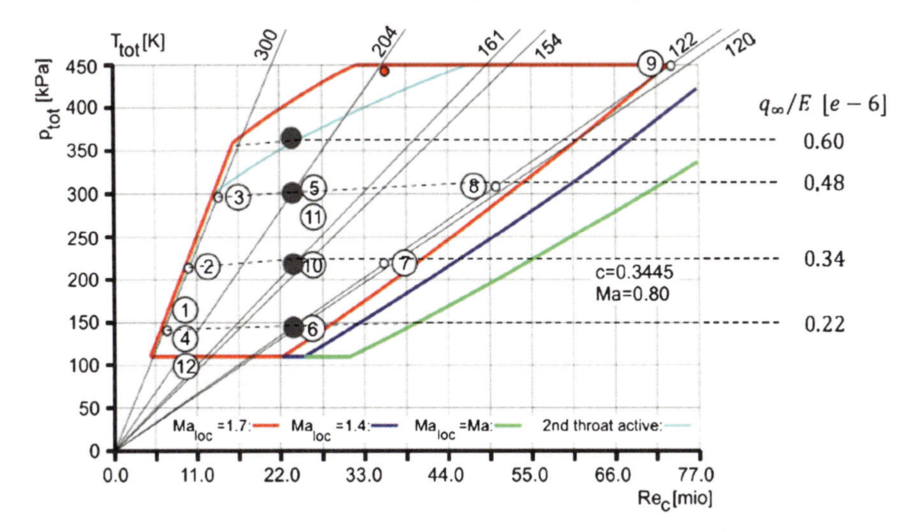

Fig. 7. ETW test envelope with experimental conditions of the HIRENASD project denoted as numbered circles and conditions of the ASDMAD project as black dots

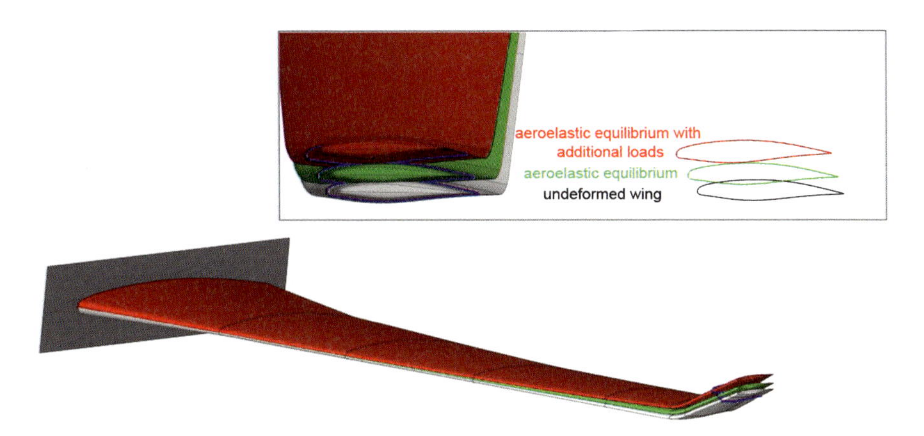

Fig. 8. Different initial AECs of the wing with and without subjecting the wing with additional loads

steady aeroelastic simulations, resulting in a modified aeroelastic equilibrium as shown in Fig. 8.

After the aeroelastic equilibrium is obtained, the additional loads are suddenly removed, initiating the unsteady simulation. The resulting imbalance between inner structural reaction forces and outer aerodynamic forces leads to vibrational motions. Fig. 9 exemplarily shows time histories of vertical displacement u_{tip} and pure structural torsion $\phi_{T,tip}$ at wing tip, the latter denoting the rotation of cross-sections oriented perpendicular to the elastic

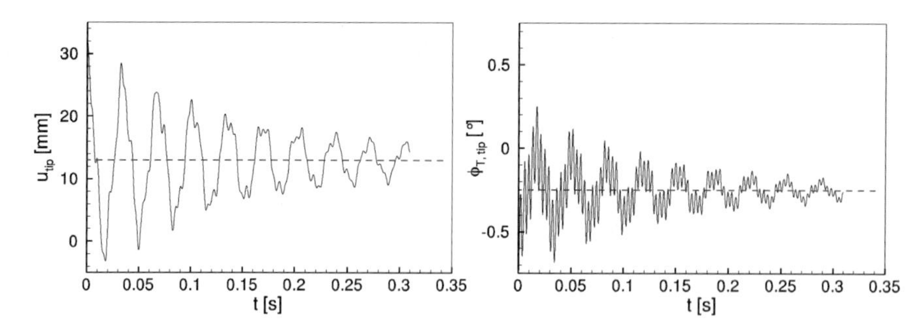

Fig. 9. Time histories of displacement u_{tip} and $\phi_{T,tip}$ on the left and right, respectively, during vibration decay

axis. The initial amplitude of bending deformation is about $20mm$, the torsional vibration starts with an amplitude of 0.4^o. Both vibrational motions reveal positive aerodynamic damping at imposed flow conditions, whereas the torsional tip motion oscillates with higher frequency than the translational one.

The test campaign of the new wind tunnel experiments will only consider two of the three independent wind tunnel conditions in the HIRENASD project, i.e. the variation of Mach number Ma and loads factor q/E. Some of the simulation results regarding the influence of these two quantities are shown in the following. In Fig. 10 the frequencies are depicted in the left column and the related aerodynamic damping factors in the right column for the 1st, 2nd and 5th eigenmode. The graphs represent numerical results for the redesigned wing with winglet compared to the original HIRENASD wing for the highest considered loads factor of $q/E{=}0.60 \cdot 10^{-6}$. The comparison shows distinct higher frequencies of the wing-winglet model compared to the HIRENASD model for the 1st and 2nd eigenmodes and a lower one for the 5th eigenmode. The wing-winglet model has higher damping values under all examined conditions. Considering only the frequency of the 1st eigenmode, both distributions remain almost constant in the diagram. The corresponding damping factors decrease at first and increase after $Ma{=}0.83$, whereas the ASDMAD model has a higher gradient. In the middle row of Fig. 10 the frequencies and damping factors of the 2nd eigenmode increase monotonously. However, the values vary only slightly indicating a weak dependence on Mach number changes in the investigated range. The frequencies and damping factors of the 1st torsion dominated eigenmode show the opposite behaviour, while one magnitude increases monotonously the other decreases after $Ma{=}0.78$. Nevertheless, the dependence on Mach number variations remains small.

In Fig. 11 the ratios of the 1st, 2nd and 5th eigenmode frequencies and damping factors of the simulation results of the ASDMAD and HIRENASD wing are depicted over the Mach number. Furthermore, the influence of the

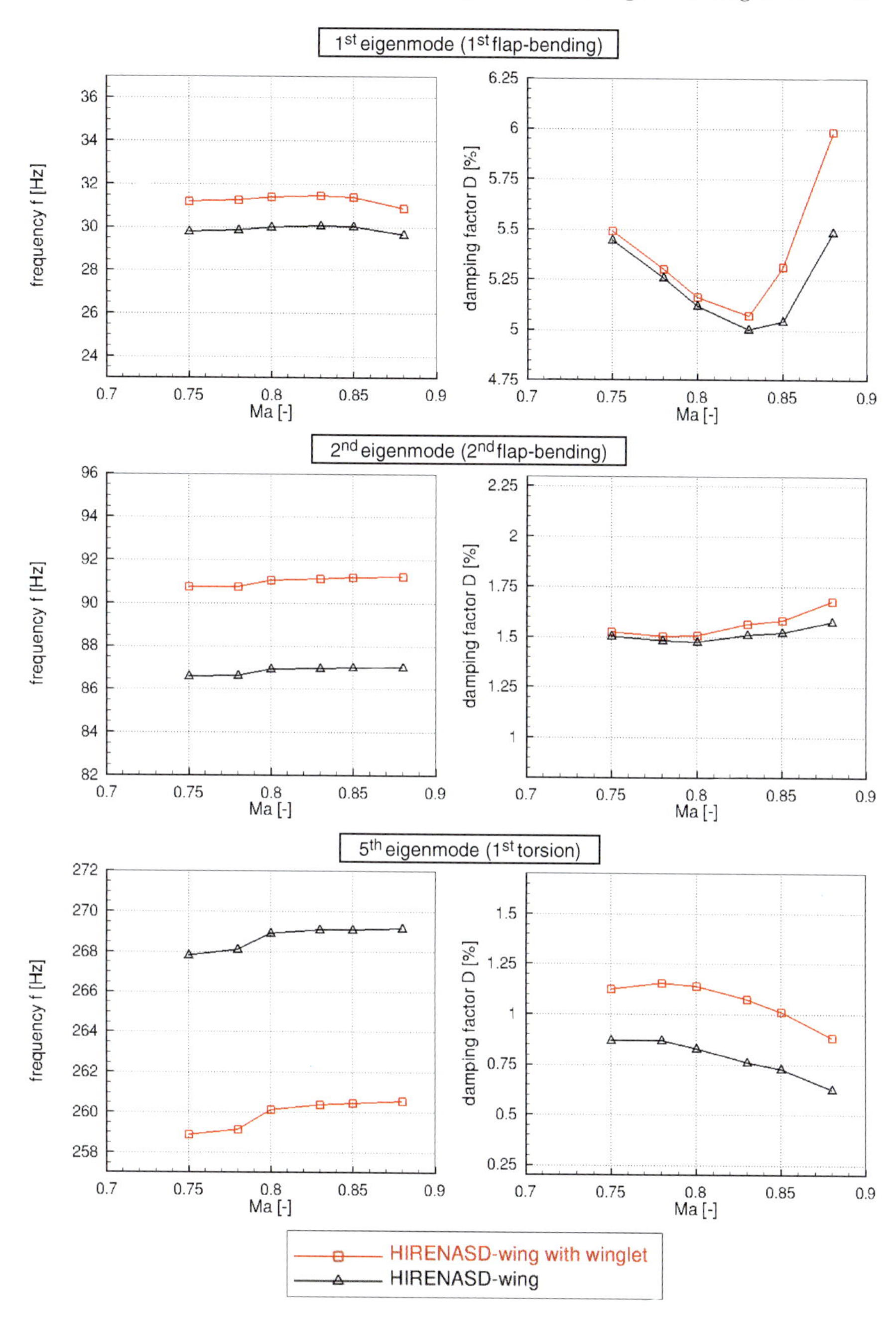

Fig. 10. Influence of Mach number Ma on frequencies and aerodynamic damping factors of 1st, 2nd and 5th eigenmode of the wing-winglet model in comparison with the original HIRENASD model (Re=$23.5 \cdot 10^6$, q/E=$0.60 \cdot 10^{-6}$ and α=$0.0°$)

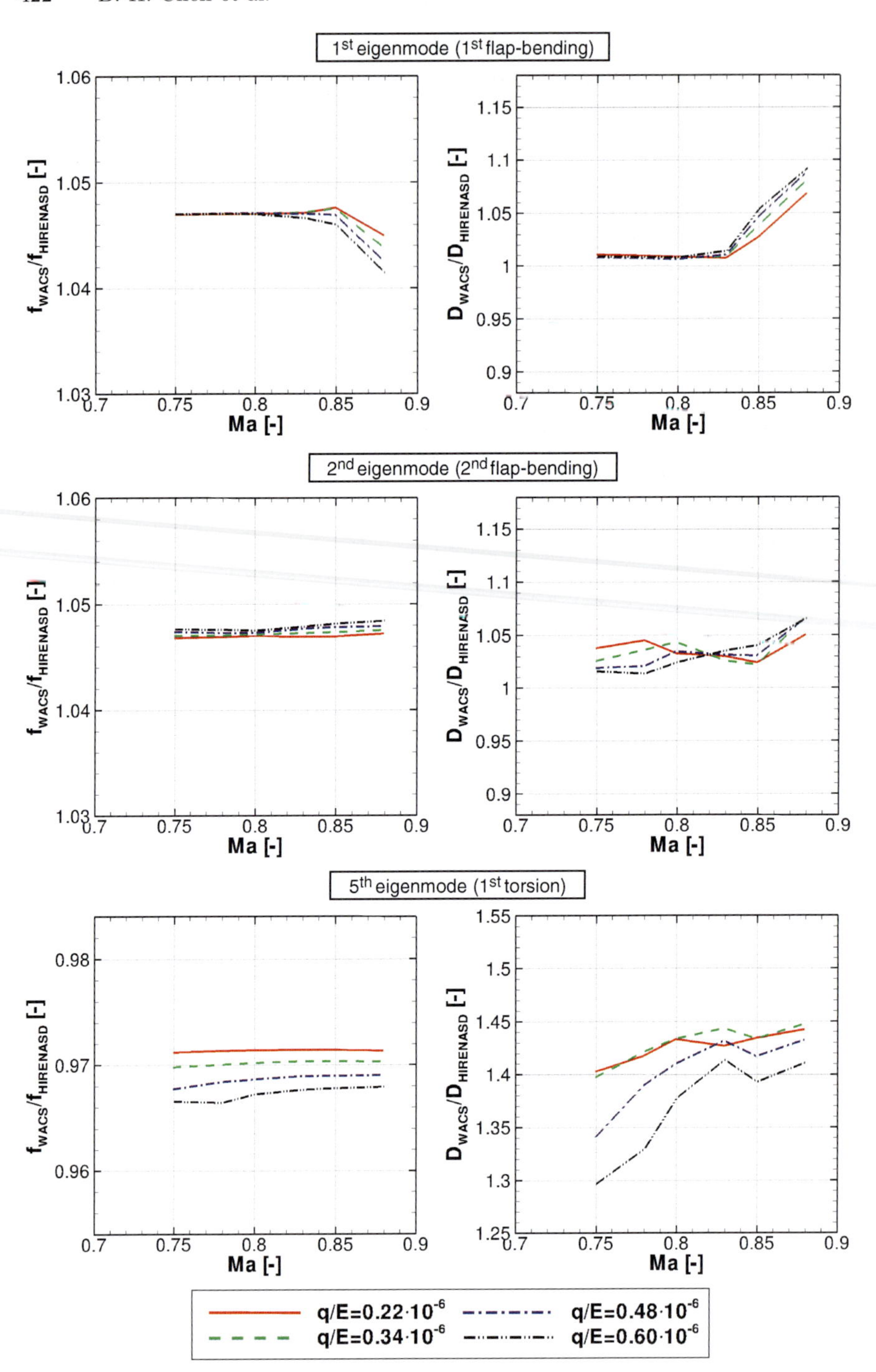

Fig. 11. Influence of Mach number Ma and loads factor q/E on frequencies and aerodynamic damping factors of 1st, 2nd and 5th eigenmode (Re=$23.5 \cdot 10^6$, α=$0.0°$)

loads factor variation (q/E=$0.22\cdot10^{-6}$, $0.34\cdot10^{-6}$, $0.48\cdot10^{-6}$ and $0.60\cdot10^{-6}$) is identified herein. In the top left diagram of Fig. 11 the distributions of the frequency ratios for the 1st eigenmode are illustrated. In the Mach number range Ma=0.75 to 0.83 the eigenfrequencies of the wing-winglet configuration appear completly unaffected to changes of Ma and q/E. However, above Ma=0.85 the influence of both quantities is clearly noticeable. Furthermore, the gradient of the drop increases with growing loads factor. The top right diagram of Fig. 11 shows the ratio of the damping factors for the 1st eigenmode, in which the opposite effect in comparison with the frequencies is visible. The middle row of Fig. 11 is related to the 2nd flap-bending dominated eigenmode. Both ratios remain almost constant, hence show only minor Ma and q/E dependency, which is in accordance to the results (Fig. 10) examined previously. The bottom row of Fig. 11 highlights the results for the 5th eigenmode. The frequency ratios remain almost constant within the regarded Mach number range and are shifted due to loads factor changes. The differences of the damping factor ratios on the other hand show dependancy to Ma and q/E.

The results presented here were conducted at a constant angle of attack α=0.0^o. But the dependence of characteristic vibrational parameters on angle of attack was studied by additional simulations for α=2.0^o and 4.0^o as well. These revealed only a small impact on frequency and aerodynamic damping of the 1st bending dominated eigenmode. The higher modes remained almost unaffected by varying angles of incidence, as shown exemplarily in Fig. 12.

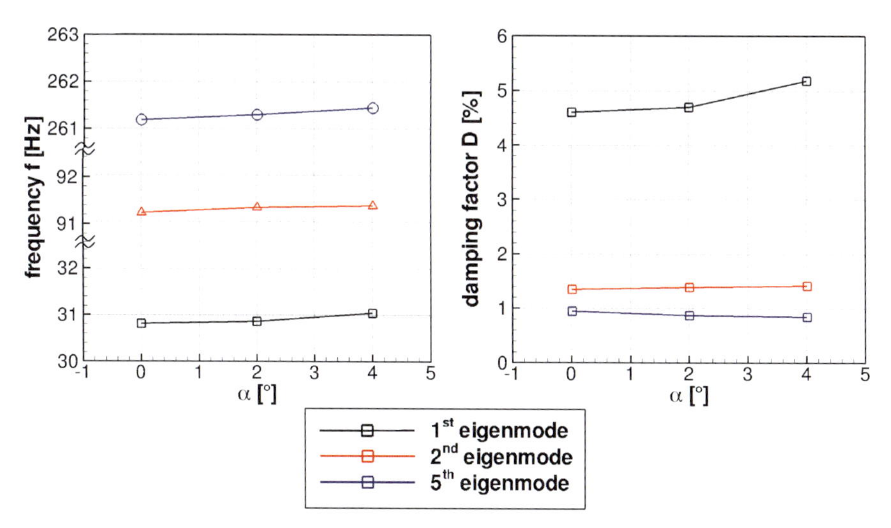

Fig. 12. Influence of the angle of attacke α on frequencies and aerodynamic damping factors of 1st, 2nd and 5th eigenmode (Ma=0.80, Re=$23.5\cdot10^{6}$ and q/E=$0.48\cdot10^{-6}$)

4 Conclusions

In this paper dynamic aeroelastic simulations of a wing-winglet model were presented, using the CAE package SOFIA. The investigations considered different variations of flow parameters, namely Mach number, loads facor and angle of attack, while others remained constant. The results of the vibrational decay test simulations of a wing-winglet model were compared with the original HIRENASD wing, by which mneans the influence of the additional winglet on the dynamic behaviour of the wing was established. The overall dynamic response behaviour of the ASDMAD model for the frequencies and damping factors of the 1st, 2nd flap-bending and 1st torsion-dominated mode shapes resembles qualitatively the HIRENASD wing. Yet, distinct curve shifts and different gradients of the frequency and damping factor distributions can be identified between these two configurations, due to the additional winglet.

5 Code Performance

The aeroelastic simulations were performed using Timoshenko beam elements for the structural idealisation, which are beneficial in terms of low additional CPU effort. The FLOWer code was developed for different computer types and is highly optimised for vector computers. FEAFA and MUGRIDO do not support vectorisation. A comparison of the average CPU time for a static and dynamic aeroelastic simulation of the wing-winglet configuration on a computational mesh of about 1.6 million grid points solving the full Navier-Stokes equations is shown in Fig. 13 for the HLRS cluster using 16 NEC SX-8 processors and the RZ Aachen using 16 Xeon processors. It reveals that the vector-based NEC SX-8 processors are three times faster than the Xeon processors on static simulations and five times faster for dynamic simulations. The total CPU time on HLRS for one aeroelastic simulation comprised the computation of the flow field which required more than 97.5% of the CPU time and the computation of the structural and mesh deformations, which required a little less than 2.5%. The average real time was less than 25 minutes for a static aeroelastic simulation, compared to 1 hour and 15 minutes on the

	static		dynamic	
	HLRS	RZ Aachen	HLRS	RZ Aachen
# iterations	2249	2249	107249	21746
real time (sec)	1439	4526	82467	86400
iterations/sec	1.5629	0.4969	1.3005	0.2517

Fig. 13. Comparison of CPU times used for static and dynamic aeroelastic simulations conducted on NEC SX-8 processors (HLRS) and Xeon processors (RZ Aachen)

Xeon processors for the same simulation. The dynamic simulation on HLRS of a vibrational decay test for this configuration takes about 23 hours for a completed run with 107249 iterations made, whereas the Xeon processors only accomplish 21746 iterations in 24 hours. The code performance on HLRS had an average of about 2480 MFLOPS and a vector operation ratio of more than 97%. The vector time was about 43% of the real run time.

References

1. Braun, C.: *Ein modulares Verfahren für die numerische aeroelastische Analyse von Luftfahrzeugen.* Doctoral thesis, Aachen University, 2007,
2. Ballmann, J., Dafnis, A., Korsch, H., Buxel, C., Reimerdes, H.-G., Brakhage, K.-H., Olivier, H., Baars, A., Boucke, A.: *Experimental Analysis of High Reynolds Number Aero-Structural Dynamics in ETW.* 46th AIAA Aerospace Sciences Meeting and Exhibit, Paper AIAA-2008-841, Reno, Nevada, United States of America, 2007
3. Reimer, L., Braun, C., Chen, B.-H., Ballmann, J.: *Computational Aeroelastic Analysis and Design of the HIRENASD Wind Tunnel Wing Model and Tests.* In proceedings of the International Forum on Aeroelasticity and Structural Dynamics (IFASD) 2007, Paper IF-077, Stockholm, Sweden, 2007
4. J. Ballmann et al.: *The HIRENASD project: High Reynolds number aerostructural dynamics experiments in the European Transonic Windtunnel (ETW)*, International Congress of the Aeronautical Sciences 2006, Hamburg, ICAS 2006-5.11.2
5. Ballmann, J.: *Experiments in the European Transonic Windtunnel (ETW)*, International Forum on Elasticity and Structural Dynamics (IFASD) 2007, Stockholm, 2007, invited keynote lecture
6. HIRENASD project homepage: https://heinrich.lufmech.rwth-aachen.de/index.php?lang=de&pg=home
7. Saad, Y.: *SPARSKIT – A basic tool-kit for sparse matrix computations (Version 2).* User guide, available at http://www-users.cs.umn.edu/~saad/software/SPARSKIT/sparskit.html, 1994
8. Bramkamp, F. D., Lamby, Ph., Müller, S.: *An adaptive multiscale finite volume solver for unsteady and steady flow computations.* Journal of Computational Physics, Vol. 197, pp. 460–490, 2004
9. Kroll, N., Faßbender, J. K. (Eds.): *MEGAFLOW – Numerical Flow Simulation for Aircraft Design.* Notes on Numerical Fluid Mechanics and Multidisciplinary Design, Vol. 89, Springer-Verlag, 2005
10. Kroll, N., Schwamborn, D, Becker, K., Rieger, H., Thiele, F. (Eds.): *MEGADESIGN and MegaOpt – Aerodynamic Simulation and Optimization in Aircraft Design.* Notes on Numerical Fluid Mechanics and Multidisciplinary Design, Springer-Verlag, 2008
11. Boucke, A.: *Kopplungswerkzeuge für aeroelastische Simulationen.* Ph.D. thesis, RWTH Aachen, 2003
12. Hesse, M.: *Entwicklung eines automatischen Gitterdeformationsalgorithmus zur Strömungsberechnung um komplexe Konfigurationen auf Hexaeder-Netzen.* Ph.D. thesis, RWTH Aachen, 2006

Transport and Climate

Prof. Dr. Ch. Kottmeier

Institut für Meteorologie und Klimaforschung, Karlsruher Institut für Technologie, Wolfgang-Gaede-Str. 1, 76131 Karlsruhe

There is an obvious tendency in HPC modelling of large environmental system like the ocean or the atmosphere to go towards Regional Earth system models (RESM) by coupling more and more complex submodels. The requirement for HPC will even increase in the near future, since processes resolvable at high resolution only, like convection and cloud processes, tend to limit the diagnostic and prognostic skills of RESMs. The CPU time requirements of such models are continuously increasing, since in oceanic and atmospheric models there are energy-containing processes in the mesoscale (10–1000 km, the convective scale (100 m – 10 km), and turbulent scales (down to mm). Processes on such scales interact considerably and progress in grid resolution means that the fields can be directly resolved better. Approximations such as parametrizations can be avoided and net effects of small scales are calculated at grid resolution. Another development concerns the requirement of ensemble runs, e.g. for numerical weather forecasting and scenarios of climate change. Reports from two ambitious projects have been selected for this publication. Other projects have been described in previous annual reports or occur still a bit premature for publication.

Several projects performed within the Baden-Württemberg programme "Herausforderung Klimawandelmade use of the HLRS in Stuttgart and the SCC in Karlsruhe. A model restart possibility needed to be implemented to allow for decadal climate runs. The study aims at an assessment of the capabilities of regional climate models in simulating the observed climate of the last decades in a highly structured mountaineous region, and based on such validated model members to estimate future climate change at high resolution. In the project RESTER the influence of climate change on the strength and the probability of occurrence of winter storms and the expected effects on forest stands in Baden-Württemberg is investigated. Within the ReSiPrec project future changes of the frequency of occurrence and intensity of heavy precipitation events are investigated. Available climate simulations of the global climate model ECHAM5, the regional climate model (RCM) REMO are complemented for a reference time period from 1971 to 2000 by COSMO-

CLM runs of IMK at Karlsruhe Institute of Technology (KIT). By comparing these results with results for a projected time period from 2011 to 2040, where the projected changes in the radiative forcing are taken from SRES scenario A1B, the effects of the resulting changes of winter storms can be determined. From these results, an assessment of the loss potential of forest stands for the current and the future climate is performed and adaptation/mitigation measures can be developed. The third project PArK aims at the probabilistic assessment of climate change for decades in the near future (2011 to 2030). Results based on complementary methods (probabilistic-dynamical and statistical downscaling techniques with COSMO-CLM) and data sources (regional climate models driven by different realisations of global model simulations) will be produced and analysed in order to quantify the most likely climate development, including its uncertainties. The quantification of uncertainties is an essential aspect in the assessment of climate change projections.

The project AMMA (African Monsoon Multidisciplinary Analysis; IMK-TRO at KIT) focusses on the precipitation events in West Africa. The dramatic change in the region of the West African Monsoon (WAM) from wet conditions in the 50s and 60s to much drier conditions from the 70s to the 90s represents one of the strongest inter-decadal signals on the planet in the 20th century. Marked inter-annual variations in the recent decades have resulted in extremely dry years with devastating environmental and socio-economic impacts. The atmospheric processes critical for precipitation are the mesoscale convective systems (MCS), whose frequency and period of occurrence decides on wet and dry years. Within the AMMA project at the HP XC 4000 at the SCC, work was performed on two different topics. On the one hand the life cycle of a typical MCS over West Africa is modeled with respect to the soil conditions. On the other hand the focus lies on comparison between the simulated MCSs over West Africa and over the Eastern Atlantic. Both projects provide highly new results recently published in peer reviewed journals.

Modelling Regional Climate Change in Southwest Germany

Hans-Jürgen Panitz, Gerd Schädler and Hendrik Feldmann

Institut für Meteorologie und Klimaforschung, Forschungszentrum Karlsruhe/Universität Karlsruhe
panitz@imk.fzk.de

1 Introduction

In its Fourth Assessment Report (AR4) the Intergovernmental Panel on Climate Change (IPCC) states that it is very likely that many land regions, especially on the Northern Hemisphere, will warm during the 21st century due to the general climate change, which itself is caused by the increase of anthropogenic greenhouse gases [7]. On the global scale the IPCC estimates that for the next two decades a warming of about 0.2° C per decade can be expected on the basis of the SRES (Special Reports on Emission Scenarios, [11]) emission scenarios. Even if the concentrations of all greenhouse gases were kept at the year 2000 level, a further warming of about 0.1° C per decade can be expected. For many land regions of the globe it is estimated that the annual mean temperature increase will be higher than the global mean. This variability of climatic change persists over all scales, from global to regional, and more or less for all climatological observables. On the spatial scale of global climate models (about 200 km) for example, the largest warming in Europe is likely to happen in the northern part in winter and in the Mediterranean area in summer [2]. Annual precipitation is very likely to increase in most of northern Europe and decrease in most of the Mediterranean area. In Central Europe, precipitation is likely to increase in winter but decrease in summer, but the agreement between the results of various models is quite low there. Extremes of daily precipitation are very likely to increase throughout Europe [1]; compared to Northern and Southern Europe, however, climatic change for Central Europe is more difficult to assess due to sometimes conflicting tendencies.

From time series analyses of observations it is known that climatic trends may vary significantly on spatial scales in the order of tens of kilometres. However, rather little is known about the impact of climate change on these scales for Germany or even smaller regions like the Federal State of Baden-Württemberg. In order to study the possible impacts of climate change in

W.E. Nagel et al. (eds.), *High Performance Computing in Science and Engineering '09*, DOI 10.1007/978-3-642-04665-0_31,

Baden-Württemberg under various aspects and to quantify also the uncertainties of these projections, the research program "Herausforderung Klimawandel Baden-Württemberg" was initiated in 2006 by the Ministry of Environment of Baden-Württemberg. The aim of the research program is to asses the impacts and risks of climate change for the area of Baden-Württemberg. It is intended not only to quantify the changes of meteorological parameters like temperature, precipitation, and wind, but also the environmental impacts of these changes, for example on ecosystems and human health.

The Institute for Meteorology and Climate Research (IMK) of the University of Karlsruhe and the Karlsruhe Research Centre (FZK) contributes to three projects within the research program.

In the RESTER project the influence of climate change on the strength and the probability of occurrence of winter storms and the expected effects on forest stands in Baden-Württemberg is investigated. For this purpose, climate simulations of the global climate model ECHAM5 and the regional climate model (RCM) REMO are analysed for a reference time period from 1971 to 2000. By comparing these results with results for a time period from 2011 to 2040, where the projected changes in the radiative forcing are taken from SRES scenario A1B, the effects of the resulting changes of winter storms can be inferred. For selected storm events, high-resolution simulations with the model COSMO of the German Weather Service (DWD) are performed. From these results, an assessment of the loss potential of forest stands for the current and the future climate can be estimated and serve as a basis for the development of adaptation and mitigation measures.

Within the ReSiPrec project future changes of the frequency of occurrence and intensity of heavy precipitation events are investigated in order to estimate the impact of climate change for the state of Baden-Württemberg for the time period 2011–2040. For this purpose, an ensemble of simulations is analysed statistically in order to derive changes in extreme values and trends. Besides existing results of present day and future climate simulations with the RCMs REMO and COSMO-CLM (CCLM, which is the climate version of the weather forecast model COSMO of the German Weather Service, DWD) the results of our own, finer resolution (7 km) regional climate simulations with CCLM at HLRS are used. In the model evaluation, statistical comparisons of the simulation results with observations (e.g. high and low percentiles of distributions) for a control period form 1971 to 2000 are performed to assess how realistically precipitation climatologies, especially the distributions of extreme events, can be reproduced by the model simulations. Future scenario precipitation climatologies are established for the period from 2011 to 2040.

The third project PArK aims at the probabilistic assessment of changes in maen values of temperature and precipitation during the decades from 2011 to 2030. Results based on complementary methods (probabilistic-dynamical and statistical downscaling techniques with COSMO-CLM) and data sources (regional climate models driven by different realisations of global model simulations) are produced and analysed with Bayesian statistics in order to quantify the most likely climate development, including its uncertainties. The quantification of uncertainties, again based on ensemble statistics, is an essential aspect of this project.

In order to assess the impact of climate change in small scale regions like Baden-Württemberg it is necessary to perform model simulations with a high spatial resolution less than 10 km. The typical resolution of existing global climate model simulations is about 150 km, which is much too coarse since features like urban areas, valleys and mountains that can have considerable influence on regional climate are not "seen" or "seen properly" by the model at this resolution. Within the framework of the three projects mentioned above, CCLM simulations with a spatial resolution of about 7 km are performed for the periods from 1968 to 2000 and 2007 to 2040 (this includes an initial spin-up time of 3 years). To reach the goals of all projects and to assess the uncertainties of the climate projections, it is necessary to create multi-model, multi-member ensembles of climate simulations. Driving data from global climate simulations with ECHAM5 and other global models in various realisations are used to initialise and drive the regional climate model. It is desirable that an ensemble includes as many members as possible; the number which can be realised depends very much on the computer power available. It is obvious that such simulations require high computational power and large storage capacity for the results.

2 The CCLM Model

The regional climate model CCLM is the climate version of the operational weather forecast model LM (Lokal-Modell) of the German Weather Service (DWD). The two model lines have been merged in 2007, i.e. physics and numerics are identical for both, the forecast and the climate application. Due to new developments at DWD, the names of the model modes have changed. The generic name is now COSMO, the climate version is named COSMO-CLM (or CCLM for short).

The CCLM is a three-dimensional non-hydrostatic model. Due to the non-hydrostatical formulation spatial resolutions far finer than 10 km (which is considered the limit for hydrostatic models) are possible. The model solves prognostic equations for wind, pressure, air temperature, different phases of atmospheric water, soil temperature and soil water content.

Further details on COSMO can be found in [3], on the web-page of the COSMO consortium (http://www.cosmo-model.org) and in [10].

The model is coded in Fortran 90, making extensive use of the modular structures provided in this language. Code parallelisation is done via MPI (message passing interface) on distributed memory machines using horizontal domain decomposition with a 2-grid halo. The model runs on various platforms, including LINUX clusters, vector computers like NEC SX and Fujitsu VPP series as well as MPP machines like IBM-SP.

3 Regional Climate Simulations Using the HLRS Facilities

3.1 Simulation Setup and Downscaling Chain

As described in [10], dynamical downscaling is used to transfer large scale information to the regional scale. Basically, this method is a nesting of the regional climate model (RCM) into large-scale global climate model (GCM) projections or reanalyses. This means that the model is initialized once with a state derived from the large scale information and that this information is updated at the lateral boundaries of the regional model domain at regular time intervals.

Large scale driving data for CCLM used so far are the results of the global climate model ECHAM5 model and the ERA-40 ([16]) global reanalysis. The reanalysis, which includes assimilated observations, covers the period form 1957 until 2001. Therefore, it is supposed to represent a real picture of the climate of the past decades, so that, the CCLM simulations driven by the ERA40 data can be used to evaluate the model.

In order to assess the impact of climate change, data of the ECHAM5 model have been used to drive the CCLM model. Two realizations of the ECHAM5 model that take into account anthropogenic and natural radiative forcing (denoted as full forcing; [12], [13], have been chosen as forcing data for the CCLM control runs from 1968 to 2000. To assess the future development of climate for the near future, two ECHAM5 realizations based on the SRES scenario A1B have been used ([14], [15]).

The horizontal grid sizes of the global data sets are larger than 100 km (about 125 km for ERA40 and 150 km for ECHAM5). To avoid a too large step in the grid sizes, a double nesting technique is used for the dynamical downscaling. In a first step the large scale data are used to drive a CCLM simulation with a horizontal grid size of 0.44 deg (about 50 km in our latitudes). The chosen model domain for this first nest is shown in Fig. 1. In the West-East direction the domain consists of 118 grid-points, in the South-North direction of 110 grid-points. In the second step the results of this 50 km simulation are used to drive the CCLM simulations with grid spacing of 0.0625 deg (about 7 km). The corresponding model domain of this second nest is illustrated in Fig. 2, the number of horizontal grid-points is 124*140. The simulation domains and periods were chosen in such a way that the results can be used simultaneously for the three projects described above.

3.2 HPC Aspects

Based on the five global data sets described above and due to the double nesting strategy, altogether ten CCLM simulations have been carried out using the NEC SX-8 high performance computer at the High Performance Computing Centre Stuttgart (HLRS). They are summarized in Table 1. A summary of the

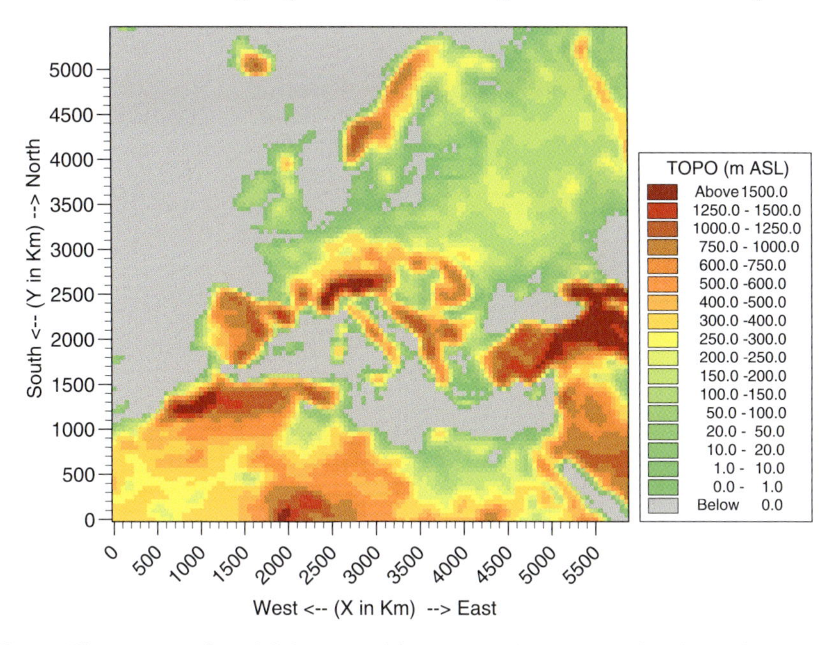

Fig. 1. Illustration of model domain of first nest. The topography shown has a mesh size of about 50 km

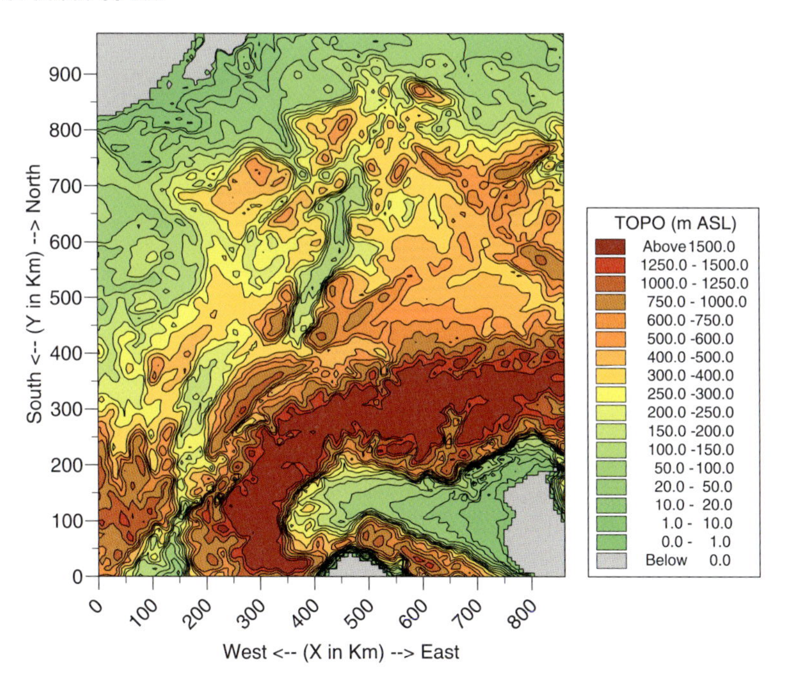

Fig. 2. Illustration of model domain of second nest. The topography shown has a mesh size of about 7 km

Table 1. Summary of ten CCLM simulations performed on NEC SX-8 at HLRS

Global Data	Period	Run No.	Nesting	External Forcing	Resolution (Deg)	Status
ERA40	1968-	1	Nest 1	ERA40	0.44	finished
	2001	2	Nest 2	CLM_50km	0.0625	finished
ECHAM5_20C3M full	1968-	3	Nest 1	ECHAM5	0.44	finished
forcing, Realization 1	2000	4	Nest 2	CLM_50km	0.0625	finished
ECHAM5_20C3M full	1968-	5	Nest 1	ECHAM5	0.44	finished
forcing, Realization 3	2000	6	Nest 2	CLM_50km	0.0625	finished
ECHAM5_A1B,	2007-	7	Nest 1	ECHAM5	0.44	finished
Realization 1	2041	8	Nest 2	CLM_50km	0.0625	finished
ECHAM5_A1B,	2007-	9	Nest 1	ECHAM5	0.44	finished
Realization 3	2041	10	Nest 2	CLM_50km	0.0625	finished

Table 2. Wall-clock and CPU-times needed for CCLM simulations on NEC SX-8 at HLRS as well as performance indicators and numbers of CPUs used. The wall-clock times do not include waiting times in the queue and system maintenance times

Run No	Wall-Clock Time (d)	CPU-Time (d)	Average vector length	Average vector ratio (%)	Average MFLOPS	CPUs used
1	9.5	143	111.8	86.1	1370.4	16
2	31	500	not recorded for this run			16
3	9	137	not recorded for this run			16
4	22	509	121.4	93.2	1683.7	24
5	9	135	112.4	86.3	1417.1	16
6	20	499	121.4	93.2	1705.7	24
7	9	137	112.4	86.2	1414.7	16
8	23	541	121.4	93.2	1681.3	24
9	9	142	112.5	86.3	1428.4	16
10	22	522	121.5	93.2	1696.8	24

mean computing requirements and performances is given in Table 2. Each of the simulations of nest 1 needed a mean wall-clock time (without waiting time in the queue and times of system maintenance) of about 9 days and a mean total CPU-time of about 140 days using 16 CPUs. For the simulations of nest 2 the corresponding times were considerably larger although 24 CPUs have been used (except run 2 using 16 CPUs). The reasons are the larger number of grid points and the short numerical time-step of 40 sec (instead of 240 sec for nest 1) that, in combination with the high spatial resolution of about 7 km (instead of 50 km for nest 1), was necessary to fulfil the Courant-Friedrichs-Levy stability criterion [9].

Altogether, wall-clock times of about 164 days and CPU-times of 3265 days were required. The first run (run No. 1) was submitted in autumn 2007, the last one (run No. 10) finished in March 2009. Thus, the real time from the

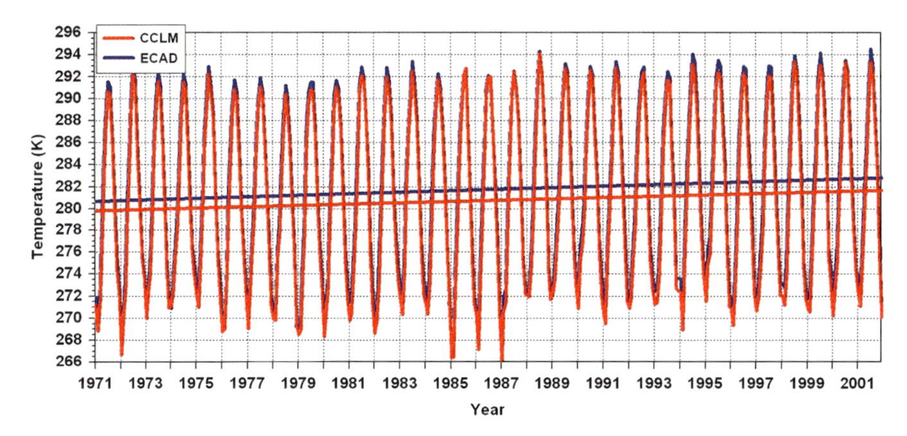

Fig. 3. Monthly means of T_2M and 31-years temperature trend for the period 1971 until 2001. Shown are the results of CCLM evaluation run (red curves) and the ECAD observational data (blue curves)

very beginning of the simulations until their end was about 1.5 years. Much of the necessary work (e. g. preparation of forcing data, CCLM simulations themselves) was carried out in parallel. Otherwise, the real time for all simulations would have at least doubled. These few numbers indicate that high performance computing is indispensable for climate simulations, and that the capacities should even be extended.

4 Results

4.1 Temperature at 2 m Height

The air temperature at 2 m height (T_2M) calculated in the evaluation run (run no.1, Table 1, Table 2) has been compared with the ECAD (ECAD: European Climate Assessment and Data; [5]) gridded data set; this daily data set based on observations matches exactly the 50 km-grid of the CCLM simulations. Discarding a model spin-up time of three years, the comparison has been performed for the period from 1971 until 2001. Fig. 3 shows the time series of T_2M monthly means for the simulation (CCLM) and the observational data (ECAD). The data are temporally averaged over each month and also spatially over all land-points of the model domain (Fig. 1). Although the visual agreement appears satisfying, the simulation shows a cold bias of about 1K on the average. However, the trend of the temperature anomaly is nearly identical for simulation and observation (Fig. 3). The anomaly plot (Fig. 4) suggests that at the present stage, simulation of changes may be more reliable than simulation of absolute values.

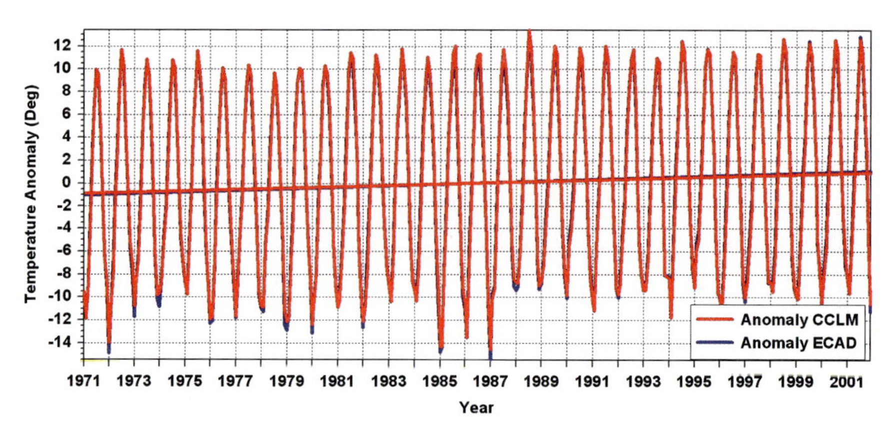

Fig. 4. Anomaly (T-Tavg) of monthly T_2M data and its trend for the period 1971 - 2001
Red curve: CCLM anomaly, Tavg = 280.72 K
blue curve: ECAD anomaly, Tavg = 281.70 K

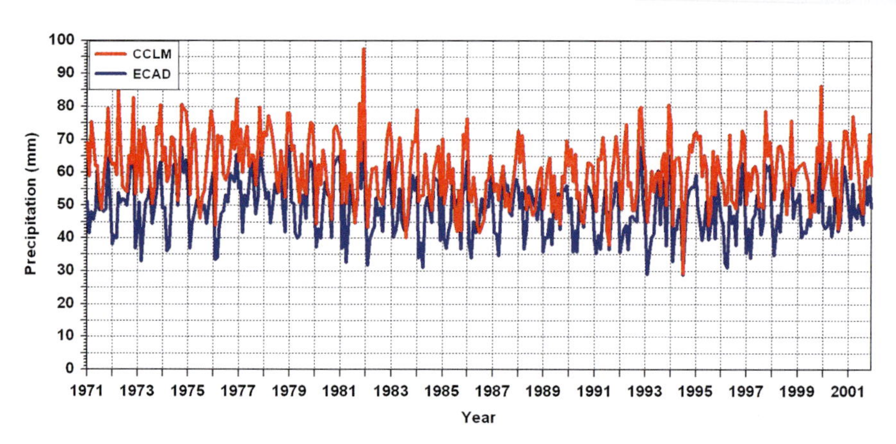

Fig. 5. Mean monthly sum of total precipitation for the period 1971 until 2001. The red curve shows the results of the CCLM evaluation run, the blue curve represents the ECAD observational data

4.2 Precipitation

The ECAD gridded data set also contains daily precipitation data. In contrast to the temperature, precipitation is overestimated by the model (Fig. 5). On the average the model precipitation is about 11 mm too high. This overestimation seems to be caused by a complex interplay of model numerics and model physics and it depends also on the region considered. The causes of the overestimation are currently analysed.

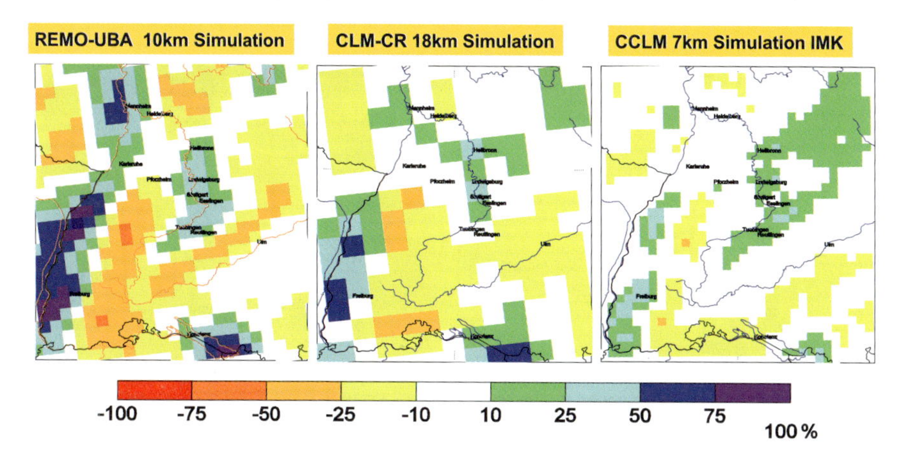

Fig. 6. Effect of horizontal resolution and improved model physics. Shown are relative deviations of annual mean precipitation between model results and observations. The model data are normalized with respect to the observed mean annual precipitation. Left: REMO UBA simulations, 10 km resolution, ECHAM5 forcing Middle: CLM Consortium Runs (CR, CLM Version 3), 18 km resolution, ECHAM5 forcing Right: CCLM simulations (CLM Version 4), 7 km resolution, improved microphysics, ECHAM5 full forcing

4.3 Added Value of High Resolution and Ensemble Modelling

Although there are still discrepancies between the model results and observations which require further investigations, the advantages of high resolution climate modelling are obvious. Due to the ability of non-hydrostatic models like CCLM to use horizontal mesh sizes less than 10 km, regional orographic features and small-scale processes can be represented much more realistically. This is demonstrated in Fig. 6 (H. Feldmann, pers. communication).

The figure shows the normalized difference between simulated and observed annual precipitation over Baden-Württemberg for the period 1971-2000, calculated by three different model runs with different horizontal resolutions [4]. It can be seen that the model REMO [8] and an older CCLM version (CLM-CR, [6]) both produce too much precipitation west of the Black Forest and too little precipitation east of it. REMO used a horizontal grid mesh of 10 km, the CLM-CR a resolution of about 18 km, and both model versions do not consider the horizontal transport of precipitation. Both simulations used boundary data derived from the global ECHAM5 model. The observed precipitation patterns are reproduced more realistically by the CCLM simulation using the 7 km resolution. This simulation is based on ECHAM5 full forcing global data for the present day climate (run 4, Table 1) and was run at HLRS. This considerably better representation of the regional patterns is a combination of the high spatial resolution and improvements in CCLM with respect to the parameterization of micro-physical processes including the transport of

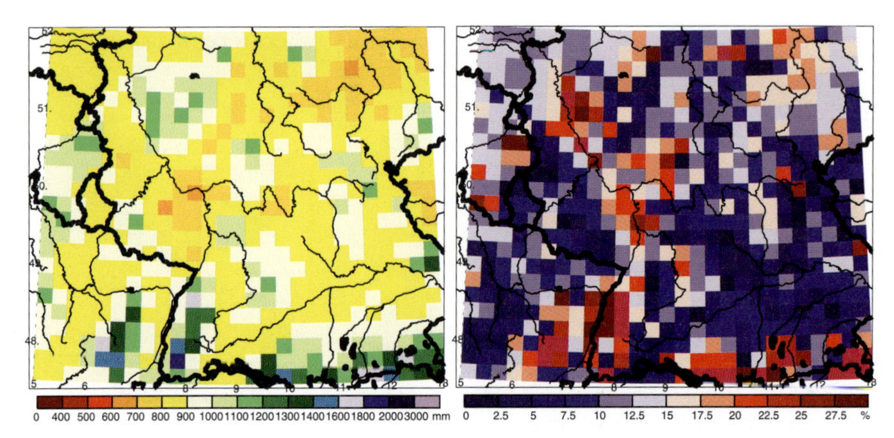

Fig. 7. Ensemble evaluation of mean annual precipitation 1971–2000.
Left: ensemble mean (area mean=927 mm, area min=685 mm, area max=1935 mm) right: ensemble standard deviation relative to annual mean (area mean=12.4%, area min=0.93%, area max=56.2%)

rain and snow by the mean wind. Figure 7 illustrates the added value which can be gained from ensemble simulations: in this case, the ensemble consists of 6 members (REMO, CLM consortial runs and CCLM HLRS runs). The figure shows that the spatial patterns of precipitation are well reproduced (left), but that the ensemble uncertainty, measured in terms of ensemble standard deviation, varies considerably (right). This ensemble standard deviation can be exploited to identify regions where climate change is likely to have a significant impact; work in this direction is presently under way.

4.4 Further Sensitivity Studies

Meissner and Schädler [10] examined the impact that different driving data (in their case the ERA40 and NCEP (http://www.cdc.noaa.gov) reanalysis data), different grid resolutions, and different soil initializations have on the results of the regional model CCLM. They identified an appropriate model configuration with 7 km grid size, climatological soil initialization, and ERA40 reanalysis data as driving data. Meanwhile, a further study has been carried out changing the numerical time integration scheme.

In this study the evaluation run (run 1, Table 1, Table 2) served as the reference. It was repeated with the same model configuration except for the numerical time integration scheme and the simulation periods which lasted from 1968 until 1985 instead of 1968 until 2001. For the reference run the Runge-Kutta method had been used. The alternative for the sensitivity study was the Leapfrog scheme. Either of these methods can be chosen by the user of the model. The time step has not been changed since it fulfilled the CFL-criterion for both schemes.

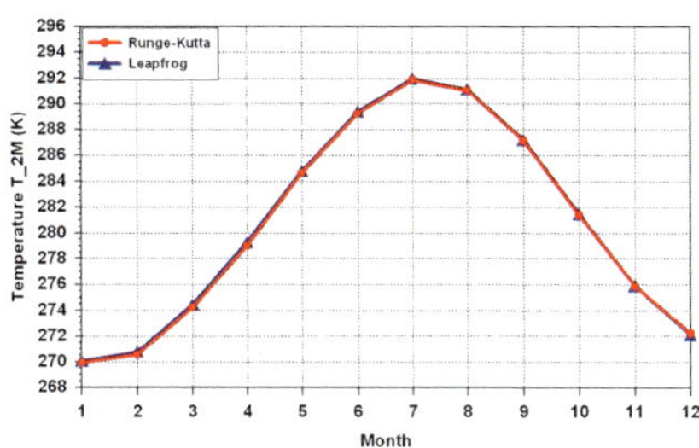

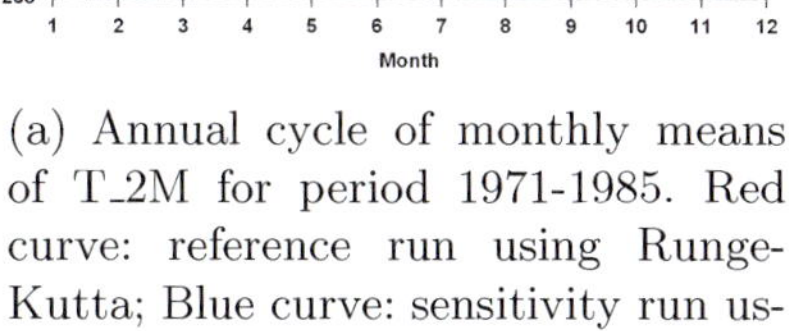

(a) Annual cycle of monthly means of T_2M for period 1971-1985. Red curve: reference run using Runge-Kutta; Blue curve: sensitivity run using Leapfrog

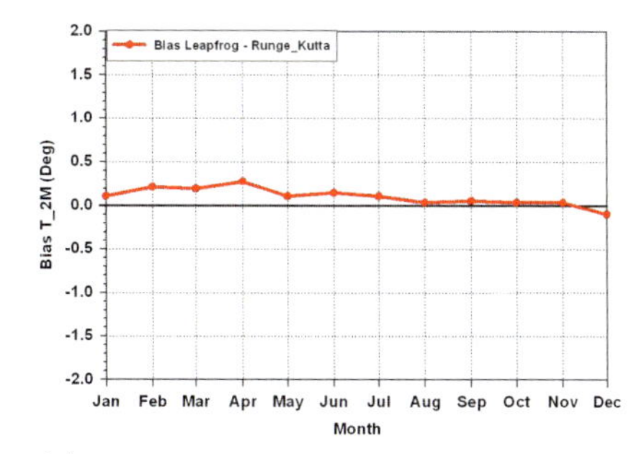

(b) Annual cycle of bias of T_2M monthly means for period 1971-1985. Difference between sensitivity run with Leapfrog and reference run with Runge-Kutta is shown

Fig. 8. Sensitivity related to different time integration schemes: T_2M

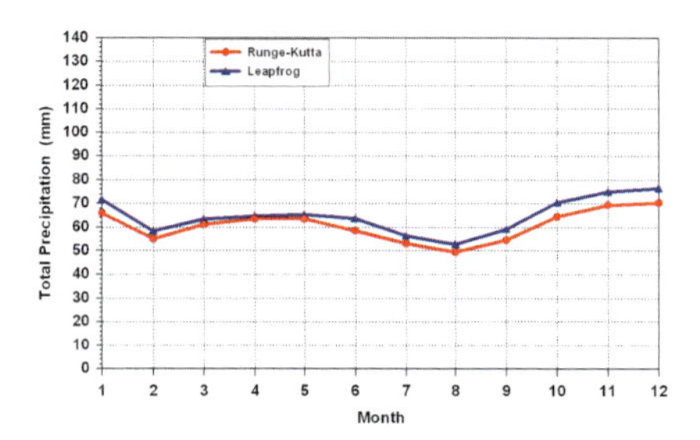

(a) Annual cycle of mean monthly sums of total precipitation for period 1971-1985. Red curve: reference run using Runge-Kutta; Blue curve: sensitivity run using Leapfrog

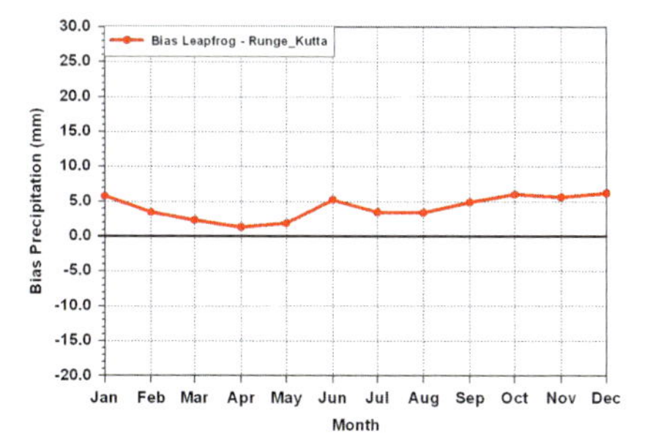

(b) Annual cycle of bias of mean monthly sums of precipitation for period 1971-1985. Difference between sensitivity run with Leapfrog and reference run with Runge-Kutta is shown

Fig. 9. Sensitivity related to different time integration schemes: Total precipitation

As an example of the results the annual cycle of the monthly means of T_2M are shown in Fig. 8a, the difference of the results between the sensitivity and the reference run is given in Fig. 8b. The temperatures obtained with the Leapfrog scheme tend to be slightly higher, in the mean about $0.1°$. With respect to the comparison with observations (Fig. 3), using the Leapfrog scheme does not improve the model results. The precipitation also increases slightly (in the mean by about 4 mm) when applying the Leapfrog method (Fig. 9a, Fig. 9b). These figures can be taken as an indication of the maximum accuracy which can be expected from the numerical point of view.

5 Future Work

As already pointed out, the ensemble aspect is very important in climate modelling. The present ensemble is still too small for reliable statistical investigations. Therefore, it is intended to increase the number of ensemble members using different realizations of further GCM climate scenarios as forcing data. Simulations using data from the HadCM3 GCM of the British Hadley Centre for Climate Prediction and Research, Bracknell, Berks, UK are presently carried out.

Furthermore, it is planned to contribute to international activities for the Fifth Assessment Report (AR5) of IPCC with focus on Africa, and to continue the investigations for Europe using the SRES scenarios to provide climate data for further impact studies. Variability, uncertainty, climate change - in particular that of extremes -, and quality of produced climate data for present day conditions are still major issues to be investigated by the new simulations. Therefore a multi model approach was favoured - at least 2 RCMs (REMO and and the recent version 4.8 of CCLM) driven by data of different GCMs (3-5 seem possible) for 1-3 scenarios.

References

1. F. Boberg, P. Berg, P. Thejll, W. J. Gutowski, and J. H. Christensen (2008): Improved confidence in climate change projections of precipitation evaluated using daily statistics from the PRUDENCE ensemble, Clim. Dyn., doi:10.1007/s00382-008-0446-y.
2. J.H. Christensen, B. Hewitson, A. Busuioc, A. Chen, X. Gao, I. Held, R. Jones, R.K. Kolli, W.-T. Kwon, R. Laprise, V. Magaña Rueda, L. Mearns, C.G. Menéndez, J. Räisänen, A. Rinke, A. Sarr, and P. Whetton (2007): Regional Climate Projections, in: Climate Change 2007: The Physical Science Basis.Contribution of Working GroupI to the Fourth Assessment Report of the Intergovernmental Panel on Climate Change. Eds. S. Solomon, D. Qin, M. Manning, Z. Chen, M. Marquis, K.B. Averyt, and M. Tignor, Cambridge University Press, Cambridge, United Kingdom and New York, NY, USA
3. G. Doms, and U. Schättler (2002) A description of the nonhydrostatic regional model LM, Part I: Dynamics and Numerics. COSMO Newsletter, 2, pp. 225–235.
4. H.Feldmann, B. Früh, G. Schädler, H.-J. Panitz, K. Keuler, D. Jacob, and Ph. Lorenz (2008): Evaluation of the Precipitation for Southwest Germany from High Resolution Simulations with Regional Climate Models. Meteorologische Zeitschrift, 17, 455–465.
5. M. R. Haylock, N. Hofstra, A. M. G. Klein Tank, E. J. Klok, P. D. Jones, and M. New (2008): A European daily high-resolution gridded data set of surface temperature and precipitation for 1950-2006. J. Geophys. Res., 113, D20119, doi: 10.1029/2008JD010201.
6. H.-D. Hollweg, U. Böhm, I. Fast, B. Hennemuth, K. Keuler, E. Keup-Thiel, M. Lautenschlager, St. Legutke, K. Radtke, B. Rockel, M.Schubert, A. Will,

M. Woldt, and C. Wunram (2008): Ensemble Simulations over Europe with the Regional Climate Model CLM forced with IPCC AR4 Global Scenarios. Technical Report No.3, pp 152Model & Data GroupMax Planck Institute for Meteorology, Hamburg, http://www.mad.zmaw.de/projects-at-md/sg-adaptation, ISSN 1619-2257
7. IPCC (2007) Summary for Policymakers, in: Climate Change 2007: The Physical Science Basis.Contribution of Working GroupI to the Fourth Assessment Report of the Intergovernmental Panel on Climate Change. Eds. S. Solomon, D. Qin, M. Manning, Z. Chen, M. Marquis, K.B. Averyt, and M. Tignor, Cambridge University Press, Cambridge, United Kingdom and New York, NY, USA
8. D. Jacob, H. Göttel, S. Kotlarski, Ph. Lorenz, and K. Sieck (2008): Klimaauswirkungen und Anpassung in Deutschland? Phase 1: Erstellung regionaler Klimaszenarien für Deutschland.a Abschlussbericht zum UFOPLAN-Vorahben 204 41 138. pp 154, Umweltbundesamt, Dessau-Roßlau, www.umweltbundesamt.de, ISSN 1862-4359
9. T. N. Krishnamurti, and L. Bounoua (1996): An Intorduction to Numerical Weather Prediction Techniques. CRC Press, Boca Raton, London, New York, Washington D.C., pp 291, ISBN 0-8493-8910-0.
10. C. Meissner, and G. Schädler (2007): Modelling the Regional Climate of Southwest Germany: Sensitivity to Simulation Setup, in: High Performance Computing in Science and Engineering '07. Eds. W. E. Nagel, D. Kröner, and M. Resch, ISBN 978-3-540-74738-3, Springer Berlin Heidelberg New York.
11. N. Nakicenovic et al.(2000). Special Report on Emissions Scenarios: A SpecialReport of Working GroupIII of the Intergovernmental Panel on Climate Change, Cambridge University Press, Cambridge, U.K., 599 pp. Available online at: http://www.grida.no/climate/ipcc/emission/index.htm
12. E. Röckner (2005a): PCC MPI-ECHAM5_T63L31 MPI-OM_GR1.5L40 20C3M_all run no.1: atmosphere 6 HOUR values MPImet/MaD Germany. World Data Center for Climate. CERA-DB "EH5-T63L31_OM_20C3M_1_6H" http://cera-www.dkrz.de/WDCC/ui/Compact.jsp?acronym=EH5-T63L31_OM_20C3M_1_6H
13. E. Röckner (2005b): IPCC MPI-ECHAM5_T63L31 MPI-OM_GR1.5L40 20C3M_all run no.3: atmosphere 6 HOUR values MPImet/MaD Germany. World Data Center for Climate. CERA-DB "EH5-T63L31_OM_20C3M_3_6H" http://cera-www.dkrz.de/WDCC/ui/Compact.jsp?acronym=EH5-T63L31_OM_20C3M_1_6H
14. E. Röckner, M. Lautenschlager, H. Schneider (2006a): IPCC-AR4 MPI-ECHAM5_T63L31 MPI-OM_GR1.5L40 SRESA1B run no.1: atmosphere 6 HOUR values MPImet/MaD Germany. World Data Center for Climate. doi: 10.1594/WDCC/EH5-T63L31_OM-GR1.5L40_A1B_1_6H
15. E. Röckner, M. Lautenschlager, H. Schneider (2006b): IPCC-AR4 MPI-ECHAM5_T63L31 MPI-OM_GR1.5L40 SRESA1B run no.3: atmosphere 6 HOUR values MPImet/MaD Germany. World Data Center for Climate. doi: 10.1594/WDCC/EH5-T63L31_OM-GR1.5L40_A1B_3_6H
16. A. J. Simmons, J. K. Gibson (2000): The ERA-40 Project Plan. ERA-40 Project Report Series, 1, 62 p

Modelling Convection over West Africa

Juliane Schwendike, Leonhard Gantner, Norbert Kalthoff, and Sarah Jones

Institut für Meteorologie und Klimaforschung, Universität Karlsruhe/
Forschungszentrum Karlsruhe/Karlsruhe Institute for Technology,
Wolfgang-Gaede-Str. 1, 76131 Karlsruhe, Germany
sarah.jones@imk.uka.de

1 Introduction

The dramatic change in the region of the West African Monsoon (WAM) from wet conditions in the 50s and 60s to much drier conditions from the 70s to the 90s represents one of the strongest inter–decadal signals on the planet in the 20th century. Marked inter-annual variations in the recent decades have resulted in extremely dry years with devastating environmental and socio-economic impacts. Vulnerability of West African societies to climate variability is likely to increase in the next decades as demands on resources increase due to the rapidly growing population. The situation may be exacerbated by the effects of climate change, land degradation, water pollution and biomass burning. Furthermore, the WAM has an impact on the downstream tropical Atlantic by providing the seedling disturbances for the majority of Atlantic tropical cyclones and on the global climate as one of the world's largest source regions of mineral dust and of fire aerosol. Motivated by the need to develop strategies to reduce the socio–economic impacts of climate variability and change in the WAM, the integrated European project African Monsoon Multidisciplinary Analysis (AMMA)[1] aims to improve our knowledge and understanding of the WAM on daily to interannual timescales and thus improve our ability to forecast the weather and climate in the West African region.

The WAM system in northern summer is characterised by the interaction of the African easterly jet (AEJ), the African easterly waves (AEWs), and the Saharan air layer (SAL) as well as by the low-level monsoon flow and the mesoscale convective systems (MCSs). The latitudinal position of the AEJ, the inter-tropical discontinuity, MCS tracks, and, hence, precipitation patterns in the West African subcontinent are strongly correlated with the movement of the intertropical convergence zone (ITCZ) which reaches its northernmost

[1] www.amma-eu.org

W.E. Nagel et al. (eds.), *High Performance Computing in Science and Engineering '09*, DOI 10.1007/978-3-642-04665-0_32,

position in August at about 11 °N and moves back to the equator during the northern hemisphere's wintertime.

The AEJ can be observed between 10 °-12 °N and at a height of about 600 hPa in northern summer (e.g., [27]). It develops as a result as a result off the reversed meridional temperature gradient generated by the monsoon layer and the SAL. The monsoon layer is located near the surface and characterised by moist and relatively cool air from the Gulf of Guinea. It extends into the Sahel and decreases in depth towards the north. Within the monsoon layer, a diurnally varying planetary boundary layer (PBL) develops. Usually, the atmosphere in the PBL is close to dry adiabatic and close to pseudoadiabatic above [29]. The hot and dry SAL, which becomes shallower towards the equator; is located above the monsoon layer as a layer of low static stability and low potential vorticity. Above the SAL, the free atmosphere prevails, which is almost pseudoadiabatic again and contains small baroclinicity. The AEJ has a typical wind speed of about $12\,\mathrm{m\,s^{-1}}$ [4, 29] and fulfils the necessary conditions for barotropic-baroclinic instability [8].

The AEWs also exist at low levels north of the AEJ (Pytharoulis and Thorncroft 1999). These synoptic-scale disturbances propagate westwards across tropical West Africa to the eastern Atlantic and the eastern Pacific. The AEWs have propagation speeds of $7\text{-}8\,\mathrm{m\,s^{-1}}$, a period of 2-5 days, and a wavelength of about 2500-4000 km [6, 5, 3]. They modulate West African rainfall and are the main precursors of Atlantic and East Pacific tropical cyclones [1].

Convective systems are usually defined as MCSs when they reach $100 \times 100\,\mathrm{km}^2$ in horizontal extension at least and are well-developed vertically [15]. They are convectively driven weather systems. The conditional instability, vertical wind shear in the lower troposphere caused by the near-surface southwest winds and the AEJ, and a moist PBL with dry air at mid-levels are favourable conditions for MCSs to form. The instability is usually released by convective scale motions that are triggered by surface heating, soil moisture inhomogeneity with a spatial variation of energy transformation, and/or topography [33, 31, 2, 22]. Large-scale convergence, such as that in AEWs, then favours the evolution of deep convection. In the African Sahelian (12 °N-18 °N) and Sudanian climate zone (9 °N-11 °N) convective systems contribute about 80-90% and about 50%, respectively, to the annual rainfall [21, 24]. Thus, they play a key role in the water cycle of West Africa. Their rainfall, however, is highly variable in space as well as in time.

In the scope of the AMMA project we will investigate the development of MCSs, the sensitivity of their life cycle to differing surface properties, the role of larger-scale weather systems (AEWs, the Saharan Heat Low) in their development, and the development of tropical cyclones out of such systems. In addition, we will investigate the interaction of the SAL with African monsoon weather systems.

2 Numerical Model

The COSMO (COnsortium for Small scale MOdelling)[2] model, which was formerly known as the Lokal-Modell (LM) [36, 37] is used for the following studies. COSMO is an operational weather forecast model used by several European weather services, e.g. the German Weather Service (DWD), and is therefore well optimised. It is a fully compressible non–hydrostatic model suitable for forecasting atmospheric processes down to the meso-gamma scale. The Arakawa C–grid is used for horizontal differencing on a rotated latitude/longitude grid. In the vertical a hybrid system with 50 layers up to about 28 km is applied. The height of the model was increased compared to the standard setup to allow for deep tropical convection. The basic equations are solved using the time–splitting technique of [20]. The comprehensive physics package of COSMO includes a turbulence and surface layer scheme using a prognostic turbulent kinetic energy equation with a 2.5 order closure by [26] with extensions by [30] and a two-category bulk model cloud microphysics scheme [11, 23]. Precipitation formation is treated by a Kessler–type bulk microphysics parametrisation [18] including water vapour, cloud water, rain and cloud ice with column equilibrium for the precipitating phase. The subgrid–scale clouds, on the other hand, are parametrised by an empirical function depending on relative humidity and height. The spatial resolution allows for the explicit treatment of deep convection. A δ–two–stream radiation scheme [32] is used for the short and longwave fluxes and the full cloud radiation feedback. The surface layer is parametrised by a stability–dependent drag–law formulation of momentum, heat and moisture fluxes according to similarity theory. The initial and boundary conditions for all runs are taken from 6–hourly European Centre for Medium Range Weather Forecasts (ECMWF) analyses.

COSMO is run operationally at several European weather services on different platforms. The code with the operational settings is well tested and optimised to produce cost effective forecasts. We will also use COSMO coupled with the aerosol and reactive trace gases module (COSMO-ART) to investigate the interaction of the SAL with African monsoon weather systems. COSMO-ART [44, 45] was developed in Karlsruhe and computes the emission and the transport of mineral dust. Computational details for three typical model setting on the HP XC 4000 can be found in Table 1.

Within the AMMA project at the HP XC 4000 at the Steinbruch Centre for scientific super computing, we focus on two different topics. On the one hand the life cycle of an MCS over West Africa is modelled with respect to the soil conditions. On the other hand the focus lies on comparison between the simulated MCSs over West Africa and over the Eastern Atlantic. Both parts are explained in more detail in the following two sections.

[2] www.cosmo-model.org

Table 1. Three typical model settings are shown here. Example I represents a typical COSMO run, Example II represents a COSMO run adapted for budget calculations, and Example III a COSMO-ART run

Parameter	Example I	Example II	Example III
Horizontal resolution	0.025 °	0.025 °	0.25 °
Vertical levels	50	50	50
Simulation time	145 h	72 h	144 h
Tasks and nodes	32 tasks running on 8 nodes	32 tasks running on 8 nodes	64 tasks running on 16 nodes
Sum of CPU-time over all processors (d-hh:mm)	8-17:57	35-05:20	31-21:20
Elapsed time (hh:mm:ss)	6:36:15	22:03:00	5:10:00
Maximum physical memory by any process (in MB)	1403	4136	1508
Maximum virtual memory by any process (in MB)	4135	6892	4675
Gridpoints	561×361	1000×500	231×132

3 Modelled Life Cycle of an MCS over West Africa and Its Sensitivity to Soil Conditions

The mechanism that determines the life cycle of precipitating convective systems in West Africa are still not well understood. Surface and convective boundary layer (CBL) processes play an important role in the initiation of convection as do mid- and upper-tropospheric forcing [16]. Important parameters are the spatial distribution and temporal development of water vapour in the CBL. Besides advective processes, water vapour is made available in the atmosphere locally through evapotranspiration from soil and vegetation. Soil moisture affects the energy balance via the albedo and emissivity of the surface, the conduction of heat in the soil and the stomata resistance of vegetation. Many research findings show that the soil moisture exerts greater influence on the CBL than vegetation (e.g. [46, 12]).

Horizontal gradients of soil moisture have been linked to the development of secondary mesoscale circulations [35, 34, 43]. In regions of convergence this leads often to a release of convective available potential energy (CAPE). West Africa is identified as a hot spot, i.e. one of the regions where a strong coupling between soil moisture and precipitation is evident [17]. Both positive and negative feedbacks are observed between soil moisture and convective precipitation. A negative feedback is described by [41] for the initiation of convection in the Sahel. [40] and [42] show from observational data that a positive feedback between soil moisture and rainfall exists in the Sahelian zone, when MCS storms pass over contrasts in surface evaporation. Although several authors report feedback mechanisms between soil moisture and moist convection or rainfall, it is still not clear if or under which circumstances this feedback is positive or negative.

Within the scope of the AMMA-project we study the sensitivity of the life cycle of an MCS to surface conditions and contribute to furthering the knowledge of the issues mentioned above. The analysis is based on simulations with the COSMO model of a real MCS event on 11 June 2006. This time of year is in the pre-onset phase of monsoon when vegetation cover is low and the impact of soil moisture is assumed to be dominant. Special emphasis is placed on the sensitivity of initiation, intensification and modification of this MCS to the soil properties. In particular, we consider whether there is a positive or negative feedback between soil moisture and precipitation.

3.1 COSMO Model Initialisation and Observed and Simulated Convective Systems on 11 June 2006

The COSMO model was initialised on 11 June 2006 at 00 UTC and the boundary conditions were updated every 6 hours. Different conditions for soil moisture were applied for initialisation of TERRA-ML. Standard initialisation is applied by feeding the model pre-processor with data from ECMWF operational analysis. The run based on the COSMO soil type distribution and on original ECMWF fields was denoted with MOI. However, comparison with AMSR-E satellite data (Figure 1a) showed that the MOI field contained too much soil moisture in the upper surface layer. To apply a proper scaling factor for the reduction of the soil moisture, the following factors were taken into consideration. The soil type in ECMWF is loam with fairly high values of permanent wilting point (PWP=0.171) and field capacity (FC=0.323). This led to values of soil moisture that were too high for the Sahel regions. According to the COSMO land use distribution the dominant soil types are sand and loam. These are characterised in TERRA-ML by a mean PWP of 0.076 and a mean FC of 0.268. Thus, the ratio between the medium soil moisture from COSMO of 0.210 and ECMWF of 0.246 is about 0.7. Therefore we reduced the volumetric soil moisture content in all layers by 35% compared to the initial conditions of MOI. This resulted in a similar soil moisture content in the uppermost level of TERRA-ML, compared to the soil moisture values of about 12% derived by the AMSR-E satellite and in-situ measurements of about 18% for the uppermost 5 cm taken at Dano (3 °W and 11 °N) [19] for the region around 11 °N, where the MCS was observed. The resulting soil moisture distribution is shown in Figure 1b and the corresponding simulation was designated as CTRL experiment. The AMSR-E observation, shown in Figure 1a, illustrated a north-south gradient in soil moisture with some pronounced patterns in the west–east direction. Some of these patterns were also represented in the model initial fields (Figure 1b). There was a tongue of dry soil extending southward from 2 °W to 1 °E and a region around the border of Burkina Faso and Benin, where moist soil extended further northward.

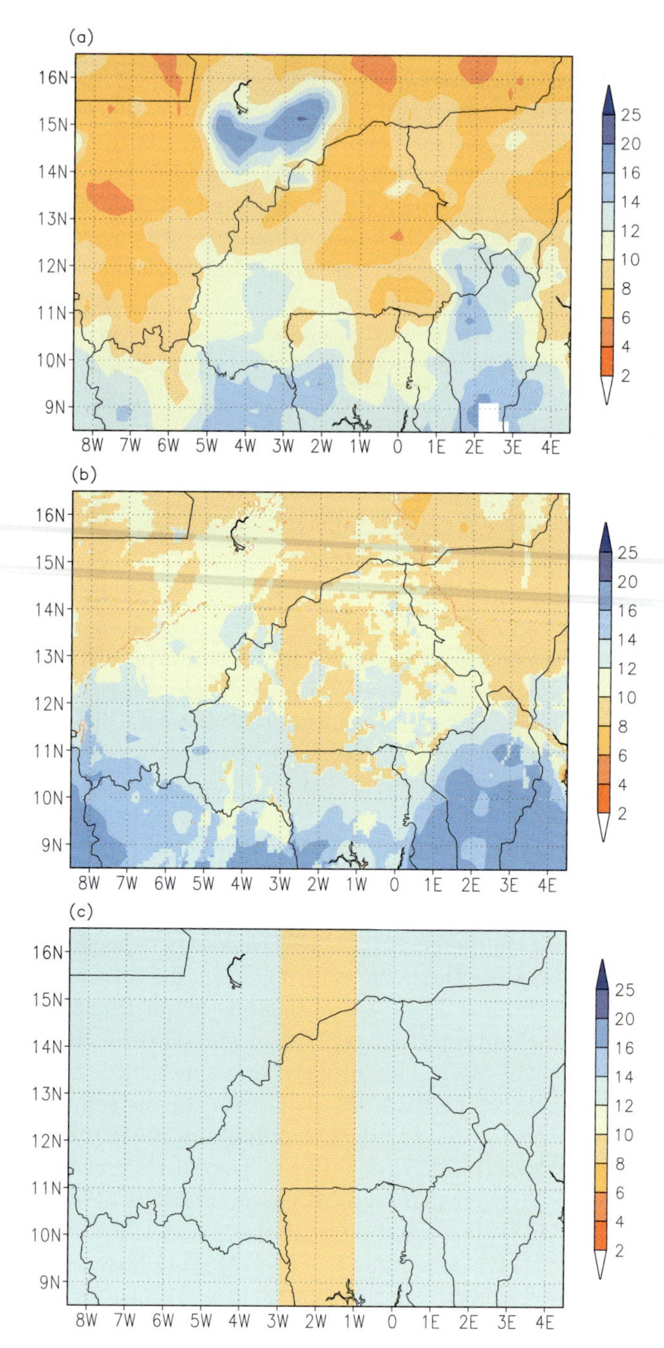

Fig. 1. Volumetric soil moisture in % (colour coded) on 11 June 2006 as seen by AMSR-E [28] (a) and COSMO initial volumetric soil moisture in % in the uppermost one cm (colour shaded) on 11 June 2006 at 00 UTC for the cases CTRL (b), and BAND (c). Taken from [14]

To eliminate the effect of spatial soil moisture variability on the initiation of convection, an additional simulation with a homogeneous (HOM) distribution of soil moisture and soil texture was performed. In this case the volumetric soil moisture was specified as a mean value of the volumetric soil moisture in the CTRL experiment along 11 °N from 4.5 °W to 4.5 °E. This resulted in a value of 12.7% in the top layer. The type loam used here was classified as medium texture, so that a volumetric soil moisture content of 12.7% corresponds to 37% of the field capacity. In the CTRL run, a dry tongue roughly positioned between 1 °W and 3 °W extended further towards the south (Figure 1b) and influenced the precipitation pattern of a convective system. Therefore, to investigate the effect of dry regions on convective systems, where the soil moisture structure is less complex than the conditions present in the CTRL run, a dry band of 2 degree longitudinal extension was inserted into the homogeneous soil moisture field. In this band the volumetric soil moisture was reduced by 35% compared to the homogeneous environment (Figure 1c), such that the volumetric moisture content in the top layer amounted to 8.3%. This value roughly agreed with the moisture content of the dry tongue in the CTRL run. The corresponding experiment was denoted as BAND.

On 11 June 2006 at 12 UTC the trough axis of an AEW was located at about 0 °W in the ECMWF analysis. This AEW formed over the Darfur mountains (24 °E, 14 °N) on 05 June and propagated over West Africa and the tropical Atlantic, where its structure was more distinct due to the reduced number of convective cells. On 11 June the axis of the AEJ calculated using the method of [3] lay approximately along 11-13 °N. Between 700 and 500 hPa the AEJ dominated the flow pattern. At lower levels, a southerly to south-westerly monsoon flow was present. The northern border of the monsoon flow meandered roughly between 12 and 15 °N. Convection was initiated predominantly in the region from Ghana to Benin (4 °W-2 °E, 7-11 °N) at about 14 UTC. This was the time at which the conditional instability was maximum. The convective systems intensified quickly, and moved rapidly westwards. The northernmost cell located at the border of Benin, Togo, and Burkina Faso (1 °W, 11 °N) experienced the strongest development and grew into a mature MCS. The centre of the MCS moved west along the southern border of Burkina Faso. New storm cells developed ahead of the strong gust front of the system. According to surface and satellite data the accumulated precipitation reached values of up to 70 mm [39]. The speed of the MCS increased from values of about 8 m s^{-1} in the afternoon of 11 June, to about 14 m s^{-1} at 20 UTC, and to 20 m s^{-1} at midnight. When it was close to its maximum intensity it had an extension of about 500×500 km^2.

The first precipitation of westward moving convective cells was simulated in CTRL between 1 and 2 °E at about 17 UTC on 11 June. The initiation of convection was east of the trough axis of the AEW. Its axis was positioned at about 2 °W at this time. Thus, the area of initiation of the convective cells agreed fairly well with the observations while the time of initiation was about 3 hours later than observed. The spatially averaged precipitation of the mature cell reached a first maximum of about 6 mm h^{-1} at 0.5 °W about 2 hours before midnight. Precipitation decreased when the system passed the region between 2 to 4 °W. Subsequently it increased again and a maximum rainfall amount of more than 6 mm h^{-1} was simulated during the morning of 12 June. The speed of the mature system was about 19 m s^{-1}. The propagation speed of the AEW was approximately 13 m s^{-1}.

3.2 Investigation of the Different Phases of Convection

The spatial distribution of CAPE and convective inhibition (CIN) of the different model experiments showed similar large scale patterns before the first convective cells developed. Figure 2 shows CAPE and CIN distributions of the CTRL run at 15 UTC. In most parts of the domain the CAPE ranged between 1000 and 2000 J kg^{-1}, while values reached up to 3000 J kg^{-1} in the central part of Burkina Faso and south–western area of Niger. Lower values were seen only north of 15 °N and in the south–western corner of the domain. CIN increased from zero in the southeast to 150 J kg^{-1} in northwest. Thus, the conditions for the initiation of convection were most favourable in the south–eastern part of the model domain, where CAPE was high and CIN was zero. In the satellite diagram and the Hovmöller plot of the CTRL case (not shown), this reafion could be identified as the area where most intense convection developed on this day.

A closer look at the CTRL case shows that three separate cells were initiated in the south–eastern part of Burkina Faso. Precipitation of up to 6 mm h^{-1} was simulated at 17 UTC (Figure 3a). The south–westernmost cell developed in the lee of an area with orographically induced upward motion. Triggering of convection often occurs in this way in West Africa, and as seen in observations and modelling studies (e.g. [10]). Two further cells developed in the east, at 1.9 °E and 11.9 °N and at 2.2 °E and 11.6 °N, but no pre–existing vertical motion was observed there in the area before the initiation of convection. In Figure 3a the precipitation pattern of the CTRL case at 17 UTC is overlaid on the soil moisture distribution in the uppermost layer at 15 UTC. This figure shows that all three cells developed in the transition zone from a wetter to dryer surface, while the centres of the precipitating cells were positioned over the dryer surface. In comparison to the CTRL case, only two precipitating cells had developed at 17 UTC in the HOM case (Figures 3b). These two cells were observed at roughly the same locations as the most intensive cells in the CTRL case. However, the precipitation of both cells was less intense than that of the CTRL case at the same time.

In the MOI case the area with CIN values of zero was slightly smaller and more restricted to a region in the south-east (not shown). Hence, the favourable conditions with high CAPE and low CIN values were more limited in space than in the other cases. In addition, the surface temperature in the region of interest in the MOI case was about 3 °C lower than in the CTRL

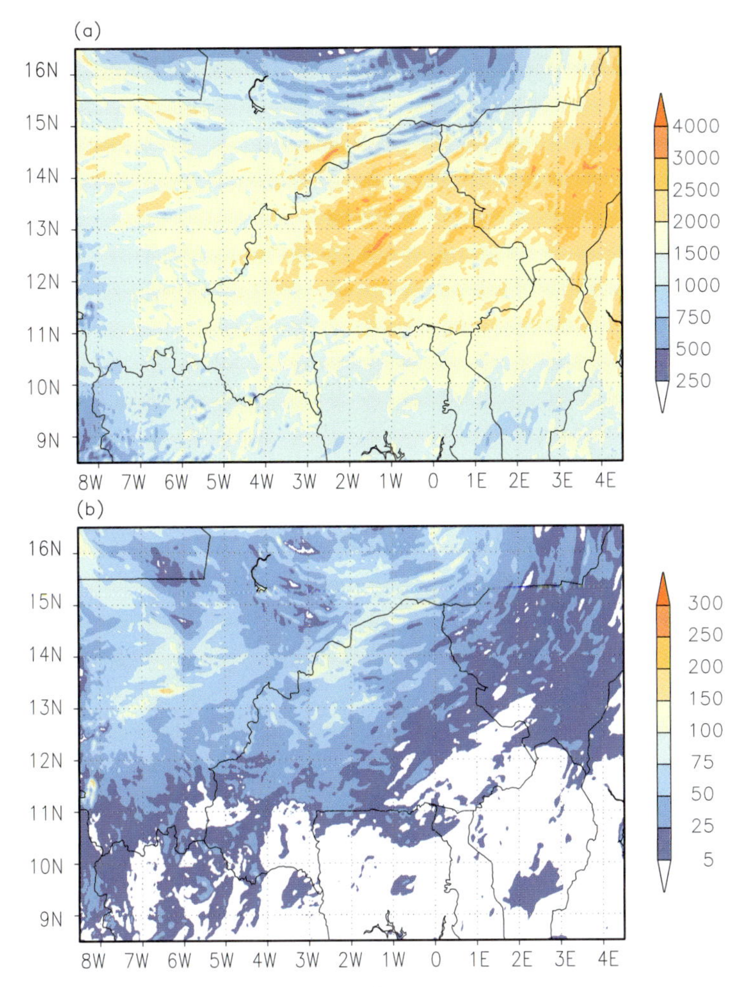

Fig. 2. CAPE (a) and CIN (b) in J kg^{-1} for CTRL on 11 June 2006 at 15 UTC. Taken from [14]

case. Under these conditions only one weak precipitating cell had developed at 17 UTC.

In conclusion, precipitating convective cells developed in all model simulations in an area where convective instability was sufficiently high combined with low convective inhibition. In addition, in the CTRL case, with the most intense convection, triggering over regions with spatial differences in soil moisture and orographically induced convergence was present. The meteorological conditions in the MOI case, with colder surfaces than in the CTRL and HOM run, were less favourable for the triggering of convection on this day.

Once triggered, the convective cells developed quickly in the CTRL and HOM case and moved with the AEJ towards the west. At 19 UTC, i.e. about

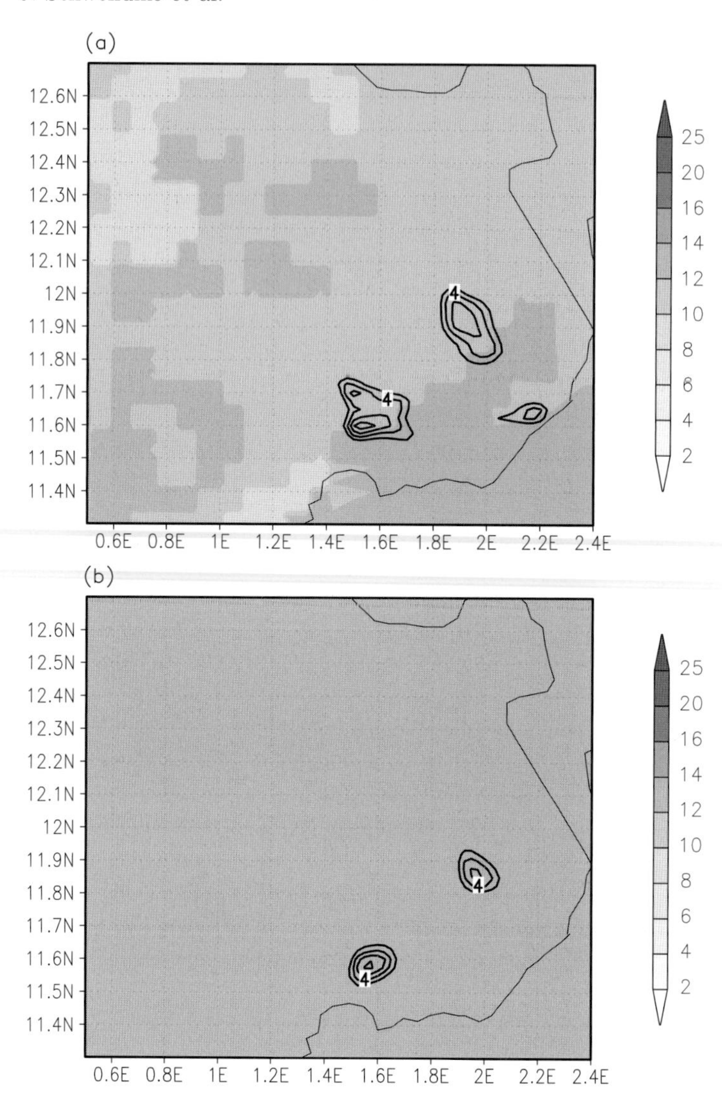

Fig. 3. Volumetric soil moisture in % at 15 UTC in the uppermost layer (grey shaded) and pre-cipitation in mm h-1 (isolines, interval 2) on 11 June 2006 at 17 UTC for CTRL (a) and HOM (b) case. Taken from [14]

two hours after its initiation, convective cells were found about 100 km westwards of its area of origin. In the CTRL run the cells had already organised into a MCS with updraughts up to -20 Pa s^{-1} and a gust front with gusts of 20 m s^{-1} which was about 130 km in length. In the HOM case three separate cells could still be distinguished at 19 UTC, which were less intense than in the CTRL case.

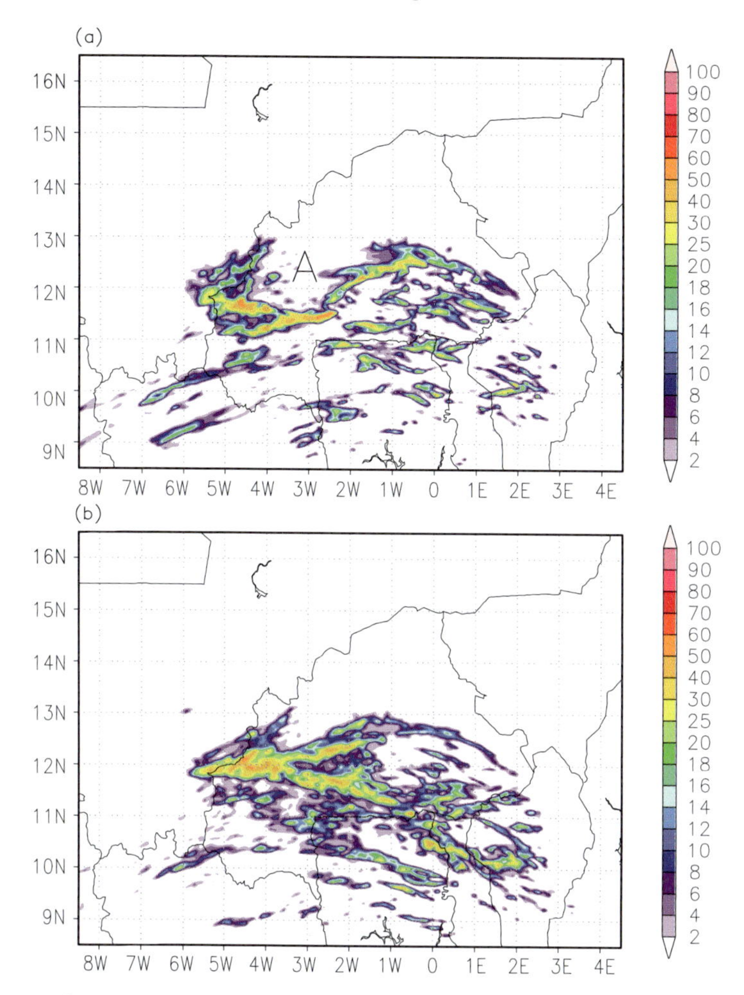

Fig. 4. 24-h accumulated precipitation in mm (colour coded) starting at 06 UTC on 11 June 2006 from CTRL (a) and HOM (b) case. In (a) the letter A indicates the area with reduced precipitation. Taken from [14]

The MOI run showed only weak convective activity at 19 UTC with some isolated cells. The convective activity of these cells weakened, while moving to the west. It appeared that the weak convective activity of the cells was completely suppressed when the cells reached the area with enhanced convective inhibition in the northwest.

During the night the MCSs of the CTRL and HOM runs moved further westwards. Figures 4a and 4b show the accumulated precipitation of both simulations from 11 June 06 UTC to 12 June 06 UTC. The precipitation patterns of the MCSs consisted of smaller precipitation bands, rather than a homogeneous rainfall distribution. These small bands were caused by the convective cells embedded in the MCS. Obviously this is a characteristic feature of convective systems [9]. Maximum values of about 60 mm h^{-1} were simulated, which was quite similar to the observations [39]. In the CTRL run a bigger gap in the precipitation pattern was simulated, roughly located between 2.5 and 4 °W and 11.5 and 13 °N (Figure 4a). This area, with little or no precipitation, was characterised by CIN values of up to 200 J kg^{-1} and slightly lower total water content, compared to the values to the east and west of it as well as soil moisture inhomogeneities (Figure 1b). Obviously, the updraughts of the MCS could not overcome the strong convective inhibition. To the west of the area with enhanced CIN the MCS gained strength again and precipitation reached former values. In the HOM run the southward oriented tongue with higher CIN and lower total water content were less pronounced, so that the MCS remained strong and the precipitation high throughout the observation period (Figure 4b).

The impact of the dry band with a volumetric moisture content of 8.3%, surrounded by a homogeneous moisture content of 12.7% (Figure 1c) on the modification of a mature MCS is shown in Figures 5a and 5b. Precipitation was significantly reduced between 0 and 2 °W, i.e. shifted by about one degree to the east of the dry band. The Hovmöller diagram reveals that the precipitation was interrupted, when the MCS approached the dry band, but re–generated when the system reached the western part of the band (Figure 5b). The figure also reveals that convection was not only initiated in the primary region along the border of Burkina Faso and Benin at about 17 UTC, but also in the western part of the dry band (around 3 °W) at about 19 UTC. At 22 UTC of the cloud cluster had reached a diameter of about 100 km; they were accompanied by significant rainfall and moved to the west with a similar speed as the cells developed earlier in the east.

To understand the decrease of the original convective system, which weakened when approached from the east, CAPE and CIN (Figure 5c) were observed. At 18 UTC an area with lower CAPE values had developed within the dry band, while CIN increased to the east and west of it. East of the band (0.83 °W, 12.2 °N) CIN further increased during the subsequent hours and was about 190 J kg^{-1} at 22 UTC, just before the MCS reached the area. Inside the band (1.8 °W, 12.2 °N) CIN was only 40 J kg^{-1} at this time. The lower CAPE inside the dry band resulted mainly through lower near–surface humidity. East of the dry band a lower near-surface temperature led to a higher CIN, which inhibited the convection of the approaching MCS. The increase of CIN inside the dry band later that night yielded from the nocturnal decrease of near-surface temperature and the passage of the MCS in the surrounding.

3.3 Summary and Conclusions

The simulations showed that convection was initiated in all model experiments, regardless of the initial soil moisture distribution. The area where convection was initiated in the simulations corresponded roughly with the obser-

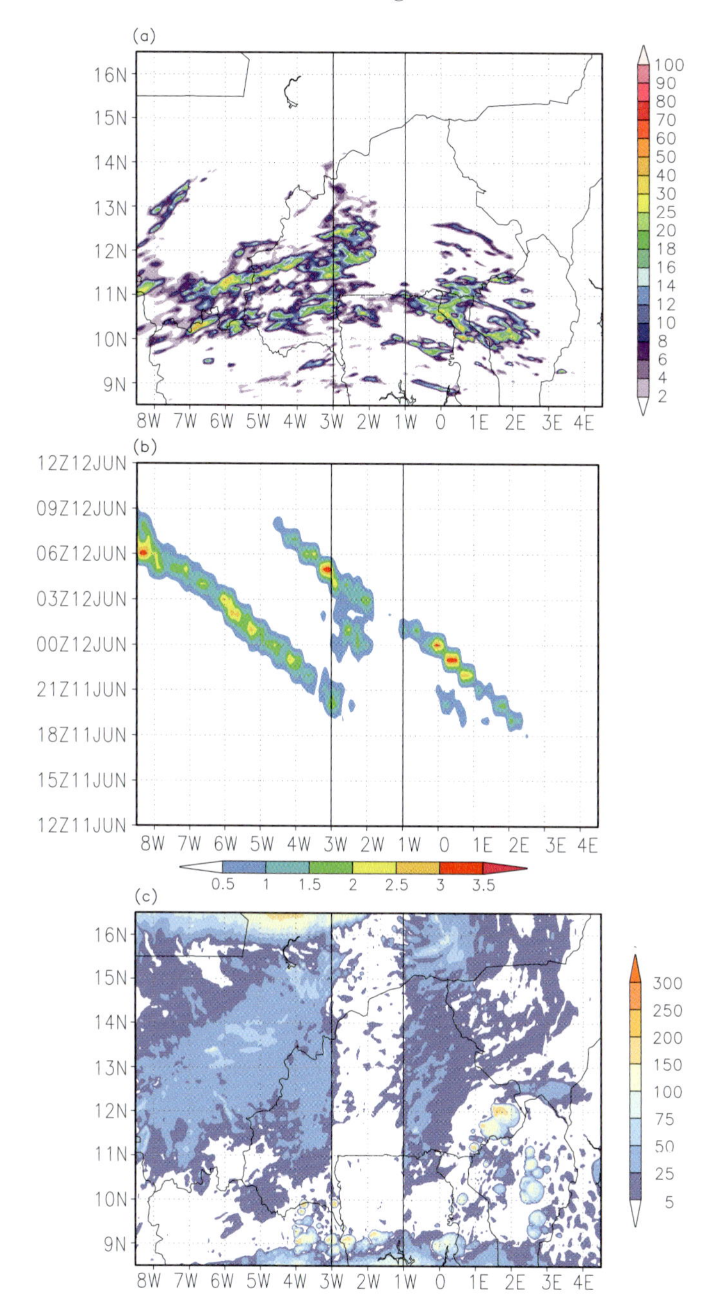

Fig. 5. 24-h accumulated precipitation in mm (colour shaded) starting from 06 UTC on 11 June 2006 (a), Hovmöller diagram of precipitation in mm h^{-1} (colour shaded) averaged between 10.5 and 13.5 °N on 11 and 12 June 2006 (b), and CIN (c) in J kg^{-1} (colour shaded) on 11 June 2006 at 18 UTC for BAND case. The solid lines enframe the area of the dry band. Taken from [14]

vations. The development was delayed by about three hours. In the CTRL case all three cells were initiated along soil moisture inhomogeneities and shifted towards the dry surface. [7] found a similar behaviour. Triggering of convection and optimal evolution was simulated in areas with low CIN, high CAPE and low soil moisture content ($< 15\%$) or soil moisture inhomogeneities. In CTRL and HOM the cells developed quickly and merged into organised mesoscale systems while moving westwards. The MCS in the CTRL run experienced a significant modification. The precipitation disappeared when the MCS reached a region which was characterised by high CIN values ($> 150\,\mathrm{J\,kg^{-1}}$), reduced total water content and soil moisture inhomogeneities. In the HOM case no significant reduction of precipitation was simulated.

In conclusion, triggering of convection on this day was favoured by drier surfaces and/or soil moisture inhomogeneities, while a mature system weakened in the vicinity of a drier surface. This means that a positive feedback between soil moisture and precipitation existed for a mature system whereas a negative feedback was found for triggering of convection. The two mechanisms are illustrative for the complexity of soil–precipitation feedbacks in an area where high sensitivity of precipitation on soil moisture was proven and reproduced by several models (e.g. [17]).

4 Comparison of Convection over Land and Ocean

The aim of this study is to investigate convective systems over West Africa and the eastern Atlantic embedded in the same AEW. The case study we have chosen is the AEW out of which Hurricane Helene (2006) developed. An MCS over land was initiated in the afternoon hours on 9 September 2006 ahead of the trough of this AEW which developed on 2 September 2006. The convective system increased quickly in intensity and size and developed three westward moving arc–shaped convective systems. As the system decayed, new convective bursts occurred in the remains of the old MCS. This lead to structural changes of the form of a squall line crossing the West African coastline at around midnight on 11 September. During the next 24 hours, intense convective bursts occurred over the eastern Atlantic. These convective bursts were embedded in a cyclonic circulation which intensified and became a tropical depression on 12 September 2006, 12 UTC. The largest and longest lived convective burst in this intensification period was analysed here and compared to the convective system over land. The convective systems modify their environment and these changes can be related to the structure of the system itself.

COSMO runs were conducted for different regions for this study and they are given in Figure 6 and Table 2. The coordinates are not rotated. A series of high–resolution runs with 2.8 km horizontal resolution were carried out such that the first model area (area 2 in Figure 6) was centred around the MCS over Burkina Faso. As the system moved across West Africa the position of the model region was adjusted for each subsequent run. All the runs are 72–

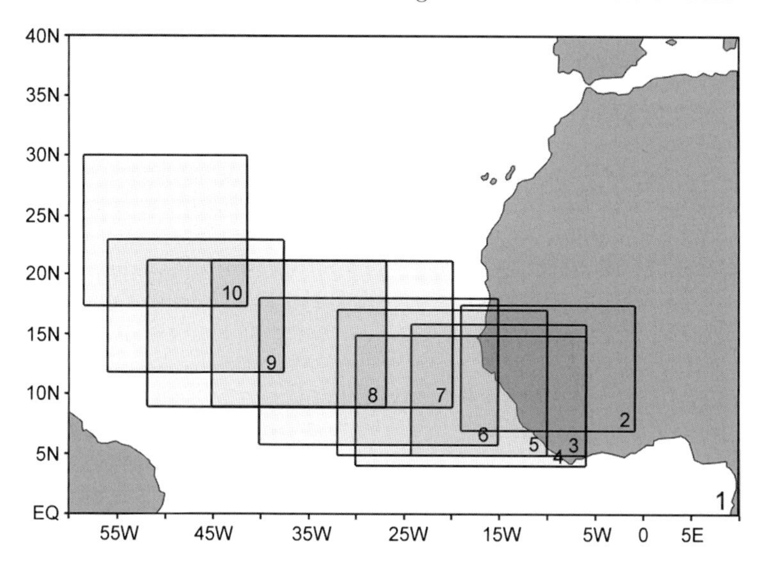

Fig. 6. The different model regions are displayed here. Further description can be found in Table 2. Taken from [38]

Table 2. List of all COSMO runs, which are 72-h forecasts. Taken from [38]

COSMO run	Initialisation	Longitude	Latitude	Resolution
1	9 Sep 2006, 12 UTC	60 °W-12 °E	1-40 °N	28 km
2	9 Sep 2006, 12 UTC	19.5-1.025 °W	7-17.5 °N	2.8 km
3	10 Sep 2006, 12 UTC	24.5-6.025 °W	5-15.975 °N	2.8 km
4	11 Sep 2006, 00 UTC	30-6.025 °W	4-14.975 °N	2.8 km
5	12 Sep 2006, 00 UTC	32-10.025 °W	5-16.975 °N	2.8 km
6	13 Sep 2006, 00 UTC	40-15.025 °W	6-17.975 °N	2.8 km
7	14 Sep 2006, 00 UTC	45-20.025 °W	9-20.975 °N	2.8 km
8	15 Sep 2006, 00 UTC	52-27.025 °W	9-20.975 °N	2.8 km
9	16 Sep 2006, 00 UTC	56.5-38.275 °W	12-22.975 °N	2.8 km
10	17 Sep 2006, 12 UTC	58.5-41.525 °W	18-29.975 °N	2.8 km

h in duration and the model region is always centred around the convective system. For the high–resolution model runs the parametrisation of convection is switched off. The model source code was adapted to provide information for moisture, temperature and momentum budgets [13].

4.1 COSMO Simulations

COSMO is able to reproduce the AEW as seen in the ECMWF analysis (not shown). The Rapid Developing Thunderstorm (RDT) Product by Météo France [25] was used to assess the ability of the high–resolution model runs to simulate the convective system. The vertically integrated cloud ice, water vapour and humidity (total water), which indicates the convective updraught cores, is compared to the convective regions, highlighted by the orange–red colours (Figure 7). A large MCS, which was initiated at around 15 UTC on 9 September 2006 over Burkina Faso, is evident in the COSMO run initiated at 9 September 2006 at 12 UTC. After the initiation the system intensified quickly and its leading edge can be seen in Figure 7b as a north south orientated band. The near–surface winds depict the mostly westerly monsoon inflow and the weak easterly inflow into the system. The regions with maximum total water content correspond to the dark orange and black regions in the RDT images where a circularly shaped MCS is visible. The MCS reached its peak intensity close to midnight on 9 September (Figure 7b). In the following hours the system decayed, and several convective bursts occurred within the remains of the old MCS (Figure 7d). The COSMO run captures these structural changes well. Four regions with high total water content are evident. The two regions of high total water within the box marked B correspond to the position of the leading edge of the convective region in the RDT image. In the following hours the convective systems furthest to the east decayed and the three remaining systems can be seen within the box in Figure 7f. It is not possible to distinguish between the convectively active regions in the RDT image, but a circle–shaped MCS is seen. This MCS was decaying and, as it approached the Highlands of Guinea, new convective cells were initiated and the structure of the system changed from a circular shape to a comma shape (not shown). In the evening hours of 10 September the system was a squall line, that crossed the West African coastline at around 11 September, 00 UTC. This change in structure and intensity was observed also in the model simulation. The main difference is that the modelled system moved more slowly than in reality. It reached the same position as in the RDT image but about 6 hours later and not with the same intensity.

When the convective system reached the eastern Atlantic the convection decayed. This was simulated by the model as well, but it occurred several hours later. As the system passed the West African coast line, a new convective system was initiated which grew rapidly into a mature MCS. When it crossed the coastline it decayed over the ocean (Figure 7h). Again, this process was delayed by about 9 hours in the model compared to the observations (Figure 7g). Over the Atlantic the character of the convective system changed. It was no longer one large MCS with either a circular or comma shape, but several large convective bursts embedded in a mesoscale circulation occurred. Most of the time the convective bursts were initiated in the northwest part of the circulation and grew very rapidly. As they moved around the circulation centre towards the southwest they decayed more slowly. While the old system decayed, a new convective burst occurred in the north–northeast which moved in the same manner. When the MCSs over West Africa reached the Atlantic, the pre-existing mid–level circulation had reached the surface (see Figure 7j). The circulation intensified and the ambient southwesterly inflow on the western and southwestern side of the circulation was much stronger than in the

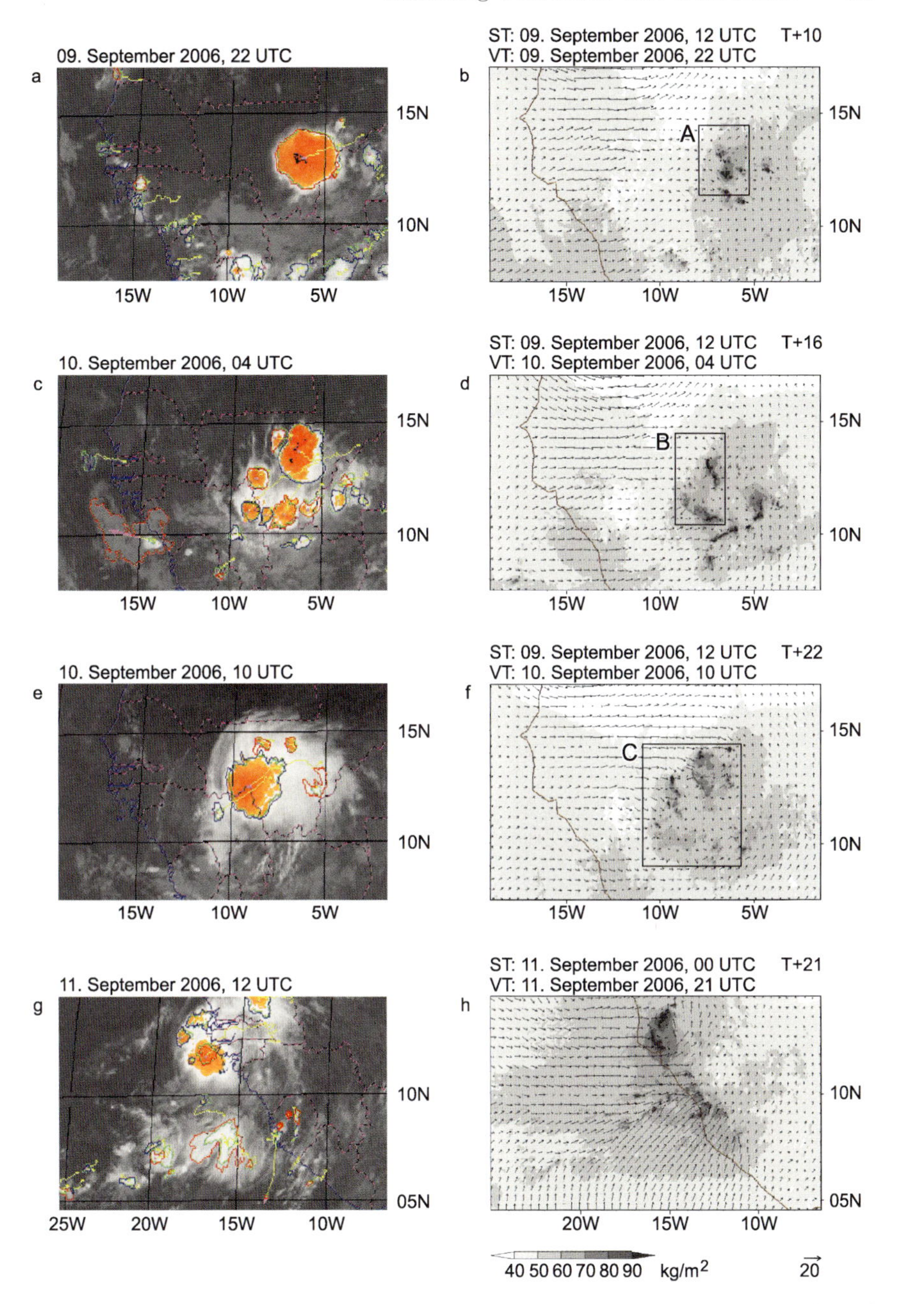

Fig. 7. Comparison between the Rapid Developing Thunderstorm Product (RDT) (left) and COSMO model runs (right). Convective objects are superimposed over the Meteosat infrared images using shadings of grey up to -65 °C, orange–red colours between -65 ° and -81 °C, and black above -81 °C in the RDT images. Continues on next page. Taken from [38]

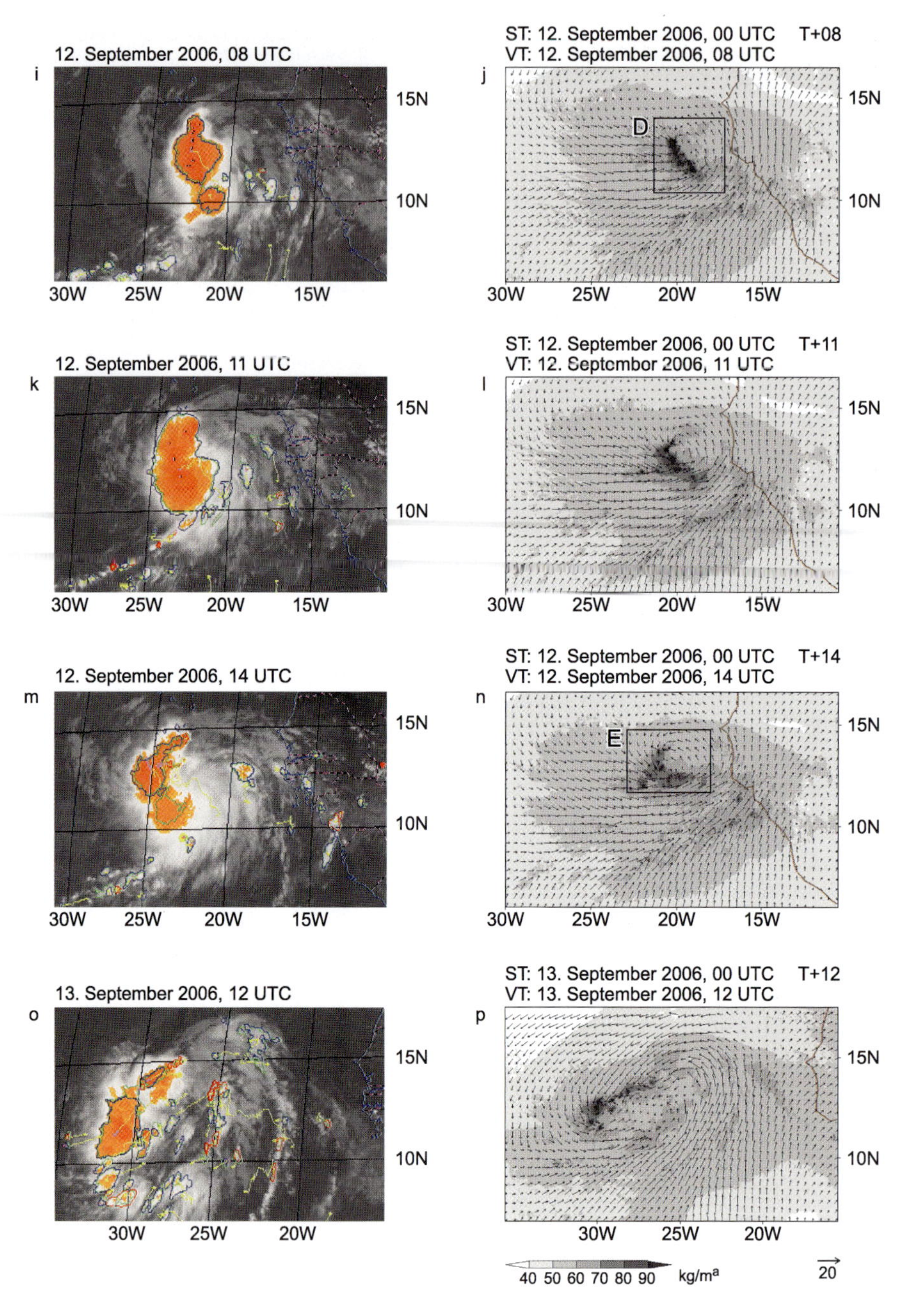

Fig. 7. (Continued) The colour contour defines the edges of the observed cloud systems. Yellow displays the triggering of the system, red the growing, violet the mature stage of the system, and the blue contour represents the decrease. Additionally, the green contour displays the edge of the system in the previous satellite image, whereas the yellow line shows the track of the system. Courtesy of Météo France. The vertical integral of cloud water, cloud ice and humidity (kg m^2) from the 28-km COSMO run is shown in the right column. Taken from [38]

Table 3. Details of boxes over land used for budget calculations giving longitudinal and latitudinal extent, time of simulation over which averaging was performed, corresponding real time, and number of grid points in a horizontal plane within the box. Taken from [38]

Box	Longitude	Latitude	Forecast Hour	Verification Time	Grid points
A	10.5-14.0 °N	8.0-5.5 °W	9-11	21-23 UTC on 9 September	14.000
B	10.4-14.5 °N	9.5-6.4 °W	15-17	03–05 UTC on 10 September	20.336
C	9.0-14.5 °N	11.0-5.8 °W	21-23	09–11 UTC on 10 September	45.760

north or the east. One example for these big convective bursts is shown in Figure 7l, 7n and a second one is shown in Figure 7p. After 12 September, 12 UTC, the convective systems embedded in the circulation were organised enough for the circulation to be classified as a tropical depression. The model was able to capture these structure and intensity changes very well.

4.2 Structure of the Convective Systems

We identified three stages in the life cycle of the MCS over West Africa: the developing, the mature, and the decaying phase. To analyse the structure of these convective systems in more detail, east–west cross sections (Figure 9) are drawn through the convective system shown in Figure 2d. The results are based on a model run initiated on 09 September 2006 at 12 UTC (see Figure 7, Table 2 and Table 4). During the mature phase, low–level convergence occurs between a strong westerly inflow and an easterly inflow from behind the system and is partly enhanced due to the descending air from the rear system. The associated ascent region extends up to about 200 hPa and is tilted eastwards with height (Figure 8a). The convergence continues up to 700 hPa. At this stage strong downdraughts have developed around 7.5 °W just behind the the low–level convergence zone that is located under the tilted updraught region. The area of the westerly inflow reaches deep into the system up to about 700 hPa and the AEJ has its maximum around 600 hPa. Divergence associated with the upper–level outflow can be seen at around 200 hPa with very strong easterly winds ahead of the system and westerly winds in the rear. Thus, there is strong mid-level convergence east of the tilted updraught. The maximum heating (Figure 8b) is collocated with the ascent region. At upper–levels, cyclonic relative vorticity occurs over a large area to the rear of the convective system and anticyclonic vorticity occurs ahead of the system. The inverse situation occurs at low–levels where a positive relative vorticity maximum is evident in the updraught region and a minimum in negative relative vorticity occurs in the region of the strong downdraughts. Below 400 hPa weaker descent occurs.

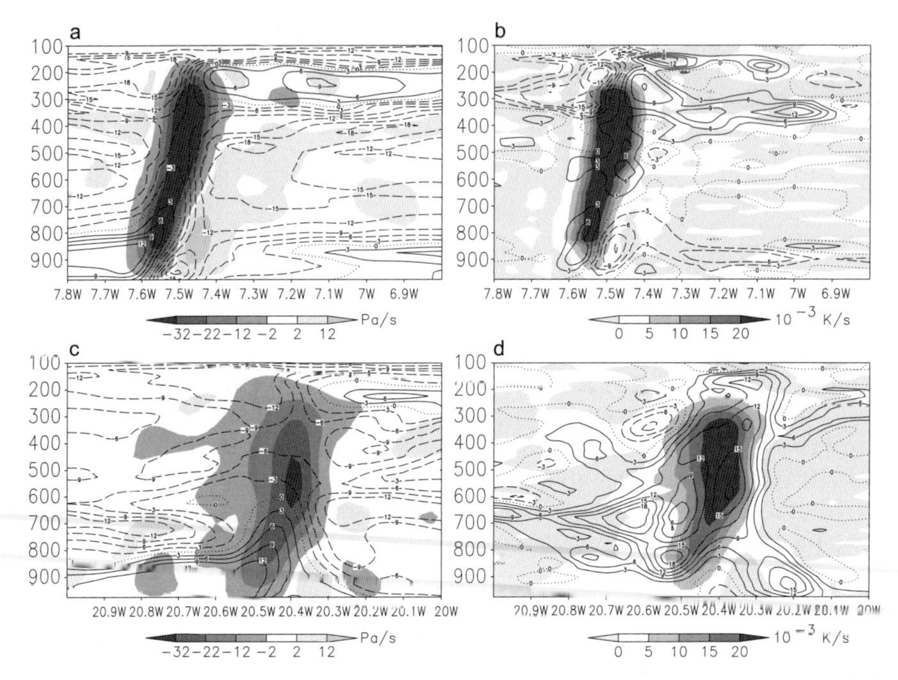

Fig. 8. Cross sections through the convective system of vertical velocity (shaded) and zonal wind (contour interval is $3\,\mathrm{m\,s^{-1}}$) in a, c as well as the total diabatic heating rate (shaded) and the vertical component of the relative vorticity (contour interval $3 \times 10^4\,\mathrm{s^{-1}}$) in b, d. The cross sections a and b are along 13.1 °N on 10 September 2006, 05 UTC, and c and d along 12.8 °N on 12 September 2006, 09 UTC. Taken from [38]

The cross section through a convective system embedded in the developing tropical depression over the eastern Atlantic is shown in Figure 8c and 8d. On 12 September at about 12 UTC, the system was classified a tropical depression. The basis for this analysis is a model run that was initialised on 12 September 2006 at 00 UTC. A major difference between the MCS over West Africa and the convective bursts that are embedded in the circulation over the Atlantic is their lifetime. The MCS over land lasted for about 3 days and the succession of MCSs over the ocean only for about 6 to 24 hours. The region of maximum heating and ascent is vertical and not tilted as in the convective system over land. The region of ascent and heating is collocated with a deep tower of positive vertical velocity, which could be described as a vortical hot tower. The AEJ seems to occur at around 700 hPa west of the convective system. This is about 100 hPa lower than for the MCS over West Africa where the air is accelerated due to air that exited the updraught core at lower levels and slightly descends. It is also apparent from the cross section that the downdrafts are not as strong as for the MCS over land. Furthermore, the low-level convergence covers a much broader region than over the continent.

Table 4. As Table 3 but over water. Taken from [38]

Box	Longitude	Latitude	Forecast Hour	Verification Time	Grid points
D	10.5-14.0 °N	21.5-17.5 °W	7-9	7-9 UTC on 12 September	22.400
E	11.5-14.5 °N	23.0-18.0 °W	13-15	13-15 UTC on 12 September	24.000

4.3 Potential Temperature Budget

To quantify the difference between the convective systems over land and over water and to assess the influence of the convective systems on the environment, potential temperature, relative vorticity, and potential vorticity budgets for regions encompassing the convective region were calculated. Details for the selected regions can be found in Figure 7 and Table 3 as well as Table 4. The budgets were then calculated over the whole box. Additionally, we defined regions within the boxes with strong ascent and descent as well as with weak ascent and descent according to the vertical velocity at 650 hPa. With the help of this partitioning, the structure of the convective systems is more evident than in averages over the entire boxes. This enables us to interpret the total potential temperature and potential vorticity tendencies better. The results for the budget calculations can be found in [38].

For this study we carried out high-resolutionn (2.8 km) COSMO model runs over large model domains (1000 × 500 gridpoints). This enabled us to identify structural features of convective systems over West Africa and the Eastern Atlantic and to analyse and discuss their differences. In future studies the analysis could be applied to other cases in order to generalise these results.

5 Outlook

This study will be extended as follows. Calculations of the heat and moisture budget for the lower troposphere will be performed to investigate the contributions of the different components to the budgets and to determine which processes were responsible for the main differences between the cases especially in triggering of convection. We will conduct a series of high resolution COSMO-ART runs in the same manner as illustrated in Figure 6 to analyse the effect of the Saharan air layer on the convective systems over land and over water as well as on Hurricane Helene.

Acknowledgement. This project received support from the AMMA-EU project. Based on a French initiative, AMMA was built by an international scientific group and is currently funded by a large number of agencies, especially from France, UK, US and Africa. It has been the beneficiary of a major financial contribution from the

European Community's Sixth Framework Research Programme. Detailed information on scientific coordination and funding is available on the AMMA International web site http://www.amma-international.org.

References

1. L. A. Avila and R. J. Pasch. Atlantic tropical systems of 1993. *Mon. Wea. Rev.*, 123:887–896, 1995.
2. G. J. Berry and C. D. Thorncroft. Case study of an intense African easterly wave. *Mon. Wea. Rev.*, 133:752–766, 2005.
3. G. J. Berry, C. D. Thorncroft, and T. Hewson. African easterly waves during 2004 – Analysis using objective techniques. *Mon. Wea. Rev.*, 135:1251–1267, 2007.
4. R. W. Burpee. Characteristics of North African easterly waves during the summers of 1968 and 1969. *J. Atmos. Sci.*, 31:1556–1570, 1974.
5. T. N. Carlson. Some remarks on African disturbances and their progress over the tropical Atlantic. *Mon. Wea. Rev.*, 97:716–726, 1969.
6. T. N. Carlson. Synoptic histories of three African disturbances that developed into Atlantic hurricanes. *Mon. Wea. Rev.*, 97:256–276, 1969.
7. W. Y. Y. Cheng and W. R.Cotton. Sensitivity of a cloud-resolving simulation of the genesis of a mesoscale convective system to horizontal heterogeneities in soil moisture initialization. *J. Hydromet.*, 5:934–958, 2004.
8. J. Charney and M. Stern. On the stability of internal baroclinic jets in a rotating atmosphere. *J. Atmos. Sci.*, 19:159–172, 1962.
9. D. B. Clark, C. M. Taylor, A. J. Thorpe, R. J.Harding, and M.E. Nicholls. The influence of spatial variability of boundary-layer moisture on tropical continental squall lines. *Q. J. R. Meteorol. Soc.*, 129:1101–1121, 2003.
10. A. Diongue, J.-P. Lafore, J.-L. Redelsperger, and R. Roca. Numerical study of a Sahelian synoptic weather system: initiation and mature stages of convection and its interaction with the large-scale dynamics. *Q. J. R. Meteorol. Soc.*, 128:1899–1927, 2002.
11. G. Doms and U. Schättler. A description of the nonhydrostatic regional model LM. Part I: Dynamics and Numerics. Cosmo documentation, Deutscher Wetterdienst, Offenbach, Germany, www.cosmo-model.org, 2002.
12. S. W. Franks, K. J. Beven KJ, P. F. Quinn, and I. R. Wright. On the sensitivity of soil-vegetation-atmosphere transfer (SVAT) schemes: equifinality and the problem of robust calibration. *Agric. For. Meteorol*, 86:63–75, 1997.
13. C. M. Grams, S. C. Jones, J.H. Marsham, D. J. Parker, J. M. Haywood, and V. Heuveline. The Atlantic inflow to the Saharan heat low: observations and modelling. *Q. J. R. Meteorol. Soc.*, page DOI: 10.1002/qj.429, 2009.
14. L. Gantner and N. Kalthoff. Sensitivity of a modelled life cycle of a mesoscale convective system to soil conditions over West Africa. *Q. J. R. Meteorol. Soc.*, page DOI: 10.1002/qj.425, 2009.
15. R. A. Houze. Mesoscale Convective Systems. *Rev. Geophys.*, 42:1–43, 2004.
16. N. Kalthoff, B. Andler, C. Bartlott, U. Corsmeier, S. Crewell, K. Träumer, C. Kottmeier, A. Wieser, V. Smith, and P. Di Girolamo. impact of convergence zones on the initiation of deep convection: A case study from COPS. *Atmos. Res.*, 93:680–694, doi:10.1016/j.atmosres.2009.02.010, 2009.

17. R. D. Koster, P. A. Dirmeyer, Z. Guo, G. Bonan, E. Chan, P. Cox, C. T. Gordon, E. Kowalczyk, S. Kanae, D. Lawrence, P. Liu, C.-H. Lu, S. Malyshev, B. McAvaney, K. Mitchell, D. Mocko, T. Oki, K. Oleson, A. Pitman, Y. C. Sud, C. M. Taylor, D. Verseghy, R. Vasic, Y. Xue and T. Yamada (the GLACE Team). Regions of strong coupling between soil moisture and precipitation. *Science*, 35:1070–1096, 2004.
18. E. Kessler. On the distribution and continuity of water substance in atmospheric circulation models. *Meteor. Monographs*, 10:Americ. Meteor. Soc. Boston, MA., 1969.
19. M. Kohler, N. Kalthoff, and C. Kottmeier. The impact of soil moisture modifications on CBL characteristics in West Africa: A case-study from the AMMA campaign. *Q. J. R. Meteorol. Soc.*, page DOI: 10.1002/qj.430, 2009.
20. J. Klemp and R. Wilhelmson. The simulation of three-dimensional convective storm dynamics. *J. Atmos. Sci.*, 35:1070–1096, 1978.
21. H. Laurent, N. D'Amato, and T. Lebel. How important is the contribution of the mesoscale convective complexes to the Sahelian rainfall. *Phys. Chem. Earth.*, 23:629–633, 1998.
22. A. Laing and J. M. Fritsch. Mesoscale convective complexes over in Africa. *Mon. Wea. Rev.*, 121:2254–2263, 1993.
23. A. Gaßmann. Numerische Verfahren in der nichthydrostatischen Modellierung und ihr Einfluß auf die Güte der Niederschlagsvorhersage. *Berichte des Deutschen Wetterdienstes*, 221:1–96, 2002.
24. V. Mathon, H. Laurent, and T. Lebel. Mesoscale convective system rainfall in the Sahel. *J. Appl. Met.*, 127:337–406, 2002.
25. C. Morel and S. Sénési. A climatology of mesoscale convective systems over Europe using satellite infrared imagery. I: Methodology. *Q. J. R. Meteorol. Soc.*, 128:1953–1992, 2002.
26. G. L. Mellor and T. Yamada. Development of a turbulence closure model for geophysical flow problems. *Rev.Geophys. Space Phys.*, 20:831–875, 1982.
27. S. Nicholson and J. Grist. The seasonal evolution of the atmospheric circulation over West Africa and equatorial Africa. *J. Climate.*, 16:1013–1030, 2003.
28. E. G. Njoku. AMSR-E/Aqua daily L3 surface soil moisture,interpretive parameters, and QC EASE-Grids V002. *National Snow and Ice Data Center, Digital media, Boulder, Colorado, USA*, 2006.
29. D. J. Parker, C. D. Thorncroft, R. Burton, and A. Diongue-Niang. Analysis of the African easterly jet using aircraft observations from the JET2000 experiment. *Q. J. R. Meteorol. Soc.*, 131:1461–1482, 2005.
30. M. Raschendorfer. The new turbulence parameterization of LM. *COSMO Newsletter*, 1:89–97, 2001.
31. J.-L. Redelsperger, A. Diongue, A. Diedhiou, J.-P. Ceron, M. Diop, J.-F. Gueremy, and J.-P-Lafore. Multi-scale description of a Sahelian synoptic weather system representative of the West African monsoon. *Q. J. R. Meteorol. Soc.*, 128:1229–1257, 2002.
32. B. Ritter and J.-F. Geleyn. A comprehensive radiation scheme for numerical weather prediction models with potential application in climate models. *Mon. Wea. Rev.*, 120:303–325, 1992.
33. D. Rowell and J. Milford. On the generation of African squall lines. *J. of Climatol.*, 6:1181–1193, 1993.
34. M. Segal and R. W. Arritt. Nonclassical mesoscale circulations caused by surface sensible heat-flux gradients. *Bull. Am. Meteorol. Soc*, 73:1593–1604, 1992.

35. G. Schaedler. Triggering of atmospheric circulations by moisture inhomogenities of the earth's surface. *Bound.-Layer Meteorol.*, 51:1–29, 1990.
36. J. Steppeler, G. Doms, U. Schättler, H. W. Bitzer, A. Gassmann, U. Damrath, and G. Gregoric. Meso-gamma scale forecasts using the nonhydrostatic model LM. *Meteorol. Atmos. Phys.*, 82:75–97, 2003.
37. U. Schättler, G. Doms, and C. Schraff. A description of the nonhydrostatic regional model LM, part VII: User's guide. *Deutscher Wetterdienst, www.COSMO-model.org.*, 2008.
38. J. Schwendike and S. C. Jones. Convection in an African Easterly Wave over West Africa an the eastern Atlantic: a model case study of Helene (2006). *Q. J. R. Meteorol. Soc.*, page in revision, 2009.
39. J. Schwendike, N. Kalthoff, and M. Kohler. The impact of mesoscale convective systems on the surface and boundary layer structure in West Africa during the AMMA campaign. *Q. J. R. Meteorol. Soc.*, page submitted, 2009.
40. C. M. Taylor, F. Saï d, and T. Lebel. Interactions between the land surface and mesoscale rainfall variability during HAPEX-Sahel. *Mon. Wea. Rev.*, 125:2211–2227, 1997.
41. C. M. Taylor and R. J. Ellis. Satellite detection of soil moisture impacts on convection at the mesoscale. *Geophys. Res. Lett.*, 33:L03404. DOI:10.1029/2005GL025252, 2006.
42. C. M. Taylor and T. Lebel. Observational evidence of persistent convective-scale rainfall patterns. *Mon. Wea. Rev.*, 126:1597–1607, 1998.
43. C. M. Taylor, D. J. Parker, and P. P. Harris. Satellite detection of soil moisture impacts on convection at the mesoscale. *Geophys. Res. Lett.*, 33:L03404. DOI:10.1029/2005GL025252, 2007.
44. B. Vogel, C. Hoose, H. Vogel, and C. Kottmeier. A model of dust transport apllied to the Dead Sea area. *Meteorologische Zeitschrift*, 15:611–624, DOI: 10.1127/0941–2948/2006/0168, 2006.
45. B. Vogel, H. Vogel, D. Bäumer, M. Bangert, K. Lundgren, R. Rinke, and T. Stanelle. The comprehensive model system COSMO-ART – radiative impact of aerosol on the state of the atmosphere on the regional scale. *Atmospheric Chemistry and Physics (ACP)*, 9:14483–14528, 2009.
46. P. J. Wetzel and J.-T. Chang. Evapotranspiration from nonuniform surfaces: A first approach for short term numerical weather prediction. *Mon. Weather Rev.*, 116:600–621, 1988.

Miscellaneous Topics

Univ.-Prof. Dr.-Ing. Wolfgang Schröder

Institute of Aerodynamics, RWTH Aachen University, Wüllnerstr. 5a, 52062 Aachen, Germany
office@aia.rwth-aachen.de

The research areas which have been tackled in the chapters above are among different fields physics, solid state physics, fluid mechanics, aerodynamics, flows with chemical reactions, thermodynamics, structural mechanics, and so forth. The following contributions even widen the field where numerical simulations represent useful tools to gain novel results and as such to improve the scientific knowledge in the various areas. The papers evidence the close link between applied and fundamental research. In other words, the close relationship between mathematics, computer science and the ability to develop scientific models is clearly shown. Compact mathematical descriptions will be solved by highly sophisticated and efficient algorithms on a high-end computer. This interdisciplinary collaboration between several scientific fields defines the extremely intricate numerical challenges and as such drives the progress in fundamental and applied research. The subsequent articles represent an excerpt of the vast amount of projects being linked with HLRS. They corroborate the statement that today's numerical simulations can no longer be considered number crunchers. The computations are not only used to obtain some quantitative results but to confirm fundamental physical models and to even derive new theoretical approaches. Nevertheless, it has to be emphasized that it will always be wise to substantiate numerical simulations by experimental investigations and analytical solutions. To follow just one of the aforementioned routes represents too high a risk to take. The close collaboration between different approaches is without any doubt the more successful solution.

The first contribution from the Chair of Thermodynamics and Energy Technology of the University of Paderborn, the Institute of Technical Thermodynamics and Thermal Process Engineering of the University of Stuttgart, and the Institute of Thermodynamics of the Technical University of Kaiserslautern deals with the molecular modeling of hydrogen bonding fluids. A major challenge for molecular modeling consists in the definition of unlike interaction potentials. A variety of combination rules were proposed in the past. In a broad study on fluid mixtures, it was recently shown that none of them is clearly superior and that all are suboptimal in many cases when accurate predictions of properties like the mixture vapor pressure are needed.

The authors continue the line of research where unlike parameters are adjusted to data on the contact angle formed between the wall and a vapor-liquid interface. On the molecular level, the precise position of the vapor-liquid phase boundary is defined by a cluster criterion. If the cohesion of the liquid phase is partly due to hydrogen bonds, successful molecular models for pure fluids can often be developed on the basis of an ab initio study of the charge distribution as well as the equilibrium position of the nuclei. This sterically realistic approach combined with adjusting the Lennard-Jones (LJ) potential parameters to vapor-liquid equilibrium (VLE) data leads to empirical models that correctly reproduce and predict thermophysical fluid properties over a wide range of conditions. This approach is applied to mixtures containing hydrogen bonding components in this article.

The subsequent paper of the Institute of Geosciences of the University of Jena introduces a basic concept of a new dynamical model of the thermal and chemical evolution of Mars. It focuses on extensions of the software Terra which allow the improvement of the solutions of the convection differential equations at strongly varying viscosity. Although these extensions have been partly tested already, some considerations on the chronology of the early evolution of Mars and on magma ocean solidification are described since they lead to a structural model of the early Mars. This is important as a starting a priory assumption for a dynamical solution of the martian evolution similar to a former model which derives the essential features of the Earth's mantle's history. Currently, there is no preliminary-reference-Earth-model analogon neither for the present time nor for the start of the solid-state creep in the martian mantle. Mars has not only a topographical and crustal dichotomy but also a chemical dichotomy. Several mechanisms are discussed which could generate not only these structures but also an early strong magnetic dipole field which vanishes after 500 Ma at the latest.

The final article of the Institute of Geodesy of the University of Stuttgart addresses numerical considerations in the framework of the main inverse problem in geodetic research, i.e., in determining the terrestrial gravitational potential. The computational analyses consist of strategies to solve the underlying systems of equations and to estimate the variance-covariance information. To be more precise, the computational ideas of the contribution are related to the Gravity field recovery and steady-state Ocean Circulation Explorer (GOCE) satellite mission which was scheduled to be launched in March 2009. This mission was supposed to upgrade the present knowledge about the Earths gravitational structure. The missions major concern was to recover the short-scale features of the terrestrial gravity field and the gravitational potential or geopotential with an intended accuracy of about 2 cm and a spatial resolution around 100 km.

From the computational point of view, modern space-borne geopotential recovery involves the solution of large-scale systems of equations with tens of millions of observations and tens to hundred of thousands of model parameters. Consequently, high-performance computing is part of modern satellite geodesy. Tailored analysis techniques have to satisfy both quality and runtime issues. Hence, black box algorithms are no efficient solutions to geodetic problems. This makes the adoption of high-performance computing techniques for geopotential determination highly demanding.

Molecular Modeling of Hydrogen Bonding Fluids: Vapor-Liquid Coexistence and Interfacial Properties

Martin Horsch[1], Martina Heitzig[2], Thorsten Merker[3], Thorsten Schnabel[2], Yow-Lin Huang[1], Hans Hasse[3], and Jadran Vrabec[1]

[1] Lehrstuhl für Thermodynamik und Energietechnik (ThEt), Universität Paderborn, Warburger Str. 100, 33098 Paderborn, Germany[†]
[2] Institut für Technische Thermodynamik und Thermische Verfahrenstechnik (ITT), Universität Stuttgart, Pfaffenwaldring 9, 70569 Stuttgart, Germany
[3] Lehrstuhl für Thermodynamik (LTD), Technische Universität Kaiserslautern, Erwin-Schrödinger-Str. 44, 67663 Kaiserslautern, Germany

1 Introduction

A major challenge for molecular modeling consists in optimizing the unlike interaction potentials. A broad study on fluid mixtures [1] recently showed that among the variety of combination rules that were proposed in the past, none is clearly superior. In many cases, all are suboptimal when accurate predictions of properties like the mixture vapor pressure are needed. The well known Lorentz-Berthelot rule performs quite well and can be used as a starting point. If more accurate results are required, it is often advisable to adjust the dispersive interaction energy parameter which leads to very favorable results [1, 2, 3, 4, 5].

A similar approach should be followed for effective pair potentials acting between fluid particles and the atoms of a solid wall. They can only be reliable if fluid-wall contact effects are taken into account, e.g., by adjusting unlike parameters to contact angle measurements [6]. Teletzke *et al.* [7] used continuum methods to examine the dependence of wetting and dewetting transitions on characteristic size and energy parameters of the fluid-wall dispersive interaction. MD simulation can be applied for the same purpose, leading to a consistent molecular approach.

On the molecular level, the precise position of the vapor-liquid phase boundary is defined by a cluster criterion. Many different criteria are known and it is not immediately obvious which of them leads to the most accurate

[†] Author to whom correspondence should be addressed: Prof. Dr.-Ing. habil. J. Vrabec. E-mail: jadran.vrabec@upb.de.

W.E. Nagel et al. (eds.), *High Performance Computing in Science and Engineering '09*, DOI 10.1007/978-3-642-04665-0_33,

results [8]. In nanoscopic systems, minute absolute differences can lead to comparably large relative deviations. Therefore, the viability of several criteria is compared in the present study with the purpose of excluding errors due to an inaccurate detection of the interface.

If the cohesion of the liquid phase is partly due to hydrogen bonds, successful molecular models for pure fluids can often be developed on the basis of an *ab initio* study of the charge distribution as well as the equilibrium position of the nuclei. This sterically realistic approach, combined with adjusting the Lennard-Jones (LJ) potential parameters to vapor-liquid equilibrium (VLE) data, leads to empirical models that correctly reproduce and predict thermophysical fluid properties over a wide range of conditions [9]. The present work applies this approach to mixtures containing hydrogen bonding components. Often potential parameters determined for one fluid carry over to a derivative with different substituents, opening the possibility of creating generic molecular models. Such a model is presented for benzyl alcohol.

The following publications in peer-reviewed international journals contribute to the present project:

- Schnabel, T., Vrabec, J. & Hasse, H. Molecular simulation study of hydrogen bonding mixtures and new molecular models for mono- and dimethylamine. *Fluid Phase Equilib.* **263**: 144–159 (2008).
- Eckl, B., Vrabec, J. & Hasse, H. An optimized molecular model for ammonia. *Mol. Phys.* **106**: 1039–1046 (2008).
- Eckl, B., Vrabec, J. & Hasse, H. Set of molecular models based on quantum mechanical ab initio calculations and thermodynamic data. *J. Phys. Chem. B* **112**: 12710–12721 (2008).
- Vrabec, J., Huang, Y.-L. & Hasse, H. Molecular models for 267 binary mixtures validated by vapor-liquid equilibria: a systematic approach. *Fluid Phase Equilib.* **279** 120–135 (2009).
- Huang, Y.-L., Vrabec, J. & Hasse, H. Prediction of ternary vapor-liquid equilibria for 33 systems by molecular simulation. Submitted.
- Horsch, M., Heitzig, M., Dan, C., Harting, J., Hasse, H. & Vrabec, J. Contact angle dependence on the fluid wall dispersive energy. In preparation.

It would exceed the scope of the present report to give a full exposition of these articles. Instead, a few points are emphasized and arranged as follows: Firstly, mixture properties are explored for binary systems containing hydrogen bonding components. Secondly, vapor-liquid interface cluster criteria and contact angles are discussed and remarks on computational details are given. Finally, a sterically accurate generic model for benzyl alcohol is introduced and evaluated.

2 Fluid Mixtures with Hydrogen Bonding Components

Vapor-liquid equilibria of 31 binary mixtures consisting of one hydrogen bonding and one non-hydrogen bonding component were studied. All models are of the rigid united-atom multi-center LJ type with superimposed electrostatic sites in which hydrogen bonding is described by partial charges. The hydrogen bonding components of the studied binary mixtures are: monomethylamine (MMA) and dimethylamine (DMA), methanol, ethanol and formic acid. The non-hydrogen bonding components are: neon, argon, krypton, xenon, methane, oxygen, nitrogen, carbon dioxide, ethyne, ethene, ethane, propylene, carbon monoxide, diflourodichloromethane (R12), tetraflouromethane (R14), diflourochloromethane (R22), diflouromethane (R32), 1,1,1,2-tetraflouroethane (R134a) and 1,1-diflouroethane (R152a).

To obtain a quantitative description of the mixture vapor-liquid equilibria, one state independent binary interaction parameter was adjusted to a single experimental data point of either the vapor pressure or the Henry's law constant. Throughout, excellent predictions were found at other state points, i.e., at other compositions or temperatures as well as for the Henry's law constant, if it was adjusted to the vapor pressure, or vice versa. Figures 1 and 2 show methane + methanol as a typical example.

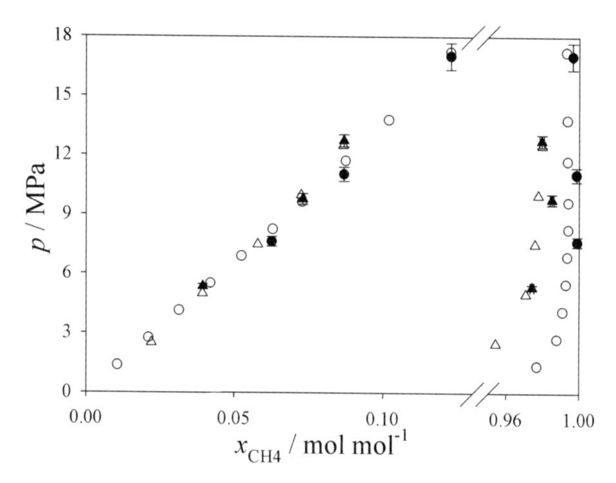

Fig. 1. Simulation data and vapor-liquid equilibria of methanol + methane: simulation data (•) and experimental data (○) at 310 K [10]; simulation data (▲) and experimental data (Δ) at 338.15 K [11]

Furthermore, a set of 78 pure substances from prior work [22, 23] was taken to systematically describe all 267 binary mixtures of these components for which relevant experimental VLE data are available. Again, per binary system, the single state independent binary interaction parameter in the energy term was adjusted to only one experimental value of the vapor pressure. The unlike energy parameter was thereby altered usually by less than 5% from the Berthelot rule. The mixture models were validated regarding the vapor pressure at other state points as well as the dew point composition, which is a fully predictive property in this context. In almost all, i.e., 97% of the

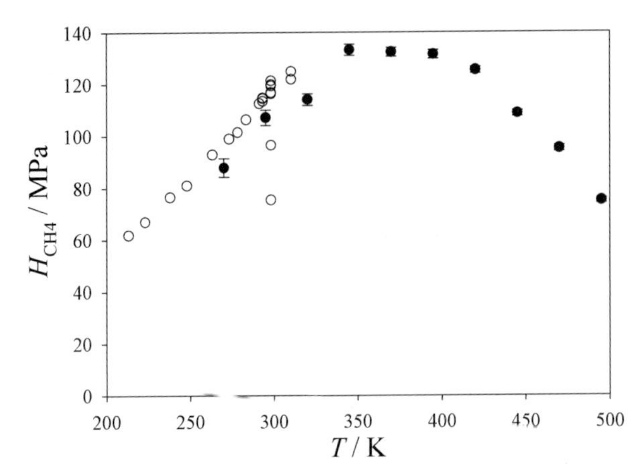

Fig. 2. Henry's law constant of methane in methanol: simulation data (•) and experimental data [12, 13, 14, 15, 16, 17, 18, 19, 20, 21] (○)

cases, the molecular models give excellent predictions of the mixture properties. Compared to other works in the literature, this is by far the largest investigation in this direction. It was facilitated by the extensive computing equipment at the High Performance Computing Center Stuttgart.

In the next step, all 33 ternary mixtures of these 78 components for which experimental VLE data is available were studied by molecular simulation. No adjustment to ternary data was carried out at all so that the calculations were strictly predictive. By comparing to experimental data, it was found that these predictions are very reliable as there was practically always an excellent match. As an example, Fig. 3 shows the ternary system consisting of methane, ethane, and carbon dioxide. Again, the computational effort was substantial, publications in the literature by other groups in this field typically cover one to two mixtures only.

3 Vapor-Liquid Interface Cluster Criteria

A suitable cluster criterion should achieve two goals: on the one hand, it needs to distinguish the bulk liquid and the bulk vapor successfully in every case – even when the vapor is supersaturated, such as in the vicinity of a droplet, or the liquid is undersaturated, such as in the vicinity of a bubble. On the other hand, the cluster criterion should also minimize noise fluctuations of the detected clusters to emphasize the signal.

The following criteria for carbon dioxide were compared for this purpose using a rigid two-center LJ plus point quadrupole (2CLJQ) model [22]:

- Stillinger [25]: all molecules with a distance of r_{St} or less from each other are liquid and belong to the same cluster (i.e., the same liquid phase or the

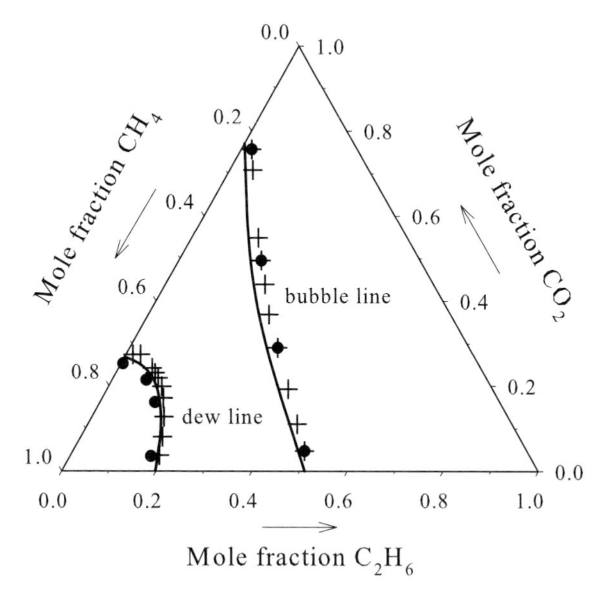

Fig. 3. Ternary vapor-liquid equilibrium phase diagram for the mixture CH_4 + CO_2 + C_2H_6 at 230 K and 4650 kPa: simulation data (•), experimental data (+) of Knapp *et al.* [24], and Peng-Robinson equation of state (—)

same droplet). The Stillinger radius was set to $r_{St} = 5.7$ Å for the present simulations.

- Ten Wolde and Frenkel [26]: all molecules with at least k neighbors within a distance of r_{St} belong to the liquid. They belong to the same cluster as all other liquid molecules within a distance of r_{St}.
- Arithmetic mean, k neighbors: a molecule is liquid if the density in the sphere containing its k nearest neighbors exceeds $(\rho' + \rho'')/2$, where ρ' amd ρ'' are the saturated liquid and vapor density, respectively. The molecule belongs to the same cluster as all other liquid molecules within the radius r_k, which defines a sphere with the volume occupied by $k + 1$ molecules at a density of $(\rho' + \rho'')/2$.
- Geometric mean, k neighbors: analogous, with a density threshold of $(\rho'\rho'')^{1/2}$.

The simulations were conducted with the massively parallel MD program *ls1 mardyn* (the precursor implementation *ls1 moldy* is described by Bernreuther and Vrabec [27]). Figure 4 shows that all of the discussed criteria are applicable. The Stillinger criterion and the geometric mean density criterion with two neighbors lead to the best results. It should be noted that at high temperatures, i.e., near the critical point, the Stillinger criterion becomes less reliable in distinguishing the liquid from a supersaturated vapor than the geometric mean density criterion.

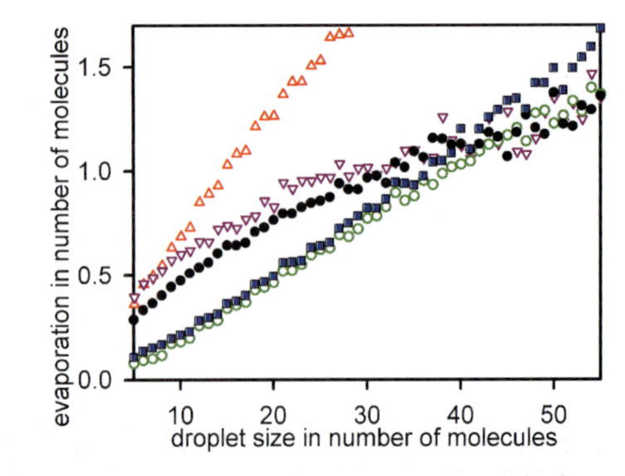

Fig. 4. Average number of molecules evaporating during a detection interval of 30 fs from droplets in a supersaturated vapor of carbon dioxide with T = 237 K, ρ = 1.98 mol/l, and N = 1,050,000, determined according to various cluster criteria: Stillinger (∘), ten Wolde and Frenkel with k = 4 (•), arithmetic mean with two (Δ) and eight neighbors (∇) as well as geometric mean with two neighbors (□)

4 Contact Angle Dependence on the Fluid-Wall Interaction

In cases where no hydrogen bonds are formed between the wall and the fluid, vapor-solid and liquid-solid interfacial properties mainly depend on the fluid-wall dispersive interaction, even for hydrogen bonding fluids. The truncated and shifted LJ (TSLJ) potential [28]

$$u_{ij}^{\mathrm{TSLJ}}(r_{ij}) = \begin{cases} 4\epsilon\left[\sigma^{12}(r_{ij}^{-12} - r_{\mathrm{c}}^{-12}) + \sigma^{6}(r_{\mathrm{c}}^{-6} - r_{ij}^{-6})\right] & r_{ij} < r_{\mathrm{c}} \\ 0 & r_{ij} \geq r_{\mathrm{c}}, \end{cases} \tag{1}$$

with a cutoff radius of $r_{\mathrm{c}} = 2.5\ \sigma$ accurately reproduces the dispersive interaction if adequate values for the size and energy parameters σ and ϵ are specified. Due to the relatively small cutoff radius, all computations are accelerated, while the descriptive power of the full LJ potential (without a cutoff) is retained even for systems with phase boundaries [29].

For the present series of contact angle simulations, the TSLJ model – with the potential parameters for methane, σ = 3.7241 Å and ϵ/k = 175.06 K – was studied, extending a previous investigation of the vapor-liquid interface of the TSLJ fluid [29]. The TSLJ potential with the size and energy parameters $\sigma' = \sigma$ as well as

$$\epsilon' = \xi\epsilon, \tag{2}$$

was also applied for the unlike interaction between the fluid molecules and the wall atoms, with the same cutoff radius as for the fluid. The wall was modeled as a system of coupled harmonic oscillators with different spring constants for transverse and longitudinal motion, adjusted to simulation results for graphite that were obtained with a rescaled version of the Tersoff [30] potential.

The simulations were carried out with the *ls1 mardyn* program. Vapor and liquid were independently equilibrated in homogeneous simulations for 10 ps. This was followed by 200 ps of equilibration for the combined system, i.e., a liquid meniscus surrounded by vapor, with a graphite wall consisting of four to seven layers, cf. Fig. 5. A periodic boundary condition was applied to the system, leaving a channel with a diameter of 27 σ between the wall and its periodic image. The contact angle was determined from the density profiles by averaging over at least 800 ps after equilibration.

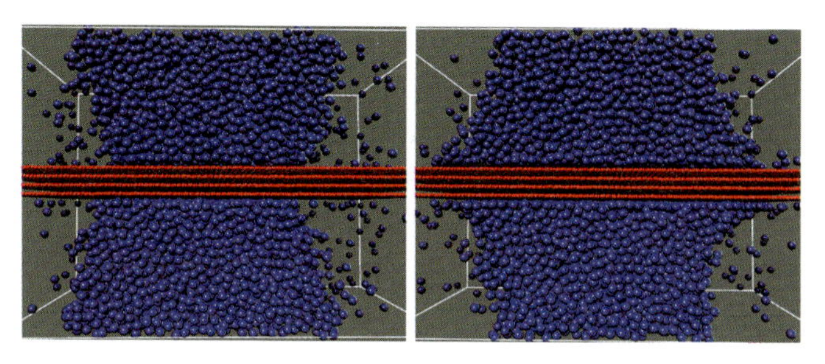

Fig. 5. Simulation snapshots for a reduced fluid-wall dispersive energy ξ of 0.09 (left) and 0.16 (right) at a temperature of 0.73 ϵ/k.

Liquid menisci between graphite walls were simulated for a reduced fluid-wall dispersive energy ξ between 0.07 and 0.16 at temperatures of 0.73, 0.88, and 1 ϵ/k. Note that the triple point temperature of the TSLJ fluid is about 0.65 ϵ/k [31] while the critical temperature is $T_c = 1.0779$ ϵ/k [29], so that the entire regime of stable vapor-liquid coexistence was covered.

High values of ξ correspond to a strong attraction between fluid and wall compoents, leading to a contact angle ϑ smaller than 90°, i.e., to partial ($\vartheta > 0°$) or full ($\vartheta = 0°$) wetting of the surface. For a higher fluid-wall dispersive energy, the extent of wetting increases, cf. Fig. 5. The transition from obtuse to acute contact angles, i.e., $\vartheta(T, \xi_0) = 90°$, occurs at a temperature independent value $\xi_0 \approx 0.113$ of the fluid-wall dispersive energy, as can be seen in Fig. 6. Furthermore, the symmetry law

$$\cos\vartheta(T, \xi_0 - \Delta\xi) = -\cos\vartheta(T, \xi_0 + \Delta\xi), \quad (3)$$

is valid over the whole relevant range of temperatures and magnitudes of the fluid-wall dispersive energy. Figure 6 also shows that there is a narrow range of ξ values that lead to the formation of a contact angle, as opposed to total dewetting or wetting. As the temperature decreases and the vapor-liquid surface tension $\gamma(T)$ increases, the contact angle approaches 90°.

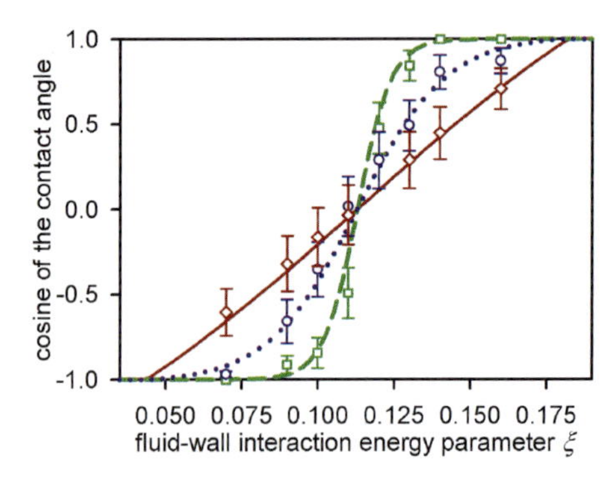

Fig. 6. Simulation results and correlation for the contact angle in dependence of the reduced fluid-wall dispersive energy ξ at temperatures of 0.73 (diamonds and solid line), 0.88 (circles and dotted line) as well as 1 (squares and dashed line) ϵ/k

5 A Sterically Accurate Generic Benzyl Alcohol Model

Benzyl alcohol (C_6H_5–CH_2OH) is widely used as a solvent for paints and inks. However, it is classified as a harmful substance (Xn) and should not be inhaled, nor used at high temperatures where it exhibits a high vapor pressure.

The basis for a new rigid molecular model of benzyl alcohol was determined by *ab initio* calculations with the GAMESS (US) quantum chemistry package [32], obtaining the quadrupole moment as well as the equilibrium positions of the nuclei as illustrated in Fig. 7.

Further potential parameters were taken from accurate empirical molecular models for related fluids, leading to a sterically accurate generic model, cf. Tab. 1, that can be used as a starting point for parameter optimization. In particular, point charges as well as σ and ϵ parameters for the hydroxyl group were taken from the ethanol model of Schnabel *et al.* [33]. Moreover, the σ and ϵ values of the corresponding LJ interaction site of the Merker *et al.* [34] cyclohexanol model were used for the CH_2 group, while the LJ parameters for the CH and C centers were set according to the Huang *et al.* [35] models of benzene and phosgene, respectively.

VLE properties of the generic model were calculated using the ms2 program, leading to an overall satisfactory first approximation, considering that all a posteriori adjustments were absent. The vapor pressure is more accurate at high temperatures, cf. Fig. 8.

6 Computing Performance

The scalability of the *ls1 mardyn* program was measured on the *cacau* supercomputer for simulation scenarios involving methane, represented by the TSLJ fluid, as well as graphite, modeled by a rescaled version of the Tersoff

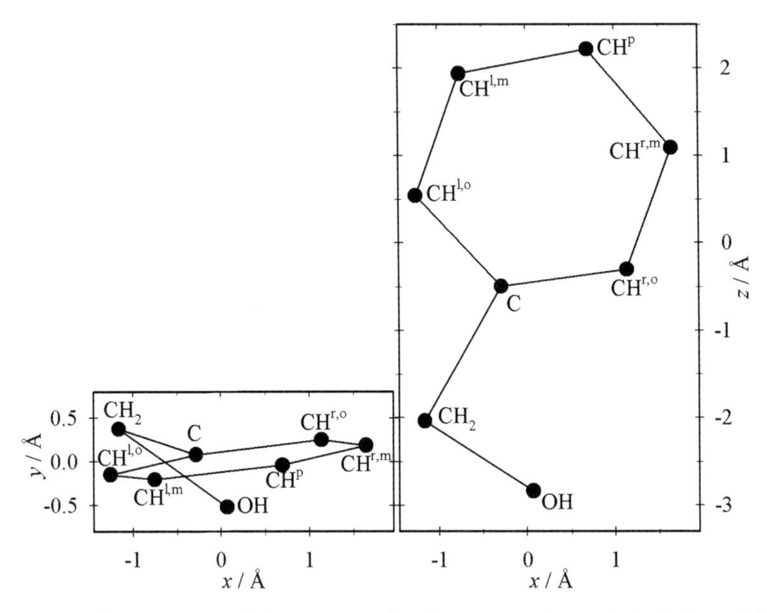

Fig. 7. Coordinates of the LJ sites for the present benzyl alcohol model

Table 1. Coordinates and parameters of the LJ sites, and the point charges, and the point quadrupole for the present benzyl alcohol model. Bold characters indicate represented atoms

Interaction site	x Å	y Å	z Å	σ Å	ϵ/k_B K	q e	Q DÅ
CH^{p}	0.695	-0.037	2.218	3.247	88.97		
CH^{l,m}	-0.753	-0.204	1.941	3.247	88.97		
CH^{r,m}	1.638	0.189	1.090	3.247	88.97		
CH^{l,o}	-1.255	-0.149	0.543	3.247	88.97		
CH^{r,o}	1.134	0.250	-0.304	3.247	88.97		
C	-0.285	0.080	-0.497	2.810	10.64		
CH$_2$	-1.160	0.372	-2.039	3.412	102.2	+0.2556	
OH	0.070	-0.516	-2.835	3.150	85.05	-0.6971	
O**H**	0.191	-1.430	-2.614			+0.4415	
Benzyl	0	0	0				2.534

[30] potential. MPI parallelization was applied according to a spatial domain decomposition scheme with equally sized cuboid subdomains and a cartesian topology based on linked cells [27].

Often the best solution is an isotropic decomposition that minimizes the surface to volume ratio of the spatial subdomains. For the simulation of homogeneous systems, this approach is quite efficient. That is underlined by

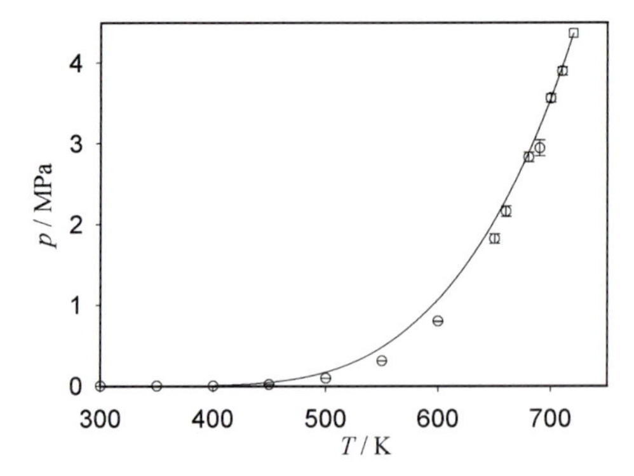

Fig. 8. Vapor pressure of benzyl alcohol according to a correlation [36] based on experiments (—) as well as present molecular simulation data (○)

the weak and strong scaling behavior of *ls1 mardyn* for typical configurations, shown in Fig. 9 (left), in cases where supercritical methane ('fluid') at a density of 10 mol/l and solid graphite ('wall') were considered with a system size of up to 4,800,000 interaction sites, representing the same number of carbon atoms and methane molecules here. Graphite simulations, containing only carbon atoms, scale particularly well, due to a favorable relation of the delay produced by communication between processes to the concurrent parts, i.e., the actual intermolecular interaction computation, which is much more expensive for the Tersoff potential than the TSLJ potential.

The simulation of combined systems, containing both fluid and solid interaction sites, is better handled by a channel geometry based decomposition

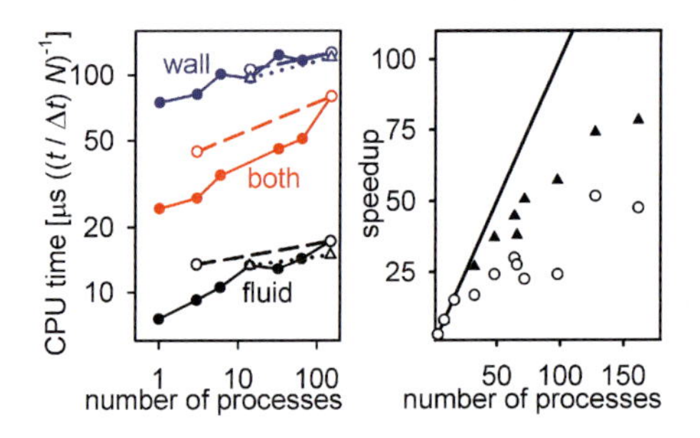

Fig. 9. Left: Total CPU time, i.e., execution time multiplied with the number of parallel processes, per time step and interaction site for weak scaling with 3,000 (dashed lines / circles) and 32,000 (dotted lines / triangles) interaction sites per process as well as strong scaling with 450,000 interaction sites (solid lines / bullets), using isotropic spatial domain decomposition. Right: speedup for a system of liquid methane between graphite walls with 650,000 interaction sites, where isotropic (circles) and channel geometry based (triangles) spatial domain decomposition was used; the solid line represents linear speedup

scheme, where an approximately equal portion of the wall and a part of the fluid is assigned to each process, cf. Fig. 9 (right). In the general case, where spatial non-uniformities do not match any cartesian grid, a flexible topology has to be used. An approach based on k-dimensional trees [37], implemented in a version of *ls1 mardyn*, showed clearly improved results with respect to the scaling of inhomogeneous systems.

7 Conclusion

The intermolecular interaction of small hydrogen bonding molecules like mono- and dimethylamine can be described by simple LJ based united-atom molecular models with point charges. Such computationally efficient models were applied to binary mixtures with non-hydrogen bonding components regarding VLE properties. Accurate predictions, covering a broad range of temperatures and compositions, were obtained, regardless whether the state independent binary interaction parameter was adjusted to Henry's law constant or vapor pressure.

For a two-center LJ plus quadrupole model of carbon dioxide, a comparison of cluster criteria with the purpose of accurately detecting the vapor-liquid phase boundary gave overall support to the geometric mean density criterion applied to the sphere consisting of a molecule and its two nearest neighbours. The Stillinger criterion was found to be particularly adequate at low temperatures. For the TSLJ fluid, the contact angle formed between the vapor-liquid interface and a wall was determined by canonical ensemble MD simulation while the mangitude of the dispersive fluid-wall interaction was varied. Over the whole temperature range between triple point and critical point, the contact angle dependence on the fluid-wall dispersive energy obeys a simple symmetry law.

A sterically realistic model for benzyl alcohol was presented, showing within the framework of LJ sites with point polarities and electric point charges that the generic molecular modeling approach can lead to a good starting point for parameter optimization with respect to VLE properties such as the vapor pressure.

The scalability of the *ls1 mardyn* program was assessed and found to be acceptable. MD simulations of methane confined between graphite walls with up to 4,800,000 interaction sites, i.e., carbon atoms and methane molecules, were conducted to demonstrate the viability of the program.

The authors would like to thank Martin Bernreuther, Martin Buchholz, Domenic Jenz, and Christoph Niethammer for contributing to the *ls1 mardyn* code, Stephan Deublein, Bernhard Eckl, Gimmy Fernández Ramírez, Gabriela Guevara Carrión, and Sergei Lishchuk for contributing to the ms2 code, Ioannis Pitropakis for performing simulation runs of binary mixtures, as well as Calin Dan, Franz Gähler, Sebastian Grottel, Jens Harting, Martin Hecht, Ralf Kible, and Guido Reina for frank discussions and their persistent support.

References

1. Th. Schnabel, J. Vrabec, and H. Hasse. Unlike Lennard-Jones parameters for vapor-liquid equilibria. *J. Mol. Liq.*, 130:170–178, 2007.
2. J. Vrabec, J. Stoll, and H. Hasse. Molecular models of unlike interactions in fluid mixtures. *Mol. Sim.*, 31:215–221, 2005.
3. Ph. Ungerer, C. Nieto-Draghi, B. Rousseau, G. Ahunbay, and V. Lachet. Molecular simulation of the thermophysical properties of fluids: From understanding toward quantitative predictions. *J. Mol. Liq.*, 134:71–89, 2007.
4. Th. Schnabel, J. Vrabec, and H. Hasse. Molecular simulation study of hydrogen bonding mixtures and new molecular models for mono- and dimethylamine. *Fluid Phase Equilib.*, 263:144–159, 2008.
5. J.Vrabec, Y.-L. Huang, and H. Hasse. Molecular models for 267 binary mixtures validated by vapor-liquid equilibria: A systematic approach. *Fluid Phase Equilib.*, 279(2):120–135, 2009.
6. T. Werder, J. H. Walter, R. L. Jaffe, T. Halicioglu, and P. Koumoutsakos. On the water-carbon interaction for use in molecular dynamics simulations of graphite and carbon nanotubes. *J. Phys. Chem. B*, 107:1345–1352, 2003.
7. G. F. Teletzke, L. E. Scriven, and H. T. Davis. Wetting transitions. II. First order or second order? *J. Chem. Phys.*, 78:1431–1439, 1982.
8. J. Wedekind and D. Reguera. What is the best definition of a liquid cluster at the molecular scale? *J. Chem. Phys.*, 127:154516, 2007.
9. Th. Schnabel, M. Cortada, J. Vrabec, S. Lago, and H. Hasse. Molecular model for formic acid adjusted to vapor-liquid equilibria. *Chem. Phys. Lett.*, 435:268–272, 2007.
10. J. H. Hong, P. V. Malone, M. D. Jett, and R. Kobayashi. The measurement and interpretation of the fluid-phase equilibria of a normal fluid in a hydrogen bonding solvent: the methane-methanol system. *Fluid Phase Equilib.*, 38:83–96, 1987.
11. N. L. Yarym-Agaev, R. P. Sinyavskaya, I. I. Koliushko, and L. Y. Levinton. *J. Appl. Chem. USSR*, 58:154–157, 1985.
12. M. G. Levi. *Gazzetta Chimica Italiana*, 31:513–541, 1901.
13. A. S. McDaniel. The absorption of hydrocarbon gases by nonaqueous liquids. *J. Phys. Chem.*, 15:587–610, 1911.
14. L. S. Bezdel and V. P. Teodorovich. *Gazovaya Prom.*, 8:38–43, 1958.
15. F. L. Boyer and L. J. Bircher. The solubility of nitrogen, argon, methane, ethylene and ethane in normal primary alcohols. *J. Phys. Chem.*, 64:1330–1331, 1960.
16. A. Lannung and J. C. Gjaldback. The solubility of methane in hydrocarbons, alcohols, water and other solvents. *Acta Chem. Scand.*, 14:1124–1128, 1960.
17. E. R. Shenderei, Y. D. Zelvenskii, and F. P. Ivanovskii. *Gazovaya Prom.*, 6:42–45, 1961.
18. J. Tokunaga and M. Kawai. Solubilities of methane in methanol-water and ethanol-water solutions. *J. Chem. Eng. Jap.*, 8:326–327, 1975.
19. J. Makranczy, L. Rusz, and K. Balogmegyery. Solubility of gases in normal alcohols. *Hung. J. Ind. Chem.*, 7:41–46, 1979.
20. P. Tkaczuk. Experimentelle Bestimmung der Gaslöslichkeit von Methan in Ketonen und Vergleich mit vorausberechneten Ergebnissen. Technical report, Technische Universität Dortmund, 1986.

21. S. Bo, R. Battino, and E. Wilhelm. Solubility of gases in liquids. 19. solubility of He, Ne, Ar, Kr, Xe, N_2, O_2, CH_4, CF_4, and SF_6 in normal 1-alkanols n-$C_\ell H_{2\ell}OH$ ($1 \leq \ell \leq 11$) at 298.15 K. *J. Chem. Eng. Data*, 38:611–616, 1993.
22. J. Vrabec, J. Stoll, and H. Hasse. A set of molecular models for symmetric quadrupolar fluids. *J. Phys. Chem. B*, 105:12126–12133, 2001.
23. J. Stoll, J. Vrabec, and H. Hasse. A set of molecular models for carbon monoxide and halogenated hydrocarbons. *J. Chem. Phys.*, pages 11396–11407, 2003.
24. H. Knapp, X. Yang, and Z. Zhang. Vapor-liquid equilibria in ternary mixtures containing nitrogen, methane, ethane and carbondioxide at low temperatures and high pressures. *Fluid Phase Equilib.*, 54:1–18, 1990.
25. F. H. Stillinger. Rigorous basis of the Frenkel-Band theory of association equilibrium. *J. Chem. Phys.*, 38:1486–1494, 1963.
26. P. R. ten Wolde and D. Frenkel. Computer simulation study of gas-liquid nucleation in a Lennard-Jones system. *J. Chem. Phys.*, 109:9901–9918, 1998.
27. M. Bernreuther and J. Vrabec. Molecular simulation of fluids with short range potentials. In M. Resch et al., editors, *High Performance Computing on Vector Systems*, pages 187–195, Heidelberg, 2006. Springer.
28. M. P. Allen and D. J. Tildesley. *Computer Simulation of Liquids.* Clarendon, Oxford, 1987.
29. J. Vrabec, G. K. Kedia, G. Fuchs, and H. Hasse. Comprehensive study on vapour-liquid coexistence of the truncated and shifted Lennard-Jones fluid including planar and spherical interface properties. *Mol. Phys.*, 104:1509–1527, 2006.
30. J. Tersoff. Empiric interatomic potential for carbon, with applications to amorphous carbon. *Phys. Rev. Lett.*, 61:2879–2882, 1988.
31. J. A. van Meel, A. J. Page, R. P. Sear, and D. Frenkel. Two-step vapor-crystal nucleation close below triple point. *J. Chem. Phys.*, 129:204505, 2008.
32. M. W. Schmidt, K. K. Baldridge, J. A. Boatz, S. T. Elbert, M. S. Gordon, J. H. Jensen, Sh. Koseki, N. Matsunaga, K. A. Nguyen, Sh. Su, Th. L. Windus, M. Dupuis, and J. A. Montgomery Jr. General atomic and molecular electronic structure system. *J. Comp. Chem.*, 14:1347–1363, 1993.
33. Th. Schnabel, J. Vrabec, and H. Hasse. Henry's law constants of methane, nitrogen, oxygen and carbon dioxide in ethanol from 273 to 498 K: Prediction from molecular simulation. *Fluid Phase Equilib.*, 233:134–143, 2005. See also *Fluid Phase Equilib.* **236**: 272 (2005) and **239**: 125–126 (2006).
34. Th. Merker, G. Guevara-Carrión, J. Vrabec, and H. Hasse. Molecular modeling of hydrogen bonding fluids: New cyclohexanol model and transport properties of short monohydric alcohols. In Wolfgang E. Nagel et al., editors, *High Performance Computing in Science and Engineering '08, Transactions of the High Performance Computing Center, Stuttgart*, pages 529–541, Heidelberg, 2009. Springer.
35. Y.-L. Huang, H. Hasse, and J. Vrabec. Molecular modeling of industrially relevant aromatic fluids. 2009.
36. R. L. Rowley, W. V. Wilding, J. L. Oscarson, Y. Yang, N. A. Zundel, T. E. Daubert, and R. P. Danner. *DIPPR R Data Compilation of Pure Compound Properties.* AIChE, New York, 2006.
37. J. L. Bentley. Multidimensional binary search trees used for associative searching. *CACM*, 18(9):509–517, 1975.

Towards a Dynamical Model of Mars' Evolution

Uwe Walzer, Thomas Burghardt, Roland Hendel, and Jonas Kley

Institut für Geowissenschaften, Friedrich-Schiller-Universität, Burgweg 11, 07749 Jena, Germany
u.walzer@uni-jena.de

Summary. We present the basic conception of a new dynamical model of the thermal and chemical evolution of Mars. Therefore new enlargements of the code Terra are necessary which allow to improve the solutions of the convection differential equations with strongly varying viscosity. These enlargements have been partly tested already. We describe considerations on the chronology of the early evolution of Mars and on magma ocean solidification since they lead to a *structural* model of the *early* Mars. This is important as a starting presupposition for a *dynamical* solution of the martian evolution similar to [122] which derives the essential features of the Earth's mantle's history. At present there is no PREM[39]-analogon neither for the present time nor for the start of the solid-state creep in the martian mantle. Mars has not only a topographical and crustal dichotomy but also a chemical dichotomy. We discuss different mechanisms which could generate not only these stuctures but also an early strong magnetic dipole field that vanishes after 500 Ma at the latest. Section 7 presents recent and future numerical improvements of the code Terra. Section 8 gives results on performance and scalability.

1 Introduction

The formation of terrestrial planets occurs stepwise. However, these steps are probably not simultaneously taking place in all solar distances. The condensation of dust grains [128] is followed by the accumulation of kilometer-sized planetesimals [126, 129, 137] which is assumed to have occured in $\leq 10^4$ a at 1 AU [27]. Gravitational attraction and gas drag generate collisions of the planetesimals inducing the growth of Moon- to Mars-sized planetary embryos within 1 Ma [90, 130]. The larger terrestrial planets are formed by further collisions. The majority of Moon- to Mars-sized planetary embryos will be consumed by these collisions within 10–100 Ma [3, 24, 26, 28]. It is well-known that the Hf-W chronometry is an excellent tool to constrain the duration of the magma ocean (MO) solidification. Based on own and other papers, Kleine et al. [64] (cf. their Table 1) estimate the following Hf/W ratios in planetary materials.

W.E. Nagel et al. (eds.), *High Performance Computing in Science and Engineering '09*, DOI 10.1007/978-3-642-04665-0_34,

Chondrites	Hf/W $= 0.6 - 1.8$
Bulk silicate Mars	Hf/W $= 3 - 4$
Bulk silicate Earth	Hf/W $= 17 \pm 5$
Bulk silicate Moon	Hf/W $= 26 \pm 2$
Basaltic eucrites	Hf/W $= 27 \pm 8$

Tungsten is siderophile. In the case of Mars, W did not possess enough time to go into the core whereas, in the case of the Earth, we observe an effective removal of radiogenic ^{182}W into the Earth's core. Therefore one might guess that Mars is essentially a single leftover planetary embryo [27]. With respect to the ^{142}Nd/^{144}Nd ratio, the Earth's mantle is nearly homogeneous. This suggests a long geological history. However, Mars has a heterogeneous ^{142}Nd/^{144}Nd ratio distribution. This points to the conservation of very early differentiated silicate reservoirs. Caro et al. [25] emphasize that Mars has preserved the largest ^{142}Nd effects of all known planetary bodies. This indicates in the same direction. Therefore and for other reasons, we propose to develop firstly simple models and algorithms, later more elaborated dynamic ones for the thermal evolution and the chemical differentiation of Mars where the numerics of the more complex code will be rather ambitious. It will be based on the program Terra [122, 123, 125]. It is clear that Terra and similar codes are suitable only for solid-state convection. If we add certain additional equations we are able to include also the chemical differentiation in small, partially melted regions of the mantle [122].

2 Chronology of the Early Evolution of Mars

2.1 General Remarks

The determinations of the time of accretion, core formation, mantle differentiation, and MO solidification mainly stem from isotope chemistry, but partly also from relatively simple models which have not been calculated in a really *geodynamic* way. It is important to have an appropriate idea at least on the order of magnitude of these times, t. All time data given here refer to the generation of Ca, Al-rich inclusions (CAIs) at an age, τ, of 4568.3 $\pm$ 0.7 Ma [64] if not noted otherwise. The absolute age of CAIs, however, is not known exactly enough but it seems to be somewhere between 4567.11 $\pm$ 0.16 Ma [4, 5] and 4568.5 $\pm$ 0.5 Ma [18]. The times of a certain process of Mars are between the times of the corresponding process of the mesosiderite parent body (MPB), of the eucrite parent body (EPB) and of the angrite parent body (APB) on the one hand and of Earth on the other hand. The accretion of MPB, EPB and APB has been largely completed within t $=$ 0.8 Ma [17]. The *initiation* of basaltic magmatism was at 2.56 Ma for MPB, at 3.05 Ma for EPB and at 3.31 Ma for APB according to [17], where these data refer to a CAI age of τ $=$ 4569.5 $\pm$ 0.2 Ma, or at about 3 Ma for MPB, EPB and APB according to [64]. On the other hand, the core formation of the Earth was finished to 63% after 11 Ma according to [134] and to 63% after about 10 Ma according to [53], Yin et al. [134] estimate the timing of core formation of the Earth at 29.5 $\pm$ 1.5 Ma using the Hf-W chronometry with a two-stage model, at 11 $\pm$ 1 Ma with a MO model. They believe that the MO segregation model is the most realistic estimate. Kleine et al. [63] determine the age for core formation on Earth at 33 $\pm$ 2 Ma. This is supplemented by [62], where they conclude from the W isotope evolution of the bulk silicate Earth that the segregation of metal into the Earth's core cannot have ceased earlier than $t = 30$ Ma. Also at present, a general review [64] shows that the tungsten model ages for core formation in Earth range from $t = 30$ Ma to $t > 100$ Ma after CAIs. These estimations show considerably higher time values in comparison to Mars.

2.2 Accretion and Core Formation

The time of *core formation of Mars* is estimated at 1 Ma in the case of a runaway growth [28] using the isotope chronometry. In this case, a significant heating contribution by short-lived isotopes such as ^{26}Al is to be expected. Jacobsen [56] derived a martian core formation age of 3.3 Ma. Kleine et al. [61] concluded an age of 12 $\pm$ 5 Ma using Hf-W evidence from martian meteorites. Instantaneous martian core formation timescales range from 0 to 10 Ma due to uncertainties in the Hf/W ratio of the martian mantle [85]. Metallic core formation in Mars is constrained to $t = 7 - 15$ Ma from Hf-W chronometry on SNCs according to [45]. Summarizing we conclude that the martian core formation is very little constrained and that its t is somewhere between 0 and 20 Ma [64] but in any case *considerably* below the t of the Earth's core formation.

If the main part of *Mars* evolved already as a planetary embryo, we have to expect a time of 0.1 – 1 Ma for its accretion [27]. This is the result of a computer modeling of the collisions and gravitational interactions between planetesimals. Ghosh et al. [50] propose that Mars accreted 3 Ma after CAI formation. That time is necessary to account for a deep martian MO. Hf/W chronometry: For a two-stage model, Mars reached 90% of its final mass at $t = 6$ Ma. For another model, $t = 20$ Ma is possible for completion of 90% of martian accretion. A runaway growth would lead to 1 Ma [53]. Hf-W chronometry cannot currently distinguish between the mentioned scenarios of *martian accretion* [64].

2.3 The Magma Ocean Solidification of Mars

This process is influenced by the solidus as a function of depth in relation to the adiabatic temperature as a function of depth. Although a crust is evolving soon at the surface it will be quickly subducted into the liquid MO. Because of the different slopes of the two curves, the MO freezes essentially from below.

Mg-silicates are enriched in the solid layer which develops just above the core-mantle boundary (CMB). Iron- and incompatible-element rich fluid silicates dominate more and more in the remaining MO above the solid mantle. So, a gravitationally instable chemical layering evolves which causes a Rayleigh-Taylor (R-T) overturn by solid-state creep after solidification. According to the model by Elkins-Tanton [40], these two processes occur in the first 5–10 Ma of the existence of Mars. The process of martian mantle solidification alone is 98% complete in less than 5 Ma if we presuppose small initial volatile contents (0.05 wt% H_2O, 0.01 wt% CO_2). [So, this estimate takes already into account the influence of an early dense martian atmosphere.] For a low-volatile martian MO, however, this 98%-solidification occurs in less than 0.1 Ma. An earlier estimate [41] for the 98% martian MO solidification is 0.1 Ma, evidently without taking into consideration the effects of volatiles. Taking into account the atmosphere, Zahnle et al. [138] determined the final cooling of MO to 300 °C as about 5 Ma.

This MO process could admittedly be protracted by later impacts. This is the reason why Debaille et al. [35] think about a duration of martian MO of about 100 Ma. They applied the short-lived [$t_{1/2}$ = 9 Ma] ^{146}Sm–^{142}Nd and the long-lived [$t_{1/2}$ = 106 Ga] ^{147}Sm–^{143}Nd chronometer to a suite of shergottites. Kleine et al. [64] show that there is an essential ϵ^{182}W-difference between nakhlites (N) and Chassigny (C) on the one hand and the shergottites (S) on the other hand. According to ϵ^{142}Nd, shergottites show an additional division into three groups. Therefore the early martian mantle differentiation must have taken place earlier than 60 Ma. The time of the differentiation of the NC source has been estimated at $t = 40 \pm 18$ Ma [25, 45]. *It is evident that there is a certain discrepancy between the isotope chronometry and the mineralogical and physical modeling. This problem is to be solved.*

3 The Generation of the Dichotomy of Mars

3.1 Observational Constraints

Mars shows a distinct *bipartition of its surface* into heavily cratered southern highlands and relatively smooth northern lowlands which cover about 42% of Mars' surface. The histogram with respect to percentage of surface area of the topography, i.e. the hypsometric curve, has two distinct maxima in a distance of 5.2 km [100, 127]. This difference has the same order of magnitude which we obtain for the hypsometric curve of the Earth.

The martian *crustal thickness* [81, 127] histogram has also two clear peaks, the difference of which amounts to 25 km. Unfortunately there are no seismometers on Mars up to now. Therefore there are no *seismic* determinations of the crustal thickness. That is why but also for an exact determination of the core-mantle-boundary (CMB) depth and other structural features, it is extremely important to promote the installation of broad band seismometers on Mars [71]. The previous determinations of the martian crustal thickness are based on inversions of free-air gravity, geoid and topography [81, 82, 131, 144], on considerations on the movement of inertia [15, 103], and on mass-balance constraints [117, 131]. According to [131] the martian crust is between 33 and 81 km thick where the smaller thicknesses belong to the northern plains. Nimmo [82] estimates the span of martian elastic thicknesses to values between 37 and 89 km whereas Hoogenboom and Smrekar [55] derive values between only 10 and 25 km for the elastic thickness *of the northern lowlands*. Assuming crustal densities between 2700 and 3400kg/m^3, Belleguic et al. [10] derive thickness values between 30 and 90 km for the whole martian crust. The cumulative frequency curves of the quasi-circular-depression (QCD) population of the northern plains have the same distribution as the curves of the southern-highlands population. Some authors deduce from it that the ages of the northern and southern units of the dichotomy might be equal. If this conclusion would be *exactly* true we run into difficulties regarding the generation of the magnetic stripe anomalies of the southern highlands [31]. These stripes are considerably broader (about 10°) than the magnetic anomalies which run parallel with the mid-oceanic ridges of the Earth. The variations in the crustal magnetic field of Mars show an association with major faults of the martian highlands crust [31]. In the northern lowlands, however, the remanent crustal magnetism of the early Noachian is only hardly to distinguish. It is entirely missing in Hellas, Argyre, Isidis, Utopia and also in the younger Tharsis region. Therefore Stanley et al. [112] try to prove the possibility of a single-hemisphere dynamo. Perhaps it would be more natural to try to prove that the age of the northern lowlands is somewhat smaller than that of the southern highlands. Although the areal QCD density of the lowlands has the same order of magnitude as that of the southern highlands, it is lower than that of those parts of the highlands that include highland QCDs [46, 47, 127]. Therefore the northern lowlands crust seems to be somewhat younger than the southern highlands crust [46, 47, 127].

There is also a *chemical dichotomy* of the martian crust. The dominant type is probably rich in silica, aluminum, and the main heat-producing elements, namely U, Th and K. It has a relatively low density. The second type is a dense basalt [87] which resurfaced large portions of the northern plains. Up to now it is not clear whether the insinuated chemical dichotomy refers only to the crust. Bennett et al. [11] speculate whether the *Earth's* mantle could be laterally different regarding the chemical composition in spite of 4500 Ma of thermal convection. There are contemporaneous Greenland and Australian Eoarchean terranes with 35 ppm ^{142}Nd excess and 23 ppm ^{142}Nd excess, respectively. Since some authors tend to assume a lower stirring intensity for Mars in comparison to the Earth, we should not exclude such a type of model for Mars. Novel geochemical models of this kind [11] would, of course, strengthen the argumentation of [112].

Assuming that the lithosphere is laterally isotropic, Hoogenboom and Smrekar [55] investigated the admittance signature, the ratio of gravity to topography data as a function of wave number, for four areas in the northern lowlands. They found topographic power spectra similar to highlands regions. They [55] conclude that the Noachian highland and lowland basement formed at *similar* times and speculate whether the low magnetization of the lowlands is induced by demagnetization. Solomon et al. [106] also suggest that hydrothermal circulation along deep faults may have led to oxydation of magnetic carriers and a reduction in magnetization preferably in the lower-lying areas of major drainage basins. A gamma ray spectrometer has revealed the distribution of iron, thorium and potassium at the martian surface. Dohm et al. [37] suggest that the higher iron abundance in the northern lowlands is induced by solution transport in acidic brines that may have leached iron from the highlands and deposited it in the lowlands. This conclusion corroborates [106]. It could, however, also be that the lateral differences in the Fe, Th and K concentrations have a causal relation with the lateral chemical differences mentioned by [87].

3.2 Is the Martian Dichotomy Induced by Endogenic Mechanisms?

Only one item is clear up to now: Cratering evidence shows that the martian dichotomy is an ancient feature (age $\tau > 3.5$ Ga). It is probably related to crustal formation by chemical differentiation. It is not clear whether the generation of the dichotomy and the generation of the early magnetic dipole field can be traced back to the same mechanism, but it is possible.

The *first* possibility to explain the dichotomy would be the assumption that there was a degree-1 pattern of thermal convection *with plate tectonics* in the very early times where the southern highlands crust clustered above the downwelling current. The physically necessary conditions for the possibility of the plate-tectonics mechanism on Earth and other terrestrial planets have been derived by [12, 13, 88, 113–115, 118, 124]. Sleep [99] was presumably the first one who proposed an early martian plate tectonics. Nimmo and Stevenson [86] thought that martian plate tectonics took place for the first 500 Ma and caused an increased surface heat flow especially above the northern lowlands. This again augmented the core heat flow and produced an episodically reversing magnetic dipole field for the first 500 Ma. However, they did not propose any explanation for the cessation of martian plate tectonics. Lenardic et al. [68] also proposed early martian plate tectonics. They believe that the end of this plate-producing mechanism is a natural consequence of the growth of the southern highlands. Also Breuer and Spohn [22] and Spohn [109] discuss the possibility of early plate tectonics on Mars. The early liquid water ocean seems to be a presupposition for plate tectonics on Mars [8]. Fairén and Dohm [43] assume a potential plate-tectonic phase in the Early to Middle Noachian. This hypothesis based on the observation that the basement of the northern lowlands is younger than the basement of the southern highlands, but older than the material exposures on the cratered highlands. Parts of the highlands crust are highly magnetized [31] so that the lowlands basement probably postdates the shut off of the dynamo. The northern plains show various episodic flood inundations and more than one ocean as a function of time [37, 44].

A *second* possibility for the production of the martian crustal dichotomy by an endogenic mechanism is a degree-1 mantle convection *without plate tectonics*. This kind of convection is generated by *strong radial viscosity variations* [75, 94, 136, 143].

A *third* possibility is a *superplume* induced by destabilization of a lower thermal boundary layer [59]. This but also other enumerated proposals could be connected with large flood basalts in the northern plains. All possibilities, up to now mentioned in Subsection 3.2, are based on the finding that subsolidus martian mantle convection is an essential prerequisite for an appropriate numerical estimation of the thermal evolution of Mars [97].

A *fourth* possibility is a *Rayleigh-Taylor overturn* after early magma ocean crystallization.

3.3 Is the Martian Dichotomy Caused by Impact(s)?

It is rather uncontested that the Utopia, Acidalia, and Chryse quasi-circular depressions are generated by an impact. Furthermore, the Tharsis region is not representative for the northern plains [81]. Therefore there have been early proposals that the northern basin was produced by a giant impact [132] or a series of multiple impacts [48]. Already early McGill and Squyres [74] disputed the multiple-impact hypothesis since it is improbable that the impacts prefer one hemisphere because of Mars' rotation. Andrews-Hanna et al. [6] investigated the MGS mission and found an elliptical northern basin if the younger Tharsis additions are stripped off. The present-day irregular boundary will become quite elliptical in this way measuring about 10,650 by 8,520 km, centered at 67° N, 200° E. Marinova et al. [73] model impacts with energies of $(3\text{-}6) * 10^{29}$ J, at low impact velocities (6-10 km/s) and oblique impact angles (30-60°) which generated structures similar in size and ellipticity of the martian lowlands basin. We want to check also this option especially regarding subsequent hydrodynamic and thermodynamic mechanisms. Admittedly we should expect that a high mountain rim is generated by the impact at the margin of the northern basin. There is no trace to be seen of it now. Can erosion explain the non-existence of the rim? Nimmo et al. [84] used a high-resolution, 2D, axially symmetric hydrocode to model vertical impacts. It is to be expected that it is not possible to obtain the elliptic form of the corrected [6] basin. Melosh [76] summarizes the recent contributions to the giant impact hypothesis. He concludes, however, that all these recent studies are not the last word on the impact origin of the Borealis basin.

4 Interior Structure Models

We explain in Section 6.2 why structural models of Mars are important as a presupposition to be able to calculate a *dynamic* martian evolution model using the code Terra since there are no shell models of Mars based on measured seismic data. Early martian structural models have been presented by [16, 72, 98]. Sohl and Spohn [103] proposed two end-member structural models: The first one was optimized to satisfy chemical SNC data whereas the second model was optimized to fulfill the *polar* moment of inertia factor. Spohn [108] reviewed the principles which are necessary to derive the internal structure of a solid-state planet. Sohl et al. [102] emphasized that the *mean* moment of inertia is required for constructing spherically symmetric structure models and used a mean moment-of-inertia factor of $I/M_P R_P^2 = 0.3635 \pm 0.0012$. They derived crustal-thickness–relative-core-radius curves for different mantle densities,ρ_m, molar magnesium numbers, Mg/(Mg+Fe), and magnesium-to-silicon ratios, Mg/Si. Zharkov and Gudkova [139] calculated similar multitudes of martian structural models, however, with more complex assumptions. Zharkov et al. [140] refined these models abandoning the spherical symmetry and using mainly new results of Konopliv et al. [65]. Verhoeven et al. [119] derive structural models of Mars combining electromagnetic, geodetic and seismic properties of the relevant materials. Their one-dimensional models depend on eight parameters: crustal thickness, crustal mean density, bulk volume fraction of iron in the mantle, olivine volume fraction of the mantle, pressure gradient, temperature profile, core radius, core mass. They select eight models fitting the observational data.

5 Tectonic Episodicity

The tectonic activities of Earth and Mars are very differently distributed as functions of time. The observations on *Earth* show an episodic continental growth [30, 60] which is distributed over the whole time axis. The integrated activity declines, but only slowly. Parman [89] showed that not only the production peaks of terrestrial ocean island basalts (OIB) but also those of the terrestrial mid-oceanic ridge basalts (MORB) can be correlated to the zircon-age peaks. These observations argue for episodic mantle melting and crustal growth on Earth. We [121, 122, 125] developed numerical models to explain the mechanism behind these terrestrial observations. The bulk of the *martian* crust formed, however, about 4.5 Ga ago, presumably from a magma ocean [87]. Later additions to the martian crust were volumetrically minor. Dohm et al. [37] observe episodes of the formation of Oceanus Borealis which took place very early though. Nevertheless, the geologic history of Mars is far from static. The tectonism decreases. It becomes concentrated near the large shield volcanoes. The Tharsis volcanoes did not stay at the same place but they migrated in the earliest epochs [83, 141, 142] similar to the Pacific volcanic chains until they came to a standstill near the margin of Borealis basin. Neukum et al. [78–80] found an appearance of episodicity of martian resurfacing events at ages of 3.5 Ga, 1 to 1.5 Ga, 300 to 600 Ma and 200 Ma. So, the geologic activity in the martian crust was highly active > 4 Ga ago and has strongly decreased in magnitude through time. However, this activity curve was not continuous but episodic.

6 Our New Model of Martian Evolution

6.1 Explaining Introduction

This part deals with physical and chemical considerations on the development of a *new martian stuctural model*. In contrast with the most published models of that kind, e.g. Sohl and Spohn [103], the new model should be derived from the evolution of the martian MO. As soon as the MO is essentially solidified, our version of the code Terra is applicable, i.e, we can dynamically calculate the solid-state creep and a simplified chemical differentiation. In other words, we can solve the differential equations of the balance of momentum, energy, mass etc. similar to the method used by Walzer and Hendel [122].

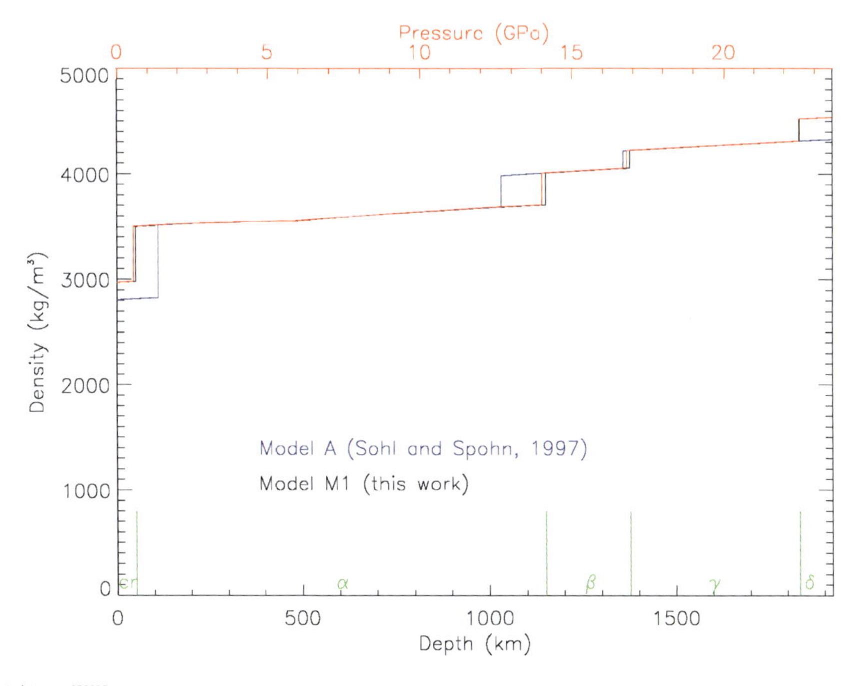

Fig. 1. A comparison of the density profile for the stuctural model [103] and our preliminary structural model of the present-day *martian* mantle. The red line refers to the pressure abscissa, the black line to the depth abscissa of our model. α, β, and γ signify the mineralogical phases of olivine. The crust is denoted by cr, a hypothetical thin perovskite layer by δ

Section 7 refers to *numerical improvements of the code Terra* including ideas developed in the doctoral theses [66, 77]. After the development of the new structural model of Mars, we intend to use it as an initial configuration for the newly developed Terra. We refer to Müller [77] to show that the numerical improvements are essentially more precise and comprehensive than that what we can show in this paper because of lack of space.

6.2 Towards a New Structural Model of the Early Mars

There is no seismic model of the martian mantle based on direct seismic observations since there are no martian seismic stations up to now. So, we have no analogon to PREM [39] which applies for Earth and which we have used in [122]. To apply Terra, we need a PREM-analogon for Mars but also other physical quantities as the viscosity, η, and the thermal expansivity, α, at least as a function of depth. Therefore we want to derive a chemical and mineralogical distribution model of the martian mantle similar to [103] or [52], however, not for the present time but for the time immediately after solidification of the MO and after a possible early R-T overturn.

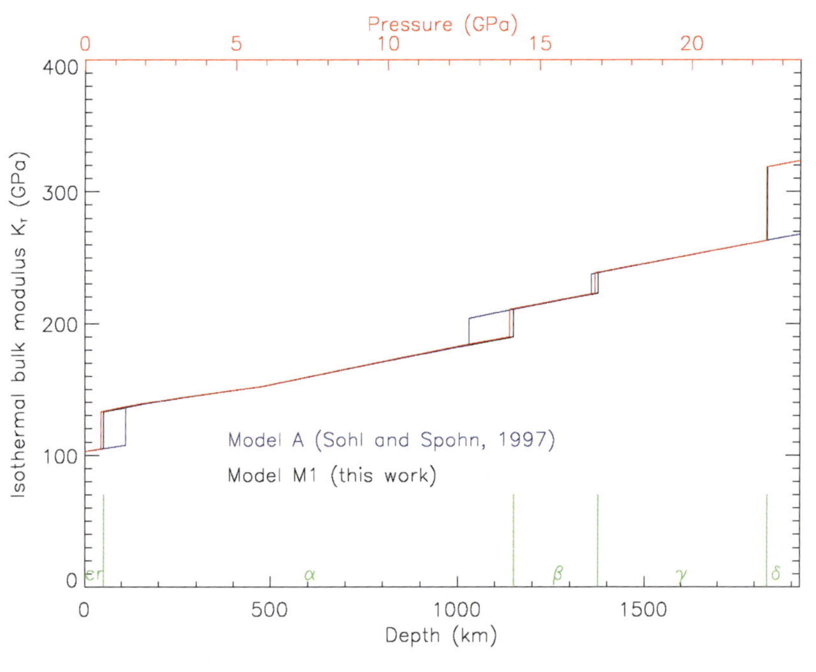

Fig. 2. A comparison of the isothermal bulk modulus profile, $K_T(r)$, for the stuctural model [103] and our preliminary stuctural model of the present-day *martian* mantle. Further remarks cf. Fig. 1

Beforehand, different chemical models of Mars [38, 70, 96] and, if available, new chemical models should be checked and compared. With regard to core formation, the partitioning of FeO is especially important. Rubie et al. [95] have determined the partitioning of FeO between liquid iron and magnesiowüstite at pressures, P, up to 70 GPa and temperatures, T, up to 3500 K and used also data of [7] for higher pressures. If there would not appear essentially new findings, we would start, for the time being, assuming 18 wt% FeO [38] in the martian MO. For the Earth, only 8 wt% FeO are to be supposed. Even for the less realistic case of a *homogeneous* chemical distribution in the martian mantle, we need a phase diagram analogous to [57, 58], to determine the *mineralogical* layering for the solid state and to know the depths of phase transitions. In every case, the phase boundaries of [57, 58] have to be shifted in dependence on the FeO- and H_2O-abundance. The group of Falko Langenhorst works to get experimentally phase diagrams for minerals [49, 101] which are relevant for the martian mantle.

Walzer et al. [120] derived a *preliminary* martian structural model related to [103] and [38] assuming a chemically homogeneous martian mantle model *for the present time*. The Figures 1 to 4 show some results of this model. We

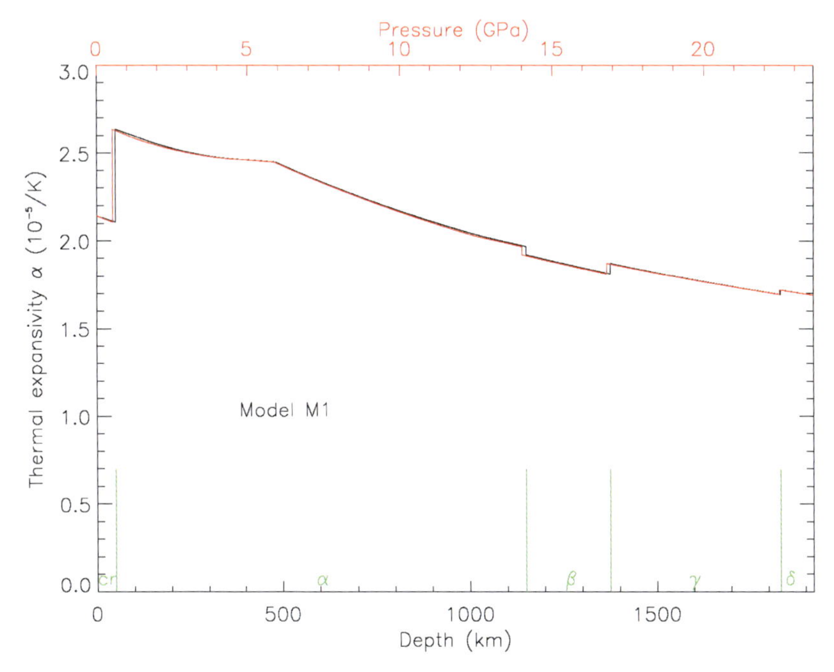

Fig. 3. The thermal expansivity, α, as a function of depth for a chemically homogeneous structural model of the present-day *martian* mantle. Further remarks cf. Fig. 1

do not consider these pictures as definitive but we want to improve them, especially the physical base of the derivation of Grüneisen's parameter, γ, (cf. Fig. 4), and to apply the whole system to proposed chemical layerings [14, 40–42] for the temporally first solid martian silicate mantle. This solidified mantle develops after about 5 Ma according to Elkins-Tanton [14, 40–42]. An unresolved question is whether we obtain a chemical layering or a compositional gradient so that the chemical composition is gradually varying as a function of radius. We want to critically test the Elkins-Tanton mechanism. As a first step, we intend to apply the physical theory to the end-member case of a chemically homogeneous martian mantle. However, from investigations of the ^{176}Lu–^{176}Hf and ^{147}Sm–^{143}Nd of shergottites, we have got evidence for a very early differentiation and a less efficient homogenization in the martian mantle [25, 36]. To check this conclusion, it is necessary to study the MO and its solidification. At first, we will start assuming an initially hot and entirely molten martian mantle which probably existed, based on geochemical evidence [20, 29, 54, 61, 93]. Kleine et al. [61] conclude from the chondritic ^{182}W/^{184}W ratio that the core-mantle-crust differentiation is constrained to 7–15 Ma.

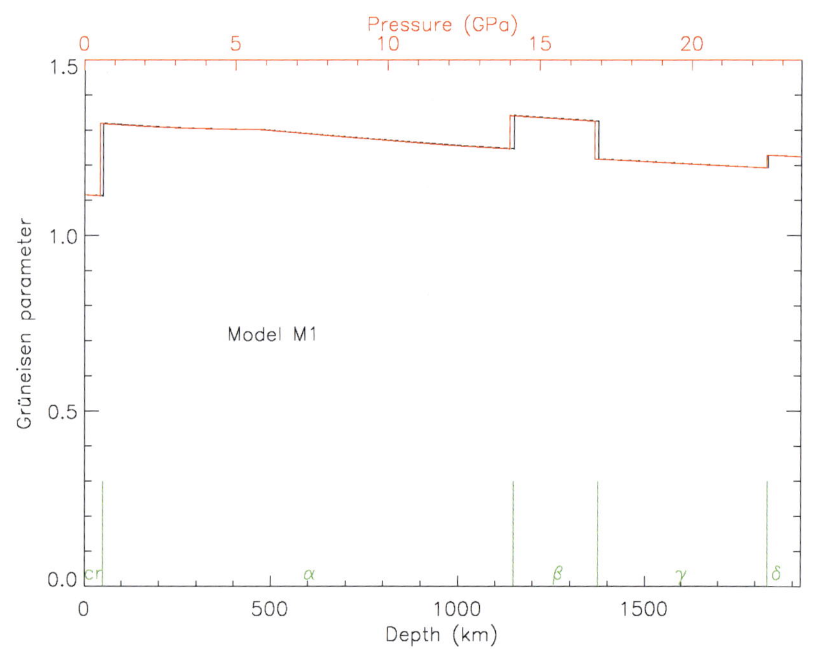

Fig. 4. The Grüneisen parameter, γ, as a function of depth for a chemically homogeneous structural model of the present-day *martian* mantle. Further remarks cf. Fig. 1

Our actual starting point will be the model by Elkins-Tanton [40–42] which, however, is to be revised in essential, but also in minor items. But the idea is nevertheless interesting. Which part of the intended revision is the most important one is not a priori known. Here we continue some considerations of Section 2. Since it is meanwhile clear that not only the highlands but also the northern plains have a high age, it might be that the basic features of each of the two peaks in the percentage of surface area of the topography [100] and the crustal thickness [81, 127, 144] are induced by very early processes.

- Layered convection, delineated by mineral septa, is unlikely in a planetary-scale martian MO. This statement [40] seems to be right because of the high Rayleigh number of the MO. This Rayleigh number is high because, e.g., Liebske et al. [69] demonstrated that the viscosity of a synthetic peridotite liquid varies between 0.019 and 0.13 Pa·s for a pressure interval of 2.8 to 13 GPa and a temperature span between 2043 and 2523 K. The viscosity of this magma is governed by a Vogel-Fulcher-Tamman equation. So, it is non-Arrhenian. Therefore the magmatic viscosity rises as a function of pressure for $P = 0 - 8$ GPa whereas it decreases with growing pressure for $P > 8$ GPa. So, the MO is expected to be less viscous than water and to convect with high turbulence. Therefore it is very improbable that there is a chemical differentiation before the emergence of the first crystals. In spite of these arguments, we will check this conclusion by the results of [92, 104, 105].
- The solidus and liquidus curves of Longhi et al. [72] have been used in [41]. We want to replace them by new curves which are to be derived together with a new phase diagram for 18 wt% FeO as an analogon to [57, 58].
- Presumably the adiabates will be steeper than the newly derived or measured solidi, also for future phase diagrams. If so, the solid layer will grow from the CBM towards the top. Elkins-Tanton et al. [40–42] *presuppose* that the growing solid mantle *is waiting* until it touches the preliminary crust near the surface. They think that the solid-state creep begins to work not before the whole martian mantle is solidified. Furthermore they do *not dynamically* calculate the R-T overturn which comes into being by the iron enrichment of the uppermost layers of the mantle. It should be possible for us to check by the code Terra whether the solidified lower part of the martian mantle begins to flow by a R-T instability **before** the whole mantle is solidified.
- One mineralogy model has been proposed for the first solidified whole-mantle martian MO, another mineralogy model for the mantle just after the R-T overturn [40–42]. We intend to check whether these findings are appropriate if we use newer results [6] etc.
- We should be able to calculate the dynamic process of the R-T overturn by an improved Terra. We will solve the full set of differential equations which guarantees the conservation of mass, energy, momentum, angular momentum and the sums of atoms of the four most relevant radionuclides

plus their corresponding daughter nuclides. Is the result really a simply reversed zero-pressure density profile given by Elkins-Tanton or is it a sequence of chemically homogeneous layers or a more complex structure?

- We can systematically vary the mineralogical assumptions and look for systematic variations in the dynamic results. Perhaps we can divide the results into classes.
- Since we observe a very early division of the mantle (cf. [64], their Fig. 10(b)) into a shergottite reservoir and a NC reservoir as well as a further division by chemical differentiation into a depleted-shergottite source (DS reservoir) with ^{180}Hf/^{184}W ≈ 8 and an enriched-shergottite source (ES reservoir) with ^{180}Hf/^{184}W ≈ 2, we should numerically solve the problem on what conditions these reservoirs are preserved until an age of 1.3 Ma, yet, and probably even until the present time.
- It is in no way evident whether the reservoir classification of the martian mantle, mentioned in the previous item, is connected with the iron enrichment in the northern lowlands. If so, then it would be an indication that the essential features of the martian dichotomy stem from the earliest history of the planet. We want to investigate if, firstly, this idea is appropriate according to observations and, secondly, if we can model it by Terra. (18–24 wt% FeO have been observed in the lowlands, 14–18 wt% FeO in the highlands. An alternative explanation is, however, the idea that the FeO-dichotomy might be due to acidic aqueous activity which could leach iron from the surface via hydrologic activity [19].) A mega-impact could be another alternative explanation of the formation of the martian crustal and chemical dichotomy [6, 46, 73, 84]. Also Watters et al. [127] think that it is most probable that the martian dichotomy evolved very early. But it is an open question whether this dichotomy came into being by one impact, by a series of impacts, or by a rapid endogenic mechanism like R-T overturn or a thermal-convection mechanism. We want to test the two last-mentioned possiblities by a Terra-based model.
- Agee [2] showed by a sink/float method that the crystal-liquid pressure dependence of density for a primitive martian melt has a crossover at $P =$ 9 GPa corresponding to a martian depth of $h = 750$ km in the dry case. However, the crossover will appear deeper and deeper for increasing initial water abundance. We have to check the influence of such findings to the Elkins-Tanton FeO-distribution before and after the R-T overturn. The altered starting distribution of Fe and Mg should have different effects on the initial solid-state creep that we can calculate dynamically using Terra. The results are expected to vary with varying water abundance.
- The following item is very important. Elkins-Tanton et al. [40–42] maintain that a stable compositional stratification after the R-T overturn at $t = 5$ Ma would inhibit thermally driven convection. So, they propose that the martian mantle is perpetually resistant to thermal convection. We cannot believe it for the following reasons. Intensive East-West striking magnetic stripe anomalies have been discovered in the southern high-

lands [33, 34, 111]. So, the early Mars had a strong magnetic dipole field [1]. The present-day martian CMB is a solid-fluid boundary. This is a result of satellite-track investigations [135]. The orders of magnitude of the obliquity of the rotational axis and of the length of day of Mars are very similar to those of the Earth. Therefore we conclude from dynamo theory that, if the present-day temperature gradient at CMB were high enough, we should expect a strong present magnetic dipole field: The Coriolis force would bring the numerous little vortices of the fluid core into line with the rotation axis. Since the CMB heat flow is controlled by the surface heat flow we presume: The one-plate nature of the present-day martian lithosphere is the cause for the non-existence of a significant present magnetic field. On the other hand, Mars' crust is ten times more intensely magnetized than the Earth's crust. Mars *had* an intense global magnetic field in the first few hundred Ma [1, 32]. Therefore we should expect thermally driven, probably layered martian mantle convection in those first few hundred Ma, at least, presupposed that the R-T overturn occured so quickly as mentioned above [40–42]. Of course, the present-day highly effective thermal screening is mainly caused by the thick crust and the thick lithosphere [21, 81, 107, 110].

- The previous items can only be optimally put into effect after the improvements of Section 7.

7 Numerical Improvements of Terra

Müller [77] and Köstler [66] work on improvements of Terra which, however, could not be used yet in our latest simulations. Müller analyzed the discretization of the Stokes problem and found a local grid refinement with hanging nodes which is inf-sup stable. Beate Sändig augmented the block size of the Jacobi smoother in the solver and Köstler adapted the grid transfer because of the small irregularities of the icosahedric grid. However, the influence of these alterations on the convergence behavior proved to be small. Further studies on the Krylov subspace method and multigrid method have been carried out and are continued to apply the latest results of numerical research to the Stokes solver in Terra. Especially a multigrid solver of the coupled Stokes problem, proposed by Larin and Reusken [67], should be developed to augment the convergence rates and to improve the pressure correction. We intend to implement such a solver and a MINRES procedure. As a latest result, John Baumgardner got the new free-slip boundary treatment debugged and working correctly. It is to be expected that he can move forward now on a set of benchmarks and get a Terra benchmark paper.

The computational model Terra uses a discretization of the sphere which consists of 10 diamonds evolving from the projection of the regular icosahedron onto the sphere [9]. The lateral extension of one diamond edge is an angle of about 63 degrees and the depth of this domain can be chosen arbitrarily. The size of such a segment is very appropriate to model special features of the martian crust. Because parallelization in Terra is done in lateral directions, it can be applied very efficiently to a regional domain which is a wide aspect ratio spherical shell segment. For instance on an SGI Altix 4700 machine with Intel Itanium2 Montecito Dual Core processors we could achieve an efficient cache usage with subdomains of 33x33x129 grid points which would give an overall regional grid spacing of 4 km at the surface and 3 km at the martian CMB if we use 1024 processors on one diamond with a depth of 500 km. This is 4 times less the spacing we could achieve in the whole spherical shell. Communication overhead would be much less than 10% for this configuration. A similar approach can be done with the convection code CitComS in [116], where a 2 times finer regional grid is used.

Nevertheless, very strong gradients in material parameters, in particular viscosity, have to be taken into account. Therefore we want to further advance the Terra code with respect to stability, local grid refinement, and solver techniques. Specifically, we intend to investigate an inf-sup-stable grid modification that enables us to use grids with hanging nodes. Thus we will be able to adapt the grid resolution rapidly towards a heavily refined region with only a few successively refined layers. We will change the smoother of the multigrid solver, the preconditioner, the algorithm for the solution of an included Stokes problem and the method of time integration. We will start refactoring the solver to achieve greater flexibility with regard to algorithm changes. We created a test framework that automates the build process, starts automatically a series of test cases on different machines in different resolutions and checks the results. This way the response time to changes and errors in the code is noticeably reduced. That means, we have facilities assisting even radical changes of the code. We can provide a fully featured parallel multigrid solver for the convection in the domain of interest, including parallel implemented tracers for chemical differentiation by a modification of the existing code (cf. Fig. 5). In addition to the adaption described, we plan to implement refined solving strategies to meet the requirements of strongly varying viscosity due to mineral phase boundaries and to differences in temperature, pressure and water content. We will use the techniques we investigated for whole mantle convection like the local grid refinement, a special preconditioning technique as well as our existing test framework. In particular, the above-mentioned preconditioned MINRES algorithm will be implemented. The benefits achieved this way would also pay off for the new dynamic martian-mantle creep model.

Having obtained a small scale method that uses essentially the same solver we have very good preconditions to integrate the small high resolution area seamlessly in the global martian convection. The spatial grid will be refined with the described technique of hanging nodes towards the boundaries of a small scale model with only a few successive refinement steps. One has however also to couple the time stepping scheme of both methods in a very flexible manner. We will have to address both, stability and optimality with respect to the load balancing. The plan to achieve these goals is simple because we do

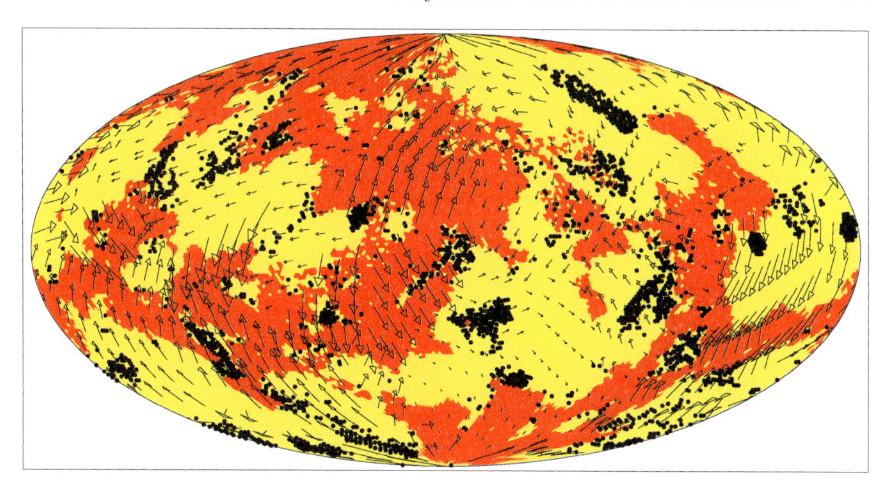

Run 819B1 σ_y = 115 MPa r_n = −0.55 Meridian 180° Time = 4490 Ma
Age = −0.1741 Ma Max vel = 0.961 cm/a Av hor vel = 0.406 cm/a

Fig. 5. A result of our convection-evolution model for the *Earth's* mantle with chemical differentiation. The tracer module used here has to be modified and strongly supplemented for the differentiation of Mars. This picture shows the distribution of continents (*red*) and oceanic plateaus (*black dots*) for the geological present of a run with yield stress $\sigma_y = 115$ MPa and viscosity-level parameter $r_n = -0.55$. The oceanic lithosphere is signified by *yellow color*

not need full adaptivity in space and time, but prescribe the spatiotemporal resolution according to optimal load balancing.

Furthermore, there exists a 2D version of Terra which can already model viscosity contrasts of up to 10^{10} between adjacent points [133]. As we use this code to study different solving strategies in two dimensions first, there is a benefit to be expected also in modeling the influences of composition and grain size. Moreover, the matrix-dependent transfer technique in the 2D code still promises an improvement of the convergence of the multigrid in the 3D code.

8 The Numerical Performance Hitherto Achieved

The Navier-Stokes equations are handled by the Finite Element Method. Pressure and velocity are solved simultaneously by a Schur-complement conjugate-gradient iteration (Ramage and Wathen, [91]). The system of linear equations is solved using a multigrid procedure in connection with matrix-dependent prolongation and restriction and with a Jacobi smoother. The temperature transport is realized by the second order Runge-Kutta method for explicit time steps.

For convergence tests we compared the results of runs with 1,351,746 and 10,649,730 nodes. The deviations concerning Rayleigh number, Nusselt number, Urey number, and the laterally averaged surface heat flow, qob, were smaller than 0.5%. Benchmark tests of the Terra code (Bunge et al. [23], Glatzmaier [51]) showed deviations of less than 1.5%.

Table 1. CPU-time, walltime and speedup for runs with 100 time steps on 1,351,746 nodes (a) and on 10,649,730 nodes (b). For comparison, Speedup (b) for 4 processors has been deliberately set to 4.00

Procs	CPU-time (a)	Walltime (a)	Speedup (a)	CPU-time (b)	Walltime (b)	Speedup (b)
1	00:27:13	00:27:17	1			
4	00:29:12	00:07:33	3.61	05:01:23	01:16:09	4.00
8	00:26:08	00:03:31	7.76	04:53:26	00:37:32	8.12
16	00:25:36	00:02:02	13.42	05:08:51	00:19:35	15.69
32	00:21:52	00:01:02	26.40	04:34:40	00:09:11	33.17
64				05:43:25	00:05:36	54.39
128				05:34:24	00:03:04	100.42

The Terra code is parallelized by domain decomposition according to the dyadic grid refinement and using explicit message passing (MPI). In Table 1 we present measurements of scalability and performance. Using the performance measuring tool `jobperf`, we obtained an average of 1201 MFlop/s with 8 processors, 1116 MFlop/s with 32 processors, and 935 MFlop/s with 128 processors, respectively. In both resolutions the speedup was almost linear, in some cases slightly superlinear due to cache usage. With the high resolution, at least 4 processors are necessary to make efficient use of the cache memory.

Acknowledgements. We kindly acknowledge the confidential cooperation with John Baumgardner, Markus Müller, and Christoph Köstler. We acknowledge the use of supercomputing facilities at SSC Karlsruhe, LRZ München and HLRS Stuttgart. This work was partly supported by the Deutsche Forschungsgemeinschaft under the grants WA1035/5-3 and KL495/16-1.

References

1. M. N. Acuña, J. E. P. Connerney, N. F. Ness, et al. Global distribution of crustal magnetization discovered by the Mars Global Surveyor MAG/ER experiment. *Science*, 284:790–793, 1999.
2. C. B. Agee. Static compression of hydrous silicate melt and the effect of water on planetary differentiation. *Earth Planet. Sci. Lett.*, 265:641–654, 2008.
3. C. B. Agnor, R. M. Canup, and H. F. Levison. On the character and consequences of large impacts in the late stage of terrestrial planet formation. *Icarus*, 142:219–237, 1999.

4. Y. Amelin, A. N. Krot, I. D. Hutcheon, and A. A. Ulyanov. Lead isotopic ages of chondrules and calcium-aluminum-rich inclusions. *Science*, 297:1678–1683, 2002.
5. Y. Amelin, M. Wadhwa, and G. W. Lugmair. Pb-isotopic dating of meteorites using ^{202}Pb - ^{205}Pb double spike: Comparison with other high-resolution chronometers. In *37th Annual Lunar and Planetary Science Conference (abstract #1970)*. LPI, 2006.
6. J. C. Andrews-Hanna, M. T. Zuber, and W. B. Banerdt. The Borealis basin and the origin of the martian crustal dichotomy. *Nature*, 453:1212–1215, 2008.
7. Y. Asahara, D. J. Frost, and D. C. Rubie. Partitioning of FeO between magnesiowüstite and liquid iron at high pressures and temperatures: Implications for the composition of the Earth's outer core. *Earth Planet. Sci. Lett.*, 257:435–449, 2007.
8. V. R. Baker, S. Maruyama, and J. M. Dohm. A theory of early plate tectonics and subsequent long-term superplume activity on Mars. *Superplume International Workshop, Tokyo, Abstract*, pages 312–316, 2002.
9. J. R. Baumgardner and P. O. Frederickson. Icosahedral discretization of the two-sphere. *SIAM J. Numer. Anal.*, 22:1107–1115, 1985.
10. V. Belleguic, P. Lognonné, and M. Wieczorek. Constraints on the Martian lithosphere from gravity and topography data. *J. Geophys. Res.*, 110(E11):E11005, 2005.
11. V. C. Bennett, A. D. Brandon, and A. P. Nutman. Coupled ^{142}Nd - ^{143}Nd Isotopic Evidence for Hadean Mantle Dynamics. *Science*, 318:1907–1910, 2007.
12. D. Bercovici. Generation of plate tectonics from lithosphere-mantle flow and void-volatile self-lubrication. *Earth Planet. Sci. Lett.*, 154:139–151, 1998.
13. D. Bercovici. The generation of plate tectonics from mantle convection. *Earth Planet. Sci. Lett.*, 205:107–121, 2003.
14. C. M. Bertka and Y. Fei. Mineralogy of the Martian interior up to core-mantle boundary pressures. *J. Geophys. Res.*, 102:5251–5264, 1997.
15. C. M. Bertka and Y. Fei. Density profile of an SNC model Martian interior and the moment-of-inertia factor of Mars. *Earth Planet. Sci. Lett.*, 157:79–88, 1998.
16. A. B. Binder. Internal structure of Mars. *J. Geophys. Res.*, 74:3110–3118, 1969.
17. M. Bizzarro, J. A. Baker, H. Haack, and K. L. Lundgaard. Rapid timescales for accretion and melting of differentiated planetesimals inferred from ^{26}Al-^{26}Mg chronometry. *Astrophys. J.*, 632:L41–L44, 2005.
18. A. Bouvier, J. Blichert-Toft, F. Moynier, J. D. Vervoort, and F. Albarède. Pb-Pb dating constraints on the accretion and cooling history of chondrites. *Geochim. Cosmochim. Acta*, 71(6):1583–1604, 2007.
19. W. V. Boynton, G. J. Taylor, L. G. Evans, et al. Concentration of H, Si, Cl, K, Fe, and Th in the low- and mid-latitude regions of Mars. *J. Geophys. Res.*, 112:E12S99, 2007.
20. A. D. Brandon, R. J. Walker, J. W. Morgan, and G. G. Goles. Re-Os isotopic evidence for early differentiation of the Martian mantle. *Geochim. Cosmochim. Acta*, 64:4083–4095, 2000.
21. D. Breuer. Thermal evolution, crustal growth and magnetic field history of Mars. Habilitation, W. W. Univ. Münster, 2003.

22. D. Breuer and T. Spohn. Early plate tectonics versus single-plate tectonics on Mars: Evidence from magnetic field history and crust evolution. *J. Geophys. Res.*, 108(E7):5072, 2003.
23. H.-P. Bunge, M. A. Richards, and J. R. Baumgardner. A sensitivity study of three-dimensional spherical mantle convection at 10^8 Rayleigh number: Effects of depth-dependent viscosity, heating mode and an endothermic phase change. *J. Geophys. Res.*, 102:11991–12007, 1997.
24. R. M. Canup and E. Asphaug. Origin of the Moon in a giant impact near the end of Earth's formation. *Nature*, 412:708–712, 2001.
25. G. Caro, B. Bourdon, A. Halliday, and G. Quitté. Superchondritic Sm/Nd in Mars, Earth and the Moon. *Nature*, 452:336–339, 2008.
26. J. E. Chambers. Making more terrestrial planets. *Icarus*, 152:205–224, 2001.
27. J. E. Chambers. Planetary accretion in the inner solar system. *Earth Planet. Sci. Lett.*, 223:241–252, 2004.
28. J. E. Chambers and G. W. Wetherill. Making the terrestrial planets: N-body integrations of planetary embryos in three dimensions. *Icarus*, 136:304–327, 1998.
29. J. H. Chen and G. J. Wasserburg. Formation ages and evolution of Shergotty and its parent planet from U-Th-Pb systematics. *Geochim. Cosmochim. Acta*, 50:955–968, 1986.
30. K. C. Condie. Episodic continental growth and supercontinents: a mantle avalanche connection? *Earth Planet. Sci. Lett.*, 163:97–108, 1998.
31. J. E. Connerney, M. H. Acuña, N. F. Ness, G. Kletetschka, D. L. Mitchell, R. P. Lin, and H. Reme. Tectonic implications of Mars crustal magnetism. *Proc. Natl. Acad. Sci. USA*, 102:14970–14975, 2005.
32. J. E. P. Connerney, M. H. Acuña, N. F. Ness, T. Spohn, and G. Schubert. Mars crustal magnetism. *Space Science Reviews*, 111:1–32, 2004.
33. J. E. P. Connerney, M. H. Acuña, P. J. Wasilewski, et al. Magnetic lineations in the ancient crust of Mars. *Science*, 284:794–798, 1999.
34. J. E. P. Connerney, M. H. Acuña, P. J. Wasilewski, G. Kletetschka, N. F. Ness, H. Rème, R. P. Lin, and D. L. Mitchell. The global magnetic field of Mars and implications for crustal evolution. *Geophys. Res. Lett.*, 28(21):4015–4018, 2001.
35. V. Debaille, A. D. Brandon, Q. Z. Yin, and S. B. Jacobsen. Coupled ^{142}Nd - ^{143}Nd evidence for a protracted magma ocean in Mars. *Nature*, 450:525–528, 2007.
36. V. Debaille, Q.-Z. Yin, A. D. Brandon, and B. Jacobsen. Martian mantle mineralogy investigated by the ^{176}Lu - ^{176}Hf and ^{147}Sm - ^{143}Nd systematics of shergottites. *Earth Planet. Sci. Lett.*, 269:186–199, 2008.
37. J. M. Dohm, V. R. Baker, W. V. Boynton, et al. GRS evidence and the possibility of paleooceans on Mars. *Planet. Space Sci.*, 2008. doi:10.1016/j.pss.2008.10.008.
38. G. Dreibus and H. Wänke. Supply and loss of volatile constituents during the accretion of terrestrial planets. In S. K. Attreya, J. B. Pollack, and M. S. Matthews, editors, *Origin and Evolution of Planetary and Satellite Atmospheres*, pages 268–288. Univ. Arizona Press, 1989.
39. A. M. Dziewonski and D. L. Anderson. Preliminary reference Earth model. *Phys. Earth Planet. Int.*, 25:297–356, 1981.
40. L. T. Elkins-Tanton. Linked magma ocean solidification and atmospheric growth for Earth and Mars. *Earth Planet. Sci. Lett.*, 271:181–191, 2008.

41. L. T. Elkins-Tanton, P. C. Hess, and E. M. Parmentier. Possible formation of ancient crust on Mars through magma ocean processes. *J. Geophys. Res.*, 110:E12S01, 2005.
42. L. T. Elkins-Tanton, E. M. Parmentier, and P. C. Hess. Magma ocean fractional crystallization and cumulative overturn in terrestrial planets: implications for Mars. *Meteorit. Planet. Sci.*, 38:1753–1771, 2003.
43. A. G. Fairén and J. M. Dohm. Age and origin of the lowlands of Mars. *Icarus*, 168:277–284, 2004.
44. A. G. Fairén, J. M. Dohm, V. R. Baker, M. A. de Pablo, J. Ruiz, J. C. Ferris, and R. C. Anderson. Episodic flood inundations of the northern plains of Mars. *Icarus*, 165:53–67, 2003.
45. C. N. Foley, M. Wadhwa, L. E. Borg, P. E. Janney, R. Hines, and T. L. Grove. The early differentiation history of Mars from ^{182}W - ^{142}Nd isotope systematics in the SNC meteorites. *Geochim. Cosmochim. Acta*, 69:4557–4571, 2005.
46. H. V. Frey. Impact constraints on, and a chronology for, major events in early Mars history. *J. Geophys. Res.*, 111:E8S91, 2006.
47. H. V. Frey. Impact constraints on the age and origin of the lowlands of Mars. *Geophys. Res. Lett.*, 33:L08S02, 2006.
48. H. V. Frey and R. A. Schultz. Large impact basins and the mega-impact origin for the crustal dichotomy on Mars. *Geophys. Res. Lett.*, 15:229–232, 1988.
49. G. Ganskow, F. Langenhorst, and D. Frost. Stability of hydrous ringwoodite in the $MgFeSiO_4$ - H_2O system. 5th Colloq. DFG SPP 1115 Münster, p. 8, 2008.
50. A. Ghosh, F. Nimmo, and H. Y. J. McSween. The effect of early accretion and redistribution of ^{26}Al on the thermal evolution of Mars (abstract #2011). *34rd Annual Lunar and Planetary Science Conference. CD-ROM*, 2003.
51. G. A. Glatzmaier. Numerical simulations of mantle convection: Time-dependent, three-dimensional, compressible, spherical shell. *Geophys. Astrophys. Fluid Dyn.*, 43:223–264, 1988.
52. T. V. Gudkova and V. N. Zharkov. Mars: Interior structure and excitation of free oscillations. *Phys. Earth Planet. Int.*, 142:1–22, 2004.
53. A. N. Halliday and T. Kleine. Meteorites and the timing, mechanisms and conditions of terrestrial planet accretion and early differentiation. In D. L. Lauretta and H. Y. McSween, editors, *Meteorites and the Early Solar System II*, pages 775–801. Univ. Arizona Press, Tucson, 2006.
54. A. N. Halliday, H. Wänke, J.-L. Birck, and R. N. Clayton. The accretion, composition and early differentiation of Mars. *Earth Moon Planets*, 96:197–230, 2001.
55. T. Hoogenboom and S. E. Smrekar. Elastic thickness estimates for the northern lowlands of Mars. *Earth Planet. Sci. Lett.*, 248:830–839, 2006.
56. S. B. Jacobsen. The Hf-W isotopic system and the origin of the Earth and the Moon. *Annu. Rev. Earth Planet Sci.*, 32:531–570, 2005.
57. T. Kawamoto. Hydrous phase stability and partial melt chemistry in H_2O-saturated KLB-1 peridotite up to the uppermost lower mantle conditions. *Phys. Earth Planet. Int.*, 143:387–395, 2004.
58. T. Kawamoto. Hydrous phases and water transport in subducting slabs. *Rev. Min. Geochem.*, 62:273–289, 2006.
59. Y. Ke and V. S. Solomatov. Early transient superplumes and the origin of the Martian crustal dichotomy. *J. Geophys. Res.*, 111:E10001, 2006.

60. A. I. S. Kemp, C. J. Hawkesworth, B. A. Paterson, and P. D. Kinny. Episodic growth of the Gondwana supercontinent from hafnium and oxygen isotopes in zircon. *Nature*, 439:580–583, 2006.
61. T. Kleine, K. Mezger, C. Münker, H. Palme, and A. Bischoff. ^{182}Hf-^{182}W isotope systematics of chondrites, eucrites, and martian meteorites: Chronology of core formation and early mantle differentiation in Vesta and Mars. *Geochim. Cosmochim. Acta*, 68:2935–2946, 2004.
62. T. Kleine, K. Mezger, H. Palme, and C. Münker. The W isotope evolution of the bulk silicate Earth: constraints on the timing and mechanisms of core formation and accretion. *Earth Planet. Sci. Lett.*, 228:109–123, 2004.
63. T. Kleine, C. Münker, K. Mezger, and H. Palme. Rapid accretion and early core formation on asteroids and the terrestrial planets from Hf-W chronometry. *Nature*, 418:952–955, 2002.
64. T. Kleine, M. Touboul, B. Bourdon, F. Nimmo, K. Mezger, H. Palme, S. B. Jacobsen, Q.-Z. Yin, and A. N. Halliday. Hf-W chronology of the accretion and early evolution of asteroids and terrestrial planets. *Geochim. Cosmochim. Acta*, 2008. revised version 05/11/2008.
65. A. S. Konopliv, C. F. Yoder, E. M. Standish, D.-N. Yuan, and W. L. Sjogren. A global solution for the Mars' static and seasonal gravity, Mars orientation, Phobos and Deimos masses, and Mars' ephemeris. *Icarus*, 182:23–50, 2006.
66. C. Köstler. *Iterative solvers for modeling mantle convection with strongly varying viscosity.* PhD thesis, Friedrich-Schiller-Univ. Jena, 2009.
67. M. Larin and A. Reusken. A comparative study of efficient iterative solvers for generalized Stokes equations. *Numer. Linear Algebra Appl.*, 15(1):13–34, 2008.
68. A. Lenardic, F. Nimmo, and L. Moresi. Growth of the hemispheric dichotomy and the cessation of plate tectonics on Mars. *J. Geophys. Res.*, 109:E02003, 2004.
69. C. Liebske, B. Schmickler, H. Terasaki, B. T. Poe, A. Suzuki, K.-i. Funakoshi, R. Ando, and D. C. Rubie. Viscosity of peridotite liquid up to 13 GPa: Implications for magma ocean viscosities. *Earth Planet. Sci. Lett.*, 240:589–604, 2005.
70. K. Lodders and B. Fegley. An oxygen isotope model for the composition of Mars. *Icarus*, 126:373–394, 1997.
71. P. Lognonné, D. Giardini, B. Banerdt, et al. The NetLander very broad band seismometer. *Planet. Space Sci.*, 48:1289–1302, 2000.
72. J. Longhi, E. Knittle, J. R. Holloway, and H. Wänke. The bulk composition, mineralogy, and internal structure of Mars. In H. H. Kieffer, B. M. Jakosky, C. W. Snyder, and M. S. Matthews, editors, *Mars*, pages 184–208. Univ. Arizona Press, Tucson, 1992.
73. M. M. Marinova, O. Aharonson, and E. Asphaug. Mega-impact formation of the Mars hemispheric dichotomy. *Nature*, 453:1216–1219, 2008.
74. G. E. McGill and S. W. Squyres. Origin of the Martian crustal dichotomy: Evaluating hypotheses. *Icarus*, 93:386–393, 1991.
75. A. K. McNamara and S. Zhong. Degree-one mantle convection: Dependence on internal heating and temperature-dependent rheology. *Geophys. Res. Lett.*, 32:L01301, 2005.
76. H. J. Melosh. Did an impact blast away half of the martian crust? *Nature Geoscience*, 1:412–414, 2008.

77. M. Müller. *Towards a robust Terra code.* PhD thesis, Friedrich-Schiller-Univ. Jena, 2008.
78. G. Neukum, A. T. Basilevsky, M. G. Chapman, et al. Episodicity in the geological evolution of Mars: resurfacing events and ages from cratering analysis of image data and correlation with radiometric ages of martian meteorites. *5th Colloquium DFG Priority Programme SPP 1115: Mars and the Terrestrial Planets*, pages 24–26, 2008.
79. G. Neukum, A. T. Basilevsky, B. A. Ivanov, S. van Gasselt, S. C. Werner, and W. Zuschneid. The geologic evolution of Mars: episodicity of resurfacing events and ages from geologic mapping and cratering analysis of image data. *4th Colloquium DFG Priority Programme SPP 1115: Mars and the Terrestrial Planets*, pages 33–34, 2007.
80. G. Neukum, R. Jaumann, H. Hoffmann, et al. Recent and episodic volcanic and glacial activity on mars revealed by the high resolution stereo camera. *Nature*, 432:971–979, 2004.
81. G. A. Neumann, M. T. Zuber, M. A. Wieczorek, P. J. McGovern, F. G. Lemoine, and D. E. Smith. The crustal structure of Mars from gravity and topography. *J. Geophys. Res.*, 109:E08002, 2004.
82. F. Nimmo. Admittance estimates of mean crustal thickness and density at the Martian hemispheric dichotomy. *J. Geophys. Res.*, 107(E11):5117, 2002.
83. F. Nimmo. Mars's rotating shell. *Nature Geoscience*, 2:7–8, 2009.
84. F. Nimmo, S. D. Hart, D. G. Korycansky, and C. B. Agnor. Implications of an impact origin for the martian hemispheric dichotomy. *Nature*, 453:1220–1223, 2008.
85. F. Nimmo and T. Kleine. How rapidly did Mars accrete? Uncertainties in the Hf - W timing of core formation. *Icarus*, 191:497–504, 2007.
86. F. Nimmo and D. J. Stevenson. Influence of early plate tectonics on the thermal evolution and magnetic field of Mars. *J. Geophys. Res.*, 105(E5):11969–11979, 2000.
87. F. Nimmo and K. Tanaka. Early crustal evolution of Mars. *Annu. Rev. Earth Planet Sci.*, 33:133–161, 2005.
88. C. O'Neill, A. M. Jellinek, and A. Lenardic. Conditions for the onset of plate tectonics on terrestrial planets and moons. *Earth Planet. Sci. Lett.*, 261:20–32, 2007.
89. S. W. Parman. Helium isotopic evidence for episodic mantle melting and crustal growth. *Nature*, 446:900–903, 2007.
90. R. R. Rafikov. The growth of planetary embryos: orderly, runaway or oligarchic? *Astrophys. J.*, 125:942–961, 2003.
91. A. Ramage and A. J. Wathen. Iterative solution techniques for the Stokes and the Navier-Stokes equations. *Int. J. Num. Meth. Fluids*, 19:67–83, 1994.
92. C. C. Reese and V. S. Solomatov. Fluid dynamics of local Martian magma oceans. *Icarus*, 184:102–120, 2006.
93. K. Righter and M. J. Drake. Core formation in Earth's Moon, Mars, and Vesta. *Icarus*, 124:513–529, 1996.
94. J. H. Roberts and S. Zhong. Degree-1 convection in the Martian mantle and the origin of the hemispheric dichotomy. *J. Geophys. Res.*, 111:E06013, 2006.
95. D. C. Rubie, H. Terasaki, Y. Asahara, et al. New constraints on core formation and planetary accretion. 5th Colloq. DFG Priority Programme SPP 1115, Münster, p. 1, 2008.

96. C. Sanloup, A. Jambon, and P. Gillet. A simple chondritic model of Mars. *Phys. Earth Planet. Int.*, 112:43–54, 1999.
97. G. Schubert, S. C. Solomon, D. L. Turcotte, M. J. Drake, and N. H. Sleep. Origin and thermal evolution of Mars. In H. H. Kieffer, B. M. Jakosky, C. W. Snyder, and M. S. Matthews, editors, *Mars*, pages 147–183. Univ. Arizona Press, 1992.
98. G. Schubert and T. Spohn. Thermal history of Mars and the sulfur content of its core. *J. Geophys. Res.*, 95:14095–14104, 1990.
99. N. H. Sleep. Martian plate tectonics. *J. Geophys. Res.*, 99(E3):5639–5655, 1994.
100. D. E. Smith, M. T. Zuber, H. V. Frey, J. B. Garvin, J. W. Head, D. O. Muhleman, G. H. Pettengill, R. J. Phillips, S. C. Solomon, H. J. Zwally, W. B. Banerdt, T. C. Duxbury, M. P. Golombek, F. G. Lemoine, G. A. Neumann, D. D. Rowlands, O. Aharonson, P. G. Ford, A. B. Ivanov, C. L. Johnson, P. J. McGovern, J. B. Abshire, R. S. Afzal, and X. Sun. Mars Orbiter Laser Altimeter: Experiment summary after the first year of global mapping of Mars. *J. Geophys. Res.*, 106:23689–23722, Oct 2001.
101. J. R. Smyth. A crystallographic model for hydrous wadsleyite (β-Mg_2SiO_4): An ocean in the Earth's interior? *Am. Min.*, 79:1021–1024, 1994.
102. F. Sohl, G. Schubert, and T. Spohn. Geophysical constraints on the composition and structure of the Martian interior. *J. Geophys. Res.*, 110:E12008, 2005.
103. F. Sohl and T. Spohn. The interior structure of Mars: Implications from SNC meteorites. *J. Geophys. Res.*, 102(E1):1613–1635, 1997.
104. V. S. Solomatov. Fluid dynamics of a terrestrial magma ocean. In R. M. Canup and K. Righter, editors, *Origin of the Earth and Moon*, pages 323–337. Univ. Arizona Press, Tucson, 2000.
105. V. S. Solomatov and C. C. Reese. Grain size variations in the Earth's mantle and the evolution of primordial chemical heterogeneities. *J. Geophys. Res.*, 113:B07408, 2008.
106. S. C. Solomon, O. Aharonson, W. B. Banerdt, et al. Why are there so few magnetic anomalies in martian lowlands and basins. *Lunar and Planetary Science Conference*, 34:1382.pdf, 2003.
107. T. Spohn. Mantle differentiation and thermal evolution of Mars, Mercury and Venus. *Icarus*, 90:222–236, 1991.
108. T. Spohn. Planetologie. In *L. Bergmann and C. Schaefer, 7: Erde und Planeten*, Lehrbuch der Experimentalphysik, pages 427–525. Walter de Gruyter, 2001.
109. T. Spohn. Planetary evolution and habitability. *European Planetary Science Congress Abstracts*, 3:EPSC2008–A–00602, 2008.
110. T. Spohn, F. Sohl, and D. Breuer. Mars. *Astron. Astrophys. Rev.*, 8:181–236, 1998.
111. K. F. Sprenke and L. L. Baker. Magnetization, paleomagnetic poles, and polar wander on Mars. *Icarus*, 147:26–34, 2000.
112. S. Stanley, L. Elkins-Tanton, M. T. Zuber, and E. M. Parmentier. Mars' paleomagnetic field as the result of a single-hemisphere dynamo. *Science*, 321:1822–1825, 2008.

113. C. Stein, J. Schmalzl, and U. Hansen. The effect of rheological parameters on plate behavior in a self-consistent model of mantle convection. *Phys. Earth Planet. Int.*, 142:225–255, 2004.
114. P. J. Tackley. Self-consistent generation of tectonic plates in time-dependent, three-dimensional mantle convection simulations. Part 1. Pseudoplastic yielding. *Geochem. Geophys. Geosys.*, 1:2000GC000036, 2000.
115. P. J. Tackley. Self-consistent generation of tectonic plates in time-dependent, three-dimensional mantle convection simulations. Part2. Strain weakening and asthenosphere. *Geochem. Geophys. Geosys.*, 1:2000GC000043, 2000.
116. E. Tan, E. Choi, P. Thoutireddy, M. Gurnis, and M. Aivazis. Geoframework: Coupling multiple models of mantle convection within a computational framework. *Geochem. Geophys. Geosys.*, 7:Q06001, 2006.
117. G. J. Taylor, W. Boynton, J. Brückner, et al. Bulk composition and early differentiation of Mars. *J. Geophys. Res.*, 112(E3):E03S10, 2007.
118. R. Trompert and U. Hansen. Mantle convection simulations with rheologies that generate plate-like behavior. *Nature*, 395:686–689, 1998.
119. O. Verhoeven, A. Rivoldini, P. Vacher, et al. Interior structure of terrestrial planets: Modeling Mars' mantle and its electromagnetic, geodetic, and seismic properties. *J. Geophys. Res.*, 110:E04009, 2005.
120. U. Walzer, T. Burghardt, and R. Hendel. Toward a dynamic model of the Martian mantle. 5th DFG Colloq. on Mars SPP 1115 Münster, 28-29 Feb, 2008.
121. U. Walzer and R. Hendel. Time-dependent thermal convection, mantle differentiation, and continental crust growth. *Geophys. J. Int.*, 130:303–325, 1997.
122. U. Walzer and R. Hendel. Mantle convection and evolution with growing continents. *J. Geophys. Res.*, 113:B09405, doi:10.1029/2007JB005459, 2008.
123. U. Walzer, R. Hendel, and J. Baumgardner. The effects of a variation of the radial viscosity profile on mantle evolution. *Tectonophysics*, 384:55–90, 2004.
124. U. Walzer, R. Hendel, and J. Baumgardner. Viscosity stratification and a 3D compressible spherical shell model of mantle evolution. In E. Krause, W. Jäger, and M. Resch, editors, *High Perf. Comp. Sci. Engng. '03*, pages 27–67. Springer, Berlin, 2004.
125. U. Walzer, R. Hendel, and J. Baumgardner. Whole-mantle convection, continent generation, and preservation of geochemical heterogeneity. In W. E. Nagel, D. B. Kröner, and M. M. Resch, editors, *High Perf. Comp. Sci. Engng. '07*, pages 603–645. Springer, Berlin, 2008.
126. W. R. Ward. On planetesimal formation: The role of collective particle behaviour. In R. M. Canup and K. Righter, editors, *Origin of the Earth and Moon*, pages 75–84. Univ. Arizona Press, Tucson, 2000.
127. T. R. Watters, P. J. McGovern, and R. P. Irwin III. Hemispheres apart: The crustal dichotomy on Mars. *Annu. Rev. Earth Planet Sci.*, 35:621–652, 2007.
128. S. J. Weidenschilling. Dust to planetesimals. *Icarus*, 44:172–189, 1980.
129. S. J. Weidenschilling and J. N. Cuzzi. Formation of planetesimals in the solar nebula. In E. H. Levy and J. I. Lunine, editors, *Protostars and Planets III*, pages 1031–1060. Univ. Arizona Press, Tucson, 1993.
130. S. J. Weidenschilling, D. Spaute, D. R. Davis, F. Marzari, and K. Ohtsuki. Accretional evolution of a planetesimal swarm, 2. The terrestrial zone. *Icarus*, 128:429–455, 1997.

131. M. A. Wieczorek and M. T. Zuber. Thickness of the Martian crust: Improved constraints from geoid-to-topography ratios. *J. Geophys. Res.*, 109:E01009, 2004.
132. D. E. Wilhelms and S. W. Squyres. The Martian hemispheric dichotomy may be due to a giant impact. *Nature*, 309:138–140, 1984.
133. W.-S. Yang and J. R. Baumgardner. A matrix-dependent transfer multigrid method for strongly variable viscosity infinite Prandtl number thermal convection. *Geophys. Astrophys. Fluid Dyn.*, 92:151–195, 2000.
134. Q. Z. Yin, S. B. Jacobsen, K. Yamashita, J. Blichert-Toft, P. Télouk, and F. Albarède. A short timescale for terrestrial planet formation from Hf-W chronometry of meteorites. *Nature*, 418:949–952, 2002.
135. C. F. Yoder, A. S. Konopliv, D. N. Yuan, E. M. Standish, and W. M. Folkner. Fluid core size of Mars from detection of the solar tide. *Science*, 300:299–303, 2003.
136. M. Yoshida and A. Kageyama. Low-degree mantle convection with strongly temperature- and depth-dependent viscosity in a three-dimensional spherical shell. *J. Geophys. Res.*, 111:B03412, 2006.
137. A. N. Youdin and F. H. Shu. Planetesimal formation by gravitational instability. *Astrophys. J.*, 580:494–505, 2002.
138. K. J. Zahnle, J. F. Kasting, and J. B. Pollack. Evolution of a steam atmosphere during Earth's accretion. *Icarus*, 74:62–97, 1988.
139. V. N. Zharkov and T. V. Gudkova. Construction of Martian interior model. *Solar System Research*, 39(5):343–373, 2005.
140. V. N. Zharkov, T. V. Gudkova, and S. M. Molodensky. On models of Mars' interior and amplitudes of forced nutations. 1. The effects of deviation of Mars from its equilibrium state on the flattening of the core-mantle boundary. *Phys. Earth Planet. Int.*, 172:324–334, 2009.
141. S. Zhong. Differential rotation of lithosphere for one-plate planets and its implications for the Tharsis Rise on Mars. *American Geophysical Union, Fall Meeting, abstract# P43C-1408*, 2008.
142. S. Zhong. Migration of Tharsis volcanism on Mars caused by differential rotation of the lithosphere. *Nature Geoscience*, 2:19–23, 2009.
143. S. Zhong and M. T. Zuber. Degree-1 mantle convection and the crustal dichotomy on Mars. *Earth Planet. Sci. Lett.*, 189:75–84, 2001.
144. M. T. Zuber, S. C. Solomon, R. J. Phillips, et al. Internal structure and early thermal evolution of Mars from Mars Global Surveyor topography and gravity. *Science*, 287:1788–1793, 2000.

Computational Considerations for Satellite-Based Geopotential Recovery

O. Baur and W. Keller

Universität Stuttgart, Institute of Geodesy, Geschwister-Scholl-Str. 24D, D-70174 Stuttgart
baur@gis.uni-stuttgart.de

Summary. This contribution addresses computational considerations in the framework of the main inverse problem in geodetic research, i.e., the determination of the terrestrial gravitational potential. The computational considerations comprise strategies for both the solution of the underlying systems of equations in order to resolve the model parameters and the estimation of the parameter variance-covariance information.

1 Introduction

On March 17, 2009, the GOCE (Gravity field and steady-state Ocean Circulation Explorer) satellite was lifted into a near-Sun-synchronous, low-altitude Earth orbit. The satellite mission will upgrade our present knowledge about the Earth's gravitational structure considerably. The mission's major concern is recovering the short-scale features of the terrestrial gravity field, the gravitational potential or geopotential respectively, with an anticipated geoid accuracy of about 2 cm, and a spatial resolution around 100 km [5].

The GOCE science data will complement the findings by the twin-satellite mission GRACE (Gravity Recovery And Climate Experiment), cf. Fig. 1. Since 2002 the GRACE spacecraft collect high-quality observations of the long- to medium-wavelength features of the geopotential [11]. The combined mission results stand for a dramatical improvement for understanding the complex system Earth which will strongly influence scientific progress in, e.g., geophysics, geodesy, oceanography and glaciology.

As a matter of fact, from the computational point of view, modern spaceborne geopotential recovery involves the solution of large-scale systems of equations with tens of millions of observations and tens to hundred of thousands of model parameters. Consequently, HPC (High Performance Computing) is part and parcel of modern satellite geodesy. Analysis techniques have to satisfy both quality and runtime issues. Hence, it is not advisable applying

W.E. Nagel et al. (eds.), *High Performance Computing in Science and Engineering '09*, DOI 10.1007/978-3-642-04665-0_35,

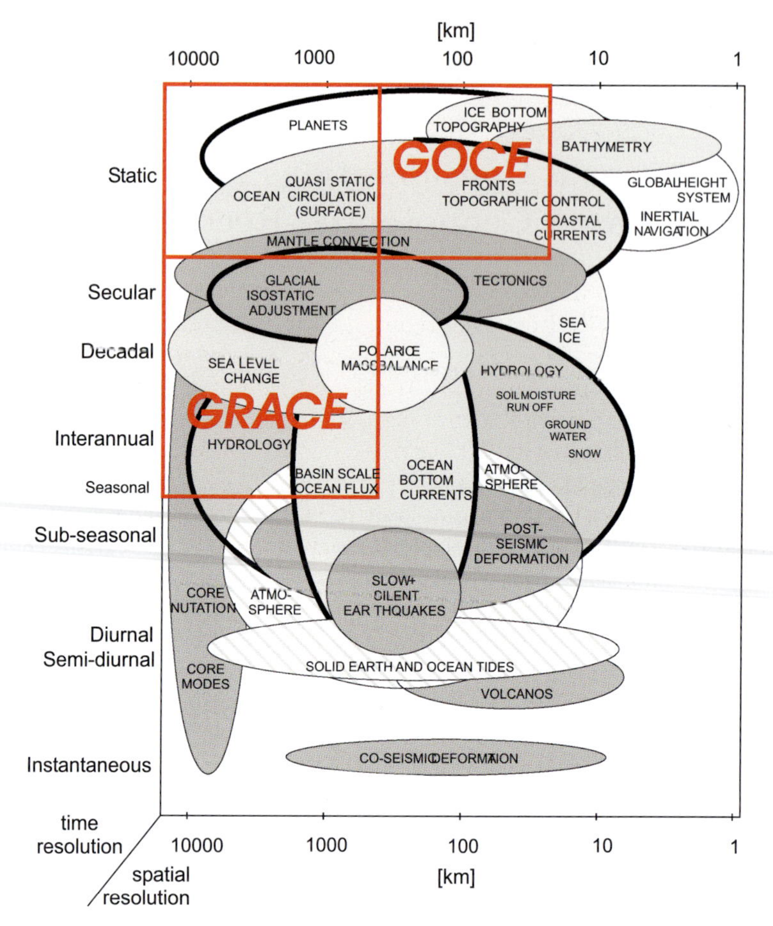

Fig. 1. GOCE and GRACE - complementary satellite missions (Credits: GOCE-Projektbüro Deutschland)

black box algorithms to geodetic problems. In fact, the challenge of tailored processing makes the adoption of HPC techniques highly demanding.

In this contribution, we highlight computational considerations in the context of the GOCE satellite mission. The next section presents the basic methodology for geopotential recovery. Section 3 briefly addresses the relevant problem dimensions. In Sect. 4 we outline two approaches to solve the resulting systems of equations. They are selected according to the HPC architectures typically available nowadays. Section 5 deals with variance-covariance information issues. Numerical examples are provided in Sect. 6. Finally, Sect. 7 summerizes the major conclusions of this contribution.

2 Generalized Methodology

The gravitational potential of the Earth is typically expressed in terms of a spherical harmonics series expansion [6] subject to

$$V(\lambda, \varphi, r) = \frac{GM}{r} \left[1 + \sum_{l=1}^{\infty} \sum_{m=-l}^{l} \left(\frac{R}{r} \right)^{l} e_{lm}(\lambda, \varphi) \bar{v}_{lm} \right]. \tag{1}$$

Polar spherical coordinates are denoted by λ (East longitude), φ (latitude) and r (radial distance from the origin). The base functions $e_{lm}(\lambda, \varphi)$ denote the 4π-normalized surface spherical harmonics. Furthermore, G indicates the Newtonian gravitational constant, M the body mass, and R the major semi-axis of a reference ellipsoid. The series coefficients $\bar{v}_{lm}$ are unknown parameters. The first series term in (1), $(^{GM}/_{r})$, specifies the gravitational potential of a homogeneous sphere. The further series terms describe the effect of geometrical deviations as well as irregularities in mass distribution of the real Earth from the idealized spherical body. It is the prime objective in satellite geodesy to determine the coefficients $\bar{v}_{lm}$ as accurate as possible. In this context, for practical applications the infinite series (1) is truncated at the maximal degree $L = l_{\max}$. The spectral resolution L depends on the sensitivity of satellite observations.

Both GOCE and GRACE do not observe the potential $V(\lambda, \varphi, r)$ directly but functionals of it. GOCE, in particular, measures so-called gravitational gradients, i.e., second partial derivatives of the geopotential. GRACE, on the other hand, detects the relative motion between the two spacecraft following each other in the same orbit. Consequently, (1) does not constitute the functional model for gravity field analysis. The observation equation for one particular measurement i may rather be written as

$$y(\lambda_i, \varphi_i, r_i) = f(\lambda_i, \varphi_i, r_i, \bar{v}_{lm}). \tag{2}$$

The unknown model parameters $\bar{v}_{lm}$ appear linearly in (2). Thus, the collectivity of observations yield a linear (overdetermined) system of equations. In standard Gauß-Markov model formulation it becomes [8]

$$\mathbf{A}\mathbf{x} = \mathbf{y} + \mathbf{r}, \tag{3}$$

$$E\{\mathbf{y}\} = \mathbf{A}\mathbf{x}, \quad E\{\mathbf{r}\mathbf{r}^T\} = D\{\mathbf{y}\} = \mathbf{\Sigma} = \sigma^2 \mathbf{P}^{-1}. \tag{4}$$

Therein, the design matrix $\mathbf{A}(n \times u)$ relates the unknown parameter vector $\mathbf{x}(u \times 1)$ to the vector of observations $\mathbf{y}(n \times 1)$. Thus, $\mathbf{A}$ assembles the partial derivatives of the functional model with respect to the parameters to be resolved. For an overdetermined system $n > u$ holds true, i.e., the right hand side of (3) is the sum of both the observation vector and residual vector $\mathbf{r}(n \times 1)$. The observation variance-covariance matrix $D\{\mathbf{y}\} = \mathbf{\Sigma}$ is the product of the variance of unit weight σ^2 and the inverse (positive definite) weight matrix $\mathbf{P}$.

For the sake of simplicity, we exclusively consider uncorrelated and equally accurate observations. In this case $D\{\mathbf{y}\} = \sigma^2\mathbf{I}$ holds true. L_2-norm minimization of the residual vector yields the target function

$$\min \|\mathbf{r}\|^2 = \min_{\mathbf{x}} \|\mathbf{A}\mathbf{x} - \mathbf{y}\|^2. \tag{5}$$

It is worthwhile to mention that in terms of a "whitening process", referred to as decorrelation, correlated systems can be reduced to formulation (5) by appropriate decorrelation filters [10].

3 Problem Dimension

Typically, data analysis in satellite-based geopotential recovery results in large-scale problems. The GOCE mission, for example, is expected to provide $n \approx 90$ million observations to resolve for $u \approx 60\,000$ unknown geopotential parameters. As outlined in Table 1, the number of unknown parameters increases roughly quadratically with increasing spectral resolution L.

Table 1. Relation between resolution and number of geopotential parameters

Spectral resolution L	Spatial resolution (km)	# geopotential parameters
50	400	2598
100	200	10 198
150	133	22 798
200	100	40 398
250	80	62 998
300	67	90 598

The problem dimensions according to Table 1 are accompanied with a huge challenge with regard to both runtime minimization for solving the target function (5) and the memory requirements involved. The peak problem dimension for future gravity field missions may even exceed the numbers considered so far. The huge amount of measurements provided by present and upcoming satellite sensors, along with high-resolution geopotential modeling can only be tackled by adopting HPC technologies.

4 Strategies for Solving the Target Function

Typically, least-squares (LS) techniques are applied to solve the target function (5). The choice of an adequate LS solver depends on both quality and computational issues. The former include regularization of ill-posed problems and the validity of the variance-covariance information on the parameter estimates. Computational issues, on the other hand, deal with efficient parallel implementation and tailored preconditioning.

Basically, two groups of LS solvers are widely adopted in geopotential research, the "brute-force" normal equation system inversion approach and a variety of Krylov-space iterative methods such as the LSQR (LS using QR decomposition) algorithm [4]. According to Table 2, both solvers have pros and cons from the computational and methodological point of view. Hence, it is worth to follow both strategies.

Table 2. Pros and cons of the "brute-force" method versus iterative LS solvers

"Brute-force" solver	Iterative solvers
• Normal matrix computation very time consuming • Huge core memory requirements • Efficient parallelization restricted to shared memory/ccNUMA systems • Straightforward estimation of the variance-covariance information	• Avoids normal matrix assembly • Well manageable core memory requirements • Independent of parallel computation platform; in general portable • Approximate estimation of the variance-covariance information

Memory considerations have led iterative LS solvers to become very popular. They avoid the explicit normal matrix assembly in favor of the row-wise processing of the design matrix. Thus, iterative solvers are characterized by well manageable core memory requirements. This, in turn, implies the independence of parallel computation platforms, i.e., they are suited well for both shared and distributed memory (DM) systems. Unfortunately, iterative solvers do not naturally provide the variance-covariance information of the parameter estimate. This is the major drawback opposed to the "brute-force" approach, cf. Sect. 5. The parallel implementation of both the "brute-force" solver and the LSQR algorithm is addressed in detail in [2, 3] and [4].

5 Strategies for Variance-Covariance Estimation

Besides the adjusted geopotential parameters, $\hat{\mathbf{x}}$, themselves, its variance-covariance matrix (VCM), $D(\hat{\mathbf{x}})$, is of utmost importance for the further use of the LS result in terms of error propagation. The "brute-force" solver provides the exact covariance information subject to

$$D(\hat{\mathbf{x}}) = \hat{\sigma}^2(\mathbf{A}^T\mathbf{A})^{-1}, \tag{6}$$

$$\hat{\sigma}^2 = \frac{\hat{\mathbf{r}}^T\hat{\mathbf{r}}}{n-u}\ , \ \hat{\mathbf{r}} = [\mathbf{A}(\mathbf{A}^T\mathbf{A})^{-1}\mathbf{A}^T - \mathbf{I}]\mathbf{y}. \tag{7}$$

Unfortunately, however, normal matrix assembly in (6), $\mathbf{N} = \mathbf{A}^T\mathbf{A}$, is very time consuming, and additionally requires huge core memory (at least one triangle of the symmetric matrix has to be stored), cf. Table 3.

Table 3. Memory requirement for normal matrix storage

Spectral resolution L	Memory requirement for $\mathbf{N}$ (MByte)
50	27
100	416
200	6500
300	33 000

In the context of the approximate computation of $D(\hat{\mathbf{x}})$, a lot of effort has been undertaken in the last few years. Monte Carlo (MC) integration techniques have been claimed to be efficient in order to provide a high-quality estimate $\hat{D}(\hat{\mathbf{x}})$ in the context of iterative LS solvers (e.g., [1]). MC integration is referred to as the approximate evaluation of integrals by random samples [7]. As outlined in [1], the VCM of the LS estimate $\hat{\mathbf{x}}$ turns out to be the mean value of M dyadic products subject to

$$\hat{D}(\hat{\mathbf{x}}) = \frac{1}{M} \sum_{i=1}^{M} \mathbf{s}_{\Delta\mathbf{x}}^{(i)} \mathbf{s}_{\Delta\mathbf{x}}^{(i)\,T} . \tag{8}$$

Each individual sample $\mathbf{s}_{\Delta\mathbf{x}}^{(i)}$, $i = 1, ..., M$ is derived from the solution of the linear system

$$\mathbf{A}\mathbf{s}_{\Delta\mathbf{x}}^{(i)} = \mathbf{s}_{\mathbf{r}}^{(i)} + \mathbf{r}^{(i)}. \tag{9}$$

The initial samples $\mathbf{s}_{\mathbf{r}}^{(i)}$ are standard normally distributed. Merging (3) and (9) means that the LS solver has to be applied to $M + 1$ right hand sides, namely $\mathbf{y}, \mathbf{s}_{\mathbf{r}}^{(1)}, ..., \mathbf{s}_{\mathbf{r}}^{(M)}$, to compute $M{+}1$ estimates for the unknown vectors $\mathbf{x}$, $\mathbf{s}_{\Delta\mathbf{x}}^{(1)}, ..., \mathbf{s}_{\Delta\mathbf{x}}^{(M)}$. The latter M estimates are used to evaluate the VCM according to (8).

6 Numerical Results

This section presents the outcome of a series of closed-loop simulation studies. The test data set consists of synthetic GOCE-like gravitational gradients. The synthetic data is generated using the EGM96 [9] gravity field model with spectral resolution $L = 300$. We analyze the data up to $L = 200$, yielding $u = 40\,398$ unknown geopotential parameters. The number of observations is $n = 518\,400$ (corresponding to an observation period of 30 days with a 0.2 Hz sampling rate).

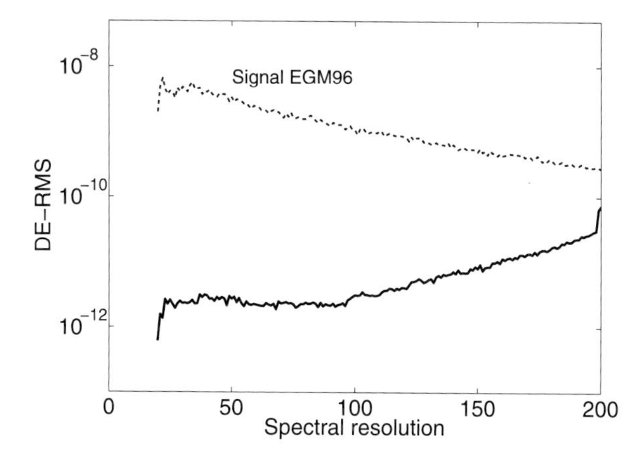

Fig. 2. "Brute-force" solver: Closed-loop simulation error in terms of degree-error RMS values, computed by $\text{DE-RMS}_l = [\sum_{m=-l}^{l} (\bar{v}_{lm}^{\text{EGM96}} - \hat{\bar{v}}_{lm})^2/(2l+1)]^{1/2}$. Synthesis: Spectral resolution $L = 300$, Analysis: Spectral resolution $L = 200$

6.1 "Brute-Force" Solver

Due to the huge memory requirements for the storage of the normal equations matrix, we implemented the "brute-force" approach for its operation on shared memory and ccNUMA (cache-coherent Non-Uniform Memory Access) systems. OpenMP is used for parallelization. The detailed parallel implementation of the method is presented in [4].

Figure 2 displays the error of the closed-loop simulation in terms of degree-error RMS (Root Mean Square) values of the coefficient estimate, $\hat{\bar{v}}_{lm}$. As the analysis of the gravitational signal is restricted to $L = 200$ (compared to $L = 300$ for data synthesis), spectral leakage effects occur. They mainly map into the estimate of high-degree coefficients. Besides spectral leakage effects, the estimate of geopotential parameters is solely affected by numerical rounding errors.

Table 4 summarizes the achieved runtime results, using up to 16 CPUs in parallel. Ranging from 80% to 93.8% of the total wall time, the computation of the normal matrix is by far the most time-consuming part of the algorithm. The computational effort for the design matrix assembly is less than 1%. Speed-up and efficiency results are satisfactory, taking into account that the parallel efficiency strongly depends on the problem dimension. For larger problems (increasing number of observations), the scaling results are expected to improve accordingly.

6.2 LSQR Solver

As a matter of fact, iterative solvers avoid the explicit normal equations matrix assembly in favor of a row-wise processing scheme of the design matrix. Thus,

Table 4. Performance of the "brute-force" approach, $n = 518\,400$ observations, $u = 40\,398$ geopotential parameters

# CPUs	1	4	8	16
Total wall time (h)	64.46	17.06	9.06	5.57
Design matrix **A** assembly (min)	31.68	8.10	4.32	2.77
Portion of total wall time (%)	0.8	0.8	0.8	0.8
Calculation of **N** (h)	61.43	15.71	8.05	4.46
Portion of total wall time (%)	93.8	92.1	88.8	80.0
Normal equations sys. inv. (min)	59.38	34.67	37.08	53.35
Portion of total wall time (%)	1.5	3.4	6.8	15.6
Speed-up	1	3.84	7.22	11.76
Efficiency (%)	100	96	90	73

they are characterized by well manageable core memory requirements. This, in turn, implies the independence of parallel computation platforms; i.e., they are suited well for both shared and distributed memory systems. We implemented the LSQR method for its operation on distributed memory systems. MPI (Message Passing Interface) is used for parallelization. The detailed parallel implementation of the method is presented in [4].

The overall computational performance of the iterative LSQR solver mainly depends on the number of iterations necessary to sufficiently approximate the exact LS solution. In order to improve the convergence behavior of the method, we apply preconditioning as outlined in [3]. In terms of quality assessment, the result obtained by the "brute-force" approach can deal as baseline accuracy. Figure 3 displays the degree-error RMS values of the closed-loop simulation using LSQR. Following this representation, from the

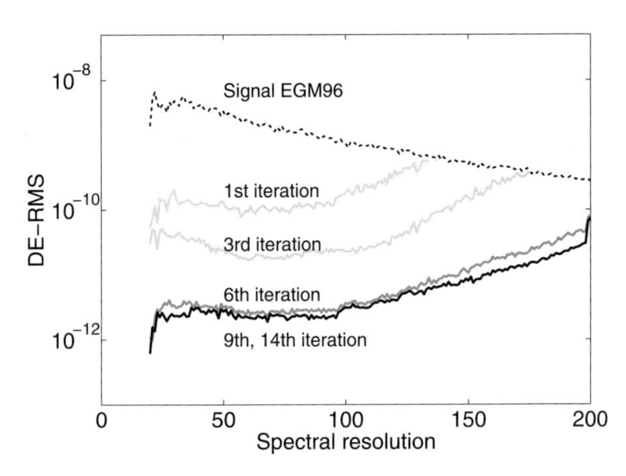

Fig. 3. LSQR solver: Closed-loop simulation error in terms of degree-error RMS values, computed by $\text{DE-RMS}_l = [\sum_{m=-l}^{l} \left(\bar{v}_{lm}^{\text{EGM96}} - \hat{\bar{v}}_{lm}\right)^2/(2l+1)]^{1/2}$. Synthesis: Spectral resolution $L = 300$, Analysis: Spectral resolution $L = 200$

ninth iteration on, no differences in the error curves become visible. Most notably, the bottom graph in Fig. 3 coincides with the baseline accuracy given in Fig. 2, proving the exact LS result being approximated very well within only a few iterations.

Table 5 highlights runtime results, dependent on the number of processing units. The speed-up is mostly close to the number of CPUs used. This demonstrates the efficient and powerful parallelization of the implementation. Again, it has to be emphasized that the parallel efficiency strongly depends on the problem dimension. Especially addressing 64 CPUs in parallel, the scaling result is expected to improve for larger problems.

Table 5. Performance of the (preconditioned) LSQR method, $n = 518\,400$ observations, $u = 40\,398$ geopotential parameters

# CPUs	# Iterations	Total wall time (h)	Speed-up	Efficiency (%)
1	14	27.65	1	100
8	14	3.70	7.5	93
16	14	2.11	13.1	82
32	14	1.20	23.0	72
64	14	0.78	35.4	55

Variance-Covariance Estimation

Adaption of MC integration according to Sect. 5 in a parallel environment requires fundamental implementational engagement. Basically, the LSQR solver has to be expanded in order to treat $M+1$ LS problems in parallel. This can either be done by distributing the $M+1$ problems on p CPUs — i.e., each CPU independently works on $(M+1)/p$ problems — or alternatively, the LS problems are successively solved by all CPUs. The former approach is much more effective with regard to memory considerations, as each CPU only addresses $(M+1)/p$ observation vectors, solution vectors respectively. On the other hand, the latter strategy is more efficient from the runtime point of view, as design matrix assembly has to be performed only once.

Figure 4 presents results applying MC integration for the approximate VCM estimation. The patterns are based on $M = 800$ random samples. We display the results in terms of relative errors

$$\epsilon_{ij} = \left| \frac{N^{-1}_{ij,\,\mathrm{approx}} - N^{-1}_{ij,\,\mathrm{exact}}}{N^{-1}_{ij,\,\mathrm{exact}}} \right| . \tag{10}$$

Therein, $\mathbf{N}^{-1}_{\mathrm{approx}}$ denotes the approximate and $\mathbf{N}^{-1}_{\mathrm{exact}}$ the exact inverse normal equations matrix, $\mathbf{N}^{-1} = (\mathbf{A}^T\mathbf{A})^{-1}$. The accuracy of the MC estimator

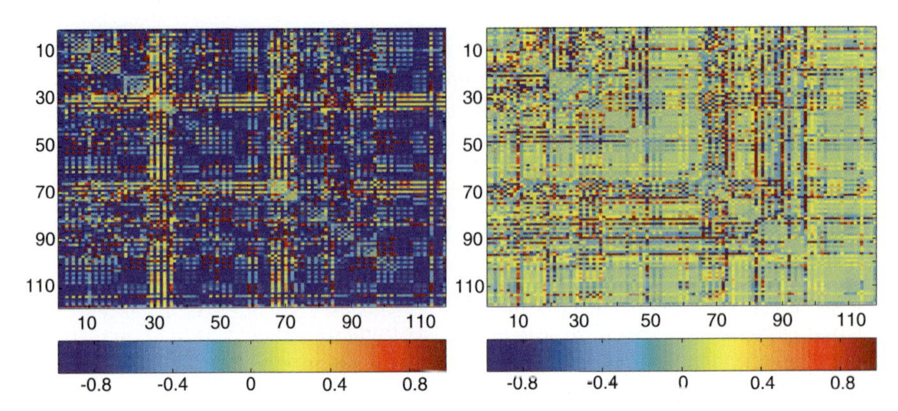

Fig. 4. Variance-covariance matrix: Relative errors ϵ_{ij} applying the MC integration method (800 samples); left: $k = 30$ iterations; right: $k = 60$ iterations

mainly depends on the number of samples used. For $M \ll \infty$ samples the MC method will never be able to reconstruct the exact variance-covariance matrix. For $M \to \infty$, $\hat{D}(\hat{\mathbf{x}}) \to D(\hat{\mathbf{x}})$ holds true. On the other hand, the numerical efficiency of the method improves with decreasing number of samples. As shown in [1], $M = 800$ samples should guarantee one significant digit of the estimated variances (main diagonal elements of the variance-covariance matrix). Fixing two significant digits would require 80 000 samples. Hence, the method becomes extremely costly from that point of view.

7 Conclusions

Satellite-based geopotential recovery requires solving large-scale (linear) systems of equations, adopting tailored algorithms. In particular, we address the "brute-force" normal equations system inversion approach and the iterative LSQR method to solve the related LS minimization problem. To a large extent, iterative solvers are not affected by restricted memory availability. Matrix-matrix and matrix-vector multiplications are avoided by means of repeated vector-vector operations. Consequently, the LSQR method is suited best for cluster systems, hence we run the algorithm on the NEC Linux Cluster (Cacau). We address the platform either massively parallel or as hybrid system. "Brute-force" approach implementation is so far realized addressing one node of the NEC SX-8 architecture, i.e., we consider the system as a shared memory platform using OpenMP for communication. With regard to hybrid parallelization we plan to extend the software accordingly.

Acknowledgement. The authors thank the High Performance Computing Center Stuttgart (HLRS) for the opportunity to use their computing facilities, most notably for the helpful technical support.

References

1. H. Alkhatib, W.-D. Schuh, (2007) Integration of the Monte Carlo covariance estimation strategy into tailored solution procedures for large-scale least squares problems, J. Geod. 81: 53–66, doi:10.1007/s00190-006-0034-z
2. O. Baur, G. Austen, W. Keller, (2006) Efficient satellite based geopotential recovery, in: W.E. Nagel, W. Jäger, M. Resch (Eds.): High Performance Computing in Science and Engineering '06, 499–514, Springer Berlin Heidelberg New York
3. O. Baur, G. Austen, J. Kusche, (2008) Efficient GOCE satellite gravity field recovery based on least-squares using QR decomposition, J. Geod. 82: 207–221, doi:10.1007/s00190-007-0171-z
4. O. Baur, (2009) Tailored least-squares solvers implementation for high-performance gravity field research, Computers and Geosciences 35: 548–556, doi:10.1016/j.cageo.2008.09.004
5. ESA SP-1233 (1999) The four candidate Earth explorer core missions - gravity field and steady-state ocean circulation mission, European Space Agency Report SP-1233(1), Granada
6. W.A. Heiskanen, H. Moritz, (1967) Physical Geodesy, W.H. Freeman and Company San Francisco
7. M. Kalos, P. Whitlock, (1986) Monte Carlo methods. Vol I: basics, Wiley New York
8. K.-R. Koch, (1999) Parameter estimation and hypothesis testing in linear models, Springer Berlin Heidelberg New York
9. F.G. Lemoine, S.C. Kenyon, J.K. Factor, R.G. Trimmer, N.K. Pavlis, D.S. Chinn, C.M. Cox, S.M. Klosko, S.B. Luthcke, M.H. Torrence, Y.M. Wang, R.G. Williamson, E.C. Pavlis, R.H. Rapp, T.R. Olson, (1998) The Development of the Joint NASA GSFC and NIMA Geopotential Model EGM96, NASA Goddard Space Flight Center, Greenbelt, 575pp
10. W.-D. Schuh, (1996) Tailored numerical solutions strategies for the global determination of the Earth's gravity field, Mitteilungen der Universität Graz 81
11. B. Tapley, J. Ries, S. Bettadpur, D. Chambers, M. Cheng, F. Condi, B. Gunter, Z. Kang, P. Nagel, R. Pastor, S. Poole, F. Wang, (2005) GGM2 – an improved Earth gravity field model from GRACE, J. Geod. 79: 467–478, doi:10.1007/s00190-005-0480-z

Simulative Analysis of Vehicle-to-X Communication considering Traffic Safety and Efficiency

O. Jetter, M. Killat, J. Mittag, F. Schmidt-Eisenlohr, J. Dinger, and H. Hartenstein

Steinbuch Centre for Computing (SCC), Karlsruhe Institute of Technology (KIT) and Institute of Telematics, Universität Karlsruhe (TH)
D-76128 Karlsruhe, Germany
hannes.hartenstein@kit.edu

Summary. Over the past several years, there has been significant interest and progress in using wireless communication technologies for vehicular environments in order to increase traffic safety and efficiency. Due to the fact that these systems are still under development and large-scale tests based on real hardware are difficult to manage, simulations are a widely-used and cost-efficient method to explore such scenarios. Furthermore, simulations provide a possibility to look at specific aspects individually and to identify major influencing effects out of a wide range of configurations. In this context, we use the HP XC4000 for an extensive and detailed sensitivity analysis in order to evaluate the robustness and performance of communication protocols as well as to capture the complex characteristics of such systems in terms of an empirical model.

1 Introduction

In recent years interest increased in improving traffic systems by enabling wireless communication between vehicles directly as well as between vehicles and infrastructure. The effort is supported by the strong interest of car manufacturers to further improve safety functionality of their products as well as by public authorities that see such technology as part of a solution to solve or reduce traffic efficiency problems. The relevance can also be seen by the reservation of a frequency band for such intelligent transportation systems (ITS) both in the U.S. and in Europe. Furthermore, this vision is fostered by the technical possibilities to build up such systems in a cost-efficient way, mainly due to the availability of low-priced communication hardware.

The introduction and operation of such wireless networks raises various research questions. Particular interest relates to scalability regarding the number of communicating vehicles with respect to the limited resources on the

W.E. Nagel et al. (eds.), *High Performance Computing in Science and Engineering '09*, DOI 10.1007/978-3-642-04665-0_36,

radio channel. Communication protocols have to be designed in a way that the network could fulfill the application requirements in all stages, i.e., from the introduction where a small number of vehicles is equipped, up to the situation where basically every vehicle has communication capabilities. In order to assess different system designs and their impact, real world experiments and measurements can be conducted in testbeds or field operational tests. However, large-scale field tests are still very expensive and thus can only be realized in a restricted way. Hence, simulations are essential and the preferred approach to study the performance of these communication systems and to design appropriate protocols. To take a wide range of possible situations into account, simulations have to be run with a significant number of varying parameters, e.g. traffic density or transmission power. We therefore make use of the high performance computation resources of the HP XC4000 to cover all these variations and follow discrete event simulation methodologies, the way communication simulations typically are performed.

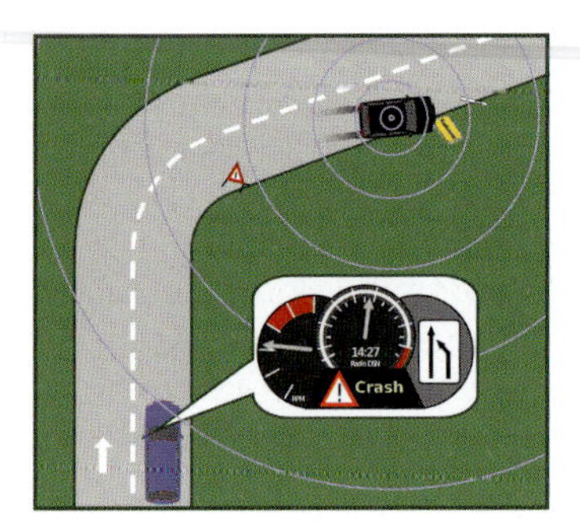

Fig. 1. Vehicle-to-Vehicle communication in a traffic safety scenario

In the envisioned Vehicle-to-Vehicle and Vehicle-to-Infrastructure – in short Vehicle-to-X (V2X)[1] – communication systems the vehicles are equipped with communication hardware; current efforts want to specifically adapt IEEE 802.11, a standard that is widely used for wireless computer networks. Planned applications can be found in the domains of traffic safety, traffic efficiency and business/infotainment. Figure 1 shows a possible traffic safety application: equipped vehicles periodically exchange information about their status, like position speed, driving direction, and acceleration. This information is spread as broadcast to vehicles in the surrounding. Each vehicle can then create a virtual map and thereby identify the current traffic situation. In critical situations, the driver can be informed or warned. V2X communication also allows to inform vehicles at further distances of upcoming dangers or critical situations, like an accident. Such information may be geographically forwarded over multiple hops. Regarding traffic efficiency, V2X communica-

[1] also known as Car-to-X (C2X) or Vehicular Ad Hoc Network (VANET)

tion can be used to dynamically collect information on current traffic situations, inform the drivers individually and thus better react on the dynamics of the traffic system. The dynamic signaling of alternative faster routes or an optimal speed advisory also considering energy efficiency would be promising applications.

In this paper we focus on the simulative investigation of traffic safety and efficiency scenarios. In Section 2 we introduce the simulation methodology that we use. In Section 3 we discuss simulation studies for traffic safety applications, while traffic efficiency applications are studied in Section 4. We conclude our paper in Section 5.

2 V2X Simulations

Simulations of wireless networks require accurate models, in particular realistic wave propagation models, to determine which vehicles either receive a message successfully, or not. For this study, we use the discrete event-based Network Simulator 2 (ns-2) [9] in version 2.33. This implementation contains substantial IEEE 802.11 model improvements [3], which we developed in collaboration with Mercedes-Benz Research & Development North America.

In a discrete event-based simulator the number of events is the dominating factor for the total simulation time and computational costs, respectively. The number of events within V2X network scenarios depends on the amount of vehicles sharing the radio channel. Basically, the number of events generated by a single transmission is linear with the number of vehicles participating in the network — one event for the transmission and an additional event for each vehicle that may receive this message. To account for all vehicles in a scenario exchanging periodic messages between each other, the number of events grows quadratically with the number of vehicles. The impact can be illustrated by the number of scheduled events for a scenario with 200 vehicles compared to a scenario with 1,000 vehicles. A rough calculation for a transmission rate of 5 Hz and a simulation time of 500 seconds results in about 6 million events for 200 vehicles compared to approximately 150 million events in case of 1,000 vehicles. Actually, the absolute number of events in a simulator is much higher due to protocol-related mechanisms. Thus, large-scale scenarios with thousands of vehicles demand a substantial amount of resources.

To improve communication protocols, sensitivity studies on the parameters are needed. The varying parameterization is a property which is very well suited for parallelization as each simulation scenario with a specific parameter set can be run independently from all other scenarios. Without the use of massive parallelization, the evaluation of different parameters would require months, sometimes even years. We used the HP XC4000 to simulate inter-vehicle communication over a wide spectrum of different scenarios, as we will explain in the following sections.

3 Traffic Safety

When considering safety-related communication, two types of messages can be identified: *periodic* and *event-driven*. By a periodic exchange of status messages – called *beacons* – which contain a vehicle's position, speed etc., a cooperative awareness between neighboring vehicles is established, which in turn will be used by safety applications to detect potentially dangerous situations, e.g. the approach of a traffic jam or a broken vehicle. Additionally, in case of the detection of a dangerous situation or an abnormal condition with respect to the environment, e.g. an exploding airbag or an icy road, vehicles are assumed to generate event-driven messages – *emergency messages* – in order to inform and warn surrounding vehicles.

While, from a safety perspective, a key challenge for direct vehicle-to-vehicle communication technologies in the market introduction phase will be to achieve a significant penetration rate of equipped vehicles, it will be challenged even more deeply in fully deployed, high density vehicular scenarios, due to the high data load on the radio channel solely caused by beaconing. With the distributed coordination of the shared wireless channel, i.e., Carrier Sense Multiple Access (CSMA), a high load on the channel is likely to result in an increased amount of packet collisions and, consequently, in a decreased 'safety level' as seen by the active safety application. Specifically, beacon messages will not be successfully decoded even when sent by a nearby vehicle and event-driven messages will show a slow and unreliable dissemination process. To tackle the issue of channel saturation, we proposed to control the generated load due to beaconing by applying a per-packet transmit power control. Thereby, the spatial reuse of the shared wireless channel and, implicitly, the ability to capture packets is increased.

Recently, we proposed a distributed transmission power control strategy called D-FPAV (Distributed Fair Power Adjustment for Vehicular environments) that controls the beaconing load under a strict fairness criterion, which has to be met for safety reasons [15]. We also developed a fast and effective, contention-based dissemination strategy for event-driven messages called EMDV (Emergency Message Dissemination for Vehicular environments) in order to deliver emergency messages to vehicles within a targeted geographical area [12]. However, an evaluation of the performance of D-FPAV and EMDV over a wide range of scenarios and parameterizations induces significantly high computational costs. The resources provided by the HP XC4000 enabled us to conduct these studies.

In this section, we will evaluate the performance of D-FPAV and EMDV over a wide range of scenarios. Due to space restrictions, we focus on the impact of different vehicle densities and different channel fading intensities and refer to [13] for a discussion of different beacon load limits and different configurations of the EMVD and D-FPAV protocol. In the following, we give a short description of the two protocols and present the obtained simulation results.

3.1 D-FPAV and EMDV

The design goals and characteristics of D-FPAV are as follows: (i) it is fully distributed and able to quickly react to the dynamic topologies of vehicular networks; (ii) it controls the beaconing load under a strict fairness criterion that has to be met for safety reasons. Strict fairness must be guaranteed since it is very important that all and not only individual vehicles have a good estimation of the state of all vehicles in their close surrounding. More specifically, a higher transmit power should not be selected at the expense of preventing other vehicles to send/receive their required amount of safety information; (iii) D-FPAV allows a clear prioritization of event-driven over periodic messages, since the transmission power of event-driven messages is not reduced.

The objective of the D-FPAV strategy is to distributively determine a power assignment for all vehicles such that the minimum of the transmit powers used for beaconing is maximized and the network load (or bandwidth consumption) experienced at each vehicle remains below a predefined threshold called MBL (Maximum Beaconing Load). By limiting the load to a MBL level below the limit of the communication channel, e.g. setting the MBL to 5.0 Mbps in a 6 Mbps channel, it is possible to reserve a portion of the bandwidth for emergency or other messages. To reach this objective, every vehicle runs FPAV (see [14]), a localized algorithm based on a 'water filling' approach as proposed in [1], to calculate the maximum common transmit power level which should be used by neighboring vehicles. For an in-depth description of the protocol and algorithm we refer to [11] and [15].

The dissemination strategy EMDV is basically a position-based forwarding approach, in which the message should generally be forwarded by a preselected optimal relay. However, in case of packet collisions and reception failure at the relay, a contention-based scheme, as proposed by [2], is used to increase the reliability of the dissemination. For a detailed description, we refer to [11] and [12].

3.2 Simulation Results

In our simulations, we assumed a 10 MHz channel and a data transmission rate of 6 Mbps which has been shown to be most suitable for safety related applications [4]. All nodes were configured to communicate according to the designated V2X communication draft standard IEEE 802.11p and to periodically transmit 400 byte packets with a beaconing rate of 10 Hz. Regarding the radio wave propagation, we assumed the probabilistic Nakagami-m model and chose either severe or moderate channel fading conditions, represented by a fading parameter of $m \in \{1, 3\}$. Table 1 provides a detailed overview of the simulation configuration. To account for different vehicle densities we used a 7 km, straight 3 lane per direction highway scenario, in which each lane is occupied by 16 vehicles/km or 20 vehicles/km.

Table 1. Simulation configuration parameters

Parameter	Value
Radio propagation model	Nakagami-m, $m \in \{1, 3\}$
IEEE 802.11p data rate	6 Mbps
Channel bandwidth	10 MHz
Preamble length	32 µs
PLCP header length	8 µs
Symbol duration	8 µs
Noise floor	-99 dBm
Carrier sense threshold	-94 dBm
SINR for preamble capture	5 dB
SINR for frame body capture	10 dB
Minimum contention window	15
Slot time	13 µs
SIFS time	32 µs
Packet size	400 bytes

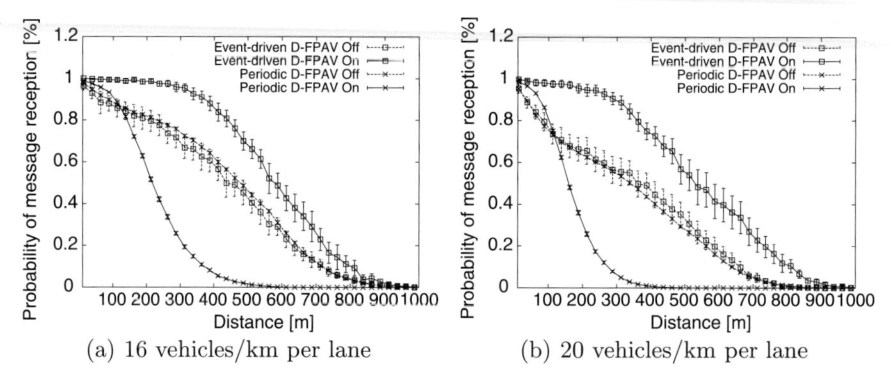

(a) 16 vehicles/km per lane (b) 20 vehicles/km per lane

Fig. 2. Probability of successful reception of periodic beacons and one-hop event-driven messages with respect to the distance to the transmitter with D-FPAV On/Off, MBL = 5.0 Mbps and vehicle densities of 16 vehicles/km per lane and 20 vehicles/km per lane

Figure 2 illustrates the impact of D-FPAV on the reception of periodic beacon and event-driven emergency messages when simulating the 7 km long and straight highway scenarios. All vehicles were configured to transmit beacon messages periodically. Additionally, one vehicle located in the middle of the scenario was selected to start the dissemination of an event-driven message, targeting all vehicles 2 km behind itself, once every second. As can be seen, D-FPAV with a MBL threshold of 5 Mbps reaches its design goals in both scenarios and provides, compared to the case without D-FPAV, higher reception probabilities for beacon messages at close distances from the sender (at the price of lower reception probabilities at further distances) and an increased reception probability for event-driven messages at all distances from

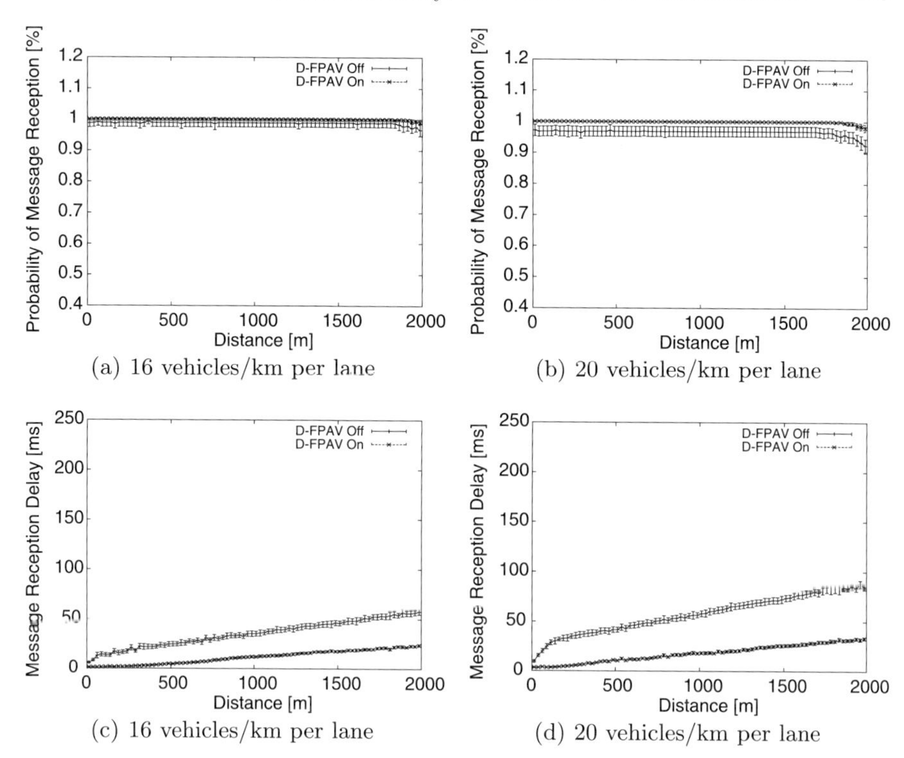

(a) 16 vehicles/km per lane

(b) 20 vehicles/km per lane

(c) 16 vehicles/km per lane

(d) 20 vehicles/km per lane

Fig. 3. Probability of message reception and reception delay with respect to the distance from the message originator (with multi-hop retransmissions) with D-FPAV On/Off, MBL = 5 Mbps and vehicle densities of 16 vehicles/km per lane and 20 vehicles/km per lane

the originator. However, the benefit of using D-FPAV in the 16 vehicles/km scenario is is not as significant as in the 20 vehicles/km scenario. We will come back to this observation later in this section.

The effect of prioritization of event-driven emergency messages over beacon messages is further illustrated in Figure 3. Combined with D-FPAV, we can see that the reliability of emergency message dissemination can be increased from 98.3%, or 96.5%, up to 100% (see Figure 3(a) and 3(b)). What initially looks like a marginal gain turns out to be quite significant: without D-FPAV, it can happen that the first and initial emergency message from the originator is not successfully received by anyone at all — critical in a safety of life situation. Apart from the reliability, the effectiveness and the speed of disseminations, in terms of average number of retransmissions and dissemination delay, are also benefiting when using D-FPAV: on the one hand, the average number of retransmissions required to cover the destination area (all vehicles located 2 km behind the originator) is reduced from 18.05 to 9.96 in the 20 vehicles/km scenario and from 11.18 to 8.48 in the 16 vehicles/km

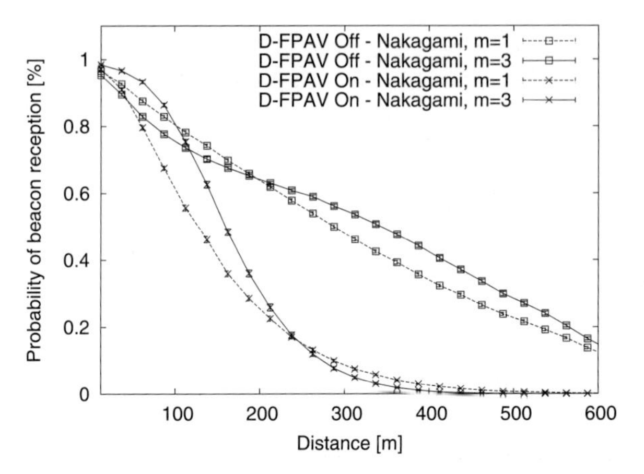

Fig. 4. Probability of successful beacon reception with respect to the distance to the transmitter for D-FPAV On and Off and different fading intensities (Nakagami $m = 1$ reflecting severe and Nakagami $m = 3$ reflecting medium fading conditions)

scenario. On the other hand, D-FPAV reduces the average reception delay with respect to the distance to the originator by nearly 50% (compare in Figure 3(c) and 3(d)). Also regarding the worst case reception delay observed in all simulations, D-FPAV achieves a reduction from 294 ms to 204 ms in the 20 vehicles/km scenario or, for the 16 vehicles/km scenario, a reduction from 248 ms to 167 ms. As before, we can notice that the benefit of using D-FPAV (in combination with EMDV) is more significant if the vehicle density is very high.

Apart from the importance of the vehicle density in the simulated scenario, we discovered that the intensity of the channel fading plays also a dominating factor with respect to protocol performance in general and possible performance gains due to the usage of D-FPAV in particular. The observation is shown by Figure 4, where the probability of successful beacon reception is illustrated when simulating a scenario with either D-FPAV enabled or disabled and applying different channel fading conditions. The results show, that transmit power control provides an increased reception probability at close distances if channel fading is assumed to be non-severe (Nakagmi $m = 3$) and no significant improvement in case of a severe channel fading (Nakagami $m = 1$). The observed behavior of the probability of beacon reception without the use of D-FPAV can be explained as follows: due to the greater variation of the received signal strengths in severe conditions, which causes a reduced number of interferences from far distances, the chance to successfully decode a message increases for close distances and decreases for far distances. Since D-FPAV has already reduced the interference from far distances by adjusting the transmit powers, a greater variation of the received signal strengths does not improve the performance of beacons from close distances and rather reduces the chance to successfully decode a message. Indeed, interference and resulting collisions are not the dominating factor anymore. In addition, the probability of reception at close distances under severe fading conditions shows almost the same performance for D-FPAV On and Off. Obviously, an intense fading (or pathloss) in signal propagation has the same effect as a transmit power adjustment and provides comparable results with respect to the probability of successful message reception. Severe fading can therefore be seen as nature's approach of how to perform congestion control in VANETs.

4 Traffic Efficiency

Considering traffic efficiency it has already been pointed out by Kawashima in 1973 that a communication network spanned over vehicles and road side infrastructure is a suitable and effective tool for a traffic control center [6]. Road operators would obtain the possibility to mitigate stressed traffic conditions by advising vehicles to appropriately adapt their driving behavior for an increased traffic throughput. An assessment whether taken measures bring the expected effect thus needs to be carried out on an evaluation unit scaled up from single vehicles to traffic flows.

As explained in Section 2, computer simulations are primarily used for evaluation purposes. In the past, advanced simulation models have been proposed which accurately determine which vehicles possess which information at which point in time. However, these models come with high computational requirements already exceeding real time computations for small numbers of communicating nodes. In this light, traffic flows which typically comprise a multitude of vehicles present a huge challenge for a traffic control center which depends on 'in time' results to choose right measures.

A remedy might be found by applying the concept of hybrid simulations. Schwetman, for example, respectively noticed in 1978 [10]:

> "By combining, in a hybrid model, discrete-event simulation and mathematical modeling, we are able to achieve a high level of agreement with the results of an equivalent simulation-only model, at a significant reduction in computational costs."

Regarding simulation studies of VANETs on vehicular traffic efficiency, a hybrid approach would offer to represent the communication system by a mathematical model. However, despite of many attempts to analytically describe communication performances in wireless communication networks according to the IEEE 802.11 standard no model is yet available which considers all fundamental influencing factors. On the other hand such simulation models do exist and thus open the door for an empirical approach: by exploiting numerous simulation studies widely covering possible scenario configurations, one obtains a lookup table providing key figures of merit on the communications system. Furthermore, the application of curve fitting techniques may help to derive a single analytical expression which then could be incorporated into a hybrid simulation architecture.

This section deals with the development of an empirical model on a key figure of merit of the communication system, namely the probability of packet reception over distance based on changing scenario parameters. We conduct an intensive simulation study to generate a suitable database, elaborate on the application of curve fitting and, finally, evaluate the derived empirical model.

4.1 Empirical Model

The capabilities of the envisioned empirical model are certainly limited to the information captured in the generated data set. In other words, the analytical expression gained from the curve fitting process only allows to interpolate between known data points. Appropriate predictions beyond the known data points, however, will not be covered. Hence, thorough considerations on the assumptions subject to the generated data set are indispensably required in advance. Our study is restricted to highway scenarios in order to exclude radio reflection effects from surrounding buildings that have been hardly explored so far.

Assumptions

The empirical model is built on assumptions in line with the discussion of Section 3. However, in the following we will use Nakagami only with $m = 3$ and when we state chosen transmission powers, we will make use of the deterministic radio wave propagation model Two-Ray Ground. When deterministic models are assumed one can identify a fixed communication range which is the largest distance at which packet receptions are still possible. Thus, when referring to the transmission power, we will state the distance ψ in meter that would correspond to the communication range in a deterministic model. Table 1 in Section 3.2 provides a detailed overview of the assumptions subject to the following simulation study.

Simulation study

A successful wireless packet reception is determined by a number of influencing factors, such as radio wave propagation and interferences issuing from simultaneous transmissions. However, if only a single sender is considered, the effects reduce to the specifics of the environment and thus to the radio propagation. Under this assumption, it was shown in [8] that the probability of message reception can be analytically derived from the Nakagami $m = 3$ model as

$$\mathcal{P}_R(x, \delta, \psi, f) = \mathcal{P}_R^{single}(x, \psi) = e^{-3\left(\frac{x}{\psi}\right)^2}\left(1 + 3\left(\frac{x}{\psi}\right)^2 + \frac{9}{2}\left(\frac{x}{\psi}\right)^4\right). \quad (1)$$

In the most plausible scenario of many senders, the probability of reception is based on a plurality of factors. For example, IEEE 802.11p MAC contentions, the capture effect and the hidden terminal problem all affect message reception but are very difficult to express analytically. For this reason, we make use of an intensive simulation study. All of the simulations were conducted according to the following scenario set-up. In order to mimic varying highway conditions, we pseudo-uniformly placed nodes on a straight line to reflect the simulated traffic density δ. In each simulation, the receptions of packets sent out by one specific node were recorded. To avoid correlations of subsequent transmissions triggered by the node of interest, we applied a relaxed transmission interval of one second to the selected node. With a scenario duration of 100 seconds and 30 seeds for each scenario, we thus captured up to 6,000 packet receptions at each distance in every scenario (note that each distance exists on both sides of the sender). The number of simulated scenarios derives from all combinations in the three dimensions

traffic density [veh/km], δ: 25, 50, 75, 100, 200, 300, 400, 500
transmission power [m], ψ: 100, 200, 300, 400, 500
transmission rate [Hz], f: 1, 2, 4, 5, 6, 8, 10

The results of the numerous simulation runs agree with the intuitive expectation that slight changes in the scenario set-up lead to only minor deviations in the probability of reception. In more detail, when a single *configuration parameter* (transmission power, transmission rate, vehicular density) varies little, then we expect no abrupt deviation in the probability of one-hop broadcast packet reception. Mathematically speaking, we suppose the probability of reception to be partially differentiable on the three-dimensional interval spanned by the aforementioned simulation parameter combinations. Figure 5 backs this supposition and exemplarily depicts how the probability of reception varies when configuration parameters change.

Model building

Instead of a lookup table based on the simulation results, we seek for a closed-form expression to be used as a model in a hybrid simulation architecture. For this purpose we follow the approach in [8] and apply general least linear curve fitting, in particular the *Levenberg-Marquardt* algorithm, to the simulation results using Expression 1 as starting point which is extended by linear and cubic terms; additionally, fitting parameter a_1 through a_4 are introduced yielding

$$\widetilde{\mathcal{P}}_R(x,\psi) = e^{-3\left(\frac{x}{\psi}\right)^2}\left(1 + \sum_{i=1}^{4} a_i \left(\frac{x}{\psi}\right)^i\right). \qquad (2)$$

Furthermore, we consider polynomial functions of fourth degree $h_i,\ i = 1\ldots4$ which approximate the gained fitting parameters $a_i,\ i = 1\ldots4$ based

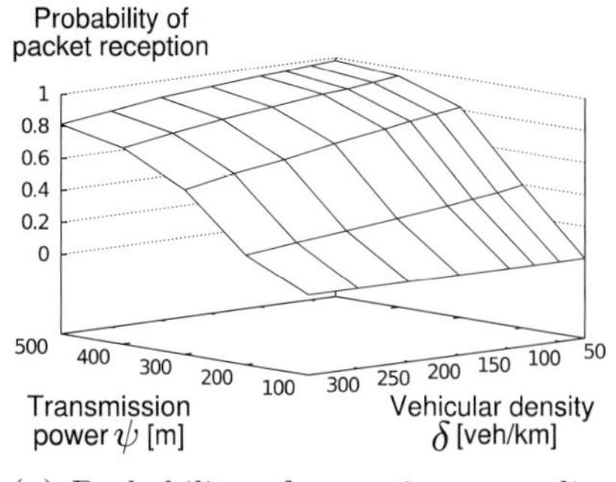

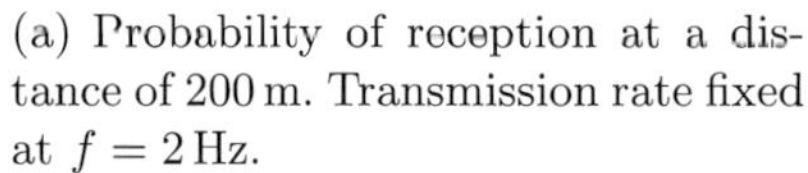
(a) Probability of reception at a distance of 200 m. Transmission rate fixed at $f = 2\,\text{Hz}$.

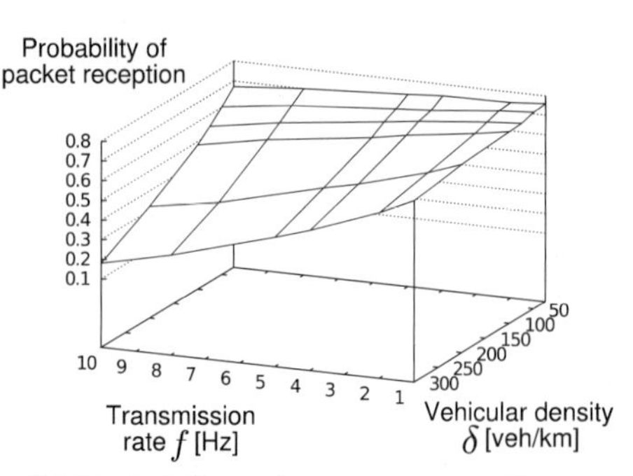

(b) Probability of reception at a distance of 75 m. Transmission power fixed at $\psi = 100\,\text{m}$.

Fig. 5. Probability of packet reception according to various configuration parameters [7]

on the configuration parameters. With $h_i,\ i = 1 \ldots 4$ again derived by curve fitting means we thus conclude with the sought empirical model given in

$$\mathcal{M}: \quad \widetilde{\mathcal{P}}_R(x, \delta, \psi, f) = e^{-3\left(\frac{x}{\psi}\right)^2} \left(1 + \sum_{i=1}^{4} h_i(\psi \cdot \delta \cdot f, \psi) \left(\frac{x}{\psi}\right)^i\right). \qquad (3)$$

The corresponding coefficients subject to the functions $h_i,\ i = 1 \ldots 4$ as well as further explanations can be found in [7].

Model evaluation

Since the empirical model given in Expression 3 is intended to capture the results of the simulation study in a single analytical term, we compare the model's 'prediction' to the simulation traces. Across all investigated scenarios we investigate the sum of squared errors (SSEs) over all distances. The maximum SSE is found in a scenario with a configured transmission power $\psi = 600\,\text{m}$, a transmission rate $f = 10$ and a vehicular traffic density of $\delta = 25\,\text{veh/km}$ and aggregates to 0.393 (cf. Figure 6(a)). The largest observed deviation between model and simulation results amounts 5.1% and is illustrated in Figure 6(b) at a distance of 139 m.

4.2 Results

From the model's evaluation we conclude its suitability to predict communication performances under changing scenario parameters. Hence, the model may serve as a valuable means in a hybrid simulation architecture which aims at accurate and efficient simulation studies on VANETs.

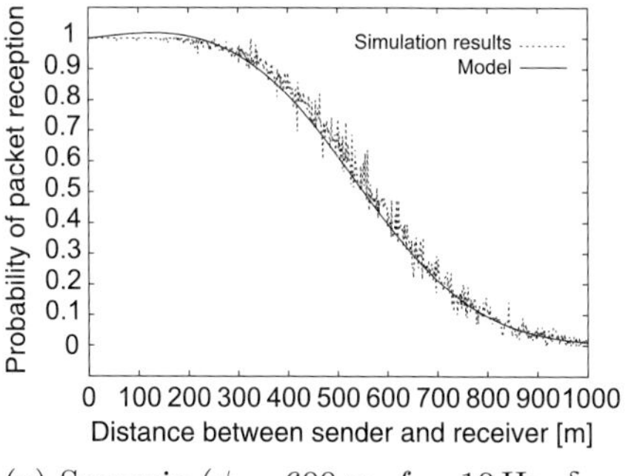

(a) Scenario ($\psi = 600\,\text{m}$, $f = 10\,\text{Hz}$, $\delta = 25\,\text{veh/km}$) for which the max. sum of squared errors has been determined.

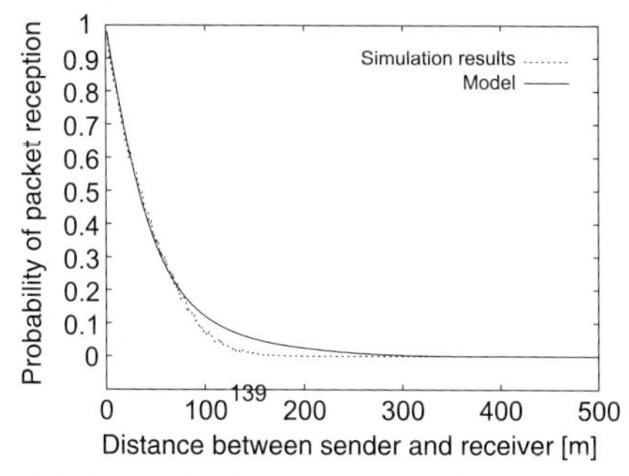

(b) Scenario ($\psi = 200\,\text{m}$, $f = 10\,\text{Hz}$, $\delta = 450\,\text{veh/km}$) for which the max. deviation has been determined.

Fig. 6. Comparison between model and simulation results [7]

5 Conclusion and Outlook

Wireless communication technology offers new promising opportunities to improve traffic safety as well as traffic efficiency. For instance, warning messages could be sent to vehicles such that drivers are informed about dangerous situations. In particular the horizon of awareness can significantly be increased up to few hundred meters or even kilometers. With respect to traffic efficiency vehicles can been regarded as a sensor and actor network that delivers accurate information about the current road conditions and also offers the opportunity to communicate with the drivers on an individual basis. Due to this promising prospects the research area of V2X communication got more and more attention in the last years.

Our research group takes part in these activities since 2003 and mainly focuses on two research areas: on the one hand improving traffic safety through the development and assessment of robust communication protocols. On the other hand evaluating the influence of V2X communication on traffic efficiency.

In terms of safety a reliable and fast message distribution is essential. We therefore developed two protocols called D-FPAV and EMDV (cf. Section 3). As there are various parameters like traffic density, different modulation schemes etc. that have impact on the performance of the protocols we made an extensive sensitivity analysis to get more insights. We therefore used the discrete event-based network simulator ns-2 which allows accurate simulations of the communication layer. Furthermore, we have been able to extract an analytical model out of the simulation runs that can be used for hybrid simulations to evaluate the influence of V2X communication on traffic efficiency (cf. Section 4).

The parallelization of the simulations has been conducted by splitting up the simulation runs on basis of the parameters. Additionally, we also had to run each scenario with different seeds for the initialization of the random number generator to get statistical significance. Hence, we achieve the maximum speed-up factor with respect to the CPU cores as the simulation runs do not have to be synchronized. However, this approach is only possible if the scenarios are small enough to run on a single core.

Besides the V2X related research, we also developed tools and methods to handle the vast amount of simulation runs and parameter sets that in particular lead to a large number of files. We improved our simulation methodology by developing a simulation management tool called Karlsruhe Eclipse Management Platform (KEMP) [5]. This tool is capable to manage simulation runs as well as scenarios and the result sets such that experiments are easily repeatable and traceable.

The presented results required a simulation amount of about 400,000 CPU hours in the last 12 months on the HP XC4000 that equals a continuous use of ca. 46 CPU cores in parallel. Apart from the CPU power itself, the HP XC4000 offered us the flexibility to use up to a few hundred cores in parallel. This gave us the opportunity to quickly get impressions of complex scenario sets what led to an accelerated research process. Finally, we were able to improve the envisioned communication protocols significantly. Moreover we presented our research results to leading European car manufacturers as member of the Car2Car Communication Consortium.

To assess the effects regarding traffic efficiency, large scale scenarios have to be examined. To reduce the computational complexity from the communication side, accurate simulations can be replaced by the empirical model that we developed. Thus, the traffic simulation part of the hybrid simulation remains the limiting factor. In the future we try to parallelize traffic simulations and to gain results from very large scale scenarios up to the whole road network of a country or continent. The simulations with respect to traffic safety are based on small to medium scale scenarios as warning messages only have a limited geographical scope. Hence, we continue our sensitivity analysis by running numerous simulations in parallel. Additionally, we are keeping track of new parallelization approaches like in the network simulator ns-3 that could lead to shorter simulation times.

Apart from the design and evaluation, the operation of the envisioned V2X communication systems will demand supercomputing capacities as well. In particular the real-time decision making process of traffic management centers will benefit from the capability to integrate and process the huge amount of traffic state information sensed by all equipped vehicles.

Acknowledgements. This work was supported by the German Research Foundation (DFG) through the Research Training Groups 895 "Information Management and Market Engineering" and 1194 "Self-Organizing Sensor-Actuator-Networks", and by Klaus Tschira Stiftung, INIT GmbH and PTV AG through the Research Group on Traffic Telematics as well as the Steinbuch Centre for Computing (SCC) that is part of the Karlsruhe Institute of Technology (KIT).

References

1. D. Bertsekas and R. Gallager. *Data Networks*. Prentice Hall, 1987.
2. L. Briesemeister and L. Schäfers. Disseminating Messages Among Highly Mobile Hosts Based on Inter-Vehicle Communication. In *IEEE Intelligent Vehicles Symp.*, pages 522–527, 2000.
3. Q. Chen, F. Schmidt-Eisenlohr, D. Jiang, M. Torrent-Moreno, L. Delgrossi, and H. Hartenstein. Overhaul of IEEE 802.11 Modeling and Simulation in ns-2. In *Proc. of the 10th ACM Symp. on Modeling, Analysis, and Simulation of Wireless and Mobile Systems (MSWiM)*, pages 159–168. ACM, 2007.
4. D. Jiang, Q. Chen, and L. Delgrossi. Optimal Data Rate Selection for Vehicle Safety Communications. In *Proc. of the 5th ACM Int'l Workshop on VehiculAr Inter-NETworking (VANET)*, pages 30–38. ACM, 2008.
5. K. Jünemann. Entwurf und Implementierung einer Simulationsmanagementlösung. Diploma thesis. Institute of Telematics, Universität Karlsruhe (TH), 2008.
6. H. Kawashima. Japanese perspective of driver information systems. *Transportation*, 17(3):263–284, 1990.
7. M. Killat and H. Hartenstein. An Empirical Model for Probability of Packet Reception in Vehicular Ad Hoc Networks. *EURASIP Journal on Wireless Communications and Networking*, 2009.
8. M. Killat, F. Schmidt-Eisenlohr, H. Hartenstein, C. Rössel, P. Vortisch, S. Assenmacher, and F. Busch. Enabling efficient and accurate large-scale simulations of vanets for vehicular traffic management. In *Proc. of the 4th ACM Int'l Workshop on Vehicular Ad hoc NETworks (VANET)*, pages 29–38. ACM, 2007.
9. NS-2. Network Simulator NS-2, Version 2.33. http://www.isi.edu/nsnam/, 2008.
10. H. D. Schwetman. Hybrid simulation models of computer systems. *Communications of the ACM*, (9), 1978.
11. M. Torrent-Moreno. *Inter-Vehicle Communications: Achieving Safety in a Distributed Wireless Environment — Challenges, Systems and Protocols (PhD Thesis)*. Universitätsverlag Karlsruhe, 2007.
12. M. Torrent-Moreno. Inter-Vehicle Communications: Assessing Information Dissemination under Safety Constraints. In *Proc. of the 4th IEEE/IFIP Conf. on Wireless On demand Network Systems and Services (WONS)*, pages 59–64, 2007.
13. M. Torrent-Moreno, J. Mittag, P. Santi, and H. Hartenstein. Vehicle-to-Vehicle Communication: Fair Transmit Power Control for Safety-Critical Information. *Accepted for publication. IEEE Trans. on Vehicular Technology*, 2009.
14. M. Torrent-Moreno, P. Santi, and H. Hartenstein. Fair Sharing of Bandwidth in VANET. In *Proc. of the 2nd ACM Int'l Workshop on Vehicular Ad hoc NETworks (VANET)*, pages 49–58, 2005.
15. M. Torrent-Moreno, P. Santi, and H. Hartenstein. Distributed Fair Transmit Power Assignment for Vehicular Ad Hoc Networks. In *Proc. of the 3rd Annual IEEE Conf. on Sensor, Mesh and Ad Hoc Communications and Networks (SECON)*, volume 2, pages 479–488, 2006.

Modelling Structural Failure with Finite Element Analysis of Controlled Demolition of Buildings by Explosives Using LS-DYNA

Georgios Michaloudis, Gunther Blankenhorn, Steffen Mattern, and Karl Schweizerhof

Institute of Mechanics, Kaiserstr. 12, Geb. 20.30, 76128 Karlsruhe
Georgios.Michaloudis, Gunther.Blankenhorn, Steffen.Mattern, Karl.Schweizerhof@ifm.uni-karlsruhe.de

1 Introduction

When buildings are not longer used, controlled deconstruction is necessary. Besides systematic dismounting with heavy machines, a very efficient way is to use controlled explosives. The main load-carrying structural parts of the building are weakened by an explosive charge, which leads to a specific collapse kinematics. This has to be predicted reliably, in order to consider the boundary conditions such as neighboring buildings and traffic loaded streets. When planning such a collapse event, it is desirable to have a reliable simulation of the complete collapse process, also considering the uncertainty of primary parameters influencing e.g. the resistance of structural elements of a building. The 'Research Unit 500' [1], funded by the German Research Foundation (Deutsche Forschungsgemeinschaft – DFG) develops a special simulation concept by subdividing the collapse analysis into several – problem specific – analyses.

The creation, simulation and evaluation of global finite element models of collapsing buildings is the main aspect of one of the research units subprojects, located at the University Karlsruhe (TH). Several reference models – academic structures and real buildings – are chosen in order to investigate modeling techniques with finite elements. Especially the prediction and efficient modeling of failure mechanisms is the focus of the analysis. The quality of the simulations is validated by comparing – as far as available – with video material from the real collapse event. An important issue of the sub-project within the research unit is the detection of structural parts, behaving like rigid bodies during the collapse. This information is used in another subproject to carry out multi-body-analyses of the reference systems, in order to reduce the computational cost of the simulations [2].

W.E. Nagel et al. (eds.), *High Performance Computing in Science and Engineering '09*, DOI 10.1007/978-3-642-04665-0_37,

In the present contribution, techniques for the simulation of failure in order to obtain a realistic kinematical behavior of the collapse are presented. The difficulties, especially concerning the determination of material parameters are discussed, and results of parametric studies on several examples, investigated within the project are presented. The used finite element program LS-DYNA [3, 4] provides different methods for modeling failure e.g. element erosion and disconnection of elements at nodes. The advantages and disadvantages of these techniques are theoretically discussed and visualized with the simulated reference structures. All computations which were necessary to develop and improve the models were performed on the HP XC6000-Cluster of the University of Karlsruhe.

2 Aspects of Numerical Simulation

2.1 General Scope

Safety and efficiency being top priorities in controlled demolition of buildings leads to the necessity of good knowledge about the kinematics of the collapse. The challenge is to produce good a-priori prognosis of the collapse, by creating representative computational tools which not only deliver reliable results but also can check the entire procedure taking into account uncertainties [5] and different possible blasting procedures. This requires a global sophisticated computational model which covers the dynamics of the entire collapse process, from the moment of the blasting until the breakdown is completed.

Within this work these aspects are being fulfilled with the application of the Finite Element Method for creating the necessary simulation models. The results of the Finite Element Method analysis can be later used also as reference solutions for judging the quality of results provided by other methods, such as the MultiBody Systems (MBS) Analysis method [6].

The results of the Finite Element Method are first validated against the real collapse process. Frames from the video and images from the analysis results are combined in one in order to estimate the agreement between them [7]. The validated results of the analysis performed are used as a guide for the proper discretization of the multibody model. They provide the necessary information concerning the identification and development of the multibody subsystems as well as their distribution within the entire model [8].

2.2 Numerical Algorithms

The entire concept of the model generation and numerical simulation consists of a set of several numerical algorithms. The problems within this work are characterized from high nonlinearities, multiple contact possibilities and necessity for discretization with continuum elements, since discretization with structural elements would be insufficient. This results in models with a very large number of elements. All the computations which are carried out within this work are performed using the Finite Element Method software LS-Dyna which applies explicit time integration. The code is highly parallelized and perfectly suited to run on clusters such as the HP XC6000.

Element Formulation

The discretization of the structures is done with 8-node hexahedral solid finite elements. These elements are reduced integrated with only one integration point at the middle. The control of the hourglass modes is achieved by applying the stabilization technique developed by Belytschko-Bindeman [9].

Simulation of Contact

An important part of the numerical modeling is the simulation of all the contact conditions which appear during a demolition, concerning contact between the building and the ground plate as well as contact between the structural parts of the building. The contact formulation used within this work is a penalty based surface-to-surface algorithm for both cases building-to-ground plate and building-to-building contact. The ground plate in all examples is modeled as a rigid body.

Modelling of Explosions

Within this work, the explosion which destroys specific structural parts leading to the collapse, is modeled by removing the according elements and the influence of the shockwave in the global investigations is neglected. This is because previous numerical tests showed that the effects of the shockwave have in such heterogeneous structures a strong local character and do not influence the global kinematics.

Material Model

One critical issue in the generation of models which should represent old buildings is the material modeling. Very often material parameters have to be quantified based on only a few data, which on the other hand do not always offer reliability because of uncertain measurements. In the computations to be presented a piecewise linear plasticity model is used (LS-DYNA MAT24) for efficiency reasons. The material model does not allow detailed modification in order to simulate the complex behaviour of reinforced concrete, however offers for many cases a sufficient level of approximation for mass dominated problems such as occurring in cases of building collapse.

The most important issue concerning the material simulation is the modeling of failure and its mechanism. This influences many aspects of the computation such as contact conditions, consistency, reliability of results and furthermore it is important for the generation of local zones of accumulated damage (hinges). Two algorithms have been chosen and tested in order to demonstrate their performance in cases of different collapse scenarios.

Element Erosion

Applying this very simple algorithm in a simulation results in removal of all elements which have reached failure satisfying a given criterion. Within the defined material law a critical value is set by the user. This value represents the critical plastic strain at failure. When during the computation the plastic strain of an element reaches this value, the element is automatically deleted and the analysis continues with the rest of the elements. The main advantage

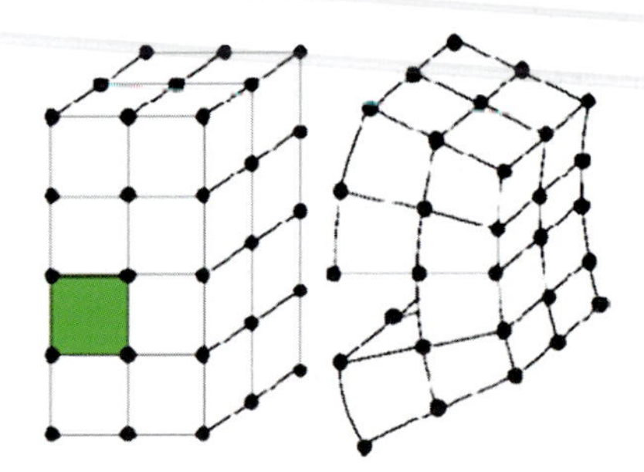

Fig. 1. Element Erosion algorithm

of this algorithm is that is directly applicable to an existing discretization, no changes or special modeling techniques are required. In addition, the algorithm does not increase the computational effort, since the number of equations to be solved remains the same, which makes this method of material failure a very efficient and straight-forward one. For problems which can be described as non vertical collapse scenarios, thus showing e.g standard breaking forming hinge lines, the element erosion algorithm can deliver reliable results and a sufficient level of representation of real demolitions can be reached. It is important to note though, that in order to do so a fairly fine mesh is required. Then the failure zone can be well captured.

However, the main characteristics of this technique impose a strong limitation on the range of its usability. The removal of the elements during the simulation results in non consistent computations in terms of volume, mass and energy. Moreover, contact conditions especially between the structural parts cannot be well described since a lot of contact surfaces are deleted due to material failure. Given the importance of the simulation of contact processes, one can expect inadequate prediction of collapse kinematics. In cases of vertical collapse scenarios this problem is more intense. Contact between the structural parts leads, as the upper parts begin to breakdown, to an excessive erosion of finite elements. As a consequence these computations are often not able to deliver useful results, as a very large part of the building is being deleted.

Node-Split

In contradiction to the failure mechanism described above, applying the so-called Node-Split algorithm does not lead to removal of elements from the computation. With this method the analyst specifies sets of nodes which are constrained to be tied together until a failure criterion is reached. Until then the coincided nodes are tightly connected. Entire regions or individual elements can be defined to be tied together.

A failure criterion is the average volume weighted plastic strain. When this is reached not all connections of the element are broken, but only the nodes of neighbor elements which carry this plastic strain are disconnected, while the other tied connections of the element remain unchanged. The nodes of the elements and not the elements themselves form the sets which are constrained to be connected until failure. Thus, the failure criterion is checked for each individual set of nodes and not for elements. E.g., in the example described in Fig. 2 the tied connection of two nodes fails. This means that after a volume weighting the plastic strain which results as an average of the plastic strains of the two elements to which the two nodes belong, has reached the critical value for failure. But at the same time, the set of nodes on the right of the previous one does not fail, although involves nodes which two of them belong to the same elements as before. This happens because the weighting procedure for this set of nodes involves two more elements, thus the resulting average plastic strain differs from the above mentioned one, which reached failure. As plastic strain develops and exceeds the critical value, the algorithm will simulate the formation of cracks which propagate through the entire structure. Concerning the definition in the model, the application of the Node-Split is not straight-forward. First an adaptation of the finite element mesh is necessary, in order to create coinciding nodes for each element. Each element, which is to be constrained with the Node-Split algorithm, independently from its connectivity conditions, consist of eight unique nodes. There are no nodes which belong at the same time to more than one element. This will increase the time needed for the creation of the model. Furthermore, the number of equations of the problem is excessively increased, leading to longer computational times in comparison to the Element Erosion technique.

In spite of the increasing of the total time of simulation, computations with node splitting demonstrate some important advantages. Since the elements even after their failure remain in the model, the simulation of the

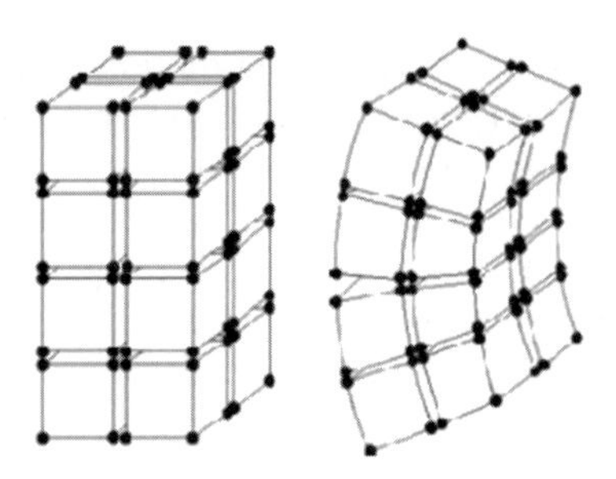

Fig. 2. The Node-Split algorithm

whole procedure remains throughout the entire time consistent in terms of mass, volume and energy. Energy is taken out of the structure only due to contact of structural parts with the ground plate. Furthermore all contact surfaces participate in the analysis, independently of the occurrence of local failure. No elements are deleted and the contact possibilities are entirely described. Thus, the kinematics are not influenced of any kind of artificial erosion and the effect of the contact procedures on the dynamical behaviour of the building during a demolition is sufficiently taken into account. The latter makes the Node-Split method much more reliable compared to the element erosion, since the evolution of the contact conditions can strongly influence the global kinematics throughout the simulation, especially during the first part of the collapse, when contact mostly occurs between the structural parts. One can realize that in cases of very tall buildings for which more often vertical collapse scenarios are designed. For a large part of the phenomenon the structural parts are under strong pressure only due to contact between them. In this case, a primitive description of these conditions would lead to pure representation of the kinematics and to overall unrealistic results. Using node splitting all surfaces are taken into consideration and all possible contacts are controlled for the entire duration of the procedure. Considering that the modeling of failure by element erosion is in such cases unable to provide useful results, a Node-Split algorithm offers a robust solution with an overall more realistic performance, which is strongly obvious in problems in which vertical collapse is the chosen method for demolition.

A technique, alternative to Node-Split with which it shares some general features, can be implemented by applying tied contact formulation which comprehends failure criteria. This is a constraint-type contact which realizes the constraining conditions by introducing elastic springs, see Fig. 3(a). A critical stress value is the failure criterion. Within this algorithm, first the slave node is projected and glued to the master surface implementing the tied contact. Then, two springs are introduced, one for the movement of the slave node to the normal direction and one for the sliding along the master surface and a stiffness value is given. With this way the geometrical constraint conditions between the slave node and master surface are accomplished. The slave node can move around the glue point within the elastic zone defined and restricted by the springs. In every step failure is checked and if the failure criterion is not fulfilled the elastic forces are computed and transferred from one body to the other through the springs. Now if the failure criterion is fulfilled the tied connection is broken and the slave node can independently move away from the glue point, since the constraining conditions are no longer active.

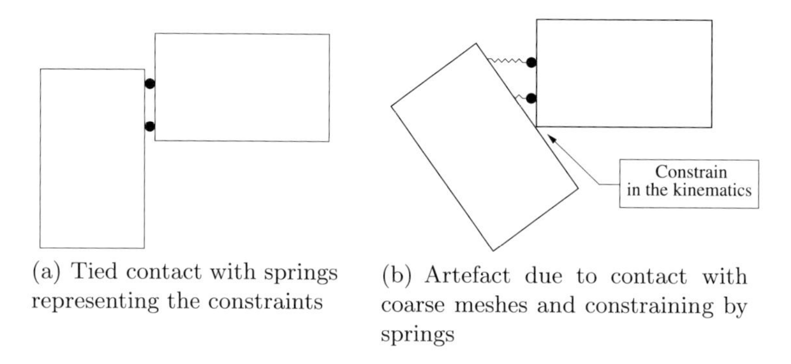

(a) Tied contact with springs representing the constraints

(b) Artefact due to contact with coarse meshes and constraining by springs

Fig. 3. Modelling local failure with tied contact

This method has some obvious disadvantages. Springs must be given a stiffness value. High stiffness values cause convergence problems concerning insufficient numerical stability. On the other hand, low stiffness values allow high penetrations between the bodies and do not satisfy adequately the geometrical constraint conditions, since the springs can open even without failure leading to a fairly soft structure. Then, the kinematics with low order elements with linear shape functions get many constraints, see Fig. 3(b) and the failure loads are too quickly reached. Moreover, wave propagation phenomena are not well described because of the constraining implementation by springs. In such a case, the slave node under the influence of the wave will penetrate because of the stiffness of the spring and in this way, it will add artificial dynamical effects, due to the additional eigenmodes created. Giving higher stiffness values to the spring would reduce this effect because then the physical interface of the elements would be better represented, however as mentioned above that leads in most cases to numerical instabilities.

3 Numerical Examples

The numerical examples which are presented in this section cover a wide range starting from simple academic examples and expanding to fairly large numerical models with up to 150.000 elements. The models of the two examples are among the 5 reference models of the research unit (FOR500) [6, 10, 11].

3.1 Academic Example

The first -simple demonstration- example is a three storey frame structure of reinforced concrete. The ground surface has dimensions 6 x 4 m and the height of the building is 9,1 m. Two rows of columns with two columns each built the vertical load carrying part of the structure. In the simulation, the collapse has been carried out by removing one row of columns of the ground floor. The collapse is then caused only by the influence of gravity. The discretization is achieved with 2330 8-node hexahedral solid finite elements, while the ground plate is modeled as a rigid body. At the simulation of the collapse

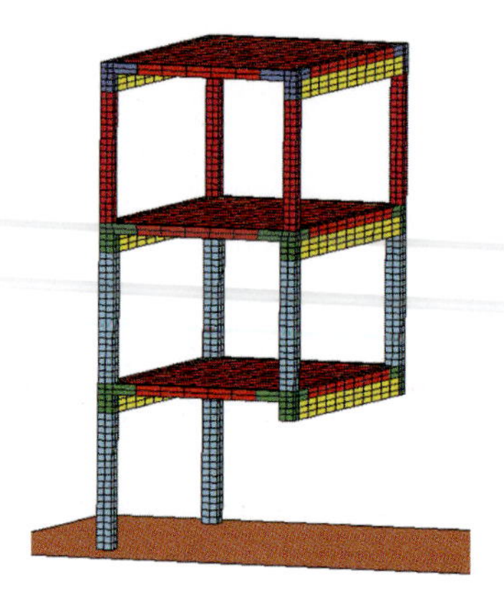

Fig. 4. Academic example - Collapse initiation: The 2 columns of the front row are removed resulting in forward bending and final collapse of the building due to gravity

of this building both algorithms concerning material failure, element erosion and Node-Split, are tested. Results concerning collapse kinematics and final rest position, which are the two major phenomena of interest, are finally compared. By comparing the figures of the collapse we can conclude that both approaches provide a similar description of the collapse kinematics at the beginning of the procedure as well as at its whole evolution. A small time difference is noted, with the model simulating the material failure with erosion of elements presenting a small delay in comparison with the model with the Node-Split algorithm. But this is not a sign of problematic behavior since the two algorithms use different procedures for determining if the failure criterion is reached within an element. Concerning the rest position of the building after the collapse the two models predict very similar results. This is due to the fact, that elements are mainly eroded at the connections forming hinges and compression is limited due to the low weight of the structure.

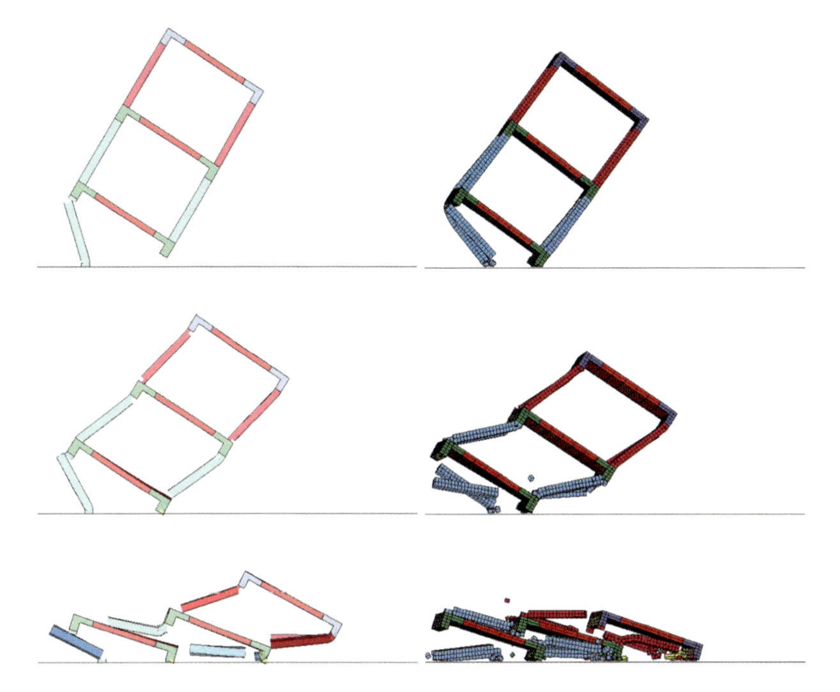

Fig. 5. Academic example - Comparison of collapse kinematics - Element Erosion (left)- Node-Split (right)

3.2 Nine Storey Building

The second example is a 9 storey building of reinforced concrete. The numerical model does not correspond to a real building, but was created as representative example of such large systems. The structural system of the building consists of frames and shear walls. It is 37,5m high and its surface has dimensions 24,4m x 12,4m.

The blasting is executed in two phases. First two rows of columns and parts of the shear walls of the 5th floor are removed. This initiates the collapse of the upper part. After 1,5 seconds two rows of columns and parts of the shear walls of the 2nd and the ground floor are destroyed aiming at a vertical collapse of the entire structure.

The finite element model consists of 146920 8-node hexahedral solid elements. As before, the ground plate is simulated as a rigid body and the building collapses under the effect of gravity. In this example the Node-Split algorithm is applied for the simulation of the material failure. The element erosion algorithm could hardly be applied in such problems, since it would result in the removal of a very large part of the structure.

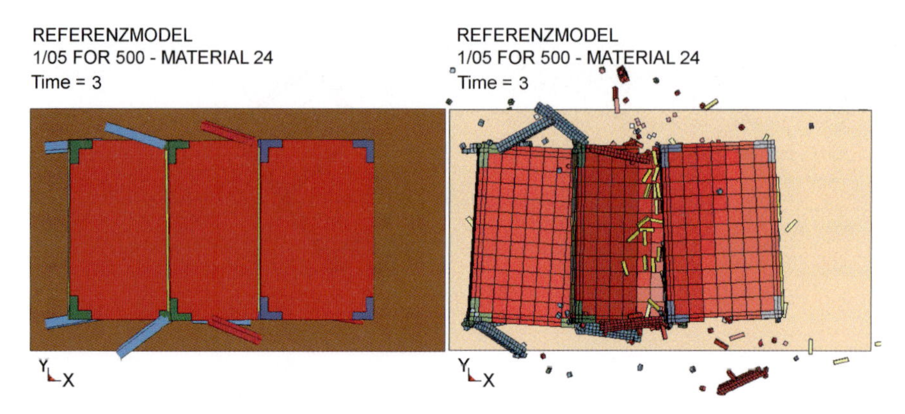

Fig. 6. Academic example - Comparison of rest position - Element Erosion (left)-Node-Split (right)

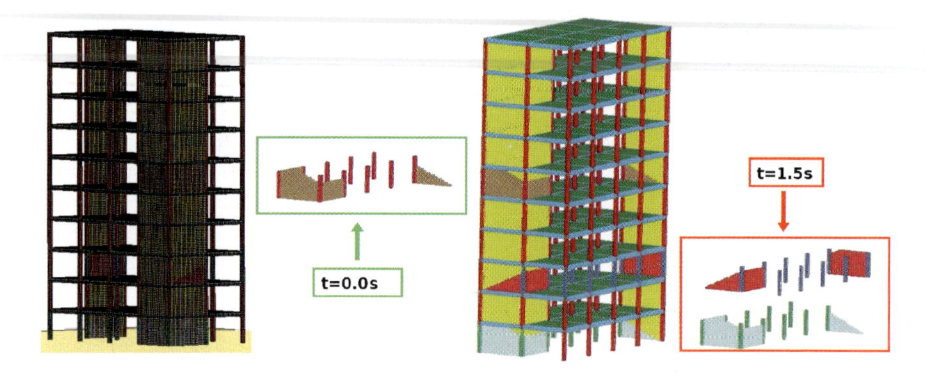

Fig. 7. Nine storey building - Computational model and blasting strategy

Simulating the blasting processes by the Node-Split algorithm provides the advantage that no loss of mass, volume or energy occurs. The energy is removed from the structural parts only by friction with the ground, resulting in a consistent analysis.

As result a fairly realistic description of the collapse kinematics as well as of the final rest position of the structure at the end of the procedure are obtained, see Fig. 8. It is shown that for a vertical collapse scenario results are achieved which are closer to reality.

Taking into account that analyses with element erosion are not consistent in terms of volume, mass and energy, the local failure approach via node splitting offers the most promising route towards the prediction of collapse kinematics and final rest position in demolition processes concerning efficiency. The fact that in the case of vertical collapse simulations with element erosion lead to intense element removal makes the application of node splitting the only eligible alternative.

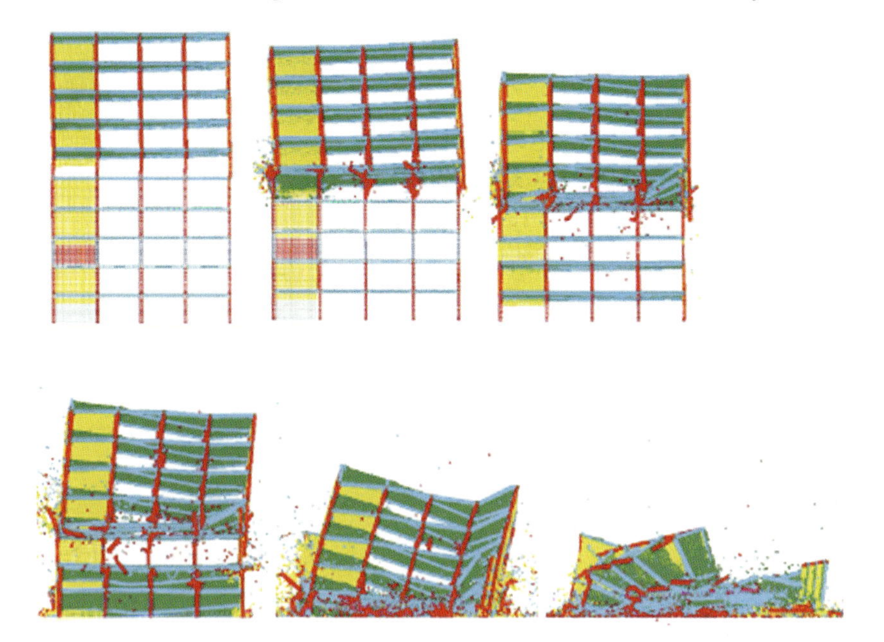

Fig. 8. Nine storey building - Evolution of collapse kinematics, Node-Split as failure mechanism

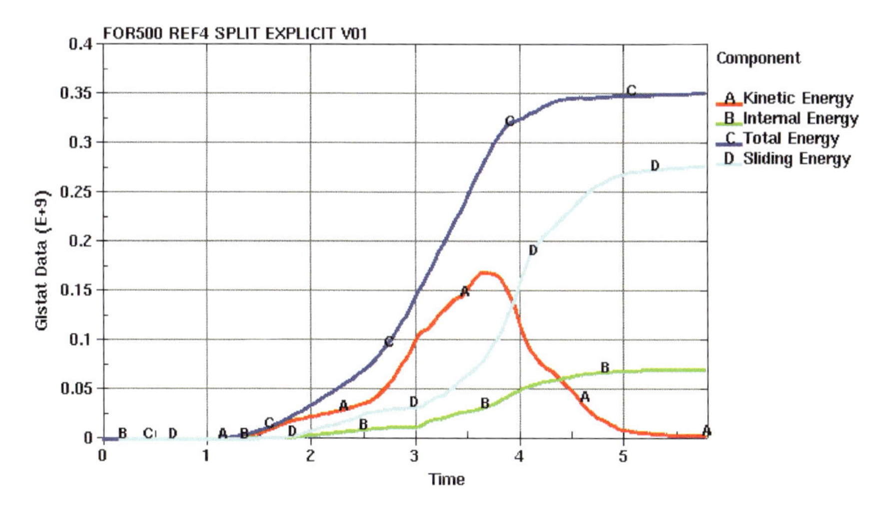

Fig. 9. Nine storey building - Evolution of the energy

The computation was performed on the HP XC6000-Cluster of the University of Karlsruhe using the MPP-Parallel version of LS-DYNA on 64 processors. The computational time for this problem of 9.393.824 unknowns, 262.579 extra constraint conditions (for the modelling of failure), computed time of

6 seconds with timestep size of approximately $2{\cdot}10^{-5}$ seconds (290.000 time steps) reaches 60 hours.

4 Conclusions and Acknowledgments

Some numerical aspects have been discussed, concerning the simulation of controlled blasting with the use of global finite element models. The goal of this study is to generate a sufficient a-priori prognosis of the collapse kinematics and final rest position of the structure, which allows taking into account uncertainties [12] and [13] and possibly different demolition strategies. The main challenge is the simulation of the material failure with high efficiency. Two different techniques have been described and tested in the numerical examples.

The element erosion algorithm is not capable of describing sufficiently all the contact surfaces which affect the collapse kinematics, especially in cases of vertical collapse. The Node-Split method increases the computational effort substantially, however modelling via node splitting does not lead to element removal out of the computation, takes into account all possible contact phenomena that may develop during the demolition and therefore offers a more robust simulation alternative, since realistic results can be achieved. However, some limitations exist due to the fairly simple failure criterion. More sophisticated failure criteria are subject of current investigations.

This work is funded by the German Research Foundation (DFG) within the frame of the research unit FOR500 'Computer aided destruction of complex structures using controlled explosives'. This support is gratefully acknowledged.

References

1. DFG Forschergruppe FOR500 www.sprengen.net 2006.
2. Mattern S, Blankenhorn G, Breidt M, Nguyen vV, Höhler S, Schweizerhof K, Hartmann D (2006) Comparison of building collapse analysis simulation from finite element and rigid body models. In: Eberhard P (ed) IUTAM Symposium on multiscale problems in multibody system contacts, IUTAM Bookseries, Springer.
3. Hallquist JO (2006) LS-DYNA Theory Manual. Livermore Software Technology Corporation.
4. Hallquist JO (2006) LS-DYNA Keyword User's Manual. Livermore Software Technology Corporation.
5. Möller B, Beer M (2004) Fuzzy Randomness - Uncertainty in civil engineering and computational mechanics, Springer, Berlin.
6. Hartmann D, Breidt M, Nguyen vV, Stangenberg F, Höhler S, Schweizerhof K, Mattern S, Blankenhorn G, Michaloudis G, Möller B, Liebscher M (2008) On a

fundamental concept of structural collapse simulation taking into account uncertainty phenomena. NATO Advanced Research Workshop 983112, Sarajevo, Bosnia and Herzegovina.
7. Mattern S, Blankenhorn G, Schweizerhof K (2006) Numerical analysis of building collapse - Case scenarios and validation. Proceedings of the NATO Advanced Research Workshop, PST.ARW981641, ed.: A. Ibrahimbegovic and I. Kozar.
8. Mattern S, Blankenhorn G, Schweizerhof K (2006) Numerical Investigation on collapse kinematics of a reinforced concrete structure within a blasting Process. Proceedings 5th German LS-DYNA Forum, Ulm, Germany.
9. Belytschko T, Bindemann LP (1993) Assumed strain stabilization of the eight node hexahedral element. Computer Methods in Applied Mechanical Engineering 105, 225:260.
10. Mattern S, Blankenhorn G., Schweizerhof K (2007) Numerical simulation of controlled building collapse with finite elements and rigid bodies - Case studies and validation. ECCOMAS Thematic Conference in Structural Dynamics and Earthquake Engineering, Rethymno, Crete, Greece.
11. Blankenhorn G, Mattern S, Schweizerhof K (2007) Controlled building collapse-Analysis and validation. Proceedings LS-DYNA Anwenderforum, Frankenthal.
12. Hartmann D, Breidt M, Nguyen vV, Stangenberg F, Höhler S, Schweizerhof K, Mattern S, Blankenhorn G, Möller B, Liebscher M (2008) Structural collapse simulation under consideration of uncertainty - Fundamental concept and results. Computers and Structures 86, 2064-2078.
13. Möller B, Liebscher M, Schweizerhof K, Mattern S, Blankenhorn G (2008) Structural Collapse under Consideration of Uncertainty - Improvement of Numerical Efficiency. Computer and Structures 86, 1875-1884.

Printing and Binding: Stürtz GmbH, Würzburg